# Climate Change Management

**Series Editor**

Walter Leal Filho, International Climate Change Information and Research Programme, Hamburg University of Applied Sciences, Hamburg, Germany

The aim of this book series is to provide an authoritative source of information on climate change management, with an emphasis on projects, case studies and practical initiatives – all of which may help to address a problem with a global scope, but the impacts of which are mostly local. As the world actively seeks ways to cope with the effects of climate change and global warming, such as floods, droughts, rising sea levels and landscape changes, there is a vital need for reliable information and data to support the efforts pursued by local governments, NGOs and other organizations to address the problems associated with climate change. This series welcomes monographs and contributed volumes written for an academic and professional audience, as well as peer-reviewed conference proceedings. Relevant topics include but are not limited to water conservation, disaster prevention and management, and agriculture, as well as regional studies and documentation of trends. Thanks to its interdisciplinary focus, the series aims to concretely contribute to a better understanding of the state-of-the-art of climate change adaptation, and of the tools with which it can be implemented on the ground.

**Notes on the quality assurance and peer review of this publication**

Prior to publication, the quality of the works published in this series is double blind reviewed by external referees appointed by the editor. The referees are not aware of the author's name when performing the review; the referees' names are not disclosed.

Javid Ahmad Parray
Editor

# Climate Change and Microbiome Dynamics

## Carbon Cycle Feedbacks

Springer

*Editor*
Javid Ahmad Parray
Department of Environmental Science
HKM Government Degree College Eidgah
Srinagar, Jammu and Kashmir, India

ISSN 1610-2002 ISSN 1610-2010 (electronic)
Climate Change Management
ISBN 978-3-031-21078-5 ISBN 978-3-031-21079-2 (eBook)
https://doi.org/10.1007/978-3-031-21079-2



This Springer imprint is published by the registered company Springer Nature Switzerland AG
The registered company address is: Gewerbestrasse 11, 6330 Cham, Switzerland

# About This Book

Climate change is a complex societal issue that we must comprehend to better deal with its challenge. Climate change has a significant impact on people's lives, energy demand, food security, etc. The book provides an overview relevant to various biological mechanisms that regulate carbon exchanges between the major components and their response to climate change. The Book will address the need to use a multifactor experimental approach to understand how soil microorganisms and their activities adapt to climate change and the implications of carbon cycle feedback. The most pressing concern is a clearer understanding of the biological factors that regulate carbon exchanges between land, oceans, and the atmosphere and how these exchanges will respond to climate change via climate–ecosystem feedbacks, which could augment or quell regional and global climate change. Terrestrial ecosystems play an important role in climate feedback as they produce and absorb greenhouse gases like carbon dioxide, methane, and nitrous oxides. The current book will focus on recent research designed to use beneficial microbes such as plant growth-promoting microorganisms, fungi, endophytic microbes, and others to improve understanding of the interaction and their potential role in promoting advanced management for sustainable agricultural solutions. Changes in climatic conditions impact all aspects of the agricultural ecosystem, including yield in terms of quantity and nutritional quality. Understanding the influence on the native microbiome, such as the distribution of methanogens and methanotrophs, nutritional content, microbial biomass, and other factors, is becoming increasingly crucial to establishing climate-resilient agriculture.

# Contents

# About the Editor

**Dr. Javid Ahmad Parray, Ph.D., PDF** GDC Eidgah Srinagar, Department of Higher Education, Jk India.

Dr. Javid A Parray holds a Master's Degree in Environmental Science and has completed his Research Programme (M.Phil. and Ph.D.) from the University of Kashmir after qualifying state level prestigious JKSLET examination. He has also done his Post Doctorate Research from the University of Kashmir. Dr. Parray was also awarded a Fast Track Young Scientist Project by SERB—DST, GoI New Delhi. He currently teaches at the Department of Environmental Science, GDC Eidgah Srinagar affiliated to Cluster University Srinagar. He has attended many courses and conferences on environmental issues and other biotechnological aspects within and outside the country like Srilanka, Indonesia, Malaysia, etc. His fundamental research interests include ecological and agricultural microbiology, climate change and microbial biotechnology, environmental microbiome, etc. He has published more than 60 high-impact research papers and book chapters in reputed journals and publishing Hubs. Dr. Parray has authored 12 books with international publishers like Elsevier, Springer, Callisto Reference USA, and Wiley-Blackwell. Dr. Parray is on the editorial board, and permanent reviewers of many journals and has been an invited speaker at various scientific meetings/conferences within India and abroad. He is also a Guest editor for a special issue on Environmental Biofilms with BioMed Research International, Hindwaii and frontiers in sustainable food system. He is member of many International and National scientific organizations and societies like Asian PGPR Society, IJMS Mumbai, Academy of Eco science, IAES Haridwar, etc. Dr. Javid was also awarded as "Emerging scientist year Gold Medal" for the year 2018 by Indian Academy of Environmental Science. Dr. Parray is national course coordinator for three (03) CeC–Swayam Moocs programme on environmental Science.

# Chapter 1
# Diversity and Biogeography of Soil Bacterial Communities

**Soheila Aghaei Dargiri and Ali Movahedi**

**Abstract** Soil microbial communities are essential for crucial soil activities such as litter decomposition, nitrogen cycling, and plant productivity, which are necessary for human health. The scientific knowledge of microbial biogeography is woefully lacking when it appears to soil bacteria, despite the widespread expectation that soil bacterial communities directly impact many ecosystem processes. Researchers are becoming increasingly interested in the global distribution of soil microbes and the influence of environmental change at the regional level. This is because of the high microbial diversity that soils contain and their important role in biogeochemical cycling. As a result, we now know that the bacterial diversity of soil is high, and the composition and diversity of soil bacterial communities change with various biological and non-biological stresses. The full range of microbial diversity can now be analyzed using ribosomal DNA. Such research could also shed light on the environmental factors influencing microbial community change. These more accurate models could anticipate the temporal-spatial dynamics of soil biodiversity and ecosystem functions in changing contexts, which could help with soil biodiversity conservation and ecological function presentation in the face of future climate change. Such knowledge could aid humans in coping with future environmental changes and increase our ability to predict microbial communities accurately and their function in a changing world. We propose the following difficulties and research opportunities for future microbial biogeographic investigations.

S. A. Dargiri (✉)
Department of Horticulture, Faculty of Agriculture and Natural Resources, University of Hormozgan, Bandar Abbas, Iran
e-mail: S.aghaei.phD@hormozgan.ac.ir

A. Movahedi
Co-Innovation Center for Sustainable Forestry in Southern China, Ministry of Education, College of Biology and the Environment, Nanjing Forestry University, Nanjing 210037, China
e-mail: ali_movahedi@njfu.edu.cn

 
J. A. Parray (ed.), *Climate Change and Microbiome Dynamics*, Climate Change Management, https://doi.org/10.1007/978-3-031-21079-2_1

## Introduction

A central purpose of microbial ecology is to link microbial distribution templates to underlying ecological activities. Developing links is significant for both fundamental knowledge and practical consequences, for instance, in manufacturing precise universal biodiversity evaluations and prioritizing management aims in the face of both native and worldwide alternation [1–3]. However, getting this critically hinges on our abilities too much specific biodiversity in the first period, with different methodological and theoretical alternations limiting our comprehensive of microbial distribution templates and their underlying ecological stimulus.

Soil bacteria are the early drivers of ecological activities [4–6]. Several bacteria depend on the manufacture or attraction of greenhouse gases such as $CO_2$, $CH_4$, and $N_2O$ [7]. Proteobacteria and Acidobacteria are the most phyla in soil bacteria [8–10]. It has been shown in the reports that the structure and diversity of soil bacteria is increased by soil characteristics and plant species. Soil pH is a significant agent in reining bacterial community manufacture [3]. Further, soil specifications as well as impact soil bacterial community combination and variety, such as nutrient accessibility [11–13] and plant variety [14–16]. The release of bacterial communities in prior studies found that the association of soil bacteria was increased by soil exclusivity, climatic, or other particular [17–19]. Comprehensive mechanisms that influence the abundance, distribution, and diversity of organisms over spatial and temporal levels are basic challenges in ecology. Several macroecological laws have been proposed for plants and animals to explain the physiological, ecological, and certain evolutionary factors that underpin these templates. Microorganisms also display spatial and temporal patterns in abundance, dispersion, and diversity [20–24]. However, it is uncertain if macroecological criteria defined for plants and animals apply to microorganisms and whether they may improve forecasts of microbe abundance, distribution, and variety.

## Definitions

### *Diversity*

The overall number of species present, i.e., species richness or abundance, and the distribution of individuals among those species, i.e., species evenness or species equitability, have been classified as biodiversity [25, 26]. Because of the necessity of observing the entirety, functionality, and long-term sustainability of both natural and managed terrestrial ecosystems, the biodiversity of soil biota is becoming more and more necessary recently [27–33]. However, our understanding of soil biota biodiversity remains hazy due to a lack of acceptable methodologies for assessing the contribution of various soil biota components to ecosystems [34].

We can distinguish the effects of various ecological processes on a community structure by quantifying and comparing biodiversity. There are numerous approaches to evaluating biodiversity [33], but they always fall into two categories: differentiation diversity or inventory diversity. Inventory diversity measurements describe diversity inside an environment (alpha diversity, according to Whittaker [35]), whereas differentiation diversity describes diversity turnover between environments (beta diversity). As a result, a community with high inventory diversity has great biodiversity within a habitat at a specific spatial scale, whereas two distinct communities with high differentiation diversity share only a few species. Numerous assessment variety statistics qualify the biodiversity based on a set of parameters. All consider the number of different taxa present in a particular sample and additional information on the evenness in relative abundance (e.g., Shannon index and heterogeneity measures). Others include the level of phylogenetic diversity (PD) within samples, which may be especially important in varied microbial communities [36, 37]. Significantly, assessment diversity characteristics may assay biodiversity on any scale. Usually, alpha diversity, also known as "native diversity" refers to diversity at the lowest spatial scale of analysis, whereas gamma diversity is a statistic for regional (landscape) diversity.

## *Biogeography*

After various decades of using molecular phylogenetic tools to study microbial community composition, we now learn that there are similarities in biogeographical templates in microbial and microbial communities [20, 38]. Biogeography is the study of the distribution of taxa through space and time, and it has provided essential insights into the mechanisms that sustain and generate species variety [39]. Numerous studies have shown that microbial communities can display biogeographic patterns, which are often qualitatively comparable to those of macroorganisms [40–42]. Understanding why microorganisms differ quantitatively in their distribution from plants and animals is crucial for various reasons. For starters, biogeographic patterns can lighten the fundamental processes governing biodiversity. Quantitative discrepancies in biogeographic patterns could imply that bacteria and larger species have different underlying mechanisms. Second, biogeography serves as the conservation and environmental management framework, including bio-prospecting. Understanding whether microbial and plant/animal biogeography follow distinct patterns is critical for developing effective management and conservation strategies [43–45]. Some argue that bacteria have weak biogeographic patterns because they differ fundamentally in ways that influence their biogeography, such as high abundance, lifespan, or dispersion capacities [9]. Others, however, have claimed that these discrepancies are byproducts of the method used to study microbial biogeography [21, 46].

Biogeographic patterns are well known to change quantitatively with geographical scale. This holds true for microorganisms [21, 47] as well as bigger organisms

[48, 49]. Environmental filtering is thought to be a more important driver of biogeographic patterns at smaller spatial scales [20, 21, 50], whereas dispersal limitation and/or diversification are supposed to be more important drivers of large-scale spatial patterns [40, 51, 52] though dispersal limitation can also play a role at local scales [21, 53].

## Changes in Soil Microbial Biogeography in the World

Soils would not be without the activity and diversity of millions of soil-inhabitant animals and microorganisms. The targets of soil microbial biogeography are to research the ecological spreads of soil microbial variety, community components, and functional properties among spaces and times from regional to worldwide measures. The research of microbial biogeography is necessary to realize further the systems that produce and preserve microbial variety and regulate key ecosystem activities, such as nutrient cycling, organic substance analysis, crop fertility, and general health [54].

### *Ecological Factor and the Global Distribution of Soil Microbial Communities*

Over the recent two decades, investigations have considerably improved our science of the deployments of soil microbial settlements from native, regional, and continental to worldwide amounts. From a classic geographical view, a negative relationship between space from the equator and the variety of plants and animals was mainly mentioned in the recent century [55]. Bacteria, protists, and planktonic foraminifera in marine habitats are negatively connected with the global latitudinal gradient [56, 57]. Nevertheless, the greatest investigations have not identified the attendance attitude of soil biodiversity worldwide in soil mechanisms. The trend of growing diversity from the poles to the orbit has been ultimately proven in the Southern Hemisphere. Environmental factors are the most important global drivers of the dispersion of soil microbial communities. Additionally, on a broad regional scale, aboveground-belowground interactions and rhizosphere-microbe relationships are important drivers of soil microbial diversity. The effects of historical factors (such as climatic legacies) (6, 26) as well as the characteristics of microorganisms themselves (such as body size, the ability to colonize, and adhesion) (41) on microbial distribution should be considered besides the effects of current environmental factors (such as climate, soil, plants, and animals) (Fig. 1.1).

Microorganisms are interdependent [58], resulting in a variety of ecologically significant but ad hoc relationships such as hostile, aggressive, mutualistic, and

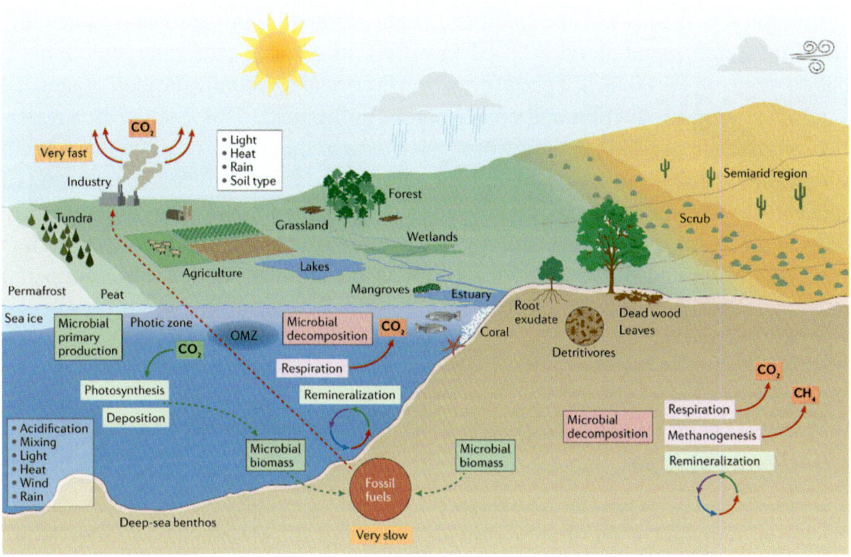


**Fig. 1.1** Changes in soil microbial biogeography in the world

predator-prey interactions [59]. This complication of the interactions between microbial partners has been prospected frequently by applying lattice analysis [60, 61]. The application of relationship networks in microbial ecology [62] has improved our valence to quantify the surface of microbial co-occurrence templates, comprehend the drivers of microbial community complex (e.g., soil carbon and pH and vegetation figures) [63–65], and know many joined taxa and keystone types [66, 67] among environmental gradients [64]. The extent of microbial networks is slightly modern and must be created based on years of experiments in researching crop and animal communities [68, 69]. Although we are yet absent a powerful document of the ecological perspective which occurs in network conclusion, that needs an experimental configuration rather in the future [70].

## *Ecosystem Function and Soil Microbial Biogeography*

Soil biodiversity displays active patterns in regulating ecological functions and ecosystem amenities [71–73]. One of the primary goals of soil microbial biogeography is to link the distribution of microbial communities to the ecological services that they're backing, which contain both single (nutrient cycling, crop fertility, and general safety) [74–77] and many (ecosystem multifunctionality) activities [72, 73]. The final experimental function identifies which microbial variety [78] and microbiome complication [79] are responsible for which ecological function. Subsequent experimental labor and worldwide projects must emphasize isolation and culture of soil microbial species and acquiring data through total-genome sequencing, proteomics, and metabolomics-based materials to allocate particular functions to specific species [78, 80–83]. This information is critical for identifying soil organisms to cultivate crop production and combat pests in the field. This data will considerably improve our current taxonomy of soil bacteria of the greatest variety that remain unidentified. Global initiatives should encourage taxonomists to devote a portion of their careers to culturing and isolating taxa, a fundamental task that is required to advance the field of microbial ecology but is often overlooked, in part because it is time-consuming and does not always result in prestigious publications, hampered researchers' early careers (Fig. 1.1).

## *Soil Biodiversity Global Atlases and Their Functions in Global -Change Scenarios*

The recent outward of the first worldwide atlases of the abundance or biodiversity of bacteria [84, 85], fungi [86], nematodes [87, 88], earthworms [89], mycorrhizal fungi, and N fixer organisms [90], highlighting possible locations including unknown species, was a major violation in soil microbial biogeography [91]. The various span of soil specifications (e.g., soil pH) and climatic situations have been used to predict and plan the worldwide dispensations of many soil organisms at zonal [92, 93], national [94, 95], and continental [96, 97], and global [86, 98] scales. These attempts have propelled the first national atlas of bacterial biodiversity among European Union (EU) member states based on the available EU-wide soil pH information [96] and the first French national atlas of soil bacterial enrichment [94]. More national tires are needed to map the dispensation of soil organisms among their territories, an effort that forms the basis for the national protection of soil biodiversity. Worldwide initiatives are needed to major study how significant land applications, such as agriculture (https://www.globalsustainableagriculture.org), adjust the global distributions of soil organisms (Fig. 1.1).

## *Biogeography of Microbial Communities*

Soil pH was the most influential environmental factor on bacterial diversity, with neutral soils having the most diverse and acidic soils having the lowest. These studies also found that taxa-area connections were poor in soil microorganisms, showing that microbial biogeography differs fundamentally from "macroorganisms." Jones [99] established the ecological features of specific populations such as Acidobacteria and validated the role of soil pH in their dispersion by applying a pyrosequencing approach to ribosomal sequences in the same soil samples. Johnson [99], on the other hand, found that changes in the genetic structure of bacterial communities from various agricultural soils were connected with soil texture and electrical conductivity rather than pH. The overall discrepancy of these results could be attributed to an insufficient sampling approach in terms of the number and representativeness of soils sampled. However, it emphasizes the need for more studies on microbial-biogeography to understand the determinism of microbial diversity better, especially since this directly affects a wide range of ecosystem functions and thus the quality of our environment.

## *Soil Bacterial Diversity*

Microorganisms are a rich source of genetic variation, but they are still poorly understood and researched [100]. Bacteria contribute significantly to this variety as one of the three domains in the evolutionary tree (Archaea, Bacteria, and Eucarya) [101]. The bacterial group has a long evolutionary history, allowing it to inhabit most terrestrial habitats. Bacteria account for the majority of biomass on Earth and are responsible for vital life processes such as the carbon, nitrogen, and sulfur cycle. As a result, there is intra-specific diversity besides bacterial species diversity. As a result, there is intra-specific diversity and bacterial species diversity. The total number of genes found in strains characterizes the bacterial genome, which can be divided into two groups: (i) the core, composed of genes found in at least 95% of strains and essential for the cell's life cycle; and (ii) the auxiliary group, found in only 5% of strains and responsible for species adaptation in different environments [102]. The core is preserved in species through speciation and vertical transmission; however, the auxiliary group does not identify the species because it is unique to each strain. This last collection of genes is also passed horizontally from strain to strain and between species [103]. This concept indicates that bacterial diversity is not static due to the high reproduction capacity linked to the short life cycle and high cell multiplication rates, which results in a high adaptability value and rapid reactions to environmental change [102, 104].

Soil bacteria are important components of soil ecosystems because they participate in the mineralization of organic matter, the biogeochemical cycling of carbon and nitrogen, and various other soil processes [105–107]. Soil qualities [9, 108], plant species [109], litter quality, and root exudates [33, 110, 111], as well as temperature and precipitation under different climatic situations, can all influence their spread [112, 113]. Microbial community study has traditionally relied on culture procedures employing a variety of culture media designed to maximize the recovery of various microbial species [114]. However, culture-dependent approaches are not commonly employed currently because it has been showed that most microorganisms cannot be cultivated in vitro [115, 116], probably because of constraints in supplying particular growth conditions in culture media [117].

## Conclusion

This chapter investigates the biogeographically distribution patterns of bacterial communities in soil. Native soil characteristics are dominant factors in shaping bacterial communities and are equally responsible for their changes. In addition, geographic distance was also an important factor in changes in bacterial communities at scale. Since soil microorganisms play an essential role in many ecosystem processes, cataloging community structures and their differences will help to predict better landscape-scale responses to environmental changes, such as erosion and soil transformations. Further work prospects include understanding the diversity patterns of another major group of soil microorganisms.

## References

1. Cameron EK, Martins IS, Lavelle P, Mathieu J, Tedersoo L, Gottschall F, Guerra CA, Hines J, Patoine G, Siebert J, Winter M (2018) Global gaps in soil biodiversity data. Nat Ecol Evol 2(7):1042–1043
2. Coyle DR, Nagendra UJ, Taylor MK, Campbell JH, Cunard CE, Joslin AH, Mundepi A, Phillips CA, Callaham Jr MA (2017) Soil fauna responses to natural disturbances, invasive species, and global climate change: current state of the science and a call to action. Soil Biol Biochem 110:116–133
3. Qiu SL, Wang LM, Huang DF, Lin XJ (2014) Effects of fertilization regimes on tea yields, soil fertility, and soil microbial diversity 74(3):333–339
4. Six J, Frey SD, Thiet RK, Batten KM (2006) Bacterial and fungal contributions to carbon sequestration in agroecosystems. Soil Sci Soc Am J 70(2):555–569
5. Bardgett R (2005) The biology of soil: a community and ecosystem approach. Oxford University Press
6. Dhuldhaj UP, Malik N (2022) Global perspective of phosphate soliloquizing microbes and phosphatase for improvement of soil, food and human health. Cell Mol Biomed Rep 2(3):173–186. https://doi.org/10.55705/cmbr.2022.347523.1048
7. Lladó S, López-Mondéjar R, Baldrian P (2017) Forest soil bacteria: diversity, involvement in ecosystem processes, and response to global change. Microbiol Mol Biol Rev 81(2):e00063-16
8. Roesch LF, Fulthorpe RR, Riva A, Casella G, Hadwin AK, Kent AD, Daroub SH, Camargo FA, Farmerie WG, Triplett EW (2007) Pyrosequencing enumerates and contrasts soil microbial diversity. ISME J 1(4):283–290
9. Lauber CL, Hamady M, Knight R, Fierer N (2009) Pyrosequencing-based assessment of soil pH as a predictor of soil bacterial community structure at the continental scale. Appl Environ Microbiol 75(15):5111–5120
10. Nemergut DR, Schmidt SK, Fukami T, O'Neill SP, Bilinski TM, Stanish LF, Knelman JE, Darcy JL, Lynch RC, Wickey P, Ferrenberg S (2013) Patterns and processes of microbial community assembly. Microbiol Mol Biol Rev 77(3):342–356
11. Broughton L, Gross KJO (2000) Patterns of diversity in plant and soil microbial communities along a productivity gradient in a Michigan old-field 125(3):420–427
12. Liu Z, Fu B, Zheng X, Liu G (2010) Plant biomass, soil water content and soil N: P ratio regulating soil microbial functional diversity in a temperate steppe: a regional scale study. Soil Biol Biochem 42(3):445–450
13. Naether A, Foesel BU, Naegele V, Wüst PK, Weinert J, Bonkowski M, Alt F, Oelmann Y, Polle A, Lohaus G, Gockel S (2012) Environmental factors affect acidobacterial communities below the subgroup level in grassland and forest soils. Appl Environ Microbiol 78(20):7398–7406

14. Stephan A, Meyer AH, Schmid B (2000) Plant diversity positively affects soil bacterial diversity in experimental grassland ecosystems 88:988–998
15. Wardle DA, Bardgett RD, Klironomos JN, Setala H, Van Der Putten WH, Wall DH (2004) Ecological linkages between aboveground and belowground biota. Science 304(5677):1629–1633
16. Kheyrodin H, Jami R, Rehman FU (2022) Cellular structure and molecular functions of plants, animals, bacteria, and viruses. Cell Mol Biomed Rep 2(1):33–41. https://doi.org/10.55705/cmbr.2022.330941.1021
17. Cho JC, Tiedje JM (2000) Biogeography and degree of endemicity of fluorescent Pseudomonas strains in soil. Appl Environ Microbiol 66(12):5448–5456
18. Zhou J, Xia B, Treves DS, Wu LY, Marsh TL, O'Neill RV, Palumbo AV, Tiedje JM (2002) Spatial and resource factors influencing high microbial diversity in soil. Appl Environ Microbiol 68(1):326–334
19. Yergeau E, Newsham KK, Pearce DA, Kowalchuk GA (2007) Patterns of bacterial diversity across a range of Antarctic terrestrial habitats. Environm Microbiol 9(11):2670–2682
20. Martiny JB, Bohannan BJ, Brown JH, Colwell RK, Fuhrman JA, Green JL, Horner-Devine MC, Kane M, Krumins JA, Kuske CR, Morin PJ (2006) Microbial biogeography: putting microorganisms on the map. Nat Rev Microbiol 4(2):102–112
21. Martiny JB, Eisen JA, Penn K, Allison SD, Horner-Devine MC (2011) Drivers of bacterial β-diversity depend on spatial scale. Proc Natl Acad Sci 108(19):7850–7854
22. Hanson CA, Fuhrman JA, Horner-Devine MC, Martiny JB (2012) Beyond biogeographic patterns: processes shaping the microbial landscape. Nat Rev Microbiol 10(7):497–506
23. Talbot JM, Bruns TD, Taylor JW, Smith DP, Branco S, Glassman SI, Erlandson S, Vilgalys R, Liao HL, Smith ME, Peay KG (2014) Endemism and functional convergence across the North American soil mycobiome. Proc Natl Acad Sci 111(17):6341–6346
24. Kivlin SN, Fei S, Kalisz S, Averill C (2020) Microbial ecology meets macroecology. Bull Ecol Soc Am 101(1):1–4
25. Borneman J, Skroch PW, O'Sullivan KM, Palus JA, Rumjanek NG, Jansen JL, Nienhuis J, Triplett EW (1996) Molecular microbial diversity of an agricultural soil in Wisconsin. Appl Environ Microbiol 62(6):1935–1943
26. Ogram A (2000) Soil molecular microbial ecology at age 20: methodological challenges for the future. Soil Biol Biochem 32(11–12):1499–1504
27. Chapman SK, Koch GW (2007) What type of diversity yields synergy during mixed litter decomposition in a natural forest ecosystem? Plant Soil 299(1):153–162
28. Costa AL, Paixão SM, Caçador I, Carolino M (2007) CLPP and EEA profiles of microbial communities in salt marsh sediments. J Soils Sediments 7(6):418–425
29. Lagomarsino A, Knapp BA, Moscatelli MC, De Angelis P, Grego S, Insam H (2007) Structural and functional diversity of soil microbes is affected by elevated [CO2] and N addition in a poplar plantation. J Soils Sediments 7(6):399–405
30. Martins-Loução MA, Cruz C (2007) microbial communities. J Soils Sediments 7(6):398
31. Winding A, Hendriksen NB (2007) Comparison of CLPP and enzyme activity assay for functional characterization of bacterial soil communities. J Soils Sediments 7(6):411–417
32. Xu Q, Jiang P, Xu Z (2008) Soil microbial functional diversity under intensively managed bamboo plantations in southern China. J Soils Sediments 8(3):177–183
33. Xu Z, Ward S, Chen C, Blumfield T, Prasolova N, Liu J (2008) Soil carbon and nutrient pools, microbial properties and gross nitrogen transformations in adjacent natural forest and hoop pine plantations of subtropical Australia. J Soils Sediments 8(2):99–105.
34. Rocha FP, Ronque MU, Lyra ML, Bacci M, Oliveira PS (2022) Habitat and host species drive the structure of bacterial communities of two neotropical trap-jaw odontomachus ants. Microb Ecol 1–14
35. Whittaker RH (1956) Vegetation of the great smoky mountains. Ecol Monogr 26(1):2–80
36. Faith DP (1992) Conservation evaluation and phylogenetic diversity. Biol Conserv 61(1):1–10
37. Sharpton TJ, Riesenfeld SJ, Kembel SW, Ladau J, O'Dwyer JP, Green JL, Eisen JA, Pollard KS (2011) PhylOTU: a high-throughput procedure quantifies microbial community diversity and resolves novel taxa from metagenomic data. PLoS Comput Biol 7(1):e1001061

38. Astorga A, Oksanen J, Luoto M, Soininen J, Virtanen R, Muotka T (2012) Distance decay of similarity in freshwater communities: do macro-and microorganisms follow the same rules?. Glob Ecol Biogeogr 21(3):365–375
39. Renner SS, Lomolino MV, Riddle BR, Brown JH (2006) Biogeography, society of systematic zoology
40. Green JL, Holmes AJ, Westoby M, Oliver I, Briscoe D, Dangerfield M, Gillings M, Beattie AJ (2004) Spatial scaling of microbial eukaryote diversity. Nature 432(7018):747–750
41. Hillebrand H, Watermann F, Karez R, Berninger UG (2001) Differences in species richness patterns between unicellular and multicellular organisms. Oecologia 126(1):114–124
42. Horner-Devine MC, Lage M, Hughes JB, Bohannan BJ (2004) A taxa–area relationship for bacteria. Nature 432(7018):750–753
43. Diamond JM (1975) The island dilemma: lessons of modern biogeographic studies for the design of natural reserves. Biol Conserv 7(2):129–146
44. Simberloff D, Abele LG (1982) Refuge design and island biogeographic theory: effects of fragmentation. Am Nat 120(1):41–50
45. Soule ME, Simberloff D (1986) What do genetics and ecology tell us about the design of nature reserves?. Biol Conserv 35(1):19–40
46. Woodcock S, Curtis TP, Head IM, Lunn M, Sloan WT (2006) Taxa–area relationships for microbes: the unsampled and the unseen. Ecol Lett 9(7):805–812
47. Franklin RB, Mills AL (2003) Multi-scale variation in spatial heterogeneity for microbial community structure in an eastern Virginia agricultural field. FEMS Microbiol Ecol 44(3):335–346
48. Nekola JC, White PS (1999) The distance decay of similarity in biogeography and ecology. J Biogeogr 26(4):867–878
49. Condit R, Pitman N, Leigh Jr EG, Chave J, Terborgh J, Foster RB, Núnez P, Aguilar S, Valencia R, Villa G (2002) Beta-diversity in tropical forest trees. Science 295(5555):666–669
50. Preston EW (1960) Time and space and the variation of species. Ecology 41(4):612–627
51. Papke RT, Ramsing NB, Bateson MM, Ward DM (2003) Geographical isolation in hot spring cyanobacteria. Environ Microbiol 5(8):650–659
52. Whitaker RJ, Grogan DW, Taylor JW (2003) Geographic barriers isolate endemic populations of hyperthermophilic archaea. Science 301(5635):976–978
53. Bell T (2010) Experimental tests of the bacterial distance–decay relationship. ISME J 4(11):1357–1365
54. Chu H, Gao GF, Ma Y, Fan K, Delgado-Baquerizo M (2020) Soil microbial biogeography in a changing world: recent advances and future perspectives. MSystems 5(2):e00803–e00819
55. Sherry TW, Kent CM (2022) Extensions and limitations of MacArthur (1958): a review of ecological and evolutionary approaches to competition and diet in the New World wood warblers (Parulidae). Ornithology 139(2):ukac010
56. Pommier T, Canbäck B, Riemann L, Boström KH, Simu K, Lundberg P, Tunlid A, Hagström Å (2007) Global patterns of diversity and community structure in marine bacterioplankton. Mol Ecol 16(4):867–880
57. Fuhrman JA, Steele JA, Hewson I, Schwalbach MS, Brown MV, Green JL, Brown JH (2008) A latitudinal diversity gradient in planktonic marine bacteria. Proc Natl Acad Sci 105(22):7774–7778
58. Graham EB, Knelman JE, Schindlbacher A, Siciliano S, Breulmann M, Yannarell A, Beman JM, Abell G, Philippot L, Prosser J, Foulquier A (2016) Microbes as engines of ecosystem function: when does community structure enhance predictions of ecosystem processes?. Front Microbiol 7:214
59. Mougi A, Kondoh M (2012) Diversity of interaction types and ecological community stability. Science 337(6092):349–351
60. Weiss S, Van Treuren W, Lozupone C, Faust K, Friedman J, Deng Y, Xia LC, Xu ZZ, Ursell L, Alm EJ, Birmingham A (2016) Correlation detection strategies in microbial data sets vary widely in sensitivity and precision. ISME J 10(7):1669–1681

61. Röttjers L, Faust K (2018) From hairballs to hypotheses–biological insights from microbial networks. FEMS Microbiol Rev 42(6):761–780
62. Barberán A, Bates ST, Casamayor EO, Fierer N (2012) Using network analysis to explore co-occurrence patterns in soil microbial communities. ISME J 6(2):343–351
63. He D, Shen W, Eberwein J, Zhao Q, Ren L, Wu QL (2017) Biochemistry, Diversity and co-occurrence network of soil fungi are more responsive than those of bacteria to shifts in precipitation seasonality in a subtropical forest. Soil Biol Biochem 115:499–510
64. de Vries FT, Griffiths RI, Bailey M, Craig H, Girlanda M, Gweon HS, Hallin S, Kaisermann A, Keith AM, Kretzschmar M, Lemanceau P (2018) Soil bacterial networks are less stable under drought than fungal networks. Nat Commun 9(1):1–12
65. Delgado-Baquerizo M, Reith F, Dennis PG, Hamonts K, Powell JR, Young A, Singh BK, Bissett A (2018) Ecological drivers of soil microbial diversity and soil biological networks in the Southern Hemisphere. Ecology 99(3):583–596
66. Banerjee S, Schlaeppi K, van der Heijden MG (2019) Reply to 'Can we predict microbial keystones?' 17(3):194–194
67. Röttjers L, Faust K (2019) Can we predict keystones?. Nature 17(3):193–193
68. Thébault E, Fontaine C (2010) Stability of ecological communities and the architecture of mutualistic and trophic networks. Science 329(5993):853–856
69. Trøjelsgaard K, Olesen JM (2013) Macroecology of pollination networks. Glob Ecol Biogeogr 22(2):149–162
70. Lv X, Zhao K, Xue R, Liu Y, Xu J, Ma B (2019) Strengthening insights in microbial ecological networks from theory to applications. MSystems 4(3):e00124-19
71. Philippot L, Spor A, Hénault C, Bru D, Bizouard F, Jones CM, Sarr A, Maron PA (2013) Loss in microbial diversity affects nitrogen cycling in soil. IME J 7(8):1609–1619
72. Wagg C, Bender SF, Widmer F, Van Der Heijden MG (2014) Soil biodiversity and soil community composition determine ecosystem multifunctionality. Proc Natl Acad Sci 111(14):5266–5270
73. Delgado-Baquerizo M, Maestre FT, Reich PB, Jeffries TC, Gaitan JJ, Encinar D, Berdugo M, Campbell CD, Singh BK (2016) Microbial diversity drives multifunctionality in terrestrial ecosystems. Nat Commun 7(1):1–8
74. Schnitzer SA, Klironomos JN, HilleRisLambers J, Kinkel LL, Reich PB, Xiao K, Rillig MC, Sikes BA, Callaway RM (2011) Soil microbes drive the classic plant diversity–productivity pattern. Ecology 92(2):296–303
75. Kanchiswamy CN, Malnoy M, Maffei ME (2015) Chemical diversity of microbial volatiles and their potential for plant growth and productivity. Front Plant Sci 6:151
76. Limmathurotsakul D, Golding N, Dance DA, Messina JP, Pigott DM, Moyes CL, Rolim DB, Bertherat E, Day NP, Peacock SJ, Hay SI (2016) Predicted global distribution of Burkholderia pseudomallei and burden of melioidosis. Nat Microbiol 1(1):1–5
77. Nelson MB, Martiny AC, Martiny JB (2016) Global biogeography of microbial nitrogen-cycling traits in soil. Proc Natl Acad Sci 113(29):8033–8040
78. Delgado-Baquerizo M, Trivedi P, Trivedi C, Eldridge DJ, Reich PB, Jeffries TC, Singh BK (2017) Microbial richness and composition independently drive soil multifunctionality. Funct Ecol 31(12):2330–2343
79. Wagg C, Schlaeppi K, Banerjee S, Kuramae EE, van der Heijden MG (2019) Fungal-bacterial diversity and microbiome complexity predict ecosystem functioning. Nat Commun 10(1):1–10
80. Martinez-Alonso E, Pena-Perez S, Serrano S, Garcia-Lopez E, Alcazar A, Cid C (2019) Taxonomic and functional characterization of a microbial community from a volcanic englacial ecosystem in Deception Island, Antarctica. Sci Rep 9(1):1–14
81. Jaswal R, Pathak A, Chauhan A (2019) Metagenomic evaluation of bacterial and fungal assemblages enriched within diffusion chambers and microbial traps containing uraniferous soils. Microorganisms 7(9):324
82. Swenson TL, Karaoz U, Swenson JM, Bowen BP, Northen TR (2018) Linking soil biology and chemistry in biological soil crust using isolate exometabolomics. Nat Commun 9(1):1–10

83. Bastida F, Crowther TW, Prieto I, Routh D, García C, Jehmlich N (2018) Climate shapes the protein abundance of dominant soil bacteria. Sci Total Environ 640:18–21
84. Delgado-Baquerizo M, Eldridge DJ (2019) Cross-biome drivers of soil bacterial alpha diversity on a worldwide scale. Ecosystems 22(6):1220–1231
85. Delgado-Baquerizo M, Oliverio AM, Brewer TE, Benavent-González A, Eldridge DJ, Bardgett RD, Maestre FT, Singh BK, Fierer N (2018) A global atlas of the dominant bacteria found in soil. Science 359(6373):320–325
86. Egidi E, Delgado-Baquerizo M, Plett JM, Wang J, Eldridge DJ, Bardgett RD, Maestre FT, Singh BK (2019) A few Ascomycota taxa dominate soil fungal communities worldwide. Nat Commun 10(1):1–9
87. Van Den Hoogen J, Geisen S, Routh D, Ferris H, Traunspurger W, Wardle DA, De Goede RG, Adams BJ, Ahmad W, Andriuzzi WS, Bardgett RD (2019) Soil nematode abundance and functional group composition at a global scale. Nature 572(7768):194–198
88. Bastida F, Eldridge DJ, Abades S, Alfaro FD, Gallardo A, García-Velázquez L, García C, Hart SC, Pérez CA, Santos F, Trivedi P (2020) Climatic vulnerabilities and ecological preferences of soil invertebrates across biomes. Mol Ecol 29(4):752–761
89. Phillips HR, Guerra CA, Bartz ML, Briones MJ, Brown G, Crowther TW, Ferlian O, Gongalsky KB, Van Den Hoogen J, Krebs J, Orgiazzi A (2019) Global distribution of earthworm diversity. Science 366(6464):480–485
90. Steidinger BS, Crowther TW, Liang J, Van Nuland ME, Werner GD, Reich PB, Nabuurs GJ, de-Miguel S, Zhou M, Picard N, Herault B (2019) Climatic controls of decomposition drive the global biogeography of forest-tree symbioses. Nature 569(7756):404–408
91. Delgado-Baquerizo M (2019) Obscure soil microbes and where to find them. ISME J 13(8):2120–2124
92. Ferrier S, Manion G, Elith J, Richardson K (2007) distributions, Using generalized dissimilarity modelling to analyse and predict patterns of beta diversity in regional biodiversity assessment. Divers Distrib 13(3):252–264
93. Bru D, Ramette A, Saby NP, Dequiedt S, Ranjard L, Jolivet C, Arrouays D, Philippot L (2011) Determinants of the distribution of nitrogen-cycling microbial communities at the landscape scale. ISME J 5(3):532–542
94. Terrat S, Horrigue W, Dequietd S, Saby NP, Lelièvre M, Nowak V, Tripied J, Régnier T, Jolivet C, Arrouays D, Wincker P (2017) Mapping and predictive variations of soil bacterial richness across France. PloS One 12(10):e0186766
95. Karimi B, Terrat S, Dequiedt S, Saby NP, Horrigue W, Lelièvre M, Nowak V, Jolivet C, Arrouays D, Wincker P, Cruaud C (2018) Biogeography of soil bacteria and archaea across France. Sci Adv 4(7):eaat1808
96. Griffiths RI, Thomson BC, Plassart P, Gweon HS, Stone D, Creamer RE, Lemanceau P, Bailey MJ (2016) Mapping and validating predictions of soil bacterial biodiversity using European and national scale datasets. Appl Soil Ecol 97:61–68
97. Jiao S, Xu Y, Zhang J, Lu Y (2019) Environmental filtering drives distinct continental atlases of soil archaea between dryland and wetland agricultural ecosystems. Microbiome 7(1):1–13
98. Tedersoo L, Bahram M, Põlme S, Kõljalg U, Yorou NS, Wijesundera R, Ruiz LV, Vasco-Palacios AM, Thu PQ, Suija A, Smith ME (2014) Global diversity and geography of soil fungi. Science 346(6213):1256688
99. Johnson MJ, Lee KY, Scow KM (2003) DNA fingerprinting reveals links among agricultural crops, soil properties, and the composition of soil microbial communities. Geoderma 114(3–4):279–303
100. Prosser JI, Bohannan BJ, Curtis TP, Ellis RJ, Firestone MK, Freckleton RP, Green JL, Green LE, Killham K, Lennon JJ, Osborn AM (2007) The role of ecological theory in microbial ecology. Nat Rev Microbiol 5(5):384–392
101. Woese CR (1994) There must be a prokaryote somewhere: microbiology's search for itself. Microbiol Rev 58(1):1–9
102. Konstantinidis KT, Ramette A, Tiedje JM (2006) Toward a more robust assessment of intraspecies diversity, using fewer genetic markers. Appl Environ Microbiol 72(11):7286–7293

103. Boucher Y, Nesbø CL, Doolittle WF (2001) Microbial genomes: dealing with diversity. Curr Opin Microbiol 4(3):285–289
104. Abby S, Daubin V (2007) Comparative genomics and the evolution of prokaryotes. Trends Microbiol 15(3):135–141
105. Bardgett RD, Freeman C, Ostle NJ (2008) Microbial contributions to climate change through carbon cycle feedbacks. ISME J 2(8):805–814
106. Burton J, Chen C, Xu Z, Ghadiri H (2010) Sediments, Soil microbial biomass, activity and community composition in adjacent native and plantation forests of subtropical Australia. J Soils Sediments 10(7):1267–1277
107. Chatterjee A, Vance GF, Pendall E, Stahl PD (2008) Biochemistry, Timber harvesting alters soil carbon mineralization and microbial community structure in coniferous forests. Soil Biol Bioche 40(7):1901–1907
108. Ushio M, Wagai R, Balser TC, Kitayama K (2008) Variations in the soil microbial community composition of a tropical montane forest ecosystem: does tree species matter?. Soil Biol Biochem 40(10):2699–2702
109. Oh YM, Kim M, Lee-Cruz L, Lai-Hoe A, Go R, Ainuddin N, Rahim RA, Shukor N, Adams JM (2012) Distinctive bacterial communities in the rhizoplane of four tropical tree species. Microb Ecol 64(4):1018–1027
110. Quideau SA, Chadwick OA, Benesi A, Graham RC, Anderson MA (2001) A direct link between forest vegetation type and soil organic matter composition. Geoderma 104(1–2):41–60
111. Grayston SJ, Prescott CE (2005) Microbial communities in forest floors under four tree species in coastal British Columbia. Soil Biol Biochem 37(6):1157–1167
112. Sowerby A, Emmett B, Beier C, Tietema A, Peñuelas J, Estiarte M, Van Meeteren MJ, Hughes S, Freeman C (2005) Biochemistry, Microbial community changes in heathland soil communities along a geographical gradient: interaction with climate change manipulations. Soil Biol Biochem 37(10):1805–1813
113. Nielsen UN, Osler GH, Campbell CD, Burslem DF, van der Wal R (2010) The influence of vegetation type, soil properties and precipitation on the composition of soil mite and microbial communities at the landscape scale. J Biogeogr 37(7):1317–1328
114. Hill GT, Mitkowski NA, Aldrich-Wolfe L, Emele LR, Jurkonie DD, Ficke A, Maldonado-Ramirez S, Lynch ST, Nelson EB (2000) Methods for assessing the composition and diversity of soil microbial communities. Appl Soil Ecol 15(1):25–36
115. Rappé MS, Giovannoni SJ (2003) The uncultured microbial majority. Annual Rev Microbiol 57(1):369–394
116. Ritz K (2007) The plate debate: cultivable communities have no utility in contemporary environmental microbial ecology. FEMS Microbiol Ecol 60(3):358–362
117. Trevors JT (1998) Bacterial biodiversity in soil with an emphasis on chemically-contaminated soils. Water Air Soil Pollut 101(1):45–67

# Chapter 2
# Microbial Consortium: A Boon for a Sustainable Agriculture

**Manoj Kaushal, Sunita Devi, Kailash Chand Kumawat, and Ajay Kumar**

**Abstract** Rhizosphere is a highly activated region in the soil where microbial number and diversity is huge. These belowground microbes are interacted with each other as well as with the plant roots and some of these interactions are beneficial for plant growth. The plant signalling molecules (like root exudates) produced by plants shapes the microbial diversity in the rhizospheric region. Some of the rhizosphere microbes are useful for the plant development and are known as plant growth promoting rhizomicrobes (PGPR). These PGPR exerted various plant growth promoting effects by various mechanisms like phosphate solubilisation, nitrogen fixation, plant growth hormones production, secretion of antimicrobial compounds etc. These PGPR are excellent substitute for chemical inputs used for increasing crop production as chemical inputs disrupt the soil biological as well as chemical property. The PGPR formulation used as biofertilizer and are generally use single microbial strain. But the application of single microbial strain biofertilizer in soil showed inconsistency in the results. Research studies have showed that application of biofertilizer containing two or more microbial strains also known as co-inoculation or consortium is more beneficial as compared to single microbial strain application. Therefore, in the present chapter the importance of biofertilizer containing microbial consortium for the application in sustainable agriculture is discussed.

M. Kaushal
Plant Production and Plant Health, International Institute of Tropical Agriculture (IITA), Dar es Salaam, Tanzania

S. Devi
Department of Basic Sciences, College of Forestry, Dr. YSP University of Horticulture & Forestry, Solan, HP 173230, India

K. C. Kumawat · A. Kumar (✉)
Department of Industrial Microbiology, Jacob Institute of Biotechnology and Bioengineering, Sam Higginbottom University of Agriculture, Technology and Sciences, Prayagraj, UP 211007, India
e-mail: ajaysamkria@gmail.com



J. A. Parray (ed.), *Climate Change and Microbiome Dynamics*, Climate Change Management, https://doi.org/10.1007/978-3-031-21079-2_2

## Introduction

With increase in human population there is demand of increasing crop production to main the food supply at equilibrium. Initially usage of chemical fertilizers achieve this target but continuous and increased usage of these chemical fertilizers have adverse affect on soil biological health which degrades the soil physico-chemical properties as well as these chemicals enter into the food chain and cause diseases in human [1]. Microorganisms are marvellous alternate for sustainable agriculture to overcome the issues encountered by the usage of chemical fertilizers. Plant associated microbiome is found to promote plant health therefore worldwide scientists trying to explore these useful microorganisms [2]. The huge diversity of these useful microbes is persisting mainly in the rhizospheric region (soil surrounding the plant root) because of exudation from plant roots which serve as nutrients for the microbes [2, 3]. These rhizospheric microbes reported to exert beneficial effect on plant growth by various mechanisms like helping in absorbing nutrients, conversion of non-usable form of nutrients to available form, protect plant from pathogens by secreting antimicrobial compounds, improve stress tolerance capacity of plants under adverse growth conditions etc. [4]. So, utilizing these beneficial microbes in agriculture not only decrease our dependence on chemical inputs but also improve soil health along with improved crop production.

The plant roots and rhizopsheric microbiome are not only connected physically but also chemically. The microbiome composition in the rhizopshere is influenced by signal molecules produced by the plant roots like root exudates. The root exudates shape the microbial composition in the rhizospheric region and microbial diversity varies with the plant species [5]. In the same harmony, rhizospheric microbes also influence the plant growth and perform other ecological cycles. These diverse rhizospheric microbes communicate in the rhizosphere using various mechanisms like quorum sensing to maintain homeostasis in this region [6].

These rhizosphere associated microbes which exerted positive effect on plant growth are known as Plant Growth Promoting Rhizomicrobes (PGPR) which includes bacteria as well as fungi. These PGPR colonize the root zone soil or may be present intracellularly within plant cell and exert positive growth affect on plant growth when applied to soil or surface of plant or seed [4]. The application of these PGPR not only improves the crop production under sustainable agriculture but also after continues use of these PGPR for 3–4 years there is no need to apply inocula of these beneficial strains as they naturally build up in sufficient quantity within soil ecosystem [7]. These PGPR are used as biofertilizers which may be phosphate solubilizers or nitrogen fixers etc. using bacteria or cyanobacteria or fungi or their combination.

Generally, biofertilizers containing single PGPR strain is applied in the agricultural soil but due to inconsistent performance of single microbial strain it is always beneficial to use mixed microbial culture or co-inoculation or consortium for the application in agriculture. This application of microbial consortium is helpful in exploiting the synergistic interaction of microbes or complimentary benefits for plant growth [8, 9]. The biofertilizer consortium basically consists of different compatible microbial strains (allochthonous) with diverse plant growth promoting attributes. The genetically different microbes in the consortium have different ability to adapt to various adverse soil conditions like pH, moisture, temperature etc. [10]. After application in the soil, these different consortium microbial strains can be activated by the root exudates or other plant physiological response in the rhizosphere region. The production of single biofertilizer strain in industry is costly as compared to the production of biofertilizer consortium [8, 10]. Also, with the application of biofertilizer consortium in the soil multiple plant growth promoting traits are activated simultaneously in the rhizospheric region. So, overall usage of microbial consortium is broad spectrum as compared to the application of single microbial strain. Therefore, to achieve improved plant growth the microbial consortium with multifarious plant growth traits are excellent tool over single microbial strain application in sustainable agriculture.

## Multifarious PGP Attributes

The growth and development of plants are influenced by PGPR through a variety of direct and indirect mechanisms [5], which may be active concurrently or sequentially at diverse phases of plant growth and development (Table 2.1). Figure 2.1 depicts each of these mechanisms, which are then detailed in depth below for a better understanding.

### *Direct Mechanisms*

The most vital nutrient in terms of plant growth and yield is nitrogen. Notwithstanding that there is over 78 percent of $N_2$ in the atmosphere, plants cannot use it. The process of biological nitrogen fixation (BNF) converts atmospheric $N_2$ into plant-available forms, with $N_2$ being converted to $NH_3$ by nitrogen-fixing microorganisms [27]. An enzyme called nitrogenase complex catalyses the $N_2$-fixation process [28]. The dinitrogenase reductase offers electrons with strong reducing power, which are then utilized by dinitrogenase for reducing $N_2$ to $NH_3$. The $N_2$-fixing mechanism differs structurally among different bacterial taxa. The enzyme, molybdenum nitrogenase that found in almost all diazotrophs, catalyses the majority of BNF [29]. Examples of diazotrophic bacteria that freely fix and supply nitrogen to a variety of plants include *Bacillus, Azospirillum, Anabaena, Azotobacter, Nostoc, Clostridium, Klebsiella, Rhodobacter*, and *Paenibacillus* [30]. Some diazotrophs, like *Herbaspirillum* spp., *Azospirillum* spp., *and*, *Azoarcus* spp., form endophytic and/or associative relationships with an array of plant roots, including cereal roots. The main *Azospirillum* species researched worldwide are *A. lipoferum* and *A. brasilense*, which are commonly used for inoculating maize, sugarcane, and rice.*A. brasilense* exhibit the potential to change the root architecture of plants by stimulating the growth and

**Table 2.1** Mechanisms of plant growth by different microbes isolated from the rhizosphere

| Biological role | Type of association | Organism involved | Mechanism | References |
|---|---|---|---|---|
| Nitrogen fixation | Free living | *Anabaena, Azotobacter,, Nostoc, Clostridium, Klebsiella* | Convert non-usable form of nitrogen into usable form and make available to plant roots | [11] |
| | Associative symbiotic | *Azospirillum, Herbaspirillum, Azoarcus, Enterobacter, Pantoea* | | [12, 13] |
| | Symbiotic | *Azolla, Anabaena, Frankia, Rhizobium* | | [14, 15] |
| Phosphate solubilisation & mobilization | Fungi | *Aspergillus, Arbuscular mycorrhiza, Glomus, Penicillium, Talaromyces, Trichoderma* | Solubilize insoluble form of phosphorus into souluble form that is absorbed by the plant roots | [16–20] |
| | Bacteria | *Bacillus, Burkholderia, Pseudomonas, Ralstonia* | | [19, 21, 22] |
| Production of plant growth promoting hormones | | *Azorhizobium, Azotobacter, Bacillus, Bradyrhizobium, Pseudomona, Rhizobium, Streptomyces* | Various plant growth hormones produced which improve the plant growth and yield | [23, 24] |
| Antifungal activity | | *Bacillus, Pseudomonas, Streptomyces* | Some microbes produce metabolites which have antifungal activity against plant pathogens | [25, 26] |

proliferation of lateral and adventitious roots, as well as root hairs [31] besides synthesizing NO via a variety of pathways. Root organogenesis, formation of root hairs along with lateral and adventitious roots, all require NO [32].

Rhizobia are the most well-known group of bacteria that exhibit the potential to fix nitrogen (>200 kg N/ha/ year) symbiotically with the plant species of *Fabaceae /Leguminosae* family in both temperate [33] and tropical [34] regions. However, two other bacterial genera, *Cyanobacteria* and *Frankia*, can also fix

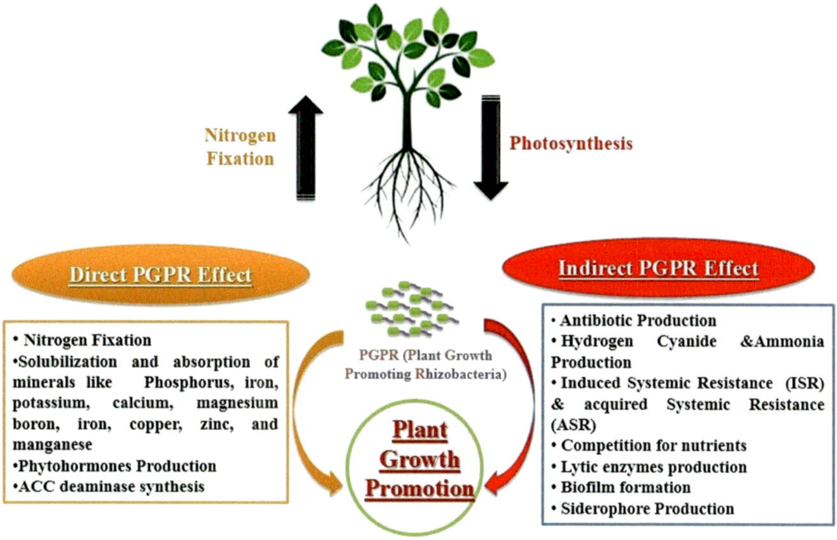


**Fig. 2.1** Direct and indirect mechanisms of plant growth-promotion

nitrogen in a symbiotic relationship with plants. For nitrogen fixation, cyanobacteria can develop a symbiotic relationship with an array of plants viz., bryophytes, gymnosperms, and angiosperms, while *Frankia* fix nitrogen by nodulating the actinorhizal plants Chang et al. [35].

Cyanobacteria serve as the main source of fixed nitrogen in the Arctic as well as terrestrial ecosystems [36]. For instance, in northern boreal forests, a high copiousness of cyanobacterial—feather moss associations contribute around 1.5–2.0 kg N/ha/year [37]. Species of the genera viz., *Anabaena, Tolypothrix, Nostoc, Aulosira, Scytonema*, and *Cylindrospermum* are found in abundance in the rice fields, all of which contribute significantly to rice fertility. Cyanobacteria have been documented to contribute approximately 20–30 kg N/ha every season, plus organic matter, which is significant for economically disadvantaged farmers who cannot afford to invest in expensive chemical nitrogen fertilizers. The amalgam of *Anabaena* (a free-living $N_2$- fixing diazotroph) with *Azolla* provides a natural way to provide nitrogen to rice plants growing under waterlogged conditions [38]. Rice biofertilization with *Anabaena* provides high nitrogen levels (up to 50 kg/ha), minimizes nitrogen loss through ammonia volatilization, and promotes the growth and development of plant [39].

The genus, *Frankia,* is comprised of aerobic, free-living, and symbiotic soil actinomycetes (family: *Frankiaceae*) that fixes nitrogen in the range of 2–300 kg N/ha/year, in harsh environments including mines, reclaimed, and degraded lands [40]. Around 200 *Frankia* strains, belonging to the genera viz., *Agromyces, Arthrobacter,*

*Corynebacterium, Micromonospora, Mycobacterium, Streptomyces* and *Propionibacteria* have been recovered from anarray of actinorhizal plant species, but not all, exhibiting $N_2$ fixing potential [41].

Phytohormone biosynthesis is also documented to encourage plant growth directly. Several species of genera *Azotobacter, Alcaligenes, Azospirillum, Bradyrhizobium, Bacillus, Brevibacillus, Enterobacter, Burkholderia, Klebsiella, Pseudomonas, Mycobacterium, Serratia,* and *Rhizobium* produce and release phytohormones viz., auxin, gibberellin, cytokinin, ethylene, and abscisic acid [42–44]. Indole-3-acetic acid (IAA) is the most commonly researched auxin in the world. Growth stimulation plus a transitory increase in IAA levels was observed in wheat seedlings upon treatment with *Bacillus subtilis* 11BM spores [45]. *Pseudomonas aeruginosa* and/or *Mesorhizobium* sp. produced IAA, which enhanced potassium and phosphate uptake in chickpea inoculated with these bacteria [46]. Species of genera viz., *Bacillus* spp., *Burkholderia cepacia, Promicromonospora* spp., and *Herbaspirillum seropedicae* are potential gibberellins (GAs) producers. *B. siamensis* is reported to enhance growth in banana plants via GA production [47]. GA3, produced by *Azospirillum* was verified to be imperative in increasing plant growth while, co-inoculation of *Pseudomonas fluorescens* plus *Azospirillum brasilense* boosted wheat biomass and yield [48].

Roots are accountable for the synthesis of 1-aminocyclopropane-1-carboxylate (ACC), which is a direct ethylene precursor. PGPR with ACC deaminase, an enzyme that converts ACC to $\alpha$-ketobutyrate and ammonium and thereby decreases ethylene levels, can metabolize ACC. Ethylene promotes the elongation process of plant root under normal and stressed environments at low concentrations. Because ACC deaminase lowers ethylene levels, modifying ACC levels in hosts may assist in alleviating the negative impacts of abiotic and biotic stressors. Besides ethylene, Abscisic acid (ABA) also regulates plant growth in stressful environments. PGPR exhibiting ABA-producing activities include *Bacillus licheniformis, Achromobacter xylosoxidans, Bacillus pumilus, Brevibacterium halotolerans, Bacillus subtilis, Pseudomonas putida,* and *Lysinibacillus fusiformis* [49].

PGPR also provide nutrients like phosphorus, and potassium to plants under nutrient-limited environs [50, 51]. Phosphorus is typically present in soil as hydroxyapatite, rock phosphate and/or calcium phosphate, and is mostly found in the form of either phytate (organic form), or insoluble phosphate (inorganic form). PGPR exhibit the potential to solubilize phosphate either via organic acid production or phytase activity [52]. Phytase producing bacteria belong to the genera viz., *Bacillus, Enterobacter, Klebsiella,* and *Pseudomonas* while, *Bacillus, Burkholderia, Erwinia, Paenibacillus, Pseudomonas, Rhizobium,* and *Serratia* genera have all been documented to solubilize phosphate- via release of organic acids like oxalate, citrate, and acetate [53, 54].

Besides phosphorus and, nitrogen, PGPR can efficiently stimulate plant growth via solubilization and absorption of other nutrients [49, 51]. For instance, a noteworthy upsurge in the uptake of potassium, calcium, and magnesium via their solubilization was observed by Ogut et al. [53] after inoculating wheat with *Bacillus* sp. or *Pseudomonas* sp. in calcareous soil without applying fertilizers. Under water-stressed conditions, *Bacillus megaterium* boosted phosphorus, calcium, boron, iron, copper, zinc, and manganese absorption as well as biomass in trefoil plants [55].

Iron is another micronutrient that plays an indispensable role in an array of metabolic activities, and its deficiency impairs key plant metabolic activities like respiration and photosynthesis. Rhizobacteria like *Pseudomonas, Bacillus, Klebsiella, Enterobacter, Rhodococcus* are known to produce siderophores which are tiny iron—chelating molecules that allow iron to be transported to root cells under iron-limiting conditions. This mechanism aids plant growth while also creating an unfavourable environment for phytopathogens that cannot thrive in iron-deficient environments [32, 56]. To demolish soilborne pathogen's cell walls, *Paenibacillus, Bacillus, Serratia, Pantoea,* and *Enterobacter* secrete lytic enzymes such amylase, chitinase, β-1, 3-glucanase, and protease [56].

## *Indirect Mechanisms*

Numerous literature sources reveal that PGPR serve the function of protective agents against soil-inhabiting pathogens [57]. Rhizobacteria can limit disease development via multiple ways, for instance, antagonistic effect of pseudomonads via synthesizing a multitude of antibiotics viz., pyoluteorin, phenazine, pyrrolnitrin, tropolone, tensin, amphisin etc. [58]; competition for nutritional substrates and ecological niches with phytopathogens plus other detrimental microbes proliferating in the rhizosphere [59]; production of cell wall degrading molecules like chitinases, β-1,3-glucanase, and biosurfactants [60], production of ammonia and hydrogen cyanide like volatile organic inhibitory molecules [61]; and induced resistance [62].

Recent scientific findings have reported that biofilm production in the rhizosphere plays a significant role in rhizobacteria's mode of action on root pathogens. The high population density of bacteria in biofilms is ascribed for the production of diverse metabolites like toxins and antibiotics in their periphery, which suppress phytopathogens in the soil. For instance, in case of *Bacillus subtilis,* biofilm is made up of surfactins, which are cyclic molecules containing lipids and amino acids that operate as potent biosurfactants with antimicrobial (antibacterial and antifungal) properties besides inducing resistance in plants [63]. The particulars of antagonist effect of *B. subtilis* strain SG6 on *Fusarium* hyphae as discerned by electron microscopic studies reveal the evident anomaly in mycelial growth that can be allied with the influence of chitinase like cell wall degrading enzymes [60]. Other toxic compounds obtained from *B. subtilis* include lipopeptide antibiotics, belong to the surfactin and iturin group that are accountable for plant disease suppression. In the rhizospheric region, antagonism encompassing competition for nutrients and space within an ecological niche is also crucial. This was demonstrated in on *B. megaterium,* a bacterium that can competently colonize roots and diminish *Rhizoctonia solani* [64].

Rhizobacteria produce siderophores as a secondary byproduct of their metabolism. These compounds exhibit the potential of sequestering $Fe^{3+}$ ions, which are mandatory for cell growth and metabolism. In this context, plant root's colonizing bacteria might display competition for the iron available in the soil, inhibiting the growth of other rhizospheric microbes. Siderophore-producing PGPR can inhibit harmful microbes from proliferating around the root [65].

To combat phytopathogens, plants possess a basal natural defensive system, but additional systems can be activated or induced to boost plant resistance [66]. Induced systemic resistance (ISR) and acquired systemic resistance (ASR) are two types of resistance induction that have been researched extensively. ISR is commenced by non-pathogenic rhizospheric microbes and does not entail the salicylic acid signalling route or synthesis of plant pathogenesis related proteins (PRPs); instead, ethylene and jasmonic acid-mediated—resistance-signaling pathway is activated [67]. In ASR, on plant's exposure to a pathogen that act as an inducing agent, defence mechanisms are activated both at the induction site that exhibits necrosis like changes as well as another distant sites, providing systematic protection to plant against subsequent infections caused by an array of pathogens [68]. ASR is followed by a rise in salicylic acid content and the build up of PRPs, which are plant defense mechanisms [69].

In nutshell, growing usage of PGPR could be envisaged amongst major avenues to maintain or enhance yield while reducing environmental imprint via explanation of many mechanisms that will assist to make these plant-beneficial rhizobacteria a valued partner in agriculture to generate future insights.

## Microbial Consortium in Agriculture (Bacteria-Bacteria and Bacteria-Fungi Consortium)

Microbes possess functional attributes that regulates the plant growth, improve the availability soil nutrients, and provides protection against stress conditions. These traits led to vast exploration of microbial strains followed by commercialization. However, in any niche area, composition and structure of microbes played crucial role in overall beneficial functions enhancement. Microbial consortia that have synergistic interactions among themselves can exhibit high level performance compared to single strains due to the diverse set of plant growth promotion attributes and biocontrol mechanisms [8]. These microbial consortia are equipped with RIDER mechanism that helps in higher nutrient uptake and ameliorating drought and salinity under extreme environments [70]. Others are crucial for maintaining soil health by nutrient assimilation, N-fixations excluding the conventional methods of agricultural production. Before developing a microbial consortium, first steps are needed. This means that the compatibility of the microorganisms used in the host plant in question, and the co-occlusion of these microorganisms, directly or indirectly affect the host. Inoculation in combination with beneficial microorganisms showed improved plant growth and yield characteristics as well as germination, nutrient absorption, plant height, number of branches, tuber formation, yield, and total crop biomass. The consortium's proposals improve the efficiency, consistency, and reliability of microorganisms in a variety of soil conditions [71]. The combination of biocontrol agents in the consortium is said to provide a higher level of protection and have the potential to control multiple plant diseases.

## *Bacteria–Bacteria Interactions*

Bacterial consortiums are usually referred to as groups of different strains of bacteria that can live together in the community. Rhizobacteria that promote plant growth (PGPR) can inhabit the soil or rhizosphere zone along with other bacterial strains [72]. Bacterial diversity has properties that promote plant growth and development, as well as general benefits that contribute to one health approach. There are many factors that influence the bacterial consortium, and interactions between consortium members are important for long-term stability. The interaction of these bacteria can be positive, negative, or neutral [73]. Positive associations include mutualism, proto-cooperation, and commensalism. Biocontrol mechanisms are the example of positive associations which employs various biological control bacterial strains having growth promoting traits to achieve desired results. These types of positive interactions require compatibility of consortium strains in soil and/ or rhizosphere zones and devoid of any kind of competition within the group. Evaluation is likely the maximum critical section for the duration of improvement of microbial consortium as it gives a know-how of its contribution in reducing stress and growing plant boom. Attempts are being made to expand microbial consortium for pests and diseases suppression and plant growth promotion. The important concept at the back of using bacterial consortiums is that an unmarried microorganism does now no longer always offer safety in opposition to a couple of pathogens, so the use of a set of microorganisms guarantees that safety in opposition to a couple of goal pathogens is provided [71, 74].

On the other hand, negative interactions bring about suppression of bacterial individuals of the consortium, disrupting network shape and characteristic. These consist of amensalism, predation, parasitoids, and competition. Competition arises whilst individuals of the bacterial consortium want the equal resources. It‘s nutrients, water, or even the space. Therefore, fast-developing strains dominate over time. Neutral interaction happens whilst the two bacterial species devour distinctive materials (nutritional differences) and do now no longer produce compounds that inhibit individuals of the consortium. In agriculture, individuals of the consortium actively have interaction whilst symbiotic associations are preferred to attain solid overall performance in long-time period cultivation for you to attain the useful outcomes anticipated whilst carried out to producing crops.

In this regard, bacterial consortium is presently most effective superficially understood. The interaction among consortium relies upon at the generation, recognition, and reaction of extracellular signaling molecules that adjust and shape bacterial populations within the consortium. In the consortium, most effective compatible bacterial strains are worried in changing plant protection responses that have an effect on plant health and production [75]. Bacterial consortium interactions are based closely on molecular signals. Among them, quorum sensing performs a critical function in bacterial compatibility in consortium formulations [76]. Of the numerous signaling molecules, the acylhomoserine lactone (AHL) signaling molecule is the maximum outstanding identified in bacterial strains [77]. On the alternative hand, AHL produced through bacterial consortium of *S. liquefaciens* and *S. phymuthica* help in root improvement and plant biomass. Other bacterial strains including *S. fredii* and *P. aananatis* form biofilm within the roots of *Oryza sativa* and *Phaseolus vulgaris* [78].

Other vital signaling compounds stated in bacterial consortia are unstable compounds called volatile organic compounds (VOCs), which are identified with bacteria–bacteria and plant–bacteria communications [79]. These compounds encompass terpenoids, alkanes, alkenes, ketones, sulfur-containing compounds, and alcohols that act as low-molecular-weight compounds. Individual and bacterial consortium of *A. brasilense* Sp7, *P. putida* KT2440, *Acinetobacter* sp. EMM02, and *Sphingomonas* sp. OF178A are the crucial examples of bacterium- maize seed interactions [80]. It was also observed that the inoculation of the bacterial consortium also improves the bacterial colonization. Bacterial colonization is predicated upon on the plant variety. The colonization of a consortium formulated with *G. diazotrophicus*, *H. seropedicae*, *H. rubrisubalbicans*, *A. amazonense*, and *B. tropica* differ in different forms of sugar cane (SP70-1143 and SP 813,250) [81].

It is essential to confirm the protection of bacterial consortium earlier than theyŕe used as biofertilizers, especially if theyŕe carefully associated with pathogenic bacterial traces. For instance, *Bacillus* sp. (RZ2MS9) and *B. ambifaria* (RZ2MS16) gift a cap potential threat because of their taxonomic proximity to pathogenic groups [82]. The coinoculation of maize with *A. brasilense* and *B. subtillis* has additionally proven more advantages than individual inoculation [83]. The maize inoculation with a consortium with *A. chrococcum* and *A. liporefum* ended in increments in shoot and seed dry weight, plant height, and yield as compared to the individual inoculation of bacterium and the control [84]. Nitrogen fertilization at 100% and the consortium plus 50% urea resulted the best increments in height, diameter, dry root weight, and grain weight compared to non-inoculated plants. These results confirmed that the bacterial consortium stimulates the growth of maize whilst a 1/2 of dose of mineral nitrogen utilized in conventional agricultural practices. In another study, the rice inoculation with a consortium (blended *Pseudomonas* culture in addition to *A. Chroococcum* and *A. brasilense*), the benefits of 50% mineral phosphorus were like the total dose of phosphorus and consortium [85, 86]. In sunflowers, the bacterial consortium (*Azotobacter* sp. and *Azospirillum* sp.), 50% nitrogen fertilization was identified in addition to the highest grain production, oil and protein levels. Most studies in which plants were inoculated with bacterial consortium found spikes in yield and biomass [87, 88].

Plants interact with indole generating and phosphate solubilizing bacteria at low nutrient situations. However, in a mild nutrient scheme, plants selectively partner with bacteria with a better potential for phosphate solubilization [89, 90]. Better plant growth and productivity with 50% urea plus the bacterial consortium could be because of the excessive phosphate solubilization functionality and indole manufacturing by few members of bacterial consortium [91]. However, it's miles important to do extra research addressing this topic, possibly the use of bacterial consortium in those mechanisms to confirm their roles in nutrient solubilization and plant growth. More research is also needed to outline the function of bacterial consortium on plant inoculation that provides an opportunity to implement sustainable agricultural practices without compromising crop yields.

## *Bacteria-Fungal Interactions*

It is now feasible to behavior studies on the character and composition of microbial interactions with plants using next-technology sequencing (NGS) techniques. Many bacterial and fungal interactions play role in plant improvement through nutrient mobilization and to cope up with numerous biotic and abiotic stresses [92]. For instance, phosphate may be solubilized through phytases secreted by soil-borne bacteria or fungi, thus favoring its uptake. Another low-molecular-weight molecule of microbial consortium are called siderophores that are the starting place with an excessive affinity for iron and contribute to solubilize iron within the rhizosphere. Biological nitrogen fixation is the most important form of symbiotic association with a microbial consortium that resolves $N_2$. The exchange of nutrients between plants, fungi (rootstock fungi) and bacteria help improve plant nutrition, including nitrogen uptake. Plant N uptake can be increased in the presence of symbiotic persistent and binding $N_2$ bacteria and mycorrhizal fungi (AMF). The minerals are taken up from the soil by mycorrhizal fungi and contribute to higher plant uptake. The minerals are then secreted by the fungal cells at the dendritic interface and picked up by the plant cells. Apart from N, the phosphatase released by bacteria associated with fungi, inorganic phosphate is absorbed by fungi and plant cells via the phosphate vector (PT). Phosphate polymers can be stored inside and outside the radical fungi at the plant roots. Polyphosphate is decomposed and inorganic phosphates are then transported to the ambient interface [93, 94].

*Rhizobium* is an alphaproteobacteria that usually causes persistent $N_2$ symbiosis with leguminous plants. This is the most characteristic process of endosymbiosis in plants containing $N_2$ bacteria. Some root species are able to induce the formation of $N_2$-fixing root nodules in the non-vegetative plant *Parasponia* sp. [95, 96]. Other blue bacteria (cyanobacteria) that dissolve $N_2$ can be associated with plants and offer $NH_4^+$ hosts without forming specialized nodules. In general, those blue bacteria that solve $N_2$ symbiotics belong to the Nostoc species. They can distinguish between specialized cells referred to as heterocysts that fix nitrogen in plants. In symbiotic rhizobia-legumes, plants benefit from reduced $N_2$ doubling even when microbes utilize carbohydrates provided by host plants [97, 98]. When there is an interaction between plants and nitrogen-fixing microorganisms, the location of the roots is rich in carbohydrates, in root exudates. In a few cases, AMF is associated with various microbes within the root area. Although these triangular interactions have not yet been accurately classified, they appear to rely heavily on food exchanges between the plant host and microbes. These exchanges include the exudate secretions with the help of fungi to facilitate access to plants [99, 100]. For example, microorganisms can be larger without problems in melting phosphates more than fungi, thus reinforcing all fungi and plants. In addition, some species of *Paenibacillus* are $N_2$ stabilizers able to dissolve phosphate and iron and secrete phytohormones [50]. Many plant-related fungi are colonized with the help of the use of endogenous diazotrophs that can present N to fungi [101–103].

Many of these tripartites may want a symbiotic status that dissolves larger green fungi and $N_2$, and there is no doubt that the use of plants will increase N acquisition. Therefore, additional studies to discover microorganisms that support the current state of symbiotic affiliation between plant life, bacterial and fungal consortium show that these three affiliations enhance plant N acquisition, especially under reduced fertilization conditions. Linking plant life to a more complex bacterial-fungal consortium is all other approaches that have the potential to improve overall plant performance. This is because fungal inoculation mixed with a bacterial consortium away from unfertilized soil promotes nutrient (N and P) uptake [104].

## Conclusions and Future Prospects

In the agricultural sector, the concern for sustainable food production that satisfies the demands of the global human population has become a critical problem. To meet present and future food demand, the development of innovative sustainable solutions to boost crop yields and quality while also restoring soil fertility is critical. Microbial consortiahave the potential to be a long-term and successful strategy for various abiotic and biotic stress conditions. Microbial consortia offer a long-term and cost-effective solution to plant productivity losses caused by changing climate variables, as well as help in the optimization of human inputs in the agro-ecosystem. The use of microbial consortium may also aid in the maintenance of agro-ecosystem ecological balance by minimizing the use of pesticides and/or heavy metals in agricultural activities. Furthermore, microbial consortium efficiency varied greatly depending on the crop and ambient circumstances. Future study should concentrate on generating more precise products, such as diving further into the interactions of the microbial strains with indigenous plant-associated microbiomes.

## References

1. Lin W, Lin M, Zhou H, Wu H, Li Z, Lin W (2019) The effects of chemical and organic fertilizer usage on rhizosphere soil in tea orchards. PLoS ONE 14(5):e0217018
2. Santos LF, Olivares FL (2021) Plant microbiome structure and benefits for sustainable agriculture. Current Plant Biol 26:100198

3. Steinauer K, Chatzinotas A, Eisenhauer N (2016) Root exudate cocktails: the link between plant diversity and soil microorganisms? Ecol Evol 6(20):7387–7396
4. Kumawat KC, Nagpal S, Sharma P (2022) Potential of plant growth-promoting rhizobacteria-plant interactions in mitigating salt stress for sustainable agriculture: a review. Pedosphere 32:223–245
5. Upadhyay SK, Srivastava AK, Rajput VD, Chauhan PK, Bhojiya AA, Jain D, Chaubey G, Dwivedi P, Sharma B, Minkina T (2022) Root Exudates: Mechanistic insight of plant growth promoting rhizobacteria for sustainable crop production. Front Microbiol 13:916488
6. Jamil F, Mukhtar H, Fouillaud M, Dufosse L (2022) Rhizosphere signaling: Insights into plant–rhizomicrobiome interactions for sustainable agronomy. Microorganisms 10:899
7. Nosheen S, Ajmal I, Song Y (2021) Microbes as biofertilizers, a potential approach for sustainable crop production. Sustainability 13:1868
8. Bradacova K, Florea AS, Bar-Tal A, Minz D, Yermiyahu U, Shawahna R, Kraut-Cohen J, Zolti A, Erel R, Dietel K, Weinmann M, Zimmermann B, Berger N, Ludewig U, Neumann G, Poşta G (2019) Microbial consortia versus single-strain inoculants: an advantage in PGPM-assisted tomato production? Agronomy 9(2):105
9. Kaur T, Devi R, Kumar S, Sheikh I, Kour D, Yadav AN (2022) Microbial consortium with nitrogen fixing and mineral solubilizing attributes for growth of barley (*Hordeum vulgare* L.). *Heliyon* 8(4):e09326
10. Sekar J, Raj R, Prabavathy VR (2016) Microbial consortial products for sustainable agriculture: commercialization and regulatory issues in India. In: Singh HB, Sarma BK, Keswani C (eds) agriculturally important microorganisms. Springer, Singapore, pp 107–131
11. Kumar S, Sindhu SS, Kumar R (2021) Biofertilizers: an ecofriendly technology for nutrient recycling and environmental sustainability. Curr Res Microb Sci 3:100094
12. Xiaomei Y, Zhi W, Yu M, Liqun W, Xu W, Qingshan X, Su P, Yu Z, Chaoling W (2018) Isolation, diversity, and growth-promoting activities of endophytic bacteria from tea cultivars of Zijuan and Yunkang-10. Front Microbiol 9:2723
13. Takada K, Tanaka N, Kikuno H, Babil P, Onjo M, Park BJ, Shiwachi H (2019) Isolation of nitrogen-fixing bacteria from Water Yam (*Dioscorea alata* L.). Trop Agric Dev 63(4):198–203
14. Mir MI, Kumar BK, Gopalakrishnan S, Vadlamudi S, Hameeda B (2021) Characterization of rhizobia isolated from leguminous plants and their impact on the growth of ICCV 2 variety of chickpea (*Cicer arietinum* L.). Heliyon 7:e08321
15. Singh RK, Singh P, Sharma A, Guo DJ, Upadhyay SK, Song QQ, Verma KK, Li DP, Malviya MK, Song XP, Yang LT, Li YR (2022) Unraveling nitrogen fixing potential of endophytic diazotrophs of different *Saccharum* species for sustainable sugarcane growth. Int J Mol Sci 23:6242
16. Chang CH, Yang SS (2009) Thermo-tolerant phosphate-solubilizing microbes for multi-functional biofertilizer preparation. Biores Technol 100:1648–1658
17. Bononi L, Chiaramonte JB, Pansa CC, Moitinho MA, Melo IS (2020) Phosphorus-solubilizing *Trichoderma* spp. from Amazon soils improve soybean plant growth. Sci Rep 10:2858
18. Doilom M, Guo JW, Phookamsak R, Mortimer PE, Karunarathna SC, Dong W, Liao CF, Yan K, Pem D, Suwannarach N, Promputtha I, Lumyong S, Xu JC (2020) Screening of phosphate-solubilizing fungi from air and soil in Yunnan, China: Four novel species in *Aspergillus*, *Gongronella*, *Penicillium*, and *Talaromyces*. Front Microbiol 11:585215
19. Elhaissoufi W, Ghoulam C, Barakat A, Zeroual Y, Bargaz A (2021) Phosphate bacterial solubilization: a key rhizosphere driving force enabling higher P use efficiency and crop productivity. J Adv Res 38:13–28
20. Fall AF, Nakabonge G, Ssekandi J, Founoune-Mboup H, Apori SO, Ndiaye A, Badji A, Ngom K (2022) Roles of arbuscular mycorrhizal fungi on soil fertility: contribution in the improvement of physical, chemical, and biological properties of the soil. Front Fungal Biol 3:723892
21. Kirui CK, Njeru EM, Runo S (2022) Diversity and phosphate solubilization efficiency of phosphate solubilizing bacteria isolated from semi-arid agroecosystems of eastern Kenya. Microbiol Insights 15:11786361221088992

22. Yu H, Wu X, Zhang G, Zhou F, Harvey PR, Wang L, Fan S, Xie X, Li F, Zhou H, Zhao X, Zhang X (2022) Identification of the phosphorus-solubilizing bacteria strain JP233 and its effects on soil phosphorus leaching loss and crop growth. Front Microbiol 13:892533
23. Jalmi SK, Sinha AK (2022) Ambiguities of PGPR-Induced plant signaling and stress management. Front Microbiol 13:899563
24. Kong Z, Liu H (2022) Modification of rhizosphere microbial communities: A possible mechanism of plant growth promoting rhizobacteria enhancing plant growth and fitness. Front Plant Sci 13:920813
25. Ali S, Hameed S, Shahid M, Iqbal M, Lazarovits G, Imran A (2020) Functional characterization of potential PGPR exhibiting broad-spectrum antifungal activity. Microbiol Res 232:126389
26. Dede A, Guven K (2022) Plant growth-promoting of olive and walnut actinobacteria: isolation, screening PGP traits, antifungal activities, identification, and hydroponic production of wheat. Arch Agron Soil Sci. 1–16
27. Shamseldin A (2022) Future outlook of transferring biological nitrogen fixation (BNF) to cereals and challenges to retard achieving this dream. Curr Microbiol 79(6):171
28. Nag P, Shriti S, Das S (2020) Microbiological strategies for enhancing biological nitrogen fixation in nonlegumes. J Appl Microbiol 129:186–198
29. Mus F, Alleman AB, Pence N, Seefeldt LC, Peters JW (2018) Exploring the alternatives of biological nitrogen fixation. Metallomics 10:523
30. Pankievicz VCS, do Amaral FP, Ane JM, Stacey G (2021) Diazotrophic bacteria and their mechanisms to interact and benefit cereals. Mol Plant-Microb Interact 34(5):491–498
31. Bashan Y, de-Bashan LE (2015) Inoculant Preparation and Formulations for *Azospirillum* spp. In: Cassan FD, Okon Y, Creus CM (eds) Handbook for azospirillum. Springer, pp. 469–485
32. Pii Y, Mimmo T, Tomasi N, Terzano R, Cesco S, Crecchio C (2015) Microbial interactions in the rhizosphere: beneficial influences of plant growth-promoting rhizobacteria on nutrient acquisition process: a review. Biol Fertil Soils 51(4):403
33. Jensen ES, Peoples MB, Hauggaard-Nielsen H (2010) Faba bean in cropping systems. Field Crops Res 115:203
34. Alves BJR, Boddey RM, Urquiaga S (2003) The success of BNF in soybean in Brazil. Plant Soil 252:1
35. Chang ACG, Chen T, Li N, Duan J (2019) Perspectives on endosymbiosis in Coralloid roots: association of cycads and cyanobacteria. Front Microbiol 10:1888
36. Rousk K (2022) Biotic and abiotic controls of nitrogen fixation in cyanobacteria-moss associations. New Phytol 235:1330–1335
37. Zackrisson O, DeLuca TH, Gentili F, Sellstedt A, Jaderlund A (2009) Nitrogen fixation in mixed *Hylocomium splendens* moss communities. Oecologia 160(2):309
38. Fosu-Mensah BY, Vlek PL, Manske G, Mensah M (2015) The influence of *Azolla pinnata* on flood water chemistry, grain yield and nitrogen uptake of rice in Dano, Southwestern Burkina Faso. J Agric Sci 7(8):118
39. Bhuvaneshwari K, Kumar A (2013) Agronomic potential of the association *Azolla-Anabaena*. Sci Res Rep 3(1):78
40. Dominguez-Nunez JA, Berrocal-Lobo M (2021) Application of microorganisms in forest plant. In: Rakshit A, Meena VS, Parihar M, Singh HB, Singh AK (eds) Biofertilizers. Woodhead Publishing, pp 265–287
41. Sellstedt A, Richau KH (2013) Aspects of nitrogen-fixing *Actinobacteria*, in particular free-living and symbiotic *Frankia*. FEMS Microbiol Lett 342(2):179
42. Kumar A, Kumar A, Pratush A (2014) Molecular diversity and functional variability of environmental isolates of *Bacillus* species. Springerplus 3:312
43. Kumar A, Soni R, Kanwar SS, Pabbi S (2019) *Stenotrophomonas*: a versatile diazotrophic bacteria from the rhizospheric soils of Western Himalayas and development of its liquid biofertilizer formulation. Vegetos 32:103–109
44. Swarnalakshmi K, Yadav V, Tyagi D, Dhar DW, Kannepalli A, Kumar S (2020) Significance of plant growth promoting rhizobacteria in grain legumes: growth promotion and crop production. Plants 9(11):1596

45. Egorshina AA, Khairullin R, Sakhabutdinova AR, Lukyantsev MA (2012) Involvement of phytohormones in the development of interaction between wheat seedlings and endophytic *Bacillus subtilis* strain 11BM. Russ J Plant Physiol 59(1):134
46. Verma JP, Yadav J, Tiwari KN, Kumar A (2013) Effect of indigenous *Mesorhizobium* spp. and plant growth promoting rhizobacteria on yields and nutrients uptake of chickpea (*Cicer arietinum* L.) under sustainable agriculture. Ecol Eng 51:282
47. Ambawade MS, Pathade GR (2015) Production of gibberellic acid by *Bacillus siamensis* BE 76 isolated from banana plant (*Musa* spp.). Int J Sci Res 4(7):394
48. Naiman AD, Latronico A, De Salamone IEG (2009) Inoculation of wheat with *Azospirillum brasilense* and *Pseudomonas fluorescens*: impact on the production and culturable rhizosphere microflora. Eur J Soil Biol 45(1):44
49. Mellidou I, Karamanoli K (2022) Unlocking PGPR-mediated abiotic stress tolerance: what lies beneath. Front Sustain Food Syst 6:832896
50. Courty PE, Smith P, Koegel S, Redecker D, Wipf D (2015) Inorganic nitrogen uptake and transport in beneficial plant root-microbe interactions. Crit Rev Plant Sci 34:4–16
51. Brown CMB, Nepomuceno RA, Anarna JA, Brown MB (2022) Plant growth promoting rhizobacteria—Advances and future prospects. In: Singh HB, Vaishnav A (eds) New and future developments in microbial biotechnology and bioengineering. Elsevier, pp 1–28
52. Devi S, Sharma P, Rana A, Pal J, Kumari A (2021) Diversity and plant growth-promoting potential of actinomycetes associated with the rhizosphere of *Arnebia euchroma* from Himachal Pradesh (India). J Environ Biol 42:964–972
53. Ogut M, Er F, Neumann G (2011) Increased proton extrusion of wheat roots by inoculation with phosphorus solubilizing microorganism. Plant Soil 339:285
54. Park Y, Solhtalab M, Thongsomboon W, Aristilde L (2022) Strategies of organic phosphorus recycling by soil bacteria: acquisition, metabolism, and regulation. Environ Microbiol Rep 14(1):3–24
55. Marulanda A, Azcon R, Chaumont F, Ruiz-Lozano JM, Aroca R (2010) Regulation of plasma membrane aquaporins by inoculation with a *Bacillus megaterium* strain in maize (*Zea mays* L.) plants under unstressed and salt-stressed conditions. Planta 232(2):533
56. Lahlali R, Ezrari S, Radouane N, Kenfaoui J, Esmaeel Q, El Hamss H, Belabess Z, Barka EA (2022) Biological control of plant pathogens: a global perspective. Microorganisms 10(3):596
57. Saeed Q, Xiukang W, Haider FU, Kucerik J, Mumtaz MZ, Holatko J, Naseem M, Kintl A, Ejaz M, Naveed M, Brtnicky M, Mustafa A (2021) Rhizosphere bacteria in plant growth promotion, biocontrol, and bioremediation of contaminated sites: a comprehensive review of effects and mechanisms. Int J Mol Sci 22(19):10529
58. Kumar V, Srivastava A, Jain L, Chaudhary S, Kaushal P, Soni R (2022) Harnessing the potential of genetically improved bioinoculants for sustainable agriculture: recent advances and perspectives. In: Soni R, Suyal DC, Yadav AN, Goel R (eds) Developments in applied microbiology and biotechnology, trends of applied microbiology for sustainable economy. Academic Press, pp 319–341
59. Ryu CM, Murphy JF, Mysore KS, Kloepper JW (2004) Plant growth-promoting rhizobacterial systemically protect *Arabidopsis thaliana* against Cucumber mosaic virus by a salicylic acid and NPR1-independent and jasmonic acid-dependent signaling pathway. Plant J 39:381
60. Zhao Y, Selvaraj JN, Xing F, Zhou L, Wang Y, Song H (2014) Antagonistic action of *Bacillus subtilis* strain SG6 on *Fusarium graminearum*. PLoS ONE 9(3):e92486
61. Kai M, Haustein MF, Petri A, Scholz B, Piechulla B (2009) Bacterial volatiles and their action potential. Appl Microbiol Biotechnol 81:1001
62. Meena M, Swapnil P, Divyanshu K, Kumar S, Tripathi YN, Zehra A, Marwal A, Upadhyay RS (2020) PGPR-mediated induction of systemic resistance and physiochemical alterations in plants against the pathogens: current perspectives. J Basic Microbiol 60(10):828–861
63. Hashem A, Tabassum B, Allah E (2019) *Bacillus subtilis*: a plant-growth promoting rhizobacterium that also impacts biotic stress. Saudi J Biol Sci 26(6):1291
64. Zheng XY, Sinclair JB (2000) The effects of traits of *Bacillus megaterium* on seed and root colonization and their correlation with the suppression of *Rhizoctonia* root rot of soybean. Biol Control 45:223

65. Kumar A, Bahadur I, Maurya BR, Raghuwanshi R, Meena VS, Singh DK, Dixit J (2015) Does a plant growth promoting rhizobacteria enhance agricultural sustainability? J Pure Appl Microbiol 9:715
66. Bonas U, Lahaye T (2002) Plant disease resistance triggered by pathogen-derived molecules: refined models of specific recognition. Currr Opin Microbiol 5:44
67. Pieterse CMJ, van Pelt JA, Knoester M, Laan R, Gerrits H, Weisbeek PJ, van Loon LC (1998) Novel signaling pathway controlling induced systemic resistance in *Arabidopsis*. Plant Cell 10:1571
68. Romeiro RS (2000) PGPR and systemic induction of resistance against plant pathogens. Summa Phytopathol 26:177
69. Moraes MG (1998) Mechanisms of acquired systemic resistance in plants. Revisao Anual de Patologia de Plantas 6:261
70. Kaushal M, Wani SP (2016) Plant-growth-promoting rhizobacteria: drought stress alleviators to ameliorate crop production in drylands. Annals Microbiol 66(1):35–42
71. Santoyo G, Guzman-Guzman P, Parra-Cota FI, Santos-Villalobos S, Orozco-Mosqueda MdC, Glick BR (2021) Plant growth stimulation by microbial consortia. Agronomy 11(2):219
72. Kaushal M, Mandyal P, Kaushal R (2019) Field based assessment of *Capsicum annuum* performance with inoculation of rhizobacterial consortia. Microorganisms 7(3):89
73. Zandbergen LE, Halverson T, Brons JK, Wolfe AJ, de Vos MGJ (2021) The good and the bad: ecological interaction measurements between the urinary microbiota and uropathogens. Front Microbiol 12:659450
74. Stubbendieck RM, May DS, Chevrette MG, Temkin MI, Wendt-Pienkowski E, Cagnazzo J, Carlson CM, Gern JE, Currie CR (2018) Competition among nasal bacteria suggests a role for siderophore-mediated interactions in shaping the human nasal microbiota. Appl Environ Microbiol 85(10):e02406-e2418
75. Bubici G, Kaushal M, Prigigallo MI, Gómez-Lama Cabanás C, Mercado-Blanco J (2019) Biological control agents against *Fusarium* wilt of banana. Front Microbiol 10:616
76. Li Z, Nair SK (2012) Quorum sensing: how bacteria can coordinate activity and synchronize their response to external signals? Protein Sci 21(10):1403–1417
77. Celine B, Schwab M, Naik T, Daude D, Chabriere E, Elias M (2018) Structural and biochemical characterization of AaL, a quorum quenching lactonase with unusual kinetic properties. Sci Rep 8:11262
78. Morohoshi T, Nakamura Y, Yamazaki G, Ishida A, Kato N, Ikeda T (2007) The plant pathogen *Pantoea ananatis* produces N-acylhomoserine lactone and causes centre rot disease of onion by quorum sensing. J Bacteriol 189(22):8333–8338
79. Kaushal M, Wani SP (2016) Rhizobacterial-plant interactions: strategies ensuring plant growth promotion under drought and salinity stress. Agr Ecosyst Environ 231:68–78
80. Stephane C, Abdul S, Hanna F, Angela S (2019) A review on the plant microbiome: ecology, functions, and emerging trends in microbial application. J Adv Res 19:29–37
81. de Oliveira ALM, de Lima C, Urquiaga S, Reis VM, Baldani JI (2006) Yield of micropropagated sugarcane varieties in different soil types following inoculation with diazotrophic bacteria. Plant Soil 284:23–32
82. Carlos MHF, Helena MVMS, Eduardo VS (2019) Promising bacterial genera for agricultural practices: an insight on plant growth-promoting properties and microbial safety aspects. Sci Total Environ 682:779–799
83. Pereira NCM, Galindo FS, Gazola RPD, Dupas E, Rosa PAL, Mortinho ES, Teixeira FMCM (2020) Corn yield and phosphorus use efficiency response to phosphorus rates associated with plant growth promoting bacteria. Front Environ Sci 8:40
84. Rozier C, Hamzaoui J, Lemoine D, Czarnes S, Legendre L (2017) Field-based assessment of the mechanism of maize yield enhancement by *Azospirillum lipoferum* CRT1. Sci Rep 7:7416
85. Sumbul A, Ansari RA, Rizvi R, Mahmood I (2020) *Azotobacter*: a potential bio-fertilizer for soil and plant health management. Saudi J Biol Sci 27(12):3634–3640
86. Molina-Romero D, Juarez-Sanchez S, Venegas B, Ortiz-Gonzalez CS, Baez A, Morales-Garcia YE, Munoz-Rojas J (2021) A bacterial consortium interacts with different varieties

of maize, promotes the plant growth, and reduces the application of chemical fertilizer under field conditions. Front Sustain Food Syst 4:616757
87. Schroder P, Beckers B, Daniels S, Gnadinger F, Maestri E, Marmiroli N, Mench M, Millan R, Obermeier MM, Oustriere N, Persson T, Poschenrieder C, Rineau F, Rutkowska B, Schmid T, Szulc W, Witters N, Sæbo A (2018) Intensify production, transform biomass to energy and novel goods and protect soils in Europe—A vision how to mobilize marginal lands. Sci Total Environ 617:1101–1123
88. Raklami A, Bechtaoui N, Tahiri A, Slimani A, Bargaz A, Meddich A, Oufdou K (2021) Co-inoculation with rhizobacteria and mycorrhizae can improve wheat/faba bean intercrop performance under field conditions. Front Agron 3:734923
89. Ding J, Jiang X, Guan D, Zhao B, Ma M, Zhou B, Cao F, Yang X, Li L, Li J (2017) Influence of inorganic fertilizer and organic manure application on fungal communities in a long-term field experiment of Chinese Mollisols. Appl Soil Ecol 111:114–122
90. Jiang Y, Qian H, Wang X, Chen L, Liu M, Li H, Sun B (2018) Nematodes and microbial community affect the sizes and turnover rates of organic carbon pools in soil aggregates. Soil Biol Biochem 119:22–31
91. Rodriguez-Andrade O, Fuentes-Ramirez LE, Morales-Garcia YE, Molina-Romero D, Bustillos-Cristales MR, Martínez-Contreras RD, Munoz-Rojas J (2015) The decrease in the population of *Gluconacetobacter diazotrophicus* in sugarcane after nitrogen fertilization is related to plant physiology in split root experiments. Rev Argent Microbiol 47:335–343
92. Kaushal M (2020) Insights into microbially induced salt tolerance and endurance mechanisms (STEM) in plants. Front Microbiol 11:1518
93. Rosewarne GM, Barker SJ, Smith SE, Smith FA, Schachtman DP (1999) A *Lycopersicon esculentum* phosphate transporter (LePT1) involved in phosphorus uptake from a vesicular-arbuscular mycorrhizal fungus. New Phytol 144(3):507–516
94. Dipta B, Bhardwaj S, Kaushal M, Kirti S, Sharma R (2019) Obliteration of phosphorus deficiency in plants by microbial interceded approach. Symbiosis 78(2):163–176
95. Op den Camp RHM, Polone E, Fedorova E, Roelofsen W, Squartini A, Op den Camp HJM, Bisseling T, Geurts R (2012) Nonlegume *Parasponia andersonii* deploys a broad *Rhizobium* host range strategy resulting in largely variable symbiotic effectiveness. Mol Plant Microb Interact 25:954–963
96. Carole S, Didier B, Claudine F (2013) Biological nitrogen fixation in non-legume plants. Ann Bot 111(5):743–767
97. Bates ST, Berg-Lyons DB, Caporosa JG, Walters WA, Knight R, Fierer N (2011) Examining the global distribution of dominant archaeal populations in soils. ISME J 5:908–917
98. Gary EH, Norman U (2019) Symbiotic root-endophytic soil microbes improve crop productivity and provide environmental benefits. Scientifica 25. Article ID 9106395
99. Reinhold-Hurek B, Bunger W, Burbano CS, Sabale M, Hurek T (2015) Roots shaping their microbiome: global hotspots for microbial activity. Annu Rev Phytopathol 53:403–424
100. Zhang L, Xu M, Liu Y, Zhang F, Hodge A, Feng G (2016) Carbon and phosphorus exchange may enable cooperation between an arbuscular mycorrhizal fungus and a phosphate-solubilizing bacterium. New Phytol 210:1022–1032
101. Torres-Cortes G, Ghignone S, Bonfante P, Schubler A (2015) Mosaic genome of endobacteria in arbuscular mycorrhizal fungi: transkingdom gene transfer in an ancient mycoplasma-fungus association. Proc Natl Acad Sci 112:7785–7790
102. Alia D, Isabelle Q, Bertrand H (2020) Beneficial soil-borne bacteria and fungi: a promising way to improve plant nitrogen acquisition. J Exp Bot 71(15):4469–4479
103. Paul K, Saha C, Nag M, Mandal D, Naiya H, Sen D, Mitra S, Kumar M, Bose D, Mukherjee G, Naskar N, Lahiri S, Ghosh UD, Tripathi S, Sarkar MP, Banerjee M, Kleinert A, Valentine AJ, Tripathy S, Sinharoy S, Seal A (2020) A tripartite interaction among the basidiomycete *Rhodotorula mucilaginosa*, $N_2$-fixing endobacteria, and rice improves plant nitrogen nutrition. Plant Cell 32:486–507
104. Hestrin R, Hammer EC, Mueller CW, Lehmann J (2019) Synergies between mycorrhizal fungi and soil microbial communities increase plant nitrogen acquisition. Commun Biol 2:233

# Chapter 3
# Overview of Soil Microbe Dynamics in Different Biosystems

**Bisma Farooq, Shahnaz Anjum, Madiha Farooq, Asma Nazir, Mohammad Umer Farooq, Shayista Yousuf, Nusrat Shafi, and Javid Ahmad Parray**

**Abstract** There are various performances done by microbes in ecosystem, which are very beneficial for microorganisms, plants and animals including soil aggregation, improved soil, water cycling and soil nutrients. Fungi, Bacteria, Protozoa, Nematodes, and Actinomycetes are few different types of microbes present in soil. In terms of soil dynamic, diversity and vegetation abundance, Plants are significant factors. The maximum rapid modifications because of soil moisture and temperature alternations or with the aid of the influx of sparkling organic depend on some stage in the numerous hours or days. They're usually associated with the microbial activity. Seasonal dynamics are resulting from annual variations in precipitation and temperature that affect the network of flora. The microbial biomass and the taxonomic composition of soil microbial communities range appreciably all through the 12 months, taking that in consideration during sample analysis and comparisons of

B. Farooq (✉)
Department of Biomedical Engineering, Sathyabama Institute of Science and Technology, Chennai 600119, India
e-mail: bismafarooq89@gmail.com

S. Anjum
Department of Botany, Lovely Professional University, Phagwara 144411, India

M. Farooq
Department of Microbiology, Asian International University, Imphal 795113, India

A. Nazir · M. U. Farooq
Department of Clinical Biochemistry, University of Kashmir, Hazratbal 190006, J&K, India

M. U. Farooq
Department of Agriculture, Lovely Professional University, Phagwara Punjab, India

S. Yousuf
Department of zoology, University of Kashmir, srinagar 190006, India

N. Shafi
Department of Chemistry, Government Degree College Eidgah Srinagar, Srinagar 190017, J&K, India

J. A. Parray
Department of Environmental Science, Government Degree College Eidgah Srinagar, Srinagar 190017, J&K, India



J. A. Parray (ed.), *Climate Change and Microbiome Dynamics*, Climate Change Management, https://doi.org/10.1007/978-3-031-21079-2_3

various soils. The lengthy-time period dynamics of microbial colonies at some stage in number one, in addition to modifications inside the taxonomic composition of microbial groups. The range of microbial communities in long-term dynamics can range in distinct ways. The longest elements of soil microbial networks are connected with changes in bioclimatic circumstances. The see of predetermination changes in soil microbial networks is suitable in explores different avenues regarding engineered changes in climatic boundaries.

## Introduction

Soil microbial networks assume a significant part in biological systems working and are on the field scale fundamental for plant nourishment and wellbeing. For a bigger scope, they add to worldwide component cycling [1, 2]. Besides, they are engaged with the turnover cycles of natural matter, the breakdown of xenobiotics and the arrangement of soil totals. An environmental condition of soils relies upon the design and movement of soil microorganisms. The consequences of soil observing in different environments in various climatic zones of Ukraine showed an unmistakable pattern for the relationship between the agroecological conditions and movement of microbiocenosis [3, 4]. The main impact of farming movement on the dirt microbiota can be seen on the inadequately soddy-podzolic and dim woods soils, where the yield development without treatment brought about a lessening in the all-out count of microorganisms by 2.2–4.5 times. Utilization of farming measures pointed toward accomplishing greatest efficiency, explicitly the mix of mineral, natural and organic composts, adds to a typical 1.3–4.1 times expansion in all out include of microorganisms in the dirt, contrasted and non-prepared variations. The dirt of regular environments is portrayed by a high all out count of the microorganisms with a reasonable construction of different natural trophic gatherings and adjusted cycles of mineralization-immobilization, natural matter decay, and humus collection [5, 6].

Soil microbial networks are impacted by base up factors like the quality and primary properties of their detrital assets. They are especially restricted by the quality and, frequently heterogeneous, spatial conveyance of their detrital assets [5–7]. The conveyance of microbial species is likewise spatially heterogeneous, on the grounds that people are separated at neighbourhood locales in view of somewhat low supplement accessibility, unforgiving ecological circumstances, or contest [8, 9]. The variety of microbial networks, which results from these neighbourhood specific tensions, makes totally different utilitarian limits across soil conditions. For instance, it was contended [10] that a few networks have major areas of strength for a field advantage, wherein they corrupt litter from their current circumstance better than unfamiliar litter. Subsequently, microbial networks probably have a huge ability to show setting subordinate changes in their utilitarian characteristics in view of the nature of their assets.

Microorganism working additionally not entirely settled by the ability to move rummaging techniques and take up natural supplements in the rhizosphere when plants discharge root exudates [11]. As a matter of fact, the biomass and exoenzyme creation of the microbial local area for the most part changes when supplements are added to the dirt [12]. There are areas of strength for obviously reliance in reactions of microbial networks to their asset base that might collaborate with hierarchical impacts to decide how soil microbial networks capability in various settings [13].

Slow eaters (detritivores/microbivores) likewise apply hierarchical consequences for organisms. High brushing tension by enormous or plentiful soil fauna can decrease microbial biomass [14], with microorganisms repaying by expanding their development rates to keep up with something very similar or higher biomass when supplements are not restricting [15, 16]. Subsequently, the greatness of the compensatory development reaction relies upon transaction between the strength of the touching effect and supplement accessibility. Microbial biomass [17] and capability [18] may stay high under brushing tension in supplement rich conditions however are bound to be discouraged in supplement unfortunate conditions. Considering that microbial biomass is connected with exoenzyme creation, microorganisms hence can intervene the flowing impacts of hunters on natural matter deterioration rate.

Regardless of the changeability in microbial networks inside soils and their reaction to natural settings, a few consensuses are starting to arise while looking at processes from the perspective of a measured methodology. In the first place, the reaction of the microbial local area to brushing pressure is exceptionally reliant upon the asset climate, with high asset conditions prompting compensatory development and low asset conditions prompting net biomass shortfall [19, 20].

Second, the impact of brushing pressure probably affects microbial local area creation and capability than on biomass essentially. Not with standing, microbial networks are seldom concentrated on utilizing this secluded point of view [21]. More observational instances of what the asset climate and nibblers mean for microbial local area collaborations are expected to construct the prescient system we are proposing.

## Soil Microbial Networks

Soil microbial networks possess the most organically assorted environments on the planet. A solitary gram of soil can uphold more than a few thousand parasitic taxa close to the root rhizosphere [22]. As referenced in different sections in this book, many elements can impact the microbial networks related with tree leaves, stems, and roots. Contrasts in have species [23], cultivar type inside an animal category, soil type, physiological status of host, and microorganism presence can impact variety in microbial networks [24, 25]. Biological equilibrium inside the related microbial local area is basic for plant wellbeing, particularly in the rhizosphere, and aggravations can cause uneven characters inside the microbial networks. Past examinations have recorded those helpful microbial connections can improve seedling power, seed germination, plant advancement, and plant development that led to higher plant efficiency, though goes after by plant microorganisms can change the microbiome construction, usefulness, and movement [26, 27].

Valuable microbial collaborations can prompt superior host opposition against pathogenic microorganisms and organisms. For instance, valuable microbial taxa can discharge different allelopathic synthetics and poisons that furnish the plant with defensive boundaries that block plant microorganisms. The rhizosphere has been displayed to contain different and complex natural networks that include microbes, growths, oomycetes, and numerous different microorganisms, for example, archaea, nematodes, and infections [28, 29]. Other tree organs, including leaves, branches, and stems, are likewise known to contain a different set-up of microbial taxa, yet by and large varieties are commonly lower than those tracked down in soils [30, 31]. Albeit microbial variety can fluctuate enormously, microorganisms can extraordinarily influence microbial networks. This section will momentarily survey the idea of path biome, how microbial networks safeguard against plant sickness, and different changes that can happen inside microbial networks within the sight of plant microorganisms. Since these exploration subjects are as of late creating in backwoods sciences, models will be gotten from editing frameworks as different as wheat, apples, and woods. True to form, microbial networks can be unfathomably different inside yearly versus enduring trimming frameworks; be that as it may, the impact of plant microorganisms on microbial networks and their biological jobs have been archived basically in different editing frameworks [32, 33].

The dirt microbial local area, which incorporates microorganisms, organisms, and archaea, gives critical biological system works and administrations [34]. The microbial local area helps abiotically in the physical organizing of the dirt through development of soil totals, expanding water maintenance and adds to natural matter arrangement and change. The dirt microbial local area is the vital driver of soil supplement cycling processes, is answerable for creation and utilization of ozone depleting substances and gives plant networks many advantages [35, 36]. These advantages incorporate direct upgrade of plant development through creation of bioactive mixtures, for example, indole acidic corrosive, and more noteworthy admittance to supplements and water through mycorrhizal symbioses. Mycorrhizal growths make establishes more open minded to stresses, for example, dry season, through a drawn out root-hyphal surface region and more impervious to bugs and microbes through actual assurance or creation of bioactive mixtures [37, 38].

Many soil processes, like disintegration and mineralization, are done by various microorganisms, and correction of upset locales with rescued soil, woods floor material or peat (or other natural changes) is presumably satisfactory for fruitful re-foundation of populaces and cycles. More testing is the compensation of miniature creatures answerable for the "thin" processes that are completed by a predetermined number of microbial species [38, 39]. Nitrogen obsession is one of these restricted cycles [40], and is answerable for the arrangement of exceptionally upset biological systems. Microorganisms engaged with mutualistic symbioses, for example, mycorrhizal growths, are likewise cornerstone living beings, accordingly numerous rebuilding projects have zeroed in on re-establishing these organic entities and affiliations (Hawkins et al., 2015), for instance in recovery of the Alberta oil sands [41, 42]. With the coming of high-throughput sequencing strategies, it is presently understood that dirt contain numerous microorganisms that we have close to zero insight into. It has been guessed that this 'uncommon' microbiome—an expected 2–28% of the absolute microbial local area—are liable for the vast majority of these 'restricted' processes [43, 44]. Proceeded with examination into recognition of these organic entities and explanation of their jobs in soil cycles will work with reclamation of soil capability on upset locales. Meanwhile, rehearses, for example, those referenced over, that energize a different soil microbial local area ought to be utilized, as high soil microbial variety builds the likelihood that these "tight" capabilities will be held following unsettling influences [12, 45].

In the AOSR, cutting edge sequencing has been utilized to think about soil microbial networks in restored soils with soils in encompassing normal boreal woods locales [46] (Fig. 13.5). 'Species lavishness (alpha variety) of prokaryotic life forms (microscopic organisms and archaea) didn't contrast among restored and normal soils, however the construction of the networks (beta-variety) varied. Copiotroph microscopic organisms (Actinobacteria, Bacteroidetes, and Proteobacteria), which flourish in supplement rich conditions and can quickly utilize an asset, were more bountiful in remade soils, while oligotrophic microorganisms (Actinobacteria, Cyanobacteria, Elusimicrobia, Firmicutes, Planctomycetes) which are better adjusted to supplement unfortunate conditions, were more plentiful in regular woodland soils. Copiotrophic microorganisms are restricted in their abilities to debase complex natural matter, which could frustrate deterioration in the recreated soils and aggregation [23, 26, 33, 47]. Nitrogen testimony, pH, earth content, and plant species were the primary factors related with the local area design of prokaryotes. Investigations of mycorrhizal organisms in the AOSR have exhibited a pattern of low quantities of mycorrhizal growths in youthful, recovered soils with expanding overflow following 15 years [48–51].

Checking of a characteristic chrono sequence (0–45 years) of post-coal-mining locales in Czechia has exhibited the progression of soil microbial networks that happens working together with soil improvement and plant progression [4–6]. During the initial 10 years when almost no vegetation was available, the dirt microbial local area was overwhelmed via autotrophic microorganisms and N2-fixing microscopic organisms like Gamma proteobacteria, Cyanobacteria and some Alpha proteobacteria. In early progression (10–20 years), the microbial local area moved from these sluggish developing oligotrophic microscopic organisms to quickly developing copiotroph microorganisms, agreeing with the presence of AMF and the advancement of trailblazer plants (spices and grasses) and arrival of root exudates. In mid-progression (20–30 years) there was fast advancement of spices and bushes, and the microbial local area was improved with rhizobacteria like Rhizobiaceae, Bradyrhizobiaceae, and Agrobacterium. The fungal: bacterial proportion was maximal at mid-progression because of the fast improvement of saprophytic micromycetes, agreeing with the gathering of natural matter through leaf litter, rhizo deposition and faunal fertilizers. In late progression (30–45 years), there was an expansion in non-cultivable microorganisms and slow-developing cultivable microbes like Firmicutes and Actinobacteria. These examinations feature the powerful transaction of biotic and abiotic factors, both over-the-ground and subterranean, that support soil cycles and capability in both normal and recreated soils [2, 4–6].

Practices to re-establish soil microbial networks following significant aggravations, for example, surface mining can be assembled into those that re-establish the circumstances that would cultivate their development, and practices focused on once again introducing either the whole local area or explicit objective creatures. Rescue and substitution of dirt gives appropriate living space and if it has not been accumulated for extremely lengthy additionally once again introduces a portion of the first microbial local area. If the dirt is to be stored for quite a while, revegetating it with wanted plant species could help with laying out propagule banks of the plants and furthermore supporting the dirt biota, in the surface layer of the reserve. Rehearses that improve soil water-holding limit and gathering of natural matter and supplements will likewise make soils more favourable for microbial expansion. Sufficient soil air circulation can be supported by staying away from compaction and cautious situation of materials [49–53].

Re-immunization of microbial networks might be vital when the surficial material is rock or overburden, or when the dirt has been stored for such a period that it as next to zero natural movement [8, 44, 51]. Arbuscular mycorrhizal organisms can stay reasonable in soil for as long as 5 months without even a trace of a host plant [54]. Some EMF spores (Wilcoxon mikolae) stay reasonable in soil for as long as 6 years, though different species, for example, Teleportal Terrestre, decline [55, 56]. Regular entrance of soil microorganisms from encompassing scenes through air, water, birds, or creatures is conceivable, however might be slow [57]. One methodology for re-establishing such destinations is the utilization of an organic soil hull, included cyanobacteria, green growth, parasites, lichens, and greeneries, that copies the normal essential progression of soil improvement on exposed rock [58, 59]. These outside layers include 70% of dryland soil surfaces around the world. Soil adjustment and water guideline and re-established the availability of the bacterial, parasitic, omnivore, and hunter food channels both over-the-ground and subterranean [59, 60].

Vaccination with nearby local soils has been demonstrated to be successful at expanding AMF and EMF disease and plant foundation and development on re-established destinations [58, 59], and may likewise give local plants an upper hand over obtrusive species [61] in correlation with "unfamiliar" soil inocula [62]. The expression "biological coordinating" has been authored to make sense of that entire AMF people group are naturally adjusted to their neighborhood have soil climate thus will work best in their local soils [62, 63]. Essentially, concentrates on looking at local wellsprings of inoculum versus business inocula, especially zeroing in on AMF, have observed that local soil inocula is generally gainful for expanding plant biomass and supplement take-up and mycorrhizal colonization on reestablished destinations [59, 60, 62, 63].

Soil microbial networks play a few significant environmental and physiological capabilities (e.g., soil natural matter deterioration and control of its cycle, guideline of mineral supplement accessibility, air nitrogen obsession, development of mycorrhiza, creation of organically dynamic substances ready to invigorate plant development) improving soil physical and compound circumstances and, subsequently, soil tenability for plants. There is a developing interest in support of agro system usefulness. It appears to be that dirt microbiota, especially its biodiversity, permits frameworks to more readily defeated normal and human-centered bothers, further developing their recuperation limit (i.e., versatility idea). Soil quality misfortune happens particularly in regions exposed to concentrated rural practices and to aimless utilization of outside input (e.g., composts, pesticides, water system water. This is the justification for why the advancement and the development of low-influence green methods ought to be worked with. Supportable practices can permit typical soil fruitfulness levels to return in the agrosystems with benefits on both soil ripeness and harvest yield quality and amount) [1, 64].

The olive tree (*Olea europaea* L.) is perhaps of the main yield in the Mediterranean Basin. In such a wide region, olive plantation the executives can be totally different relying upon pedoclimatic and financial circumstances and asset accessibility. This section reviews soil microbiological information of olive plantations exposed to various soil the board frameworks that have been applied for variable time spans under various pedoclimatic conditions. Specific consideration is given to changes in the design, elements, and intricacy of microbial networks to assess soil wellbeing status. Among the agronomic reasonable practices, the contribution of natural matter as fertilizer is perhaps of the main component influencing soil fruitfulness. Thus, cases of in situ manure creation in olive forests are examined [1, 11, 46].

Soil microbial networks possess the most organically different territories on the planet. A solitary gram of soil can uphold more than a few thousand parasitic taxa close to the root rhizosphere [65]. As referenced in different parts in this book, many variables can impact the microbial networks related with tree leaves, stems, and roots. Contrasts in have species [66], cultivar type inside an animal types, soil type, physiological status of host, and microorganism presence can impact variety in microbial networks [24]. Environmental equilibrium inside the related microbial local area is basic for plant wellbeing, particularly in the rhizosphere, and aggravations can cause uneven characters inside the microbial networks. Past investigations have archived those gainful microbial connections can improve seedling energy, seed germination, plant advancement, and plant development that lead to higher plant efficiency, while assaults by plant microorganisms can adjust the microbiome design, usefulness, and action [1, 3, 20, 23].

Advantageous microbial collaborations can prompt superior host opposition against pathogenic microorganisms and growths. For instance, valuable microbial taxa can emit different allelopathic synthetic compounds and poisons that furnish the plant with defensive hindrances that hinder plant microorganisms. The rhizosphere has been displayed to contain different and complex natural networks that envelop microbes, growths, oomycetes, and numerous different microorganisms, for example, archaea, nematodes, and infections. Other tree organs, including leaves, branches, and stems, are likewise known to contain a different set-up of microbial taxa, however in general varieties are normally lower than those tracked down in soils. Albeit microbial variety can change significantly, microorganisms can extraordinarily influence microbial networks [17, 25, 33, 60]. This section will momentarily survey the idea of path biome, how microbial networks safeguard against plant illness, and different changes that can happen inside microbial networks within the sight of plant microorganisms. Since these examination points are as of late creating in timberland sciences, models will be gotten from editing frameworks as different as wheat, apples, and woodlands. True to form, microbial networks can be immeasurably unique inside yearly versus perpetual trimming frameworks; notwithstanding, the impact of plant microorganisms on microbial networks and their environmental jobs have been reported essentially in assorted editing frameworks [21, 34, 36, 39].

The fast expansion in industrialization has prompted colossal releases of impurities into the climate. Chromium is the second most plentiful metal tracked down in most sullied locales. The most plentiful types of Cr in the climate, Cr(VI) and Cr(III), have differentiating characters. Chromium(III) is a fundamental supplement in that it adjusts glucose digestion in people. The dissolvability of Cr(III) is extremely low and for the most part hastens or edifices in normal soils pH (4–8). Interestingly, Cr(VI) is a class A cancer-causing agent, teratogen, and mutagen. The portability and solvency of Cr(VI) are far higher than that of Cr(III). In this manner, the opportunities for diffusing Cr(VI) through cell film are high, which will harm DNA. The versatility and bioavailability of these two species generally rely upon the pH and redox capability of the dirt. Of the accessible philosophies that can moderate Cr harmfulness soil flushing, sorption, decrease, EC, phytoremediation, and layer partition—remediation by decrease is viewed as the most practical procedure. Cr(VI) diminishes to Cr(III) within the sight of OC sources because of the great overflow of electrons in OC [67–70].

Soil microbial local area is extremely basic in determining the destiny of Cr in sullied soils. It has been noticed that normal weakening of Cr isn't occurring in a tannery emanating sullied site notwithstanding the site being 225 years of age. This might be because of the oxidation of Cr(III) by Mn oxides. Phytoremediation is a demonstrated compelling strategy for recuperating tainted soils. Hyperaccumulating plants offer Cr remediation from soil and oceanic media. Nonetheless, relief of Cr-polluted soil and water needs a multiscale approach, which includes the blend of physical, synthetic and organic instruments. The following are regions where future examination can zero in o [48, 53, 59, 67].

The change and elements of Cr in the dirt and sea-going media as impacted by biotic and abiotic systems to foster remediation procedures in various ecologically complex settings. The impact of heap soil properties (physical, substance and organic) and natural boundaries (precipitation and temperature) on the maintenance and versatility of Cr(VI) in various soils should be analysed under field conditions. A superior comprehension of the instrument of adsorption of Cr(VI) is expected to affirm the overall degree of inward circle and external circle complexation. This will extraordinarily assist with expanding the maintenance of Cr(VI) in tainted soil [39, 43, 48, 67–70].

In tannery emanating sullied soils, Cr(III) fixation is frequently higher than Cr(VI) species since $Cr_2\,(SO_4)_3$ is utilized as a collagen (conceal protein)-settling specialist.

In any case, the Cr(VI) focus in these destinations has been displayed to increment over the long run. Subsequently, a top to bottom review ought to be finished on the components that oxidize Cr(III) to Cr(VI) in these locales. The drawn-out strength of Cr(III) in tannery gushing tainted locales. Finding the dynamic job of electrons in Cr(VI) decrease utilizing synchrotron-based applications. Expansion of natural changes expands DOC in the dirt. The DOC is made from a few useful gatherings. The portrayal of carbon, for example, aliphatic and sweet-smelling carbon, in DOC should be evaluated. The exchange of electrons assumes a significant part in Cr(VI) decrease. Hence, redox estimations during Cr(VI) decrease should be attempted [39, 43, 48, 67–70].

Analysing the impact of carbon-based materials like dark carbon and biochar over the drawn out to decide whether reoxidation of Cr(VI) in tainted soil is conceivable. Assessing the possible worth of other minimal expense alterations, for example, chitosan-based biowaste, ocean growth and burn fluid from biochar plants on decreasing Cr(VI) in water and soil should be attempted.

## Bioclimatic Changes and Long-Term Dynamics of Soil Microbial Communities

The most long haul changes in the design of microbial networks are related with changes in climatic circumstances for a specific region. Precipitation, temperature, and the degree of insolation influence the vegetation cover and the substance and actual properties of the dirt which clearly, influences the design of the microbial local area. On account of a drawn out difference in the environment, the issue of recreating soil microbial networks of previous ages and demonstrating their progressions in what's in store emerges [68–72].

One of the ways of concentrating on soil microbial networks of different ages is the microbiological investigation of covered soils. Soils covered under regular dregs under archeological (normally earthen) developments, well as soils in the frozen state (permafrost), draw in the consideration of scientists as potential documents of microbiological data safeguarded since their internment. In covered soils, elements of the vertical circulatetion of microorganisms along the profile are preserved. Covered humus skylines are generally characterized by a larger number, biomass, and species variety of microorganisms in examination with other mineral skylines [73, 74]. However the substance of feasible microbial biomass in the covered soils might be low, they ordinarily contain a lot of microbial DNA that can be saved in soils covered at a profundity of in excess of 100 m. A few creators propose that microbiological markers in paleosols ought to be thought of as one of the types of the dirt "organic memory," which can be utilized to remake the miniature bial populace of these dirts before their covering. Be that as it may, soil entombment is seldom joined by complete preservation of microbial networks. Miniature living beings in covered soils hold their metabolic activity [75, 76].

Covered soils can some of the time be equivalent to modern surface soils as far as complete $CO_2$ emanations even in permafrost with freezing temperatures; numerous microorganisms protect their physiological movement [77]. At the point when the dirts are covered, the ordered and utilitarian design of microbial networks changes: there are fundamentally less saccharolytic microbes and more oligotrophs and anaerobic microorganisms in the covered skylines, and denitrification beats nitrification. The ordered structure of the prokaryotic local area changes significantly after entombment; specifically, the relative abundance of Verrucomicrobia is extraordinarily decreased. Specific biological highlights are ordinary for infinitesimal parasites of paleosols. The absolute biomass and length of the mycelium of growths diminishes after soil entombment and a large portion of the parasitic biomass (up to 70%) in covered soils con-sists of spores of generally little sizes [78, 79].

Among the developed structures, little spores and psychrotolerants are most bountiful. The specificities of microbial networks in bramble ied soils are brought about by changes in the ecological conditions after internment. When in doubt, covered soils are singe acterized by decreased oxygen content and expanded carbon dioxide content, and lower temperature and dampness variances. Discontinuance of the contribution of new natural matter is particularly huge for the dirt microorganisms. The substance of natural matter in paleosols consistently diminishes in the initial 100–300 years after entombment, after which the obliteration processes decelerate. As this happens, a piece of the natural matter as leftover humus (around 7% of the unique substance) can be put away in covered soils for endlessly prolonged stretch of time [9, 12, 38, 43, 68, 70].

The quantity of microbes and archaea in covered soils of the authentic period is by three-eight times lower, and the microbial bio-mass is by three-seven times lower than in the cutting edge surface soils. This proportion varies somewhat for soils covered at various times (quite a while back), and that implies that the fundamental misfortunes of microbial biomass happen during the main many years or hundreds of years after internment. In this way, the number, design, and variety of the microbial local area change essentially after soil entombment. A specific piece of the resting types of microorganisms, DNA, or other biomolecules can be inherited from the hour of soil internment, for instance, absorbed by earth minerals on a superficial level. In any case, the subject of how to isolate the microbiological markers of the "natural memory" of pale sols from the consequences of later changes in microbial networks after the dirt covering stays unsettled. Covered soils can give us information about the structure of microbial networks in the review region previously. Nonetheless, the investigation of covered soils can't is deficient to anticipate future changes in microbial networks of present day soils upon potential changes in the bioclimatic circumstances [3, 4, 10, 11].

The investigation of future changes in soil microbial networks is conceivable utilizing "artificial chronosequences"—research facility or field experiments with displaying long haul changes in the dirt and ecological circumstances. For instance, experiments on recreating environment changes, including a dangerous atmospheric devation, which ordinarily incorporate counterfeit long haul climb in temperature and changes in precipitation and insolation levels in the exploratory region. These investigations endeavor to survey how the biomass, action, and different qualities of soil microbial networks are changed because of environmental change. For instance, such analyses survey the impact of an unnatural weather change recreation on the microbiomes of icy and boreal soils. Such an effect is communicated in a reduction in the wealth of growths, an expansion in the overflow of microscopic organisms, and an adjustment of the ordered organization of the local area [80, 81].

Simultaneously, momentary changes in temperature and dampness content may not influence the design of the microbial local area at all, or the impact might show up solely after decade of the experiment and just in the surface soil layer. Frequently, it is connected with the roundabout impact of changes in the overflow and organization of the plants on the dirt microbial local area. Whether the consequences of such tests are relevant to demonstrating the genuine elements of microbial networks because of cli-mate change is a disputable issue. Notwithstanding, right now, such a recreation of an Earth-wide temperature boost stays one of only a handful of exceptional ways of foreseeing long haul changes in soil microbial networks from now on [82, 83].

Soil microbial networks change inside a gigantic scope of time: from hours to centuries. In the most limited periods, under the effect of sudden changes in soil conditions or the contribution of new natural matter, the action of microorganism's changes fundamentally. Over longer periods, the complete biomass and ordered construction of the microbial local area change due to the elements of the sythesis of plants and physical and compound properties of the dirt (particularly, pH conditions). Changes in the all-out overflow and biomass of microorganisms are normally connected with the con-tent of soil natural matter. Processes joined by the aggregation of carbon in the dirt-beginning pedo-beginning or auxiliary rebuilding progressions—normally lead to an expansion in the microbial biomass and in the parasites/microorganisms proportion. The absolute variety ($\alpha$-variety, species extravagance) of microbial networks can either increment or diminish or stay unaltered during soil cycles of totally different lengths [33, 60, 64, 83–87].

Clearly, explicit examples of changes in microbial not entirely settled by a wide range of boundaries, and it appears to be difficult to isolate a solitary general pattern. The biomass and design of microbial networks in practically all dirts and environment types are exposed areas of strength for to elements. This ought to be thought about while contrasting microbial networks of spatially far off soils, particularly those examined at different times. In any event, throughout the mid-year season, the overflow and biomass of microorganisms can change by a few times, which mutilates the consequences of near examination of various soils. Plants are vital in controlling the elements of microbial networks. For brief timeframes, the effect of plants is communicated in changes in the movement of the arrival of root exudates and, for longer periods, in changes in the overflow and structure of the plant local area during different progressions. Other factors in the elements of microbial networks—temperature, dampness, physical and synthetic legitimate ties of the dirt-may likewise influence microorganisms by implication, through the guideline of vegetation [2, 4, 6, 14, 18, 45].

## Conclusion and Future Perspectives

This study showed that dirt microbial local area is fundamental to accomplishing food security under environmental change while they moderate GHG emanations and further develop soil fruitfulness. This concentrate further rundowns microbial procedures in CSA as practical, modest, and eco-accommodating innovation that ought to be sought after. This study gave a profound comprehension of microbial innovations, soil and plant cooperation's under CSA situation. This study focused on the requirement for environmental change variation and moderation while further developing food creation in the ongoing food framework. At long last, this study adds to comprehension of what environment changes mean for soil organisms and biological system cycles, and how agrarian practices under CSA mediations can accomplish environmental change variation, GHG relief, and food security.

## References


1. Hoseini A, Salehi A, Sayyed RZ, Balouchi H, Moradi A, Piri R, Fazeli-Nasab B, Poczai P, Ansari MJ, Obaid SA, Datta R (2022) Efficacy of biological agents and fillers seed coating in improving drought stress in anise. Front Plant Sci 13:955512. https://doi.org/10.3389/fpls.2022.955512
2. Ali-Soufi M, Shahriari A, Shirmohammadi E, Fazeli-Nasab B (2017) Investigation of biological properties and microorganism identification in susceptible areas to wind erosion in Hamoun wetlands. In: Congress on restoration policies and approaches of Hamoun international wetland Zabol 2017, pp 231–240
3. Fazeli-Nasab B, Mahdinezhad N, Parray JA (2022) Metagenomics and Microbiome engineering identification of core microbiome and improvement of Rhizosphere. In: In: Core microbiome: improving crop quality and productivity
4. Ali-Soufi M, Shahriari A, Shirmohammadi E, Fazeli-Nasab B (2017) Seasonal changes biological characteristics of airborne dust in Sistan plain, Eastern Iran. In: International conference on loess research. Gorgan University of Agricultural Sciences and Natural Resources, Gorgan, Iran
5. Ali-Soufi M, Shahriari A, Shirmohammadi E, Fazeli-Nasab B (2017) Identification and isolation of associated microorganisms with airborne dust loaded over Sistan plain. In: 15th Iranian soil science congress. Isfahan University of Technology, Isfahan, Iran, Congress COI: SSCI15, Article COI: SSCI15_895
6. Ali-Soufi M, Shahriari A, Shirmohammadi E, Fazeli-Nasab B (2019) Investigation of dust microbial community and identification of its dominance species in Northern Regions of Sistan and Baluchestan Province. J Soil Sci (Sci Technol Agricul Nat Resour) 23(1):309–320. https://doi.org/10.29252/jstnar.23.1.23
7. Shaban H, Fazeli-Nasab B, Alahyari H, Alizadeh G, Shahpesandi S (2015) An Overview of the benefits of compost tea on plant and soil structure. Adv Biores 6(61):154–158. https://doi.org/10.15515/abr.0976-4585.6.1.154158
8. Driscoll SRT (2021) Using principles of seascape ecology to consider relationships between spatial patterning and mobile Marine Vertebrates in a seagrass-mangrove ecotone in Bimini. Antioch University, Bahamas
9. Pelletier N, Doyon M, Muirhead B, Widowski T, Nurse-Gupta J, Hunniford M (2018) Sustainability in the Canadian egg industry—Learning from the past, navigating the present, planning for the future. Sustainability 10(10):3524. https://doi.org/10.3390/su10103524

10. Barelli C, Albanese D, Stumpf R, Donati C, Rovero F, Hauffe H (2019) Habitat disturbance affects gut microbiota communities differently in wild arboreal and ground-feeding tropical primates. In: 56th annual meeting of the association for tropical biology and conservation: tropical biology and sustainable development, MG, p 152
11. Kaur M, Sodhi HS (2022) Combinative effect of seed priming with plant growth-promoting rhizobacteria and green chemicals on plant growth and stress tolerance. In: New and future developments in microbial biotechnology and bioengineering. Elsevier, pp 265–288. https://doi.org/10.1016/B978-0-323-85581-5.00004-5
12. Schimel JP, Weintraub MN (2003) The implications of exoenzyme activity on microbial carbon and nitrogen limitation in soil: a theoretical model. Soil Biol Biochem 35(4):549–563. https://doi.org/10.1016/S0038-0717(03)00015-4
13. Imperial MT (2005) Using collaboration as a governance strategy: Lessons from six watershed management programs. Adm Soc 37(3):281–320. https://doi.org/10.1177/0095399705276111
14. Hendrickson KL, Rasmussen EB (2013) Effects of mindful eating training on delay and probability discounting for food and money in obese and healthy-weight individuals. Behav Res Ther 51(7):399–409. https://doi.org/10.1016/j.brat.2013.04.002
15. Verseux C, Baqué M, Lehto K, de Vera J-PP, Rothschild LJ, Billi D (2016) Sustainable life support on Mars—The potential roles of cyanobacteria. Int J Astrobiol 15(1):65–92. https://doi.org/10.1017/S147355041500021X
16. Williams PJlB, Laurens LM (2010) Microalgae as biodiesel & biomass feedstocks: review & analysis of the biochemistry, energetics & economics. Energy Environ Sci 3(5):554–590. https://doi.org/10.1039/B924978H
17. Eisenhofer R, Minich JJ, Marotz C, Cooper A, Knight R, Weyrich LS (2019) Contamination in low microbial biomass microbiome studies: issues and recommendations. Trends Microbiol 27(2):105–117. https://doi.org/10.1016/j.tim.2018.11.003
18. Yan J, Wang L, Hu Y, Tsang YF, Zhang Y, Wu J, Fu X, Sun Y (2018) Plant litter composition selects different soil microbial structures and in turn drives different litter decomposition pattern and soil carbon sequestration capability. Geoderma 319:194–203. https://doi.org/10.1016/j.geoderma.2018.01.009
19. Freschet GT, Pagès L, Iversen CM, Comas LH, Rewald B, Roumet C, Klimešová J, Zadworny M, Poorter H, Postma JA (2021) A starting guide to root ecology: strengthening ecological concepts and standardising root classification, sampling, processing and trait measurements. New Phytol 232(3):973–1122. https://nph.onlinelibrary.wiley.com/journal/14698137
20. Habibishandiz M, Saghir M (2022) A critical review of heat transfer enhancement methods in the presence of porous media, nanofluids, and microorganisms. Therm Sc Eng Progr 101267. https://doi.org/10.1016/j.tsep.2022.101267
21. Rao A, Rath A, Sharma R, Meda US (2022) Microbial fuel cells and genomics: a review. ECS Trans 107(1):10729
22. Danilova N (2020) Stress response and immunity: links and trade offs. Bentham Science Publishers
23. Hough M, McCabe S, Vining SR, Pickering Pedersen E, Wilson RM, Lawrence R, Chang KY, Bohrer G, Coordinators I, Riley WJ (2022) Coupling plant litter quantity to a novel metric for litter quality explains C storage changes in a thawing permafrost peatland. Glob Change Biol 28(3):950–968. https://doi.org/10.1111/gcb.15970
24. Trivedi P, Leach JE, Tringe SG, Sa T, Singh BK (2020) Plant—Microbiome interactions: from community assembly to plant health. Nat Rev Microbiol 18(11):607–621. https://doi.org/10.1038/s41579-020 0412-1
25. Odelade KA, Babalola OO (2019) Bacteria, fungi and archaea domains in rhizospheric soil and their effects in enhancing agricultural productivity. Int J Environ Res Public Health 16(20):3873. https://doi.org/10.3390/ijerph16203873
26. Hussain A, Rehman F, Rafeeq H, Waqas M, Asghar A, Afsheen N, Rahdar A, Bilal M, Iqbal HM (2022) In-situ, ex-situ, and nano-remediation strategies to treat polluted soil, water, and air—A review. Chemosphere 289:133252. https://doi.org/10.1016/j.chemosphere.2021.133252

27. Vardhan KH, Kumar PS, Panda RC (2019) A review on heavy metal pollution, toxicity and remedial measures: Current trends and future perspectives. J Mol Liq 290:111197. https://doi.org/10.1016/j.molliq.2019.111197
28. Sarsaiya S, Jain A, Shi J, Chen J (2020) Plant root-microbe relationship for shaping root microbiome modification in benefit agriculture. In: New and future developments in microbial biotechnology and bioengineering. Elsevier, pp. 85–98. https://doi.org/10.1016/B978-0-12-820526-6.00006-3
29. Reynolds S (2021) Invasive non-native species and the management and exploitation of freshwater ecosystems. University of Cambridge
30. Sun T, Hobbie SE, Berg B, Zhang H, Wang Q, Wang Z, Hättenschwiler S (2018) Contrasting dynamics and trait controls in first-order root compared with leaf litter decomposition. Proc Natl Acad Sci 115(41):10392–10397. https://doi.org/10.1073/pnas.1716595115
31. Carper DL, Weston DJ, Barde A, Timm CM, Lu T-Y, Burdick LH, Jawdy SS, Klingeman DM, Robeson MS, Veach AM (2021) Cultivating the bacterial microbiota of populus roots. Msystems 6(3):e01306-e1320. https://doi.org/10.1128/mSystems.01306-20
32. Jiao N, Song X, Song R, Yin D, Deng X (2022) Diversity and structure of the microbial community in rhizosphere soil of Fritillaria ussuriensis at different health levels. PeerJ 10:e12778. https://doi.org/10.7717/peerj.12778
33. McKinley VL (2019) Effects of land use and restoration on soil microbial communities. In: Hurst CJ (ed) Understanding terrestrial microbial communities. Springer International Publishing, Cham, pp 173–242. https://doi.org/10.1007/978-3-030-10777-2_7
34. Xiong Y, Hou Z, Xie H, Zhao J, Tan X, Luo J (2022) Microbial-mediated $CO_2$ methanation and renewable natural gas storage in depleted petroleum reservoirs: A review of biogeochemical mechanism and perspective. Gondwana Res. In Press. https://doi.org/10.1016/j.gr.2022.04.017
35. Timmis K, Ramos JL (2021) The soil crisis: the need to treat as a global health problem and the pivotal role of microbes in prophylaxis and therapy. Wiley Online Library, pp 769–797
36. Patel HK, Kalaria RK, Vasava DK, Bhalani HN (2022) Soil microbiome: a key player in conservation of soil health under changing climatic conditions. In: Arora S, Kumar A, Ogita S, Yau YY (eds) Biotechnological innovations for environmental bioremediation. Springer Nature Singapore, Singapore, pp 53–82. https://doi.org/10.1007/978-981-16-9001-3_3
37. Parray JA, Yaseen Mir M, Shameem N (2019) Rhizosphere engineering and agricultural productivity. In: Parray JA, Yaseen Mir M, Shameem N (eds) Sustainable agriculture: biotechniques in plant biology. Springer Singapore, Singapore, pp 71–154. https://doi.org/10.1007/978-981-13-8840-8_3
38. Alamgir ANM (2018) Secondary metabolites: secondary metabolic products consisting of C and H; C, H, and O; N, S, and P Elements; and O/N heterocycles. In: Alamgir ANM (ed) Therapeutic use of medicinal plants and their extracts: volume 2: phytochemistry and bioactive compounds. Springer International Publishing, Cham, pp 165–309. https://doi.org/10.1007/978-3-319-92387-1_3
39. Manchola Rojas LA (2022) Buried wood effects on soil nutrient supply and microbial activity in different oil sands reclamation soils in Northern Alberta. University of Alberta
40. Zhang X, Zou Q, Zhao B, Zhang J, Zhao W, Li Y, Liu R, Liu X, Liu Z (2020) Effects of alternate-day fasting, time-restricted fasting and intermittent energy restriction DSS-induced on colitis and behavioral disorders. Redox Biol 32:101535. https://doi.org/10.1016/j.redox.2020.101535
41. Pouliot A (2018) Allure of Fungi. Csiro Publishing
42. Bierend D (2021) In search of mycotopia: citizen science, fungi fanatics, and the untapped potential of mushrooms. Chelsea Green Publishing
43. Pollock FJ, McMinds R, Smith S, Bourne DG, Willis BL, Medina M, Thurber RV, Zaneveld JR (2018) Coral-associated bacteria demonstrate phylosymbiosis and cophylogeny. Nat Commun 9(1):1–13. https://doi.org/10.1038/s41467-018-07275-x
44. Deel HL (2022) Microbial succession in human rib skeletal remains and fly-human microbial transfer during decomposition. Colorado State University

45. George TS, French AS, Brown LK, Karley AJ, White PJ, Ramsay L, Daniell TJ (2014) Genotypic variation in the ability of landraces and commercial cereal varieties to avoid manganese deficiency in soils with limited manganese availability: is there a role for root-exuded phytases? Physiol Plant 151(3):243–256. https://doi.org/10.1111/ppl.12151
46. Santana Martinez JC (2021) The impact of reclamation and vegetation removal on compositional and functional attributes of soil microbial communities in the Athabasca Oil Sands Region. University of Alberta
47. Ragot SA, Kertesz MA, Mészáros É, Frossard E, Bünemann EK (2017) Soil phoD and phoX alkaline phosphatase gene diversity responds to multiple environmental factors. FEMS Microbiol Ecol 93(1):fiw212. https://doi.org/10.1093/femsec/fiw212
48. Carletti P, Vendramin E, Pizzeghello D, Concheri G, Zanella A, Nardi S, Squartini A (2009) Soil humic compounds and microbial communities in six spruce forests as function of parent material, slope aspect and stand age. Plant Soil 315(1):47–65. https://doi.org/10.1007/s11104-008-9732-z
49. Kheyrodin H, Jami R, Rehman FU (2022) Cellular structure and molecular functions of plants, animals, bacteria, and viruses. Cell Mol Biomed Rep 2(1):33–41. https://doi.org/10.55705/cmbr.2022.330941.1021
50. Greacen EL, Sands R (1980) Compaction of forest soils. A review. Soil Res 18(2):163–189. https://doi.org/10.1071/SR9800163
51. Dhuldhaj UP, Malik N (2022) Global perspective of phosphate soliloquizing microbes and phosphatase for improvement of soil, food and human health. Cell Mol Biomed Rep 2(3):173–186. https://doi.org/10.55705/cmbr.2022.347523.1048
52. Zipper CE, Burger JA, Skousen JG, Angel PN, Barton CD, Davis V, Franklin JA (2011) Restoring forests and associated ecosystem services on Appalachian coal surface mines. Environ Manag 47(5):751–765. https://doi.org/10.1007/s00267-011-9670-z
53. Batey T (2009) Soil compaction and soil management—A review. Soil Use Manag 25(4):335–345. https://doi.org/10.1111/j.1475-2743.2009.00236.x
54. Druille M, Cabello MN, Omacini M, Golluscio RA (2013) Glyphosate reduces spore viability and root colonization of arbuscular mycorrhizal fungi. Appl Soil Ecol 64:99–103. https://doi.org/10.1016/j.apsoil.2012.10.007
55. Nguyen NH, Hynson NA, Bruns TD (2012) Stayin' alive: survival of mycorrhizal fungal propagules from 6-yr-old forest soil. Fungal Ecol 5(6):741–746. https://doi.org/10.1016/j.funeco.2012.05.006
56. Varenius K, Kårén O, Lindahl B, Dahlberg A (2016) Long-term effects of tree harvesting on ectomycorrhizal fungal communities in boreal Scots pine forests. For Ecol Manag 380:41–49. https://doi.org/10.1016/j.foreco.2016.08.006
57. Huang J, Nara K, Zong K, Lian C (2015) Soil propagule banks of ectomycorrhizal fungi along forest development stages after mining. Microb Ecol 69(4):768–777. https://doi.org/10.1007/s00248-014-0484-4
58. Policelli N, Bruns TD, Vilgalys R, Nuñez MA (2019) Suilloid fungi as global drivers of pine invasions. New Phytol 222(2):714–725. https://doi.org/10.1111/nph.15660
59. Ishida TA, Nara K, Tanaka M, Kinoshita A, Hogetsu T (2008) Germination and infectivity of ectomycorrhizal fungal spores in relation to their ecological traits during primary succession. New Phytol 180(2):491–500. https://doi.org/10.1111/j.1469-8137.2008.02572.x
60. Mueller RC, Scudder CM, Whitham TG, Gehring CA (2019) Legacy effects of tree mortality mediated by ectomycorrhizal fungal communities. New Phytol 224(1):155–165. https://doi.org/10.1111/nph.15993
61. Karpati AS, Handel SN, Dighton J, Horton TR (2011) Quercus rubra-associated ectomycorrhizal fungal communities of disturbed urban sites and mature forests. Mycorrhiza 21(6):537–547. https://doi.org/10.1007/s00572-011-0362-6
62. Policelli N, Horton TR, García RA, Naour M, Pauchard A, Nuñez MA (2020) Native and non-native trees can find compatible mycorrhizal partners in each other's dominated areas. Plant Soil 454(1):285–297. https://doi.org/10.1007/s11104-020-04609-x

63. Giachini AJ, Oliveira VL, Castellano MA, Trappe JM (2000) Ectomycorrhizal fungi in Eucalyptus and Pinus plantations in southern Brazil. Mycologia 92(6):1166–1177. https://doi.org/10.1080/00275514.2000.12061264
64. Fazeli-Nasab B, Sayyed RZ (2019) Plant growth-promoting rhizobacteria and salinity stress: a journey into the soil. In: Sayyed RZ, Arora NK, Reddy MS (eds) Plant growth promoting Rhizobacteria for sustainable stress management : volume 1: rhizobacteria in abiotic stress management. Springer Singapore, Singapore, pp. 21–34. https://doi.org/10.1007/978-981-13-6536-2_2
65. Banerjee S, Walder F, Büchi L, Meyer M, Held AY, Gattinger A, Keller T, Charles R, van der Heijden MG (2019) Agricultural intensification reduces microbial network complexity and the abundance of keystone taxa in roots. ISME J 13(7):1722–1736. https://doi.org/10.1038/s41396-019-0383-2
66. Gentry TJ, Pepper IL, Pierson III LS (2015) Microbial diversity and interactions in natural ecosystems. In: Environ microbiology. Elsevier, pp 441–460. https://doi.org/10.1016/B978-0-12-394626-3.00019-3
67. Miretzky P, Cirelli AF (2010) Cr(VI) and Cr(III) removal from aqueous solution by raw and modified lignocellulosic materials: a review. J Hazard Mater 180(1–3):1–19. https://doi.org/10.1016/j.jhazmat.2010.04.060
68. Zayed AM, Terry N (2003) Chromium in the environment: factors affecting biological remediation. Plant Soil 249(1):139–156. https://doi.org/10.1023/A:1022504826342
69. Chernov T, Zhelezova A (2020) The dynamics of soil microbial communities on different timescales: a review. Eurasian Soil Sci 53(5):643–652. https://doi.org/10.1134/S106422932005004X
70. Kotaś J, Stasicka Z (2000) Chromium occurrence in the environment and methods of its speciation. Environ Pollut 107(3):263–283. https://doi.org/10.1016/S0269-7491(99)00168-2
71. Zhelezova A, Chernov T, Nikitin D, Tkhakakhova A, Ksenofontova N, Zverev A, Kutovaya O, Semenov M (2022) Seasonal dynamics of soil bacterial community under long-term abandoned cropland in boreal climate. Agronomy 12(2):519. https://doi.org/10.3390/agronomy12020519
72. Costa D, Tavares RM, Baptista P, Lino-Neto T (2022) The influence of bioclimate on soil microbial communities of cork oak. BMC Microbiol 22(1):1–17. https://doi.org/10.1186/s12866-022-02574-2
73. Wang Q, Zhao X, Tian P, Liu S, Sun Z (2021) Bioclimate and arbuscular mycorrhizal fungi regulate continental biogeographic variations in effect of nitrogen deposition on the temperature sensitivity of soil organic carbon decomposition. Land Degrad Dev 32(2):936–945. https://doi.org/10.1002/ldr.3651
74. Parizadeh M, Mimee B, Kembel SW (2021) Neonicotinoid seed treatments have Significant non-target effects on phyllosphere and soil bacterial communities. Front Microbiol 11:619827. https://doi.org/10.3389/fmicb.2020.619827
75. Jerbi M, Labidi S, Bahri BA, Laruelle F, Tisserant B, Jeddi FB, Sahraoui AL-H (2021) Soil properties and climate affect arbuscular mycorrhizal fungi and soil microbial communities in Mediterranean rainfed cereal cropping systems. Pedobiologia 87:150748. https://doi.org/10.1016/j.pedobi.2021.150748
76. Ruthsatz K, Lyra ML, Lambertini C, Belasen AM, Jenkinson TS, da Silva Leite D, Becker CG, Haddad CF, James TY, Zamudio KR (2020) Skin microbiome correlates with bioclimate and Batrachochytrium dendrobatidis infection intensity in Brazil's Atlantic Forest treefrogs. Sci Rep 10(1):1–15. https://doi.org/10.1038/s41598-020-79130-3
77. Aley P, Singh J, Kumar P (2022) Adapting the changing environment: microbial way of life. In: Microbiome under changing climate. Elsevier, pp 507–525. https://doi.org/10.1016/B978-0-323-90571-8.00023-7
78. Majumder N, Biswas K (2021) Soil microflora and its role in diminution of global climate change. In: Lone SA, Malik A (eds) Microbiomes and the global climate change, Springer, Singapore, pp 225–246. https://doi.org/10.1007/978-981-33-4508-9_13

79. Medhi K, Bhardwaj R, Laxmi R (2021) Climate change with its impacts on soil and soil microbiome regulating biogeochemical nutrient transformations. In: Choudhary DK, Mishra A, Varma A (eds) Climate change and the microbiome: sustenance of the ecosphere. Springer International Publishing, Cham, pp 95–138. https://doi.org/10.1007/978-3-030-76863-8_6
80. Wei L, Li Y, Zhu Z, Wang F, Liu X, Zhang W, Xiao M, Li G, Ding J, Chen J (2022) Soil health evaluation approaches along a reclamation consequence in Hangzhou Bay China. Agricul Ecosyst Environ 337:108045. https://doi.org/10.1016/j.agee.2022.108045
81. van Rijssel SQ, Veen G, Koorneef GJ, Bakx-Schotman J, Ten Hooven FC, Geisen S, van Der Putten WH (2022) Soil microbial diversity and community composition during conversion from conventional to organic agriculture. Mol Ecol 31(15):4017–4030. https://doi.org/10.1111/mec.16571
82. Yang Y, Huang Y, Tang X, Li Y, Liu J, Li H, Cheng X, Pei X, Duan H (2021) Responses of fungal communities along a chronosequence succession in soils of a tailing dam with reclamation by *Heteropogon contortus*. Ecotoxicol Environ Saf 218:112270. https://doi.org/10.1016/j.ecoenv.2021.112270
83. Zhang Y, Cao C, Cui Z, Qian W, Liang C, Wang C (2019) Soil bacterial community restoration along a chronosequence of sand-fixing plantations on moving sand dunes in the Horqin sandy land in northeast China. J Arid Environ 165:81–87. https://doi.org/10.1016/j.jaridenv.2019.04.003
84. Verma A, Shameem N, Jatav HS, Sathyanarayana E, Parray JA, Poczai P and Sayyed RZ (2022) Fungal endophytes to combat biotic and abiotic stresses for climate-smart and sustainable agriculture. Front Plant Sci 13:953836. https://doi.org/10.3389/fpls.2022.953836
85. Javid AP, Sumira J, Azra NK, Raies AQ, Parvaiz A (2016) Plant Growth Promoting Rhizobacteria (PGPR): from physiology to genomics. J Plant Growth Regul. https://doi.org/10.1007/s00344-016-9583-4
86. Parray JA, Ali U, Mir MY, Shameem N (2021) A high throughputs and consistent method for the sampling and isolation of Endophytic bacteria allied to high altitude the medicinal plant Arnebia benthamii (Wall ex. G. Don). Micro Environer 1(01):1–6. https://doi.org/10.54458/mev.v1i01.6668
87. Mir MY, Hamid S, Parray JA (2021) Phyllosphere microbiomes: implications and ecofunctional diversity. In: Microbial diversity and ecology in hotspots, pp 81–95. Elsevier. https://doi.org/10.1016/B978-0-323-90148-2.00005-5

# Chapter 4
# Microbial Community Dynamics Due to Land Use Change: Some Circumstances in the Tropical Rain Forest of Indonesia

**Enny Widyati, Helbert, Yayan Wahyu C. Kusuma, Ragil S. B. Irianto, Neo E. Lelana, Irma Yeny, and Chao Gao**

**Abstract** This chapter discusses concerning land use shifting influences to the soil microorganisms dynamic, especially in Indonesia where the biggest tropical rain forest established. Indonesia is among the region with largest tropical rain forest in the world. The country is also rich in plants biodiversity associated with the biophysical and the climate conditions forming the tropical rain forest. The high of plant diversity of Indonesia forest is illustrated by Malik et al. (Jurnal Ilmiah Pendidikan Sains 1:35–42, 2020), in Kalimantan in a hectar of forest can be identified more than 150 species.

## Introduction

Kusmana and Hikmat [1] summarized, despite the fact large of terrestrial region of Indonesia is only about 1.3% from total of the earth, 25% of world seed plants (spermatophytes) species are distributed in Indonesia. Hence, Indonesia is positioned as the 7th world plant biodiversity with about 20,000 numbers of species. Among the 20,000 species, 40% are endemic species (origin) of Indonesia. The most abundance family is belong to Orchidaceae that is reached 4,000 species, followed by Dipterocarpaceae with 386 species numbers (70% of dipterocarps population in the world), Myrtaceae and Moraceae (each 500 species numbers); Ericaceae (737 species), involved Rhododendron and Naccinium with 287 and 239 species numbers,

---

E. Widyati (✉) · Helbert · Y. W. C. Kusuma · I. Yeny
Research Center for Ecology and Ethnobiology, Research Organization for Life Sciences and Environment, The National Research and Innovation Agency (BRIN), Jakarta, Indonesia
e-mail: enny.widyati@brin.go.id

R. S. B. Irianto · N. E. Lelana
Research Center for Applied Microbiology, Research Organization for Life Sciences and Environment, The National Research and Innovation Agency (BRIN), Jakarta, Indonesia

C. Gao
Department of Mechanical and Industrial Engineering , Norwegian University of Science and Technology , 7034, Trondheim, Norway

 
J. A. Parray (ed.), *Climate Change and Microbiome Dynamics*, Climate Change Management, https://doi.org/10.1007/978-3-031-21079-2_4

respectively. Malik et al. [2] noticed that Indonesia is the producer of 75% of world rattan.

Malik et al. [2] reviewed forest of Indonesia colonized by the world highest palm family (Arecaceae), for instance there is 122 species numbers of bamboo. Kusmana and Hikmat [1] also reported that Indonesia has a high diversity of ferns about 4,000 species numbers, rattan about 332 species numbers involved of big stems of Genera Calamus (204 species) and Genera of Daemonorops (86 species). On the other hand, Indonesia also noticed as the center of distribution of Vavilov (biodiversity of cultivated plants) such as *Syzygium aromaticum*, *Nephelium* spp., *Musa* spp., *Durio* spp., and *Myristica fragrans* (ref).

However, along with the increase of population and development, lots of forests convert into many land utilization caused a deforestation phenomenon. Referring to [3], sometimes deforestation is planned for infrastructure development even it gave several negative impacts to the environment. It was reported, forest coverage of Indonesia during in 12 years (1985–1997) was drastically reduced from 119 million ha to 95 million ha [4]. Several activities such as intensive forest management, illegal logging, mining, agriculture, transmigration, forest fire, and land grabbing are indicated as the major reason for deforestation and forest degradation [5]. In addition, the excessive oilpalm plantation and mining activities in the forest area lead to enormously increase of forest vulnerability [4].

Plant is a sessile organism hence they need assistance from various microbes living around them for reaching nutrients, growth factors, and safeguard against pathogens. Plants actively initiate in assembling a favorable environment to invite beneficial microbes colonized around their root system. On the other hand, vegetation (species, stage of growth, etc.) determine structure and composition of soil microbes [6]. Various studies shown that many plant-microbes association have a remarkable impact on germination of seeds; vigor of seedlings; plant nutrition; plant disease; as well as plant growing, development and yield [7].

Berg and Smalla [8] have summarized from earlier studies, essentially every plant species requires a set of microorganism communities in its rhizosphere, both to support its growth (nutrient availability and growth factors) and its specific health (biocontrol and anti-pathogens). Therefore, it is crucial to consider knowledge on the plant-rhizosphere community interrelation in developing strategy for soil treatments, multi-species cropping, and crop rotations. The characteristic of plant species is vital for biological control applications. Moreover, it is also important to recognize the existing specific association among plants and microbes in correlation to issues of nature conservation. It means that once a plant species distinct, soil rhizosphere community will be disturbed.

Pitman and Jorgensen [9] discovered approximately 22–47% of the world's plants are threatened with extinction. Unknown microbial diversity may be impacted when plants become extinct. Improved understanding on specific interactions among plants with microorganisms in their rhizosphere is useful for reforestation activities that include replanting degraded forests and woodlands with native tree stock. It is also reported, the interactions among microorganism and plant are crucial issues influencing the invasive species competition with the indigenous flora. Hence, the influence of climate-change on interaction among vegetations and microorganisms, i.e. on plant diseases, is also urgent to be calculated.

The activities causing to forest coverage changing is presumed to give many alterations to the underground organisms, involved soil microorganisms. Furthermore, this chapter discussing review results on the dynamic and function of forest coverage related to development and planning from previous publications (journals, IOP proceedings, books, reports, etc.), especially focused on (1) the dynamic of soil microbe under forest harvesting/tree cutting; (2) the role of soil microbes to the succession of pioneer in the secondary forest, involved to the invasive alien species distribution, (3) the alteration of soil microbes population due to land use change from natural to monoculture plantation, (4) responsibility of soil microorganisms on the mining land and the limitations to reclamation achievement.

## The Dynamic of Soil Microbes Under Forest Harvesting/Tree Cutting

Plants are the initiator in rhizosphere configuration and controlling the composition and structure of root-microbial communities by releasing diverse organic compounds from photosynthesis [10]. It is estimated at 10–30% of photo-synthate [11], collectively labeled as root exudates [12] released to the root zone, for attracting soil microorganism and creating an unique environment known as the rhizosphere [13].

The rhizosphere recognized as the confine zone around and impacted by roots, is a hotspot for a variety of organisms and is the most dynamic ecosystems [14, 15]. In the rhizosphere is colonized by nematodes, arthropods, protozoa, algae, archaea, bacteria, fungi, oomycetes, and viruses [16, 15]. Most of them compose the complexity of food web using the large proportion of nutrients supplied by the plant, involved root exudates, border cells, mucilage [7]. The root exudates is a major driving force, with functions to attract and deter soil microbes hence the structure, size, and array of rhizosphere colonization match with the types, growth, and the stage of plant development [17, 7, 6]. Berg and Smalla [8] concluded, the rhizosphere is the important area for plant nutrition, health and productivity. Rhizosphere determine nutrient cycling in terrestrial ecosystems and ecosystem functioning.

Therefore, tree harvesting is perhaps the most harmful to trees since it removes all plant portions that operate as photosynthetic patches. This is an important process in the manufacturing of root exudates. Kögel-Knabner [18] found a half portion of root exudates is released as sugars, the main source of carbon for soil microbes [19]. Furthermore, tree felling is thought to influence the rhizosphere's interaction between plants and microbes.

Earlier studies, on a larger scale, the practice of forest harvesting conducted by clear-cutting. It removes in excess of the tree bole, which remarkable decreasing the total content of soil nitrogen and biomass of microbes (Johnson and Curtis 2001). This resulted a niche selecting some sensitive taxa and alter structure of soil community [20], which can be considered as an environmental screening [21]. The loss of susceptible microbes due to tree harvesting may support the colonization of better-adapted microbes, it shift the microbial community hence modify the process of decomposition [22, 23].

Specifically, the reset of soil community is due to forest harvesting contributes large amount of soil organic compound into soil. Referring to [24] huge of available organic C should facilitate copiotrophs microorganisms. Tate [25] divided soil microbes into two groups. Copiotrophs microbes group is opportunist, when resource conditions are plentiful, they prefer to ingest unstable soil organic C pools, then aggressively grow. In contrast, oligotrophic group have slower growth speed and are incapable to compete with the copiotrophs in poor nutrient circumstances [25].

Study on short rotation coppice monoculture plantation of *Callyandra callothyrsus* in Majalengka District, Indonesia by Widyati et al. [10] found cutting decreased the below ground sugars flux by 80% and lead to decrease the soil pH rapidly. The depletion of total soil sugar is hypothesized as the strategy for *C. calothyrsus* to survive and regenerate after being cut. Sugar deficiency causes major alteration in the size and composition of rhizosphere community. Another survival strategy for limiting adjacent competitor populations in the rhizosphere of callyandra is to increase soil acidity [10].

## The Role of Soil Microbes to the Succession of Pioneer in the Secondary Forest, Involved to the Invasive Alien Species Distribution

Once forested land opened due to harvesting, fire, or other catastrophes, this is the opportunity for a new plant to occupy this new habitat. Vegetation formation in the earth is started with seed dispersal and establishment of seedlings in soil. A seed reach the new habitat by seed dispersal vectors, such as animals, wind, water, or human being. Nelson [26] reported the impact of environment and microbial interactions in plant development take place initially in germination and early growth stages. The microbiome developing throughout seed sprouting and spreads to seedlings and diverse organs of full-grown plants after a long time may contain microbes that were picked up along the way [26]. Afterwards, an extensive range of biotic (plant traits) and abiotic (soil properties) variables determine the diversity of structure and function of the microbial communities in the new rhizosphere assemblage [8].

From the seed stage onward, interactions between plants and microorganisms have been documented, the interaction is known as seed microbiome [26]. Furthermore, [26] classified seed microbiome into endophytic and epiphytic microbiota. Endophytic microbiota are microbes living inside seed tissues and inherited to its descendant through progeny process during seedlings development, while epiphytic microbiota are microbes inhabiting outside seed and may or may not be adopted to inner tissues of seeds and transmitted either vertically to their seedlings or horizontally to other plants [26]. Previous studies reported that seed-associated bacterial distinct due to species of plant [27], plant traits [28], stages of seed development [29], topographical locations [30], and the existence of plant pathogen [31]. Links et al. [27] explained seed endophytic bacteria deliver almost the entire species assemblage from where the seed microbiome recruited, it indicated that in some plant species the seed endophytic were substantially conserved. The seed endophytic microbiota is frequently dissimilar with the soil bacteria colonized the plants rhizosphere [32]. It is indicated that, the microbes colonizing the seed is predominantly brought from the parent plant environment [26], it carried away from the habitat where the origin of the host plants grow [33]. It is not clearly explained, either local site characteristics or host genotypes assembly the bacterial seed microbiome [33, 30].

Plants have an impact on soil microbial populations; every plant type is presumed to form a distinguish rhizosphere communities. Root exudates are the main force to carry out the selection process [8]. The type of vegetation determines the conformation of substances released by roots, which determines the relative abundance of microorganisms surrounding the roots [34]. To shape their own rhizosphere, plants allocate nutrients for the desired microbes, in the contrary it deliver unique antimicrobial metabolites to get rid the unwanted microorganisms.

The interested phenomenon on succession is invasive plant occupation, which has remarkable effects on the society of soil microorganisms [35]. The invasive plant species generally characterized by their capacity to grow rapidly, hence they will immediately replace the origin vegetation composition [36]. In the new ecosystem, these exotic plants will change *the net primary productivity* (NPP) and nutrient cycling processes [37]. Because there is an intently link among the plant aboveground and belowground subsystems, hence the alteration in species plant dominant in a community will simultaneously affect interactions among plants and microorganisms in the rhizosphere. Afterward, it determine the nutrient cycle processes [38]. It is due to, substances released by plant root facilitate the rhizosphere association, which in reciprocate they decompose organic matter to provide nutrients to the plants [39]. Zhang et al. [40] found more peculiar fixed carbon released in the rhizosphere of *Spartina alterniflora* Loisel., an invasive species, than it found in the native plants. In consequence, the carbon turnover effectiveness at the plant-soil boundary increase with the intention of achieving successful invasion.

Significant modification on the assembly of soil microbial associations, biomass, and their activities due to plant infiltrations determine the fundamental ecosystem behaviors such as decomposition of soil organic matter and nutrient cycling [41]. Stefanowicz et al. [42] convinced that the invader plants change belowground microbial performance significantly only in two growing seasons. The modification of soil environment is the effort of the invasive species to construct their proper niche to support the growth and successfully conquer the new habitat [42].

Stefanowicz et al. [42] summarized, the various alterations due to invasion of alien species can be classified into: impact on soil physic-chemicals (nutrients and pH), impact on soil communities (soil bacteri and fungy, ectomycorrhiza and endomycorrhizal fungi), impact on microbial activities (enzymatic and respiration). Characteristic of the plant invaders such as crown formation, rooting architecture, or chemical content of tissue define the distinction in the reactions of soil to plant invasion [42]. Root exudate is a selection tool in a rhizosphere because a root exudate with a certain composition is only suitable for the structure of a particular microbial community, otherwise, that composition can be a killing machine for other microbial groups [43]. Thus, it can be understood that the introduction of new plant species into a habitat, massively, will lead to dramatic reformation of the community of underground microbes as consequences of the powerfull reciprocal influence.

It is widely recognized that invasive plants brought negative impacts to the indigenous plant communities, even the invader often completely eliminate native species and change the habitat to a monodominant communities [44]. The shift in plant composition by exotic plant species interfere the linkage between above-ground communities [45], it modify soil chemical properties (pH, N content, N mineralization processes) due to revolution in the structure of microbial communities that control the main biogeochemical cycles in the habitat.

Every exotic species has a unique consequence to the physico-chemical characteristics of the soil in its new habitat [46]. Study on invasive species showed, they caused alteration on soil physical attributes, especially the soil porosity, temperature, water-holding capacity, and moisture [47]. This is due to the changes in the vegetation type in the habitat which has different in tissue biomass characteristics, rooting depth, leaf area index, and transpiration rate [48]. Modification in soil moisture and root exudate composition result in changes in the rhizospheric microbial flora to promotethe further invasiveness [49]. More over, the invasion also influence the chemical characteristics of soil due to the shifting of soil organic matter input, paterns of cycling of carbon and nitrogen, and soil pH. Invasive species also found to release of some allelopathic substances [50, 51, 52].

The belowground microbial community strongly determines the invasive capacity of exotic plant species [53]. Li et al. [53] reported one of most destructive invasive weeds in China, *Ageratina adenophora*, which formed a single species community rapidly. The existence of *A. adenophora* resulted in shifting of microbial composition either in the bulk soil or rhizosphere, for example *Bradyrhizobium* replace *Aeromicrobium* [53], the specific microbes rule in N-cycling processes. Li et al. [53] confirmed that *A. adenophora* change the soil pH of the rhizosphere environment to impose homogenous microbial communities. They selected appropriate microbial communities in providing their obligations in soils to encourage their invasiveness.

In Indonesia there are several invasive species incidences that caused alteration on the habitat dramatically. In Batukahu National Park, there were 10 identified invasive plant species member of 10 genera and five families [54]; which were classified as 40% herbs, and 30% each shrubs and grasses [54].

The most phenomenal invasive species in Indonesia is *Vachellia nilotica* (L.) P.J.H. Hurter & Mabb commonly known as thorny acacia, is notorious for its ability to conquer diverse environment, especially grassland (Fig. 4.1.) After being introduced for the first time in the 1969, to the Baluran National Park (BNP), Indonesia, the

**Fig. 4.1** *Vachellia nilotica* in The BNP (**a**) and *Merremia peltata* (**b**) the most remarkable invasive species in Indonesia

tree currently has invaded wild bull habitat of the national park more than half area [55]. BNP is the biggest Bull (*Bos sondaicus*) in Indonesia with 1500–2000 ha of savannah ecosystem [56]. The invasion of *V. nilotica* threatened the population of the bulls due to the invasive species eradicated the bulls feeding plants.

Another terrific invasive plant in Indonesia is *Merremia peltata* causing serious hazards to the regeneration of indigenous plant [57]. Both the opened areas and the bared land, before planting for estate and agriculture, in entire regions of Indonesia are susceptible to be invaded by this species [57]. This species has a large underground tuber. They climb and cover all over crowns of the woody plants, hence it disturb the photosynthesis process [57]. Merremia is classified into a fast-growing plant that is regenerate by rooting their nodes, or by resprouting and rooting the broken stem fragments [58]. The species dispersal also occurred by seeds that is unconsciously carried away by human activities or as a result of soil displacements [59]. Yudaputra [57] estimated that currently, *M. peltata* have influenced or perhaps destroyed the habitat of 30.4% of total terrestrial ecosystem of Indonesia.

Unfortunately, the study on the influents of invasive plant species in Indonesia on the microbial population and biogeochemical process in soil is lacking. Due to each plant species need a specific collective microbes forming their own microbiome, the gaps of the information is inspiration to conduct further studies.

## Alteration of Soil Microbes Population Due to Land Use Shifting from Natural Forest to Monoculture Plantation

In natural forests, the presence of various types of plants growing together in a site will complement each other so that nutrient absorption becomes more efficient [60]. Multispecies swards have shown a variety of diversity benefits on aboveground performance, including yield, nitrogen contents, and even soil-legacy effects on a subsequent crop [61]. Diverse plant functional attributes in multispecies vegetations resulted in complementarity of resource acquisition [62, 63], such as growing season [64], rooting depth [65], and $N_2$ fixation capability [65]. Importantly, [61] explained that these plant species diversity beneficial impacts resulted from interactions across the plant species and are thus more than merely the comparative contribution of each species (their identity effect).

The large-scale development of monoculture forest plantation will eventually replace the ecosystem's community. Plantation species are typically chosen for their highly adaptable traits, which are comparable characteristics to those of invasive plants. Monoculture cultivation's success in an ecosystem has replaced native plant dominance with exotic species. Due to changes in the content of plant root rhizodeposition into the rhizosphere, these alien species modify the network between aboveground and belowground communities in new settings [45]. Because the root exudate generated by new plantations has a different composition than the original soil environment, it alters the structure and function of the soil community of rhizosphere. Consequently, massive planting of new species as monoculture commonly drastically changes the important characteristic of soil such as pH, component of nitrogen and carbon, rate of mineralization and nitrification, and portion of essential elements such as potassium (K), calcium (Ca), and magnesium (Mg) (Table 4.1) [66].

The biotic and abiotic properties in soil can be modified by plant, and this will give impact to other plants that subsequently grow in this ecosystem. In multi-species plantation the effect of a plant type to the belowground ecosystem will be very complex [69]. It is depend on what it function and abundance in the ecosystem, it is also determined by species composition exist in the ecosystem and the characteristics of the soil [69]. Previous study carried out by Fox et al. [61] found that soil microbial community structures were highly driven by plant species identity. The difference physiology of plant species such as structures, differing root biomass, and symbiotic $N_2$-fixation induced soil physicochemical change.

**Table 4.1** Increase of deforestation from 4 important sectors during 2016–2017 (analyzed from [67, 68])

| Land use change | Year (ha) | | | | | Forest conversion until 2020 (%) |
|---|---|---|---|---|---|---|
| | 2016 | 2017 | 2018 | 2019 | 2020 | |
| IPF | 10,842,974 | 11,178,601 | 11,439,445 | 11,258,485 | 11,141,179 | 9.092 |
| OPP | 11,201,500 | 12,383,100 | 14,326,300 | 14,456,600 | 14,858,300 | 12.310 |
| RP | 3,637,300 | 3,659,100 | 3,671,100 | 3,675,900 | 3,681,300 | 3.030 |
| MO | 27,316.84 | 65,047.14 | 147,825.75 | 249,005.94 | 559,218.59 | 0.463 |

IPF: industrial forest plantation
OPP: oil palm plantation
RP: rubber plantation
MO: mining operation

Furthermore, poor species diversity on monoculture changes the rhizosphere microbial community [70]. Since, soil microbe abundance, composition, and diversity are strongly affected by plant species [71], changes in plant composition from multi- to monoculture modify the rhizosphere properties. Intensive monoculture activities over a long period lead to nutrient depletion because plants with the same growth rate in even-age forests require large amounts of the same nutrients [72]. As a result, they will release the same root exudate to invite microbes for helping grow and improving fitness. This continuous process will give negative impacts on soil function and yield sustainability due to different performance of their new rhizosphere composition.

Soil microbes have vital rule in a variety of ecological activities, including organic matter decomposition, nitrogen cycling, and plant productivity [73, 74]. The study of how different plant species and their configurations, such as forbs, grasses, and legumes, regulate their collaborated microbial association is receiving more consideration (e.g., [75, 76]). Within a particular soil type, distinct plant species found to assemblage-distinguished configuration of microbial colonization [77]. The diverse physiologies and features of different species, such as root architectures and activities, root productivity and array of rhizodeposition, are fundamental determinants of such variations [61] (Fig. 4.2).

After plants were dead or harvested, these changes in the soil microbiome mediated by plant left as "legacy" and determine the other plant species that grow subsequently (plant-soil feedbacks) [78]. The kind of soil-transferred legacy effects varies depending on different parameters, such as the prior plant, climate conditions, and soil type [79]. Rhizodeposits and litter attributes of plants determine soil microorganism [78]. Legacy effect is strongly defined by the amount and type of transmitted-persistance residue in the soil when the previous crop is removed [61]. The persistence best adapted decomposers to plant residues under the prevailing situations [80] may be assisted by the retaining of such plant excess in the soil environment, keeping this crucial macronutrient accessible in the habitat.

These kinds of legacy effects are likely to have wider ecological consequences. Plant legacy effects on the microbiome may effect on competition among plants,

**Fig. 4.2** Monoculture oil palm plantation (**a**) and natural forest (**b**). *Source* Google images

establishment and succession of plant, and the composition of the overlying plants [81]. Plants legacy can either negatively or positively effect succeeding plant species. Negatively effects occur when there are plant pathogens congregations in the soil and positively effect through the build-up of beneficial microbes [78]. So, that why one of negative effects of monoculture plantations is the occurrence of soil pathogens because its legacy may be the accumulation of pathogens.

The same species of plant has the same root system so the area of competition in the absorption of nutrients and water will be stronger [82]. The root competition of the same species of plants occurs three to five times greater than if they compete with different species [83]. To conquer the neighboring plants, they will release allellochemical, the compounds released frequently have impact either increase the growth of soil-borne pathogens or prevent the growth of advantageous microbes [84]. Similar plants will release similar allellochemicals, and there will be more buildup over time with recurrent plantation. The formation of numerous diseases known as replanting disease has been linked to one of the important chemicals in the allelopathic system: phenolic acid [85]. The phenolic acid level in soil was 400 percent higher in a continuous monoculture rye plantation than in a diverse cropping system, resulting in a decreased actinomycetes population [86]. Actinomycetes play a pivotal role in the rhizosphere, such as preventing plants from various soil-borne pathogens [87].

## Role of Soil Microbes on Mining Land and the Limitations to Reclamation Achievement

Indonesia has the biggest deposit of mineral in the world, such as second position for gold and third for nikkel of the global supply (ESDM 2016). Indonesia also has 34.8 billion tons of coal deposit (the 8th position) (ESDM 2021). In one hand, mining sectors are the enormous source for the country income. On the other hand, minning results significant ecological effects such as soil erosion, holes formation, and biodiversity loss. Soil and water on ex-mining sites contaminated due to the chemicals used in the ore purification processes. Ex-mining sites are characterized by poor in macronutrients but rich in heavy metals, acidic soil reaction and inappropriate soil texture and moisture. Nikkel, tin, and coal mining are among the harmful to the forest area, due to those are operated in opened pit mining (OPM) that remove all of soil layers above the ore deposits, included the vegetation. The removal of vegetation brings immense consequence to the elimination of the origin soil microbiome, the essential actors in soil functioning and biogeochemical cycling.

The most tremendous consequence of OPM is the incident of acid mine drainage that is much more detrimental to the environment. The OPM systems reveal layers of rock containing sulfide compounds, expose to atmospheric oxygen hence it undergoes oxidation. This oxidation process will cause the previously inert rock to become reactive and release very strong sulfuric acid to the environment. Consequently, it will quickly acidify the surrounding waters and soil. The study conducted by Widyati [88] on ex-coal mining soil in South Sumatera, Indonesia, soil pH may decrease up to 2.8. This condition may dissolves metals, immobilizes various macro elements hence they are not available to plants, which can result in the death of various aquatic biota, as well.

Referring to Akcil and Koldas [89] mining of nickel, gold, and copper, is accompanied by acid drainage problems, that is in long-term destruct water bodies and life. When sulfide-containing rocks are exposed to oxygen and water, it resulted a phenomenon called acid-mine drainage (AMD) [89]; released sulfuric acid solution that will be polluted the surface water (rainwater, pond water) and shallow subsurface water. Once AMD is happened, extremely acidic water rich in heavy metals will be continually formed and transported follow the water movement [89]. The AMD phenomenon can be illustrated in the following reactions (Fig. 4.3) [89]:

(1) $$\mathbf{2FeS_{2\,(S)} + 7O_{2\,(g)} + 2H_2O_{\,(l)} \rightarrow 2Fe^{2+}\,(aq) + 4SO_4^{2-}{}_{(aq)} + 4H^+{}_{(aq)}}$$

The initial reaction is the sulfide mineral such as pyrite ($FeS_2$) reacts with atmospheric oxygen and in the moist condition will dissolve ferrous ($Fe^{2+}$) ion. The ferrous will be immediately oxidized into ferric ($Fe^{3+}$) ion (reaction 2). AMD formation will be rapider in the region with high rainfall, like in Indonesia.

(2) $$\mathbf{4Fe^{2+}{}_{(aq)} + O_{2\,(g)} + 4H^+{}_{(aq)} \rightarrow 4Fe^{3+}{}_{(aq)} + 2H_2O_{\,(l)}}$$

The rate of acid generation is strongly determined by the chemical, biological and physical attributes of the rocks and environments. Waste rock dump permeability is particularly the important physical factors. High permeability of dumping rock facilitates excessive oxygen access, which is contributes to rapid chemical reaction rates [89]. The acid environment favor the colonization

**Fig. 4.3** AMD is characterized by forming reddish color (**a**) or torquize (**b**). The picture taken at the ex-coal mining in South Sumatra (**a**) and at the ex-cement mining land in Sukabumi West Java (**b**)

of bacteria *Acidithiobacillus ferrooxidans* and the bacteria will be most favorable when the water pH is less than 3.2 [89]. The bacteria is classified as lithotrophs ("litho" means "rock") groups that are getting energy rock weathering. It is also classified as chemotroph organisms that get energy from oxidation of inorganic compound i.e. $FeS_2$ [90]. Bacteria *A. ferrooxidans* rapidly release lead and zinc from the rocks [90]. Removing soil layers rich in soil organic matter (top soil and sub soil) due to mining excavation give advantages to the bacteria group, and rapidly colonize in the habitat.

This reactions undergo either spontaneously or being catalyzed by *A. ferrooxidans*. The cathion $Fe^{3+}$ will oxydize much more pyrite and release more ion responsible in acidifying the environment.

(3) $$\mathbf{FeS_{2\,(S)} + 14Fe^{3+}{}_{(aq)} + 8H_2O_{(l)} \rightarrow 15Fe^{2+}{}_{(aq)} + 2SO_4{}^{2-}{}_{(aq)} + 16H^+{}_{(aq)}}$$

Other problem inherited by mining operation is talling, that can be highly diverse in their physic-chemical characteristics, generally is described as sandy or silty soil, and toxic peculiarities. Tailings from ore-metal minings are constantly not only sulphidic but also rich in residual metals and metalloids (mainly Arsenic) [90]. In many places of the world, surface stabilization by revegetation (i.e. phytostabilization) is essential to decrease the negative effects of legacy tailings. However, phytostabilization of sulphidic-based metal tailings through phytoremediation is limited by the tailings' incapacity to facilitate the growth of vegetations [90]. Phytoremediation is a technology employing plant activities to absorb and eliminate elemental contaminants or decline their concentration in soil [91]. Avoidance and tolerance are two defense schemes employed by plants to deal with heavy metals poisonous in soils [92]. It is highly recommended to apply phytoremediation in ex-opened pit mining area with unsteady structure and high soil erosion, or on tailing of metal extraction [93]. The application of phytoremediation needs heavy metal detoxification as precondition process [94].

Beneficial microbes found in association with plants playing as phytoremediation activities. Earlier studies showed alteration in community structure of roots of pioneer grown in tailing containing Pb and Zn and improvement on microbial biomass [90]. Soil microbes can be engaged to assist in improving ex-mining land, directly or indirectly. Directly, microbial communities help in biogeochemical cycling of tailings. In the oxidized layer of neutralized base metal tailings can be colonized by microbial with significant biomass. However, the microbial diversity (mainly bacteria) is lower than it in the unpolluted soils [93]. The soil microbes population can be improved by inoculation. Introduction of sulphate-reducing bacteria inoculum to the ex-coal mining soils, have been improved the pH and soil nutrients [88], hence improve the seedlings planted as revegetation [95]. The bacteria reduced $SO_4{}^{2-}$ into $S_2$ that is immobile [88].

Indirectly, favorable microbes in the rhizosphere of revegetation plants facilitate the revegetation process in a variety of manners. For example, arbuscular mycorrhizas acting as a prohibiting barricade for heavy metal uptake by absorption, adsorption, or chelation process [96]. (2) Microorganisms promote immobilize the metal ions by adsorbing metals to their cell walls, creating chelators, and stimulating precipitation processes [97]. They can also help with phytostabilization by increasing root surface and depth, as well as acting as a separation barricade to protect shoots from ion translocation from roots [98]. (3) Microbes directly stimulate root multiplication, promote plant development, increase plant tolerance to heavy metal, and improve plant health.

The group of plant growth-promoting rhizobacteria (PGPR) can be employed in ex-mining revegetation because their ability to enhance plant growth and fitness, improve plant nutrition, and the most important is their protection to plants from heavy metal uptake and translocation [97]. This is performed through producing organic acids, enzymes, siderophores, antibiotics, and phytohormones, among other chemicals [97].

## Future Strategies

As one of most populated country in the world (more than 270 millions), Indonesia, will encounter food, energy, and water security in the future. The situation may be aggravated by environment destruction and climate change. It is need tight collaboration among all stakes in formulating smart strategies to deal with the challenges, included strengthen knowledge on importance of soil microbial to improve land productivity, to clean pollution, as well as to enhance land revegetation.

a. ***Optimize land utilization in food, water and energy nexus to preserve deforestation.***
   Cultivation of mixed crops that produce food, bioenergy and species that quickly increase water catchment needs to be developed to prevent expansion of deforestation and optimize land productivity. In addition, the use of local varieties needs to be expanded for restoring biodiversity, also reducing destruction of the microbiome due to "strange rhizophere assemblage" by invasive exotic species.
b. ***Rhizosphere engineered for environment friendly agriculture.***
   Plants rhizosphere can be engineered to produce substances for increasing nutrient availability, for defensing from biotic and abiotic pressures, or for promoting the growth of beneficial bacteria. Rhizosphere engineering can involve inoculation of beneficial microbial populations to the selected plants. Soil amendment can be applied to enhance the fitness of root associated bacterial communities. Hence, the rhizosphere favor selected bacteria collaborative synergically in consortia appropriate for barricading roots from pathogens. Rhizosphere engineering with various activities of PGPR improve the soil aggregation, soil health and fertility, hence facilitate plant growth better and increase the productivity.
c. ***Ex-mining rehabilitation and revegetation employ beneficial microbes***
   The crucial step in ex-mining reclamation process is soil amendment to provide favorable environment for revegetation planting. To improve revegetation succeed, both organic and inorganic ameliorants can be added to the contaminated

soil. Inorganic amendment is aimed to modify metal toxicity, reduce heavy metal bioavailability through adjustment soil reaction [99]. While, organic amendment is intended to increase the organic matter content. Those soil organic improvements add essential nutrients of the soil, improve physic, chemical and biological soil attributes, improve water-holding capacity which can benefit plant colonization in ex-mining sites. Earlier study on augmentation the ex-coal mining with material consists of raw organic matter, such as paper mills sludge, in a huge dosage (50%) successfully depleted the population of bacteria *Thiobacillus thiooxidans* in the ex-mining soil [88], that is recognized as biocatalyzer of AMD.

Another key method for maximizing the success of ex-mine land revegetation is species selection. The selected species should be tolerant to heavy metal environments, have a dense roots system and have capability to preserve soil structure, and prevent soil erosion, [92]. Qualification of selected plants for ex-mining revegetation such as fast growing for building large canopy in in a short period of time. It will assist land to modify microclimates, rapidly. They also produce lots of biomass that can be supplied to soil as organic matter. On the other hand, the selected plant should be effortless to cultivate in the field [91, 100]. The most familiar pioneer is acacias which have the ability to rehabilitate soils by absorbing and storing heavy metals like zinc (Zn), lead (Pb), copper (Cu), cadmium (Cd), and chromium (Cr) in their leaves, shoots, and roots [93]. Including microbes in ex-mining revegetation activities for example microbes enabling nitrogen fixation [101] that will improve not only soil remediation, soil amendment, but also assist plant to grow better in the severe environment.

## References

1. Kusmana C, Hikmat A (2015) The Biodiversity of Flora in Indonesia. Jurnal Pengelolaan Sumberdaya Alam dan Lingkungan 5:187–198. e-ISSN: 2460-5824. https://doi.org/10.19081/jpsl.5.2.187
2. Malik AA, Prayudha JS, Annreany R, Sari MW, Walid A (2020) Keanekaragaman hayati flora di Taman Nasional bukit Barisan SelatanResor Merpas Bintuhan Kabupaten Kaur. Jurnal Ilmiah Pendidikan Sains. 1:35–42
3. Contreras-Hermosilia A (2000) The underlying causes of forest decline. CIFOR Occasional Paper 2000, No. 30. Center for International Forestry Research, Bogor, Indonesia
4. Sumargo W, Nanggara SG, Nainggolan FA, Apriani I (2011) Potret Keadaan Hutan Indonesia. Edisi Pertama. Forest Watch Indonesia. pp 52
5. DJPKTL [Direktorat Jenderal Planologi. Kehutanan dan Tata Lingkungan] (2017) Deforestasi Indonesia Periode 2014–2015. Direktorat Inventarisasi dan Pemantauan Sumber Daya Hutan, Direktrorat Jenderal Planologi Kehutanan dan Tata Lingkungan, Kementerian Lingkungan Hidup dan Kehutanan, Jakarta
6. Widyati E (2017) Understanding the business in the rhizosphere: how do plants and soil microbes make transactions. Dee Publish. Yogyakarta Indonesia. (in Indonesian)
7. Mendes R, Garbeva P, Raaijmakers JM (2013) The rhizosphere microbiome: significance of plant beneficial, plant pathogenic, and human pathogenic microorganisms. FEMS Microbiol Rev 37:634–663. https://doi.org/10.1111/1574-6976.12028

8. Berg G, Smalla K (2009) Plant species and soil type cooperatively shape the structure and function of microbial communities in the rhizosphere. FEMS Microbiol Ecol 68:1–13. https://doi.org/10.1111/j.1574-6941.2009.00654.x
9. Pitman NCA, Jorgensen PM (2002) Estimating the size of the threatened world flora. Science 298:989
10. Widyati E, Irianto RSB, Susilo A (2022) Rhizosphere upheaval after tree cutting: soil-sugar flux and microbial behavior. Commun Integr Biol 15:105–114. https://doi.org/10.1080/19420889.2022.2068110
11. Gunina A, Kuzyakov Y (2015) Sugars in soil and sweets for microorganisms: review of origin, content, composition. Soil Biol Biochem 90:87–100. https://doi.org/10.1016/j.soilbio.2015.07.021
12. Hütsch BW, Augustin J, Merbach W (2002) Plant rhizodeposition - An important source for carbon turnover in soils. J. Plant Nut. Soil Sci. 165:397–407. https://doi.org/10.1002/1522-2624(200208)165:4(397::%3cAID-JPLN397%3e3.0.CO;2-C
13. Dundek P, Holík L, Rohlík T, Vranová V, Rejšek K et al (2011) Methods of plant root exudates analysis: a review. Acta Universitatis Agriculturae et Silviculturae Mendelianae Brunensis 3:241–246. https://doi.org/10.11118/actaun201159030241
14. Hinsinger P, Bengough AG, Vetterlein D, Young IM (2009) Rhizosphere: biophysics, biogeochemistry and ecological relevance. Plant Soil 321:117–152
15. Raaijmakers JM, Paulitz TC, Steinberg C, Alabouvette C, Mo€enne-Loccoz Y (2009) The rhizosphere: a playground and battlefield for soilborne pathogens and beneficial microorganisms. Plant Soil 321:341–361
16. Bonkowski M, Villenave C, Griffiths B (2009) Rhizosphere fauna: the functional and structural diversity of intimate interactions of soil fauna with plant roots. Plant Soil 321:213–233
17. Bakker PAHM, Berendsen L, Doombos RF, Winterman PCA, Pieterse CMJ (2014) The rhizosphere revisited: root microbiomic. Front Plant Sci 4:1–8. https://doi.org/10.3389/fpls.2013.00165
18. Kögel-Knabner I (2002) The macromolecular organic composition of plant and microbial residues as inputs to soil organic matter. Soil Biol Biochem 34:139–162. PII:S0038-07170100158-4
19. Sasse J, Matinoia E, Northen T (2018) Feed your friends: to plant exudates shape the microbiome? Trends Plant Sci 23:25–41
20. Dvorsky M, Dolezal J, De Bello F, Klimesova J, Klimez L (2011) Vegetation types of East ladakh: species and growth form composition along main environmental gradients. Appl Veg Sci 14:132–147
21. Zhang X, Liu S, Li X, Wang J, Ding Q et al (2016) Changes of soil prokaryotic communities after clear-cutting in a kars forest: evidences for cutting-based disturbance promoting deterministic processes. FEMS Microbiol Ecol 92:1–12. https://doi.org/10.1093/femsec/fiw026
22. Hernesmaa A, Björkölf K, Jørgensen KS, Haahtela K, Romantschuk M (2008) Potential impacts of clear-felling on microbial activities in boreal humus and mineral soil layers. Boreal Environ Res 13:525–538
23. Hernesmaa A, Björkölf K, Kiikkilä O, Fritze H, Haahtela K et al (2005) Structure and function of microbial communities in the rhizosphere of Scot pine after tree-felling. Soil Biol Biochem 37:777–785
24. Fierer N, Bradford MA, Jackson RB (2007) Toward an ecological classification of soil bacteria. Ecology 88:1354–1364
25. Tate R (2000) Soil microbiology. 2nd ed. John Wiley and Sons, New York, New York, USA
26. Nelson EB (2018) The seed microbiome: origins, interactions, and impacts. Plant Soil 422:7–34. https://doi.org/10.1007/s11104-017-3289-7
27. Links MG, Demeke T, Grafenhan T, Hill JE, Hemmingsen SM, Dumonceaux TJ (2014) Simultaneous profiling of seed-associated bacteria and fungi reveals antagonistic interactions between microorganisms within a shared epiphytic microbiome on *Triticum* and *Brassica* seeds. New Phytol 202:542–553

28. Barret M, Brian M, Bonneau S, Preveaux A, Valiere S, Bouchez O, Hunault G, Simoneau P, Jacques MA (2015) Emergence shapes the structure of the seed microbiota. Appl Environ Microbiol 81:1257–1266
29. Liu Y, Zuo S, Zou YY, Wang JH, Song W (2013) Investigation on diversity and population succession dynamics of endophytic bacteria from seeds of maize (*Zea mays* L., Nongda108) at different growth stages. Ann Microbiol 63:71–79
30. Klaedtke S, Jacques M, Raggi L, Préveaux A, Bonneau S, Negri V, Chable V, Barret M (2016) Terroir is a key driver of seed-associated microbial assemblages. Environ Microbiol 18:1792–1804. https://doi.org/10.1111/1462-2920.12977
31. Rezki S, Campion C, Iacomi-Vasilescu B, Preveaux A, Toualbia Y, et al (2016) Differences in stability of seed-associated microbial assemblages in response to invasion by phytopathogenic microorganisms. PeerJ 4. https://doi.org/10.7717/peerj.1923
32. van Overbeek LS, Franke AC, Nijhuis EHM, Groeneveld RMW, da Rocha UN, Lotz LAP (2011) Bacterial communities associated with *Chenopodium album* and *Stellaria media* seeds from arable soils. Microb Ecol 62:257–264
33. Johnston-Monje D, Lundberg DS, Lazarovits G, Reis VM, Raizada MN (2016) Bacterial populations in juvenile maize rhizospheres originate from both seed and soil. Plant Soil 405:337–355
34. Somers E, Vanderleyden J, Srinivisam M (2004) Rhizosphere bacterial signaling: a love parade beneath our feet. Crit Rev Microbiol 30:205–240
35. Van der Putten WH, Klironomos JN, Wardle DA (2007) Microbial ecology of biological invasions. ISME J 1:28–37
36. Liang MQ, Zhang CF, Peng CL, Lai ZL, Chen DF, Chen ZH (2011) Plant growth, community structure, and nutrient removal in monoculture and mixed constructed wetlands. Ecol Eng 37:309–316. https://doi.org/10.1016/j.ecoleng.2010.11.018
37. Guedes BS, Olsson BA, Sitoe AA, Egnell G (2018) Net primary production in plantations of *Pinus taeda* and *Eucalyptus cloeziana* compared with a mountain miombo woodland in Mozambique. Glob Ecol Conserv 15:e00414. https://doi.org/10.1016/j.gecco.2018.e00414
38. Carvalhais LC, Dennis PG, Badri D, Kidd BN, Vivanco JM (2015) Linking jasmonic acid signaling, root exudates, and rhizosphere. Microbiomes 28:1049–1058. https://doi.org/10.1094/MPMI-01-15-0016-R
39. Kulmatiski A, Beard KH, Stevens JR, Cobbold SM (2008) Plant-soil feedbacks: a metaanalytical review. Ecol Lett 11:980–992. https://doi.org/10.1016/j.soilbio.2017.06.003
40. Zhang P, Nie M, Li B, Wu J (2017) The transfer and allocation of newly fixed C by invasive *Spartina alterniflora* and native *Phragmites australis* to soil microbiota. Soil Biol Biochem 113:231–239. https://doi.org/10.1016/j.soilbio.2017.06.003
41. Zubek S, Majewska ML, Błaszkowski J et al (2016) Invasive plants affect arbuscular mycorrhizal fungi abundance and species richness as well as the performance of native plants grown in invaded soils. Biol Fertil Soils 52:879–893. https://doi.org/10.1007/s00374-016-1127-3
42. Stefanowicz AM, Stanek M, Majewska ML, Nobis M, Zubek S (2019) Invasive plant species identity affects soil microbial communities in a mesocosm experiment. Appl Soil Ecol 136:168–177. https://doi.org/10.1016/j.apsoil.2019.01.004
43. Coats VC, Rumpho ME (2014) The rhizosphere microbiota of plant invaders: An overview of recent advances in the microbiomics of new-excotic plants. Front Microbiol 5:368–377. https://doi.org/10.3389/fmicb.2014.00368
44. Sun F, Ou Q, Yu H, Li N, Peng C (2019) The invasive plant Mikania micrantha affects the soil foodweb and plant-soil nutrient contents in orchards. Soil Biol Biochem 139:107630. https://doi.org/10.1016/j.soilbio.2019.107630
45. Sanon A, Andrianjaka ZN, Prin Y, Bally R, Thioulouse J, Comte G, Duponnois R (2009) Rhizosphere microbiota interfers with plant- plant interactions. Plant Soil 321:259–278. https://doi.org/10.1007/s11104-009-0010-5
46. Gibbons S, Lekberg Y, Mummey DL, Sangwan N, Ramsey PW, Gilbert JA (2017) Invasive plants rapidly reshape soil properties in a grassland ecosystem. mSystems 2:e00178–16. https://doi.org/10.1128/mSystems.00178-16

47. Lone PA, Dar JA, Subashree K, Raha D, Pandey PK, Ray T, Khare PK, Khan ML (2019) Impact of plant invasion on physical, chemical and biological aspects of ecosystems: a review. Trop Plant Res 6:528–544. https://doi.org/10.22271/tpr.2019.v6.i3.067
48. Levine JM, Vila M, Antonio CMD, Dukes JS, Grigulis K, Lavorel S (2003) Mechanisms underlying the impacts of exotic plant invasions. Proc R Soc Lond, B, Biol Sci 270:775–781
49. Si C, Liu X, Wang C, Wang L, Dai Z, Qi S, Du D (2013) Different degrees of plant invasion significantly affect the richness of the soil fungal community. PLoS ONE 8:e85490. https://doi.org/10.1371/journal.pone.0085490
50. Ehrenfeld JG (2003) Effects of exotic plant invasions on soil nutrient cycling processes. Ecosystems 6:503–523. https://doi.org/10.1007/s10021-002-0151-3
51. Hawkes CV, Wren IF, Herman DJ, Firestone MK (2005) Plant invasion alters nitrogen cycling by modifying the soil nitrifying community. Ecol Lett 8:976–985
52. Thorpe AS, Thelen GC, Diaconu A, Callaway RM (2009) Root exudate is allelopathic in invaded community but not in native community: field evidence for the novel weapons hypothesis. J Ecol 97:641–645
53. Li Q, Wan F, Zhao L (2022) Distinct soil microbial communities under *Ageratina adenophora* invasions. (Abstract). Plant Biol 24. https://doi.org/10.1111/plb.13387
54. Mukaromah L, Imron MA (2019) Invasive plant species in the disturbed forest of batukahu nature reserve, Bali, Indonesia. Biotropia 27:22–32. https://doi.org/10.11598/btb.2020.27.1.933
55. Zahra S, Hofstetter RW, Waring K, Gehring C (2020) Review: the invasion of Acacia nilotica in Baluran National Park, Indonesia, and potential future control strategies. Biodiversitas 21:104–116
56. Radiansyah AD, Susmianto A, Siswanto W, Tjitrosoedirdjo S, Djohor DJ et al (2015) Strategi Nasional dan Arahan Rencana Aksi Pengelolaan Jenis Asing Invasif di Indonesia. Deputi Bidang Pengendalian Kerusakan Lingkungan dan Perubahan Iklim, Jakarta. pp 2–3. ISBN 978-602-72942-2-6
57. Yudaputra A (2022) Future spatial prediction of invasive plant Merremia peltata in Indonesia. 2nd ISeNREM 2021. IOP Conf Series: Earth and Environmental Science 50:012084. https://doi.org/10.1088/1755-1315/950/1/012084
58. Paynter Q, Harman H, Waipara N (2006) Prospects for biological control of Merremia peltata. Landcare Research Contract Report: LC0506/177. Landcare Research, Auckland, New Zealand
59. Kirkham WS (2005) Valuing invasions: understanding the Merremia peltata invasion in postcolonial Samoa. University of Texas at Austin, Austin, TX, USA
60. Liu CLC, Kuchma O, Krutovsky KV (2018) Mixed-species versus monoculture in plantation forestry: development, benefits, ecosystem services and perspecitves for the future. Glob Ecol Conserv 15:e00419. https://doi.org/10.1016/j.gecco.2018.e00419
61. Fox A, Lüscher A, Widmer F (2020) Plant species identity drives soil microbial community structures that persist under a following crop. Ecol Evol 10:8652–8668. https://doi.org/10.1002/ece3.6560
62. Hooper DU, Chapin FS, Ewel JJ, Hector A, Inchausti P, Lavorel S, Wardle DA (2005) Effects of biodiversity on ecosystem functioning: a consensus of current knowledge. Ecol Monogr 75:3–35. https://doi.org/10.1890/04-0922
63. Loreau M, Naeem S, Inchausti P, Bengtsson J, Grime J, Hector A, … Wardle D (2001) Biodiversity and ecosystem functioning: current knowledge and future challenges. Science 294:804–808. https://doi.org/10.1126/science.1064088
64. Husse S, Lüscher A, Buchmann N, Hoekstra NJ, Huguenin-Elie O (2017) Effects of mixing forage species contrasting in vertical and temporal nutrient capture on nutrient yields and fertilizer recovery in productive grasslands. Plant Soil 420:505–521. https://doi.org/10.1007/s11104-017-3372-0
65. Hoekstra NJ, Suter M, Finn JA, Husse S, Lüscher A (2015) Do belowground vertical niche differences between deep- and shallow-rooted species enhance resource uptake and drought resistance in grassland mixtures? Plant Soil 394:21–34. https://doi.org/10.1007/s11104-014-2352-x

66. Simba YR, Kamweya AM, Mwangi PN, Ochora JM (2013) Impact of the invasive shrub, *Lantana camara* L. on soil properties in Nairobi National Park, Kenya. Int J Biodivers Conserv 5:803–809. https://doi.org/10.5897/IJBC2013.0623
67. The Center of Statistic Council of Indonesia (BPS) (2022) Luas tanaman perkebunan Indonesia. https://www.bps.go.id/indicator/54/131/1/luas-tanaman-perkebunan-menurut-provinsi.html
68. The Ministry of Forestry of Indonesia (KLHK) (2022) Statistik Kehutanan Indonesia 2020. https://www.menlhk.go.id/site/single_post/4697/statistik-2020
69. Kostenko O, Bezemer TM (2020) Abiotic and biotic soil legacy effects of plant diversity on plant performance. Front Ecol Evol 8:87. https://doi.org/10.3389/fevo.2020.00087
70. Gajda A, Martyniuk S (2005) Microbial biomass C and N and activity of enzymes in soil under winter wheat grown in different crop management systems. Pol J Environ Stud 14:159–163
71. Ushio M, Kitayama K, Balser TC (2010) Tree species effects on soil enzyme activities through effects on soil physicochemical and microbial properties in a tropical montane forest on Mt. Kinabalu, Borneo. Pedobiologia 53:227–233
72. Acosta-Martinez V, Zobeck TM, Allen V (2004) Soil microbial, chemical and physical properties in continuous cotton and integrated crop-livestock systems. Soil Sci Soc Am J 68:1875–1884
73. Schnitzer SA, Klironomos JN, HilleRisLambers J, Kinkel LL, Reich PB, Xiao K, … Scheffer M (2011) Soil microbes drive the classic plant diversity–productivity pattern. Ecology 92:296–303. https://doi.org/10.1890/10-0773.1
74. Wagg C, Bender SF, Widmer F, van der Heijden MGA (2014) Soil biodiversity and soil community composition determine ecosystem multifunctionality. Proc Nat Acad Sci 111:5266–5270. https://doi.org/10.1073/pnas.1320054111
75. Ladygina N, Hedlund K (2010) Plant species influence microbial diversity and carbon allocation in the rhizosphere. Soil Biol Biochem 42:162–168. https://doi.org/10.1016/j.soilbio.2009.10.009
76. Zhou Y, Zhu H, Fu S, Yao Q (2017) Variation in soil microbial community structure associated with different legume species is greater than that associated with different grass species. Front Microbiol 8:1007 https://doi.org/10.3389/fmicb.2017.01007
77. Leff JW, Bardgett RD, Wilkinson A, Jackson BG, Pritchard WJ, De Long JR, Fierer N (2018) Predicting the structure of soil communities from plant community taxonomy, phylogeny, and traits. ISME J 12:1794–1805. https://doi.org/10.1038/s41396-018-0089-x
78. Hannula SE, Heinen R, Huberty M, Steinauer K, De Long JR, Jongen R, Bezemer TM (2021) Persistence of plant-mediated microbial soil legacy effects in soil and inside roots. Nat Commun 12:5686. https://doi.org/10.1038/s41467-021-25971-z
79. Anderson RL (2011) Synergism: a rotation effect of improved growth efficiency. In: Sparks DL (ed) Advances in agronomy, vol 112. Elsevier Academic Press Inc., San Diego, CA, pp 205–226
80. Allison SD, Lu Y, Weihe C, Goulden ML, Martiny AC, Treseder KK, Martiny JBH (2013) Microbial abundance and composition influence litter decomposition response to environmental change. Ecology 94:714–725. https://doi.org/10.1890/12-1243.1
81. Kardol P, De Deyn GB, Laliberté E, Mariotte P, Hawkes CV (2013) Biotic plant–soil feedbacks across temporal scales. J Ecol 101:309–315. https://doi.org/10.1111/1365-2745.12046
82. Foxx AJ, Fort F (2019) Root and shoot competition lead to contrasting competitiveoutcomes under water stress: a systematic review and meta-analysis. PLoS ONE 14:e0220674
83. Rubio G, Walk T, Ge Z, Yan X, Liao H, Lynch J (2001) Root gravitropism and below-ground competition among neighbouring plants: a modelling approach. Ann Bot 88:924–940. https://doi.org/10.1006/anbo.2001.1530
84. Pollock JA, Kogan LA, Thorpe AS, Holben WE (2011) Catechin, a root exudate of the invasive *Centaurea stoebe* Lam. (Spotted Knapweed) exhibits bacteriostatic activity against multiple soil bacterial populations. J Chem Ecol 37:1044–1053. https://doi.org/10.1007/s10886-011-0005-6

85. Wu L, Wang J, Huang W, Wu H, Chen J, Yang Y, Zhang Z, Lin W (2015) Plant-microbe rhizosphere interactions mediated by *Rehmannia glutinosa* root exudates under consecutive monoculture. Sci Rep 5:15871. https://doi.org/10.1038/srep15871
86. Liu XB, Herbert SJ, Hashemi AM, Zhang X, Ding G (2006) Effects of agricultural management on soil organic matter and carbon transformation—A review. Plant Soil Environ 52:531–543. https://doi.org/10.17221/3544-PSE
87. Bhatti AA, Haq S, Bhat RA (2017) Actinomycetes benefaction role in soil and plant health. Microb Pathog 111:458–467. https://doi.org/10.1016/j.micpath.2017.09.036
88. Widyati E (2006) Bioremediation of ex-coal mining soil use sludge of pulp and paper to enhance its land revegetation. PhD thesis. IPB University
89. Akcil A, Koldas S (2006) Acid mine drainage (AMD): causes, treatment and case studies. J Clean Prod 14:1139–1145. https://doi.org/10.1016/j.jclepro.2004.09.006
90. Li X, Bond P, Van Nostrand J, Zhou J, Huang L (2015) From lithotroph- to organotroph-dominant: directional shift of microbial community in sulphidic tailings during phytostabilization. Sci Rep 5:12978. https://doi.org/10.1038/srep12978
91. Berti WR, Cunningham SD (2000) "Phytostabilization of metals". In: Raskin I, Ensley BD (eds) Phytoremediation of toxic metals: using plants to clean-up the environment. John Wiley & Sons, Inc., New York, NY, pp 71–88
92. Yan A, Ming Y, Tan SN, Yusof MLM, Ghosh S, Chen Z (2020) Phytoremediation: a promising approach for revegetation of heavy metal-polluted land. Front Plant Sci. https://doi.org/10.3389/fpls.2020.00359
93. Yahya A, Vijayanathan J, Ishak MF, Kadir WRWA (2018) Reversing soil degradation via phytoremediation techniques in an ex-tin mine and gold mine in Peninsular Malaysia. https://www.researchgate.net/publication/327716844_Reversing_soil_degradation_via_phytoremediation_techniques_in_an_ex-tin_mine_and_gold_mine_in_Peninsular_Malaysia. Accessed 19 May 2022
94. Thakur S, Singh L, Wahid ZA, Siddiqui MF, Atnaw SM, Din MFM (2016) Plant-driven removal of heavy metals from soil: uptake, translocation, tolerance mechanism, challenges, and future perspectives. Environ Monit Assess 188:206. https://doi.org/10.1007/s10661-016-5211-9
95. Sembiring YRV, Andriyanto M, Siagian N, Widyati E, Azwir (2016) Isolasi bakteri pereduksi sulfat untuk memperbaiki sifat kimia tanah bekas tambang batubara dan pengaruhnya terhadap karet (hevea brasiliensis) di polibeg. Jurnal Penelitian Karet 34(2):165–174
96. Hall J (2002) Cellular mechanisms for heavy metal detoxification and tolerance. J Exp Bot 53:1–11. https://doi.org/10.1093/jexbot/53.366.1
97. Ma Y, Prasad M, Rajkumar M, Freitas H (2011) Plant growth promoting rhizobacteria and endophytes accelerate phytoremediation of metalliferous soils. Biotechnol Adv 29:248–258. https://doi.org/10.1016/j.biotechadv.2010.12.001
98. Göhre V, Paszkowski U (2006) Contribution of the arbuscular mycorrhizal symbiosis to heavy metal phytoremediation. Planta 223:1115–1122. https://doi.org/10.1007/s00425-006-0225-0
99. Burges A, Alkorta I, Epelde L, Garbisu C (2018) From phytoremediation of soil contaminants to phytomanagement of ecosystem services in metal contaminated sites. Int J Phytoremediat 20:384–397. https://doi.org/10.1080/15226514.2017.1365340
100. Marques AP, Rangel AO, Castro PM (2009) Remediation of heavy metal contaminated soils: phytoremediation as a potentially promising clean-up technology. Crit Rev Environ Sci Technol 39:622–654. https://doi.org/10.1080/10643380701798272
101. Suter M, Connolly J, Finn JA, Loges R, Kirwan L, Sebastià MT, Lüscher A (2015) Nitrogen yield advantage from grass–legume mixtures is robust over a wide range of legume proportions and environmental conditions. Global Change Biol 21:2424–2438. https://doi.org/10.1111/gcb.12880

# Chapter 5
# Climate Change and Microbes: Mechanisms of Action in Terrestrial and Aquatic Biosystems

**Sonal Kalbande, Arun Goud, Vishal Hivare, Mukesh Bhendarkar, and Karankumar Ramteke**

**Abstract** The most crucial issue in the contemporary environmental picture is climate change. Climate change causes changes in a variety of elements at the same time, resulting in complicated alterations in the terrestrial and aquatic microbial population. These issues develop due to rising $CO_2$ levels, greenhouse gases in the atmosphere, changing temperature trends, and global warming, which directly and indirectly affect soil microbial communities. Microbial interactions play a vital role in the worldwide fluctuations of the significant biogenic greenhouse gases (carbon dioxide ($CO_2$), nitrous oxide, and methane). They usually respond to climate change immediately. Microbes regulate terrestrial and aquatic greenhouse gas fluxes. Thus, considering microbe's intricate interactions with various biotic and abiotic variables. The promise of lowering greenhouse gas emissions by regulating terrestrial and aquatic microbial processes to combat climate change is a tempting option for the future. This environmental issue is resolved by changing the microbial community structure and composition, a key feedback response mechanism for climate change when microbial communities and their mechanisms are coupled, a good strategy for addressing climate change emerges.

S. Kalbande (✉) · M. Bhendarkar
ICAR-National Institute of Abiotic Stress Management, Baramati 4131151, India
e-mail: sonalkalbande123@gmail.com

A. Goud
College of Fisheries, Dr. Balasaheb Sawant Konkan Krishi Vidyapeeth (Agricultural University), Dapoli 415629, India

V. Hivare
International Crops Research Institute for the Semi-Arid Tropics, Patancheruvu, Telangana, India

K. Ramteke
ICAR-Central Institute of Fisheries Education, Mumbai, India



J. A. Parray (ed.), *Climate Change and Microbiome Dynamics*, Climate Change Management, https://doi.org/10.1007/978-3-031-21079-2_5

## Introduction

Currently, climate change is universally acknowledged as the most significant contemporary human threat. Based on a recent study by the Intergovernmental Panel on Climate Change [16], the situation is worsening, with 3,300 million people considered to be highly vulnerable to the effects of climate change and existing unsustainable development models increasing ecosystems and human susceptibility to climate risks. Microbes are the only life forms in specific habitats, such as deep seas and extreme environments. Microbes inhabit all the environments on earth. Microbes have been on Earth for at least 3.8 billion years, and despite any potential extinction events, they seem likely to last a long time [6].

As part of multiple processes, including the carbon and nitrogen cycles, microbes use and produce greenhouse gases such as carbon dioxide and methane. Microbes are essential to climate change models because they can respond positively and negatively to temperature. Numerous studies have demonstrated the importance of microorganisms to climate change.

It is difficult to determine their function in the ecosystem due to their diversity and the wide variety of responses to environmental change. However, microbes are rarely referred in conversations on climate change. Due to a lack of knowledge, most climate change models have not effectively accounted for microbial activity concerning climate change and its effects on the microbial population; this review aims to understand better the role of microorganisms in terrestrial and aquatic environments. The review emphasizes how vital the biosphere's microbial component is as both a "victim" and a "producer" of climate change.

## Climate Change

"Climate change" refers to long-term modifications in weather patterns and temperatures. These changes could be natural, like when the sun's cycle changes, but since the 1800s, people have been the main factor in earth's climate change. Most of the time, this is because they use fuels like coal, oil, and natural gas, which are made of carbon [29]. Changes in global temperatures and the frequency of heat waves, droughts, floods, storms, and other extreme weather occurrences are all part of this.

### *System of Climate*

The atmosphere, the oceans, the cryosphere (snow and ice), the land surface, the biosphere, and their interactions make up the very complex global system known as the climate system [12]. These interactions determine both daily weather and long-term climate averages. Natural occurrences like volcanic eruptions, solar radiation, and changes in the composition of the atmosphere by humans impact the internal dynamics of climate systems. The sun is the sole source of energy for the climate system. The equilibrium of radiation on earth can be affected by three primary factors:

1. By modifying the amount of solar energy flowing in.
2. By altering the amount of reflected solar radiation (known as "albedo").
3. By modifying the amount of long wave radiation that earth emits back into space.

Feedback mechanisms both directly and indirectly affect the climate [14].

## *Factors Leading to Climate Change*

Greenhouse gas emissions have increased significantly in recent years due to natural events such as volcanic eruptions and human activities. Carbon dioxide, methane, nitrous oxide, and halocarbons are some gases that fall under this category. Due to the gradual accumulation of these gases in the atmosphere, the concentration of these gases gradually increases over time. During the time of industrialization, all of these gases have seen prominent peaks in their concentrations. Many different factors also contribute to the acceleration of climate change. Some of these elements are beyond our ability to manage because they are naturally occurring and are not affected by human activities. Climate change has also been caused by other natural events such as meteor strikes, which dramatically impact the earth's conditions. The climate is also impacted by variations in the sun and the earth's orbit [8]. When fossil fuels are used for ignition, cooling, transportation, construction, and cement production, carbon dioxide is produced, thereby speeding up climate change.

Additionally, it is released by microbial decomposition, respiration, and deforestation. Because of fossil fuels and biomass burning, aerosols, including organic chemicals, black carbon, and sulfide compounds, have increased. Aerosols are tiny particles that vary in size, concentration, and chemical composition and are present in the atmosphere. Although some aerosols are created using different materials that are immediately discharged into the atmosphere, others are created using factory processes. There exist both naturally occurring and artificially produced materials.

As a direct result of human activities such as open-pit mining and industrial processes, there is more dust in the atmosphere. Natural aerosols are volcanic sulfate and dust aerosols, sea salt aerosols, land- and ocean-based biogenic emissions, and surface-emitted mineral dust. The biggest culprit, however, has been the explosion of $CO_2$ released into the atmosphere due to human activity, especially when seen over the previous century. Fifty percent of the world's carbon is emitted by just 10 percent of the population, according to a 2015 Oxfam research. Human activities such as fossil fuel production, distribution and combustion, landfills and garbage, animal husbandry, biomass burning, and rice agriculture contribute to methane production. Wetlands and oceans are unique producers of methane emissions due to their natural processes [23].

## Microorganisms and Climate Change

The sustained life of all higher trophic living species depends on the presence of microorganisms. Microorganisms are essential to the process of carbon and nutrient cycling, as well as to the maintenance of animal and plant health (including human health), as well as to the functioning of agriculture and global food, even though microorganisms are crucial in minimizing the consequences of climate change.

### *Role of Aquatic Microbes*

According to the Census of Marine Life, microbial biomass makes up 90% of aquatic biomass. Aquatic microbes are abundant, perform essential ecological tasks, and are the backbone of ocean food webs, which in turn support the global carbon and nutrient cycles by fixing carbon and nitrogen and remineralizing organic compounds [3].

The micro, nano, and picoplankton found in the ocean, including bacteria and archaea, are responsible for most of the ocean's carbon cycle's mechanical processes. In aquatic environments, primary microbial production plays a vital role in the sequestration of $CO_2$. As a result, aquatic bacteria release $CO_2$ into the atmosphere as they recycle nutrients for use in aquatic food chains. The aquatic ecosystem is also a considerable contributor to methane emissions into the atmosphere. Because methane is constantly escaping from holes in the ocean floor and each of these methane seeps has its unique population of methane-eating bacteria because no species are consistent over the entirety of the deep sea at these places. These microorganisms can remove approximately 75% of the newly produced methane before it is released into the atmosphere. As a result, these species play an essential role in protecting the climate by reducing greenhouse gas emissions [28].

### *Role of Terrestrial Microbes*

Around 1029 microorganisms can be found in all terrestrial ecosystems, which is similar to the number seen in marine habitats [13]. Microorganisms are the principle organic matter decomposers in a spectrum of terrestrial ecosystems, liberating nutrients for plant growth and greenhouse gases such as $CO_2$ and $CH_4$ into the soil. Microbes play a crucial role in altering the emission of greenhouse gases. The Interactions between microbes and biotic, abiotic factors lead to these alterations. It's well understood that microbes play an important role in determining the concentrations of greenhouse gases. Microbes react and impact climate change (through greenhouse gas emissions), and climate change affects microbial responses (e.g., increased $CO_2$, warming, and changes in precipitation) oxygen minimum zone (OMZ), or oxygen most community. Microorganisms in the soil control the amount of organic carbon sequestered there, and the amount returned to the atmosphere. They also indirectly affect the amount of carbon sequestered in plants and soils by providing macronutrients (nitrogen and phosphorus) that regulate plant productivity [4].

Atmospheric permafrost is the most significant terrestrial carbon sink due to the accumulation of carbon from organic matter (the remains of plants, animals, and microorganisms) [18, 21]. Terrestrial ecosystems rely mostly on higher plants for net primary production to remove $CO_2$ from the atmosphere. However, microbes also play an essential role in net carbon exchange through decomposition and heterotrophic respiration, indirectly through their roles as plant pathogens or symbionts and their influence on soil nutrient availability. The decomposition of organic matter by soil bacteria leads to an annual release of between 50 and 75 Pg. of carbon into the atmosphere; this release is 7.5–8 times more than the total amount of carbon emitted by humans across the globe [9]. This mechanism is vulnerable to the impacts of global warming, which have the potential to exacerbate atmospheric warming by creating carbon cycle-climate feedbacks. These feedbacks can be considered a positive feedback loop in which the carbon cycle influences climate.

## Climate Change: Mechanisms of Action

Temperature, precipitation, and the seasons' duration are all climate change indicators [24]. Consequently, the mechanism of action is mostly shown with variations in moisture and temperature.

### *Aquatic Microorganisms*

Climate changes have an impact on the microbial community's structure and capabilities both directly and indirectly. Climate change has speeded global warming by decomposing organic matter, thereby increasing carbon dioxide emissions into the atmosphere [26, 31]. Microbes and enzymes also stimulate warming by decomposing organic matter more efficiently, emitting toxic compounds into the environment, and averting climate change. Nearly the ocean covers 70% of the surface of the planet. has a mean depth of 4,000 m, and is diverse chemically and physically, with over 50 biomes ranging from poles to tropics and from oceanic surface to the dark abyssal zones. Microbes in the ocean account for about 98% of the global biomass; they produce 50% of the planet's oxygen and are the main processors of greenhouse gases. Marine microbes can also alleviate the effects of climate change [30]. With an evolutionary history of nearly 4 billion years, the oceanic microbes have adapted to continuously changing earth's environment and developed resiliency and physiological plasticity, which would offer some protection from artificial climate changes. At present, the rate of climate change is higher in the earth's history due to heat-trapping greenhouse gases, posing a huge threat to marine microbes [27]. An increase in green gases elevates the global temperature, thereby increasing the temperature of the sea surface. In this century, due to global warming, there is expected to be an increase in surface ocean water temperature from 2 to 6 °C [19]. This wide range of temperature fluctuations may directly affect water chemistry, thereby majorly affecting microbes' growth and biological activity.

Increasing temperature affects biological processes and reduces water density of water and thus affecting the stratification and cycling of organismal dispersal and nutrient transport. Enhance in stratification also increases the pace of future warming. Hot upper layers in deeper lakes may reduce air exchange, usually one of the processes of adding oxygen to water. Large anoxic dead zones that cannot support life may result from this. The oxygen minimum zone (OMZ) has increased due to ocean warming over the last five decades, reducing oxygen solubility [20]. Increased carbon dioxide levels could assist changes in composition and competition among algal communities. In the aquatic ecosystem, the abundance of microbes is inversely proportional to temperature. The water's important property, i.e., viscosity, also relies on temperature, and its changes significantly impact the growth rate of consumers, carrying capacity, and the mean density of apex predators. Oceanic phytoplankton multiplication and cell density are higher, and early decaying occurs at a higher temperature. Temperature and other environmental factors determine the global biogeography of phytoplankton and select species based on optimal growth potential [15]. The effects of warming on controlling the phytoplankton dynamics in aquatic systems, such as lakes and the open ocean have been reported.

Survival of phytoplankton at high temperatures depends on phenotypic domestication, mutation, and selection. Microbes can adapt to adverse conditions due to phenotypic acclimation, which results from physiological modifications. A general trend indicates that warming favors smaller phytoplankton's as they have more tolerance to increasing temperature. Nature selection toward small-size primary producers possesses a great effect on biogeochemical cycles. Both marine and freshwater microalgae growth rates are affected by temperature, showing rapid responses to climate changes. Such changes are exhibited by changing algal species in the oceanic environment. These effects on algae are useful in understanding the past and detecting current anomalies. For example, changes in red algae pigmentation indicate an irradiation state and are therefore good signs of climate change. In some micro-algal species, the increase in temperature increases metabolism and growth. And also, competition at the species and community level is altered among other sensitive species. A prominent role is played temperature in the distribution of algal species. In general, microorganisms disperse more than macroscopic organisms [2]. It is mentioned that the algal species Fucus distichus disticus is distributed to the north of 13° isotherms in Britain. A 1–2 °C increase in seawater temperatures in summer could lead to a shift by 13° isotherms northward by this species and their disappearance in Britain [11]. Because of decreasing nutrient contents and shallowing of the surface mixed layer, remote sensing data show that diatom populations dropped globally from 1998 to 2012, mainly in the North Pacific [6].

In marine microbial food webs, the Heterotrophic bacteria occupy the central position. Temperature regulates the metabolic activity of heterotrophic bacteria and their interactions with other compartments in the web. In aquatic systems, the bacterioplankton activity is mainly determined by temperature, and because of their huge numbers and significant turnover, these play an important role in biogeochemical cycles. Important ecosystem processes such as bacterial production, growth efficiency, respiration, and bacterial–grazer trophic interactions may alter in warmer oceanic water. A higher correlation is found between temperature and bacterial activity in estuarine and coastal environments than in the open ocean and freshwater environments. As temperatures increases, the grazer's predation rates are anticipated to surge in proportion to the predator's body mass. Temperature and no substrate availability limit the bacterial productivity in cooler temperate coastal regions. However, the rising ocean temperatures may favour heterotrophic bacterioplankton over phytoplankton, which may lead to substantial heterotrophic yield.

## *Terrestrial Microbes*

Soil microbes play a vital role in maintaining climate by controlling the turnover rate of soil organic matter (SOM), the biggest organic carbon pool in the lithosphere. Microbial communities found in soil carry out Carbon (C) and nutrient cycling in ecosystems. Rising atmospheric carbon dioxide levels, changing weather patterns, and global warming may affect the microbial populations in soil directly or indirectly. We have little understanding of how climate change affects soil microbes and climate. Multiple factors are altered because of climate change that brings complex changes in the soil microbial community. These alterations may have a major impact on the microbes and plants and also on the carbon balance of the soil [7]. Interactions between multiple variables of climate change factors could selectively target specific soil microbes, which could lead to changes in communities that may ultimately determine the state of ecosystems in the future [5].

Biotic and abiotic factors like temperature, litter inputs, and moisture affect microbial activities. And both abiotic and biotic factors are affected by atmospheric and climatic changes. Climate changes induce stress in abiotic factors, which may change the diversity of soil microbes and their processes [22]. Microbe's activity, processing, and turnover ability enhance with increasing temperature, which may cause the microbial community to shift towards representatives adapted to high temperatures and faster growth rates. For instance, climate change in western USA had effect on the arid topsoil cyanobacteria i.e. Microcoleus vaginatus and Microcoleus steenstrupii. As the temperature increased, the thermo-tolerant Microcoleus steenstrupii replaced and outcompeted the Microcoleus vaginatus, which is psychrotolerant. These bacteria maintain the topsoil microbial population as they control soil erosion [10]. The quantity and function of soil microbes are affected by climate change. Microorganisms that control cycles, like denitrification, nitrification, nitrogen fixation, and methanogenesis, are also affected which may affect other ecosystem processes.

Increased microbial activity because of climate change may also elevate soil respiration [32]. Changes in the structure of the microbial community, availability of substrate, quality, and quantity of plant litter, and available carbon abundance brought by an increase in temperature trigger alterations in soil respiration. Soil respiration and temperature are correlated positively and may be inhibited at high or low moisture content. And also, the enzyme production rate is affected by alterations in moisture and temperature because of its effect on the availability of substrate, enzyme, and microbial efficiency. The N-degrading enzymes are less sensitive to temperature than enzymes degrading C [25]. Soil respiration, microbial biomass turnover, and soil organic matter are greater in tropical regions when compared to temperate soils [17].

Plants are prominent biotic factors that change the soil respiration rates by emitting carbon substrates from roots and also alter temperature and soil moisture through evaporation and by giving shade and altering the amount of precipitation. Moisture also plays an important effect in soil respiration patterns in many land ecosystems. The microbial activity could also be suppressed by moisture in many environments like saltwater and soils. Moisture may have severe effects on dynamics and the emission of carbon dioxide [1].

## Climate Change Effect on Microorganisms

Climate change affects the speed in direct and indirect ways or slows down the composition of land and aquatic-based microbial groups and their roles. The following are the effects of climate change on microorganisms: biodiversity stimulation, diversity, and composition can lead to extinction or alteration, which can have beneficial or adverse effects on the reduction or effect on its physiology and the production of greenhouse gases. The architecture of the microbial community changes in response to rising temperatures, which also affects the structure of the microbial community changing with increasing temperature, which also affects the accelerating processes such as respiration, fermentation, and methane production. The resulting heat waves, wildfires, intense storms, rising floods, natural disasters, extreme heat, poor air quality, drought, injury caused by spread and emerging diseases, and death risks are all involved in the impact of global warming on biotic and abiotic elements. The effects of bacteria, fungi, algae, and archaea on: first, an acceleration of global climate change is as follows warming caused by the breakdown of the organic component; second, an increase in carbon dioxide flux into the atmosphere.

Climate change impacts terrestrial microbes through altering temperature patterns, changing precipitation, increasing carbon dioxide levels, and altering ecosystems, among other things. Climate change impacts aquatic ecosystems due to increased ocean stratification, a rise in the temperature of coastal waters, the extinction of species, and an increase in the nitrogen-fixing capacity of plants and animals. Due to the warming of the ocean, primary output has been reduced. The melting of ice, the prevalence of storms, the rising amounts of carbon dioxide, variations in particular

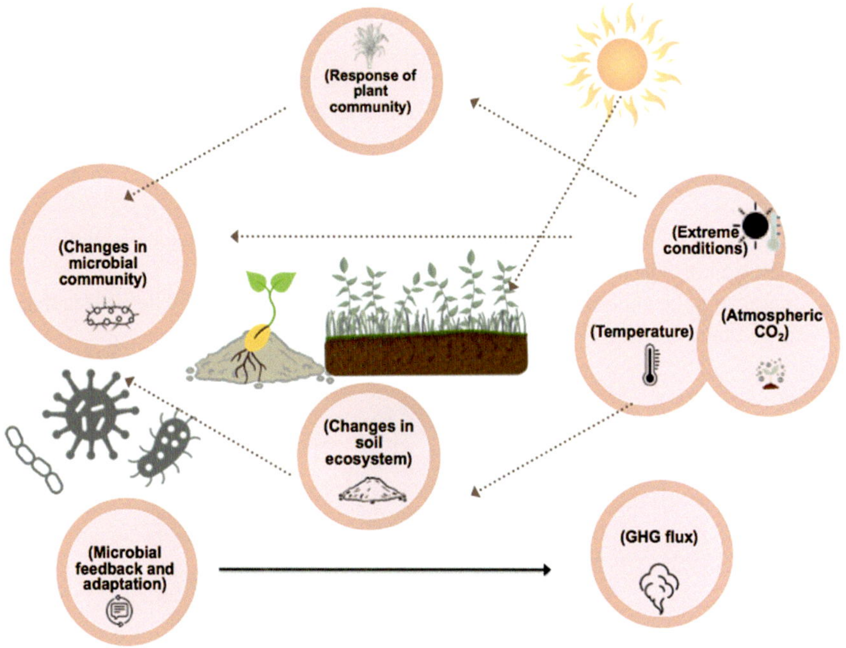


**Fig. 5.1** Effect of Climate change on microbial diversity and functioning

ocean bacteria, and increase in toxic algal blooms are all effects of climate change (Fig. 5.1).

## Mitigation Strategies

1. A better understanding of microbial interactions might help build climate change remedies.
2. Strategies to reduce emissions in agriculture are provided by an understanding of the ecophysiology of the microorganisms that convert $N_2O$ to $N_2$.
3. Reduce the usage of synthetic chemical fertilizers in agriculture and replace them with beneficial microorganisms as bio-fertilizers, which eliminate immediately all greenhouse gas emissions.
4. Rumen microbiome manipulation and breeding programs targeting host genetic variables alter microbial community responses. The United Nations' 17 Sustainable Development Goals can be addressed by implementing microbial technologies, which provide practical solutions (chemicals, materials, energy, and remediation) for these issues.
5. It is essential to introduce new species into the ecosystem regularly.
6. Improving the ability of biotic organisms to withstand drought.

7. Implementing afforestation programs on a global scale. The sequestration of carbon is then easily managed.
8. Getting people to support these actions will be much easier if they know more about microorganism's crucial roles in global warming, called "microbiology literacy.
9. Using bio-based chemicals and polymers because they reduce greenhouse gas emissions.
10. Plastic bags can be recycled and reused to reduce the impact of land-based pollution on maritime ecosystems.
11. Increasing the general public's knowledge of microbiology will help them make more environmentally responsible judgments (Fig. 5.2).

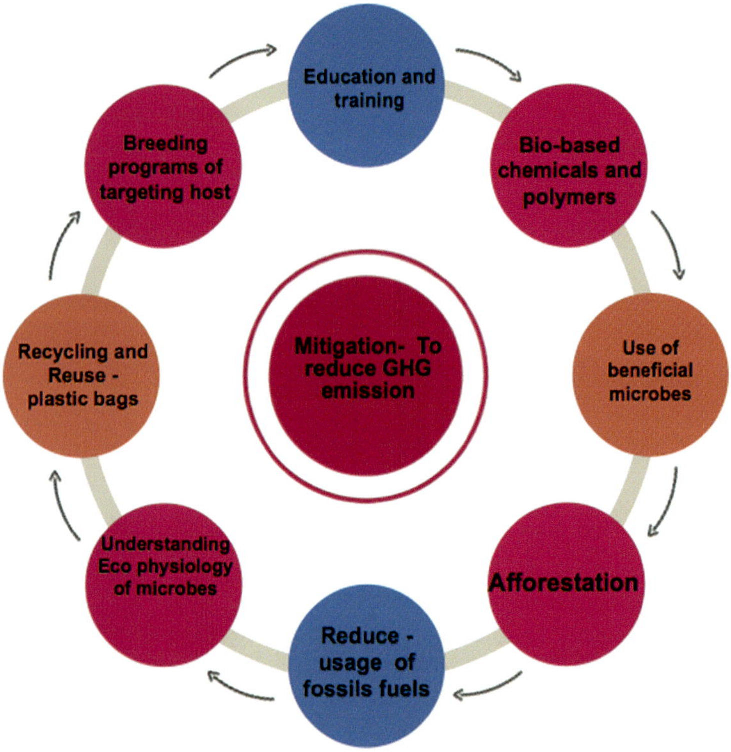


**Fig. 5.2** Mitigation strategies for climate change

## Conclusion

To the scientific community's admiration, bacteria play a vital role in determining greenhouse gas emissions. Estimates of these bacteria's long- and short-term reactions to changing climate and their direct and indirect effects can be used to determine their potential contributions. We can use microbes as a natural resource to combat global warming if they are appropriately utilized. Consequently, ignoring this could be a significant contribution to the problem's worsening. Investigating this issue, learning more about the underlying mechanisms, and using that knowledge in developing practical solutions are long overdue.

## References


1. Aanderud ZT, Schoolmaster DR, Lennon JT (2011) Plants mediate the sensitivity of soil respiration to rainfall variability. Ecosystems 14:156–216
2. Alharmoudi FA (2020) A brief analysis of microorganisms and climate change
3. Azam F, Malfatti F (2007) Microbial structuring of marine ecosystems. Nat Rev Microbiol 5(10):782–791
4. Bardgett RD, van der Putten WH (2014) Belowground biodiversity and ecosystem functioning. Nature 515:505–511
5. Castro HF, Classen AT, Austin EE, Norby RJ, Schadt CW (2010) Soil microbial community responses to multiple experimental climate change drivers. Appl Environ Microbiol 76(40):999–1007
6. Cavicchioli R, Ripple WJ, Timmis KN, Azam F, Bakken LR, Baylis M, Behrenfeld MJ, Boetius A, Boyd PW, Classen AT, Crowther TW (2019) Scientists' warning to humanity: microorganisms and climate change. Nat Rev Microbiol 17(9):569–586
7. Classen AT, Sundqvist MK, Henning JA, Newman GS, Moore JAM, Cregger MA, Moorhead LC, Patterson CM (2015) Direct and indirect effects of climate change on soil microbial and soil microbial plant interactions: what lies ahead? Ecosphere 6(8): 1–21
8. Climate change: what is it, causes, effects, and solutions by sentient media. June 11, 2020. https://sentientmedia.org/climate-change/?
9. Crowther TW, Thomas SM, Maynard DS, Baldrian P, Covey K, Frey SD, van Diepen LTA, Bradford MA (2015) Biotic interactions mediate soil microbial feedbacks to climate change. Proc Natl Acad Sci USA 112(22):7033–7038
10. DiGregorio BE (2015) Climate change affecting microbes in North America soils. American Society for Microbolgy. https://www.microbemagazine.org/index.php?option=com_contentandview=articleandid=6497:climatechangeaffectingmicrobesinnorthamericasoils. Accessed 15 Dec 2015
11. Dutta H, Dutta A (2016) The microbial aspect of climate change. Energy, Ecol Environ 1(4):209–232. https://doi.org/10.1007/s40974-016-0034-7
12. Edenhofer O, Pichs-Madruga R, Sokona Y, Seyboth K, Matschoss P, Kadner S, Zwickel T, Eickemeier P, Hansen G, Schlomer S, von Stechow C (2011) IPCC special report on renewable energy sources and climate change mitigation. Prepared By Working Group III of the Intergovernmental Panel on Climate Change, Cambridge University Press, Cambridge, UK
13. Flemming HC, Wuertz S (2019) Bacteria and archaea on Earth and their abundance in biofilms. Nat Rev Microbiol 17:247–260
14. Government of Canada Climate Change. The Climate System. Accessed 5 July 2016
15. Huertas IE, Rouco M, Lopez-Rodas V, Costas E (2011) Warming will affect phytoplankton differently: evidence through a mechanistic approach. Proc Royal Soc B: Biol Sci 278(1724):3534–3543

16. IPCC (in press) Climate change 2022: impacts, adaptation, and vulnerability. Contribution of working group II to the sixth assessment report of the intergovernmental panel on climate change. Cambridge University Press, Cambridge, United Kingdom
17. Joergensen RG (2010) Organic matter and micro-organisms in tropical soils. In: Dion P (ed) Soil biology and agriculture in the tropics. Springer, Berlin, pp 17–43
18. Lupascu M, Welker JM, Seibt U, Maseyk K, Xu X, Czimczik CI (2014) High Arctic wetting reduces permafrost carbon feedbacks to climate warming. Nat Clim Chang 4(1):51–55
19. Sarmento H, Montoya JM, Dominguez EV, Vaque D, Josep M, Gasol JM (2010) Warming effects on marine microbial food web processes: how far can we go when it comes to predictions? Philos Trans R Soc B https://doi.org/10.1098/rstb.2010.0045
20. Schmidtko S, Stramma L, Visbeck M (2017) Decline in global oceanic oxygen content during the past five decades. Nature 542:335–339
21. Schuur EA, McGuire AD, Schadel C, Grosse G, Harden JW, Hayes DJ, Hugelius G, Koven CD, Kuhry P, Lawrence DM, Natali SM (2015) Climate change and the permafrost carbon feedback. Nature 520(7546):171–179
22. Shade A, Peter H, Allison SD, Baho DL, Berga M, Burgmann H, Huber DH, Langenheder S, Lennon JT, Martiny JB, Matulich KL (2012) Fundamentals of microbial community resistance and resilience. Front Microbiol 3:417
23. Singh BK, Bardgett RD, Smith P, Reay DS (2010) Microorganisms and climate change: terrestrial feedbacks and mitigation options. Nat Rev Microbiol 8:779–790
24. Smith P, Fang C, Dawson JJC, Moncrieff JB (2008) Impact of global warming on soil organic carbon. Adv Agron 97: 1–43
25. Stone MM, Weiss MS, Goodale CL, Adams MB, Fernandez IJ, German DP, Allison SD (2012) Temperature sensitivity of soil enzyme kinetics under N-fertilization in two temperate forests. Glob Change Biol 18(3):1173–1184
26. Swati T, Ramesh S, Shaily J (2014) Effect of climate change on plant-microbe interaction: An overview. Eur J Mol Biotechnol 5:149–156
27. Tiedje JM, Bruns MA, Casadevall A, Criddle CS, Eloe-Fadrosh E, Karl DM, Nguyen NK, Zhou J (2022) Microbes and climate change: a research prospectus for the future. Mbio e00800–22
28. Trinastic J (2015) Methane-munching microbes limit global warming. http://www.nature.com/scitable/blog/eyesonenvironment/methanemunching_microbes_limit_global_warming. Accessed 15 Dec 2015
29. United Nations Department of Economic and Social Affairs. The Sustainable Development Goals Report 2018 (United Nations, 2018). Society of general microbiology
30. Vivekanandan E (2016) Role of marine microorganisms in climate change. http://www.envismadrasuniv.org/nl20134Roleofmarinemicroorganismsinclimatechange.htm Accessed 20 March 2016
31. Weiman S (2015) Microbes help to drive global carbon cycling and climate change. Microbe Mag 10:233–238
32. Wu Z, Dijkstra P, Koch GW, Penuelas J, Hungate BA (2011) Responses of terrestrial ecosystems to temperature and precipitation change: a meta-analysis of experimental manipulation. Global Change Biol 17:927–942

# Chapter 6
# Plant Exudates and Microbial Interaction—A Change in Dynamics

**Dawa Dolma Bhutia, Saroj Belbase, Jiwan Paudel, and Shrvan Kumar**

**Abstract** Plant rhizosphere encompasses a dynamic zone of interactions between microorganisms and their respective plant hosts. This phytobiome has a significant role in the growth, development and fitness of the plants that ultimately contributes in increasing the productivity since the root zone is enriched by the compounds that are being secreted by both microorganism and plants. This chapter deals with the mechanisms that drive the root exudation process and its effect on the rhizospheric microbes and overall plant health. Root system architecture is influenced by the influx and efflux of metabolites at the tip of the root and the root exudates in turn are greatly influenced by microbes as they establish a strong sink for plant carbon that increases the gradient concentration of metabolites. These root exudates that are passively lost from roots of plants (including primary metabolites—sugar, amino acids and organic acids) by diffusion, are being utilized by the rhizosphere- abiding microbes and by the plants themselves for sensing the nutrient availability and signaling to transport the nutrient through the use of nutrient transporters.

## Introduction

Plants rely on soil for water and nutrients, which are distributed unevenly and often dynamically. Plants have evolved ways to affect the physicochemical characteristics and microbial populations of the rhizosphere, i.e. the soil compartment under the influence of the roots, in order to optimize their foraging activities. This constant interplay between root-soil microbiome interactions produces new features that affect plant nutrition and health [1]. Plants achieve this by changing the design and structure of their root systems in response to environmental cues, allowing them to explore different soil layers and locate and exploit water and nutrient-rich regions [2]. Plants

D. D. Bhutia (✉) · S. Belbase · J. Paudel
Department of Mycology and Plant Pathology, IAS, BHU, Varanasi, India
e-mail: dolmaassampa@gmail.com

S. Kumar
Department of Mycology and Plant Pathology, RGSC, BHU, Varanasi, India

 
J. A. Parray (ed.), *Climate Change and Microbiome Dynamics*, Climate Change Management, https://doi.org/10.1007/978-3-031-21079-2_6

have also evolved systems to change soil physicochemical parameters and microbial populations under the influence of roots (the rhizosphere) in order to increase their foraging activities [56].

## Holobiont

Holobiont refers to the single biological entity comprising the interaction between hosts and its endosymbiont in all types of ecosystems and its genetic information (host genome and associated microbial genomes) is termed as hologenome. Lynn Margulis first introduced the term Holobiont in 1991 [3]. Holobiont is a holistic approach of defining every natural animal and plant i.e., host and diverse symbiotic microbes and viruses [4]. Microbial symbionts can be transmitted from parent to offspring by a variety of methods viz. cytoplasmic inheritance, coprophagy, direct contact during and after birth, and the environment. Also, the host-associated microbes contribute to the anatomy, behavior, physiology, fitness, innate and adaptive immunity and also to genetic variation and to the origin and evolution of holobiont [53]. The nature of universal presence of host associated microbes and their role in host ecology, biology and evolution has widened the concept of holobiont in several branches of biology. Further the development of NGS and newer molecular techniques also proved the ubiquitous nature of microorganism and their role in biological and evolutionary processes [3]. Similarly, the hologenome concept of evolution postulates that the holobiont with its hologenome is a level of selection in evolution which is likely to occur between the host and the microbes but also among microbes [55]. Acquisition of microbes and microbial genes is a powerful mechanism for driving the evolution of complexity and exhibit synergetic phenotypes that are subjected to evolutionary forces. Within the holobiont population, the phenotypes encoded by nuclear genome, beneficial, deleterious and neutral microbes in microbiome are subjected to drift and selection. Evolution proceeds both via cooperation and competition, working in parallel [55]. The change in host genome and subsequent change in symbiotic microbes genomes results in genetic variation among the hologenomes [54]. The genetic variation of the microbiome outnumbers that of the host genome, and it increases far faster than that of the host genome. Given that genetic variation is the raw material on which evolution eventually works, microbial sources of hologenomic variation are possible targets of evolution, and scientists must consider include the microbiome in the general study of evolution, despite its intrinsic complexity [53].

The plant can regulate its microbiome to adapt to its surroundings in real time. The core microbiota must be established at multiple hierarchical scales of ecology to better understand the amount of plant dependency on microbiotic components, whereas pan-microbiome research would increase characterization of the functions exhibited [4]. The change in availability of phosphorous in the soil resulted the shift in aarbuscular-mycorrhizal fungal communities in the rhizospheric area of wild grass *Holcus lanatus* that was detected through the use of metatranscriptomics. Hence, the holobint can also be determined with changes in relation with soil types [5].

## Extended Phenotype

The term "extended phenotype" is not new; it is drawn from Dawkins' (1982) proposal that an organism's phenotype should extend from its cellular components to its surroundings. Manipulation of an organism's physical environment and behavioral changes, both of which can begin at the gene level, are examples of extended phenotypes. Other evolutionary biology concepts provide a larger understanding of heredity that is shaped at numerous levels beyond the individual, with natural selection acting on ecological units other than the individual [6]. Extended phenotype in terms of community genetics perspective can be defined as the effect of genes at level higher than the population and focuses on the intraspecific genetic variation that is due to the extended phenotypic genes—a heritable character [7]. The rhizosphere is thought of as an extended root phenotype, a representation of plant genes' effects on their environment both within and outside the organism [1]. The notion of multilevel selection, often known as group selection, is maybe one of the most important among them [6]. Qualitative Trait Locus and genetic mapping technique are important in understanding the genetic basis of quantitative variation as few genes can have significant difference in phenotypic character or large number of genes can have small effect [8], which can significantly alter an extended phenotype and resulting interaction.

Extended phenotype is expected to shift in dominant and keystone species due to genotype– environment interaction [7]. The microbiota in the rhizospheric layer is shaped by the plant genotype that drives the plant phenotypic characters establishing a correlation between the plant microbiota and host plant phenotypic character. Also, the environmental factors can drive the development of plant phenotypes and the assembly of plant microbiota [9]. Soil pH, nutrient profile, environmental factors (temperature, water availability and UV radiation) altered the bacterial communities in the phyllospheric and rhizospheric bacterial communities of *Boecherastricta.* In the drought stress condition plant root sites produces root glucosinolate for culticle thickness that ultimately affects the root associated bacteria of *Boecherastricta* [10]. A plant growing in nitrogen-limited soils could gain a fitness advantage over competitors by enriching its rhizosphere for microbial communities that enhance nutrient capture and utilization capabilities [6].

Natural selection occurring on complete groups of organisms as well as individuals is known as multilevel selection. By applying the notion to the plant rhizosphere, researchers may be able to better understand the intricate interactions that occur in microbe-microbe and microbe-plant networks, which can be influenced by natural and artificial selection at multiple levels. Given the sorts of selection forces that drive microbial density-dependent rhizosphere activities like nutrient cycling, applying this notion to microbiome science might be extremely beneficial. Selection may work on numerous levels of organization across biological units to alter the observable phenotype, according to a core component of multilevel selection theory. Individual and group-level selection forces continually interact in the plant rhizosphere to shape the phenotype of the rhizosphere [6].

## Mechanism of Plant Root Exaduation

Living roots release a variety of organic substances into the soil (Known as rhizodeposits), which alter the rhizosphere's physicochemical properties [11]. Primary and secondary metabolic products, volatile organic carbon compounds, cell debris orginated from the root cap (i.e. border cells), and metabolites derived from root epidermal cell senescence as well as root turnover are all found in rhizodeposits [12]. Roots lose on average 17% of the net C fixed by photosynthesis, which is recovered via rhizosphere respiration (12%) and soil residues (5%), corresponding to 50% of the C exported by shoots to belowground [13]. Rhizodeposit amount and composition vary greatly depending on plant community diversity, plant species, genotypes, plant age, and growth circumstances. Root tips are the first plant tissue to detect changing soil surroundings and are key exudation hotspots in a variety of ways.

Some of the mechanism involved in rhizodeposition includes:

### *Root Border Cells Sloughing Off*

The tip portion of root cap of the plant roots containing the apical meristems gets sloughs off as the root wends through the soil pores. In some context, the entire cap gets slog off particularly in mature branched roots as a result of pathogen attack or normal developmental processes [14]. These sloughed off cells are generated from the cap and differentiated into statocytes which are then able to secrete mucilages [13]. These cells are also referred as border cells [15]. These border cells are viable for several days even after separation from the root tip. Its function is-

(a) Decrease in frictional resistance as the root wends in the soil.
(b) Change in rhizospheric microbial dynamics by attracting the pathogens and preventing the possible damage to root meristem.
(c) Promoting gene expression in symbiotic microorganism.
(d) Protection against heavy metal toxicity [16, 17].

### *Secretion of Mucilage by Roots*

The mucilage (polymerized sugar) are supposed to be secreted actively from root cap cells, however, being observed at the surface of roots in the form of droplets [12, 18]. As an illustration, mucilage is being secreted by root hair and fibrillar mucilage by epidermal cells in case of *Sorghum* [19]. Mucilage is composed of polymerized sugar, upto 6% proteins, sugars (glucose, fructose, xylose, galactose, arabinose) [20]. The mucilage is being generated in the endoplasmic reticulum, completed in Golgi sassules and transported to the cell surface through golgi vesicles and plasmalemma by exocytocis [13].

## Root Exudation

The root of plant excrete a wide variety of compounds which includes amino acids, simple and complex sugar, organic acids, alcohol, phenols, hormone, enzymes, protein and polypeptides. The plant-derived primary and secondary metabolites diffuse or are actively transferred from root cells to soil [1]. Concentration gradients stimulate the movement of low-molecular-weight substances including sugar, amino acids, and organic acids from root cells to the rhizosphere. In undifferentiated root tip tissues, the lack of an apoplastic barrier (i.e. Suberin, casparian strip, or sclerenchyma) favors passive diffusion of hydrophilic substances through the plasma membrane, which is mediated by specialized transporters. Transmembrane primary active transporters (ATP-dependent transporters) such as ABC transporters or secondary active transporters (associated with H + pumps) are required for the expulsion of high-molecular-weight substances such as polysaccharides, proteins, alkaloids, and phenolics [21].

## Role of Compounds in Host-Microbe and Microbe Microbe Interaction

Different studies have demonstrated the importance of various root exudates which includes polypeptides, organic acids, carbohydrates, amino acids, simple and complex sugars, sterols, phenolics that serve as a carbon source for rhizobacteria [13, 22]. The presence of various plant metabolities was discovered in lyophilized root exudates of *Brachypodium distachyon* according to metabolomics study which includes in Table 6.1.

(i) Carbohydrates and their derivatives (glucose, fructose, xylose, sucrose, trehalose, maltose, galactose, and others);
(ii) Sugar alcohols (β-mannosylglycerate, myo-inositol, galactinol, 2-deoxyerythritol, ribitol, threitol and cellobitol);
(iii) Amino acids and derivatives (glutamine, tyrosine, glutamic acid, asparagine, aspartic acid, valine, phenylalanine, isoleucine, glycine, serine, proline, leucine, tryptophan, cysteine, methionine, citrulline, and others);
(iv) Organic acids (aconitic, allantoic, γ-aminobutyric, azelaic, citric, fumaric, 2-furoic, D-glyceric, 3-hydroxypropionic, α-ketoadipic, malic, methylmalonic, nicotinic, quinic, succinic, threonic);
(v) Assorted metabolites including heterocyclic compounds, phenolics, and biogenic amines, etc. (3-hydroxypyridine, maleimide, noradrenaline, 4-hydroxy-3-methoxybenzoate, 5-methoxytryptamine, uracil, aminomalonic acid, palmitic acid, and urea) [23].

**Table 6.1** Change in rhizospheric dynamics by root exudates and its impact on plant

| Plant | Root exudates | Attract | Benefits to plants | Reference |
|---|---|---|---|---|
| *Medicago truncutala* | C-compound | Mycorrhiza | Plant gets facilitated with P availability | [47] |
| Maize | Benzoxazinoids | *P. putida* | Triggers ISR againstmaize anthracnose *Colletotrichumgraminicola* | [48, 49] |
| Legumes | Flavonoids | *Rhizobium* | Synthesis of Nod factors | [50] |
| *Arabidopsis thaliana* | Malic acid | *Bacillus subtilis* | ABA and Salicyclic Acid signaling pathway | [22] |
| Wheat | 2,4-diacetylphloroglucinol (DAPG) | Fluorescent *Pseudomonas* spp. | Control *Gaeumannomyces graminis* var. *tritici* (*Ggt*) | [51] |
| Tomato | Peroxidase, Oxylipins | *Trichoderma harzianum* | Antagonistic activity against *Phythium ultimum*, *Phytopthora capsici* | [35, 52] |

The formation of the microbial rhizosphere is very dynamic, and it is largely controlled by rhizodeposits [24], which may serve as key carbon sources for microorganisms, as well as signaling chemicals and antimicrobial compounds [1]. These compounds serve as carbon and energy sources for rhizobacteria, and the presence of the intact corresponding catabolic pathways is essential for competitive colonization of roots and disease suppression [25]. The chemo-attractants [22], osmoprotectants [23] for beneficial microorganism in different plant models esteems root colonization. The root exudates also contained osmoprotectants that may help microorganisms to persist in the rhizosphere of drought-stressed plants. *Bacillus subtilis* RR4 showed a positive response to chemotactic ability towards Malic Acid (MA) -organic acid- and induce biosynthesis of MA in rice roots [22].

Xylose, major structural component of plant cell wall is dominant constitute of root exudate in wide range of plant [26, 27]. Most of the Gram positive bacteria are capable to catabolize xylose and utilize as a sole source of carbon. In vivo expression technologies being utilized for profiling of *Pseudomonas fluorescens* SBW25 and identified xylose isomerase among genome regions whose expression is specifically induced during bacterial colonization of sugar beet seedlings [28]. The aggressive colonization of *Pseudomonas fluorescens* in xylose rich region of sugarbeet and other crops i.e., wheat, maize, pea inhibit the damping off pathogen *Pythium ultimum* [28], changing the dynamics of the rhizospheric region.A genome-wide transposon mutagenesis approach (RB-TnSeq) for screen of *Pseudomonas simiae* identified genes for the catabolism of *myo*-inositol, carbohydrate metabolism, among traits essential for the colonization of *Arabidopsis thaliana* roots [29]. Furthermore, secondary metabolites like coumarins, which are well-known iron-mobilizing exudates, influence the rhizosphere microbiome in *Arabidopsis* by acting as antimicrobials against fungal infections [30]. Other secondary metabolites from maize and legume root exudates, such as benzoxazinoids and canavanine, have been found to have similar activities. In *Brachypodium* and barley, architectural features including root type and root hairs have been discovered to have a considerable impact on the makeup of rhizosphere microbial communities [31].

Differential exudation patterns affect microbial colonization along growing roots, changing the distribution of microbial biomass along the root as well as the kinematics of root tip development across soil profiles [32]. Chemotaxis toward signaling molecules released by roots pulls microorganisms to the vicinity of root surfaces, whereas root elongation rate modulates the dynamics of root surface adherence and longitudinal transport along elongating roots. In general, a greater and more diverse number of active bacteria accumulate towards the root tip, whereas fewer microbial taxa are associated with the root extension zone. Bacterial density gradually declines from the elongation zone to the mature root zone [33]. Bacterial density gradually declines from the elongation zone to the mature root zone. This is most likely due to the fast growth of root epidermal cell size (up to 30 times in 6 h when cells transit the elongation zone), which 'dilutes' microbial cells living on the root surface until they divide and form a continuous biofilm in the maturation zone. Dispersion of rhizosphere bacteria and chemotactic motions may also influence changes in rhizosphere populations that favor the presence of bacterial decomposers [33, 34]. Lombardi et al. [35] observed the root exudates (Peroxidase and Oxylipins) released during the time of stress by *Fusarium oxysporum* in tomato triggered the number of spores of *Trichoderma harzianum.*

## Mycorrizal Association with Plants

Mycorrhizas ('fungus roots') are symbiotic relationships between plants and specialized soil fungi. There are seven different varieties of mycorrhizas, yet many of them are fairly similar. The most common kind of mycorrhiza is vesicular–arbuscularmycorrhizas (VAM, also known as arbuscularmycorrhizas) [36]. The very advantageous symbiosis between the plant root and the fungal symbiont has spurred the diversity of plant root shape as well as VAM structure and function, according to research on vesicular-arbuscularmycorrhiza (VAM) [6]. Due to the interchange of restricted energy and nutritional resources, mycorrhizal evolution would have moved from endophytic hyphae to balanced relationships where partners were interdependent [36]. The AM (Arbuscular Mycorrihizae) colonization particularly *Glomus etunicatum, G. intraradices* and *G. mosseae* around the root of *Sorghum bicolor* had increased the bacterial number in the rhizospheric soil [37].

The species composition on the soil microbes is affected by the specific selection pressure from the roots and the arbuscularmycorrhizal exudation in the mycorrhizosphere soil and through the exchange of nutritional compounds [37]. Exudates generated by the arbuscularmycorrhizal fungus *Rhizophagus irregularis*, in particular, have been demonstrated to encourage bacterial growth and affect bacterial community structure, resulting in an increase in the abundance of certain Gammaproteobacteria including a taxon within the Enterobacteriaceae [38]. Notably, the capacity of bacteria to colonize the mycosphere is linked to their ability to utilise certain carbonaceous chemicals prevalent in mycosphere exudates including L-arabinose, L-leucine, m-inositol, m-arabitol, D-mannitol, and D-trehalose through BIOLOG-based substrate utilization test [39]. The effect of presence and absence of the arbuscularmycorrhizal fungus *Glomus hoi* in the soil was studied and demonstrated that the fungus has a considerable impact on bacterial community structure, implying that nitrogen export by the fungus is a major driving force behind bacterial community shifts [40]. Fungal hyphae or fruiting bodies have long been recognized as key habitats that may be colonized by particular bacterial taxa, including *Pseudomonas* strains and bacteria from the Oxalobacteraceae, Bacillaceae, and Burkholderiaceae families, among others, as part of the mycosphere [37, 41, 42]. Hence, fungal exudates appear to have a specialized function in mycosphere colonization by promoting the development of certain bacteria or altering the structure of the bacterial population.

Plant roots produce carbon-rich rhizodeposits that contain low-molecular-weight compounds, lysates, and mucilages. These exudates feed soil microorganisms and regulate their attachment to host plants [23]. Microbes' varied strategies for cooperating and competing on plant tissues show that microbe-microbe interactions play critical roles in forming and organizing microbial networks in nature. As a result, the interaction of host-microbe and microbe-microbe is likely crucial for the creation of complex and diversified multi-kingdom plant-associated microbiota [40].

## Phytobiome in Plant Growth and Development

Varied populations of microorganisms that live on the root surface (rhizoplane) and in the endophytic compartment have an impact on plant health [43], particularly on plant growth, productivity, carbon sequestration and phytoremediation. Microbes in the rhizosphere can help plants grow and operate better by boosting their resistance to pathogens, retaining more water, absorbing and using more nutrients, and generally enhancing their development [44]. Colonization of the root occurs despite a sophisticated plant immune system, suggesting finely tuned discrimination of mutualists and commensals from pathogens [32]. This root microbiome is hypothesized to be controlled by host plant immunological function, root exudate-mediated communication and metabolic compatibility, as well as intermicrobial interactions within the rhizosphere, and is recruited from surrounding soil communities [32, 45]. These interactions are crucial for the creation of a root-associated bacterial population that is different from that of the surrounding soil, especially during the early colonization stage. Plant genetic variables, particularly immunological phytohormone pathways, have been shown to have a role in regulating bacteria's capacity to colonize plants in several investigations of plant diseases [46].

## Conclusion

Rhizodeposition representing loss of energy for plants alters the microbial dynamics through the interrelated processes like organic matter dynamics, nutrient cycling, soil-borne pathogen and inoculants dynamics, pollutants biovailability etc. [13]. Root development modifies soil structure around the root and thus contributes to the formation of the rhizosphere. Novel engineering strategies to improve biological product development, and will facilitate the mechanistic exploration of the root colonization process [29]. In order to determine the various mechanism underlying in the interaction of plant root exudates and rhizobiome dynamics, integeration of omics technique is a must. Metabolomics, metagenomics, plant transcriptomics, metatranscriptomics, and plant genetics are some of the approaches that combined can disentangle the complex interactions occurring between members of the holobiont [35].

## References


1. de la Fuente Cantó C, Simonin M, King E, Moulin L, Bennett MJ, Castrillo G, Laplaze L (2020) An extended root phenotype: the rhizosphere, its formation and impacts on plant fitness. Plant J 103(3):951–964. https://doi.org/10.1111/TPJ.14781
2. Lynch JP (2019) Root phenotypes for improved nutrient capture: an underexploited opportunity for global agriculture. New Phytol 223(2):548–564. https://doi.org/10.1111/NPH.15738
3. Simon JC, Marchesi JR, Mougel C, Selosse MA (2019) Host-microbiota interactions: from holobiont theory to analysis. Microbiome 7(1):1–5. https://doi.org/10.1186/s40168-019-0619-4
4. Vandenkoornhuyse P, Quaiser A, Duhamel M, Le Van A, Dufresne A (2015) The importance of the microbiome of the plant holobiont. New Phytol 206(4):1196–1206. https://doi.org/10.1111/nph.13312
5. Young, E, Carey M, Meharg AA, Meharg C (2018) Microbiome and ecotypic adaption of *Holcus lanatus* (L.) to extremes of its soil pH range, investigated through transcriptome sequencing. Microbiome 6(1): 48. https://doi.org/10.1186/S40168-018-0434-3/FIGURES/8
6. Garcia J, Kao-Kniffin J (2018) Microbial group dynamics in plant rhizospheres and their implications on nutrient cycling. Front Microbiol 9:1516. https://doi.org/10.3389/fmicb.2018.01516
7. Whitham TG, Young WP, Martinsen GD, Gehring CA, Schweitzer JA, Shuster SM, Wimp GM, Fischer DG, Bailey JK, Lindroth RL, Scott W, Kuske CR (2003) Community and ecosystem genetics: a consequence of the extended phenotype. Ecol Soc Am 84(3):559–573. https://doi.org/10.1890/0012-9658%282003%29084%5B0559%3ACAEGAC%5D2.0.CO%3B2
8. Leal SM (2001) Genetics and analysis of quantitative traits. Am J Hum Genet 68(2):548–549
9. Li Y, Wu X, Chen T, Wang W, Liu G, Zhang W, Li S, Wang M, Zhao C, Zhou H, Zhang G (2018) Plant phenotypic traits eventually shape its microbiota: a common garden test. Front Microbiol 9:2479. https://doi.org/10.3389/FMICB.2018.02479/BIBTEX
10. Wagner MR, Lundberg DS, del Rio TG, Tringe SG, Dangl JL, Mitchell-Olds T (2016) Host genotype and age shape the leaf and root microbiomes of a wild perennial plant. Nat Commun 7(1):12151. https://doi.org/10.1038/ncomms12151
11. Hinsinger P, Bengough AG, Vetterlein D, Young IM (2009) Rhizosphere: biophysics, biogeochemistry and ecological relevance. Plant Soil 321(1):117–152. https://doi.org/10.1007/S11104-008-9885-9
12. Sasse J, Martinoia E, Northen T (2018) Feed your friends: do plant exudates shape the root microbiome? Trends Plant Sci 23(1):25–41. https://doi.org/10.1016/j.tplants.2017.09.003
13. Nguyen C (2009) Rhizodeposition of organic C by plant: mechanisms and controls. Sustain Agri 23:97–123. https://doi.org/10.1007/978-90-481-2666-8_9
14. Varney GT, McCully ME (1991) The branch roots of Zea. II. Developmental loss of the apical meristem in field-grown roots on JSTOR. New Phytol 118(4):535–546. https://www.jstor.org/stable/2557581?seq=1#metadata_info_tab_contents
15. Hawes MC (1990) Living plant cells released from the root cap: a regulator of microbial populations in the rhizosphere? Plant Soil 129(1):19–27. https://doi.org/10.1007/BF00011687
16. Bengough AG, McKenzie BM (1997) Sloughing of root cap cells decreases the frictional resistance to maize (*Zea mays* L.) root growth. J Exp Bot 48(4):885–893. https://doi.org/10.1093/JXB/48.4.885
17. Miyasaka SC, Hawes MC (2001) Possible role of root border cells in detection and avoidance of aluminum toxicity. Plant Physiol 125(4):1978–1987. https://doi.org/10.1104/PP.125.4.1978
18. Oades JM (1978) Mucilages atthe root surface. J Soil Sci 29(1):1–16. https://doi.org/10.1111/J.1365-2389.1978.TB02025.X
19. Werker E, Kislev M (1978) Mucilage on the root surface and root hairs of *Sorghum*: heterogeneity in structure, manner of production and site of accumulation. Ann Bot 42(4):809–816. https://doi.org/10.1093/OXFORDJOURNALS.AOB.A085520

20. Bacic A, Moody S, Mccomb J, Hinch J, Clarke A (1987) Extracellular polysaccharides from shaken liquid cultures of *Zea mays*. Funct Plant Biol 14(6):633–641. https://doi.org/10.1071/PP9870633
21. Oleghe E, Naveed M, Baggs EM, Hallett PD (2017) Plant exudates improve the mechanical conditions for root penetration through compacted soils. Plant Soil 421(1–2):19–30. https://doi.org/10.1007/S11104-017-3424-5/FIGURES/8
22. Kandaswamy R, Baskar B, Srinath S, Usha B (2018) Plant-growth-promoting rhizobacteria Bacillus subtilis RR4 isolated from rice rhizosphere induces malic acid biosynthesis in rice roots. Can J Microbiol 64(1):20–27. https://doi.org/10.1139/CJM-2017-0409
23. Mavrodi OV, McWilliams JR, Peter JO, Berim A, Hassan KA, Elbourne LDH, LeTourneau MK, Gang DR, Paulsen IT, Weller DM, Thomashow LS, Flynt AS, Mavrodi DV (2021) Root exudates alter the expression of diverse metabolic, transport, regulatory, and stress response genes in rhizosphere *Pseudomonas*. Front Microbiol 12:698. https://doi.org/10.3389/FMICB.2021.651282/BIBTEX
24. Zolla G, Bakker MG, Badri DV, Chaparro JM, Sheflin AM, Manter DK, Vivanco J (2013) Understanding root-microbiome interactions. Mol Microb Ecol Rhizosphere 2:743–754. https://doi.org/10.1002/9781118297674.CH70
25. Lugtenberg B, Kamilova F (2009) Plant-growth-promoting Rhizobacteria. Annu Rev Microbiol 63(1):541–556. https://doi.org/10.1146/annurev.micro.62.081307.162918
26. Rennie EA, Scheller HV (2014) Xylan biosynthesis. Curr Opin Biotechnol 26:100–107. https://doi.org/10.1016/j.copbio.2013.11.013
27. Rovira AD (1959) Root excretions in relation to the rhizosphere effect. Plant Soil 11(1):53–64. https://doi.org/10.1007/BF01394753
28. Liu Y, Rainey PB, Zhang XX (2015) Molecular mechanisms of xylose utilization by *Pseudomonas fluorescens*: overlapping genetic responses to xylose, xylulose, ribose and mannitol. Mol Microbiol 98(3):553–570. https://doi.org/10.1111/MMI.13142
29. Cole BJ, Feltcher ME, Waters RJ, Wetmore KM, Mucyn TS, Ryan EM, Wang G, Ul-Hasan S, McDonald M, Yoshikuni Y, Malmstrom RR, Deutschbauer AM, Dangl JL, Visel A (2017) Genome-wide identification of bacterial plant colonization genes. PLoS Biol 15(9):e2002860. https://doi.org/10.1371/JOURNAL.PBIO.2002860
30. Stringlis IA, De Jonge R, Pieterse CMJ (2019) The age of coumarins in plant-microbe interactions. Plant Cell Physiol 60(7):1405–1419. https://doi.org/10.1093/PCP/PCZ076
31. Cotton TEA, Pétriacq P, Cameron DD, Meselmani MA, Schwarzenbacher R, Rolfe SA, Ton J (2019) Metabolic regulation of the maize rhizobiome by benzoxazinoids. ISME J 13(7):1647–1658. https://doi.org/10.1038/s41396-019-0375-2
32. Lundberg DS, Lebeis SL, Paredes SH, Yourstone S, Gehring J, Malfatti S, Tremblay J, Engelbrektson A, Kunin V, del Rio TG, Edgar RC, Eickhorst T, Ley RE, Hugenholtz P, Tringe SG, Dangl JL (2012) Defining the core Arabidopsis thaliana root microbiome. Nature 488(7409):86–90. https://doi.org/10.1038/nature11237
33. Massalha H, Korenblum E, Malitsky S, Shapiro OH, Aharoni A (2017) Live imaging of root–bacteria interactions in a microfluidics setup. Proc Natl Acad Sci 114(17):4549–4554. https://doi.org/10.1073/pnas.1618584114
34. Berendsen RL, Pieterse CMJ, Bakker PAHM (2012) The rhizosphere microbiome and plant health. Trends Plant Sci 17(8):478–486. https://doi.org/10.1016/j.tplants.2012.04.001
35. Lombardi N, Vitale S, Turr ÀD, Reverberi M, Fanelli C, Vinale F, Marra R, Ruocco M, Pascale A, D'Errico G, Woo SL, Lorito M (2018) Root exudates of stressed plants stimulate and attract *Trichoderma* soil fungi. Mol Plant-Microbe Interact 31(10):982–994. https://doi.org/10.1094/MPMI-12-17-0310-R/ASSET/IMAGES/LARGE/MPMI-12-17-0310-R_T3.JPEG
36. Brundrett MC (2002) Coevolution of roots and mycorrhizas of land plants. New Phytol 154(2):275–304. https://doi.org/10.1046/j.1469-8137.2002.00397.x
37. Andrade G, Mihara KL, Linderman RG, Bethlenfalvay GJ (1997) Bacteria from rhizosphere and hyphosphere soils of different arbuscular-mycorrhizal fungi. Plant and Soil 192(1):71–79. https://www.jstor.org/stable/42948049?seq=7#metadata_info_tab_contents

38. Toljander JF, Lindahl BD, Paul LR, Elfstrand M, Finlay RD (2007) Influence of arbuscular mycorrhizal mycelial exudates on soil bacterial growth and community structure. FEMS Microbiol Ecol 61(2):295–304. https://doi.org/10.1111/j.1574-6941.2007.00337.x
39. Warmink JA, Nazir R, van Elsas JD (2009) Universal and species-specific bacterial 'fungiphiles' in the mycospheres of different basidiomycetous fungi. Environ Microbiol 11(2):300–312. https://doi.org/10.1111/j.1462-2920.2008.01767.x
40. Hassani MA, Durán P, Hacquard S (2018) Microbial interactions within the plant holobiont. Microbiome 6(1):1–17. https://doi.org/10.1186/S40168-018-0445-0
41. Scheublin TR, Sanders IR, Keel C, Van Der Meer JR (2010) Characterisation of microbial communities colonising the hyphal surfaces of arbuscular mycorrhizal fungi. ISME J 4(6):752–763. https://doi.org/10.1038/ismej.2010.5
42. Warmink JA, Van Elsas JD (2009) Migratory response of soil bacteria to *Lyophyllum* sp. strain Karsten in soil microcosms. Appl Environ Microbiol 75(9):2820–2830. https://doi.org/10.1128/AEM.02110-08
43. Bulgarelli D, Schlaeppi K, Spaepen S, van Themaat EVL, Schulze-Lefert P (2013) Structure and functions of the bacterial microbiota of plants. Annu Rev Plant Biol 64(1):807–838. https://doi.org/10.1146/annurev-arplant-050312-120106
44. Olanrewaju OS, Ayangbenro AS, Glick BR, Babalola OO (2019) Plant health: feedback effect of root exudates-rhizobiome interactions. Appl Microbiol Biotechnol 103(3):1155–1166. https://doi.org/10.1007/s00253-018-9556-6
45. Bai Y, Müller DB, Srinivas G, Garrido-Oter R, Potthoff E, Rott M, Dombrowski N, Münch PC, Spaepen S, Remus-Emsermann M, Hüttel B, McHardy AC, Vorholt JA, Schulze-Lefert P (2015) Functional overlap of the *Arabidopsis* leaf and root microbiota. Nature 528:364–369. https://doi.org/10.1038/nature16192
46. Jones JDG, Dangl JL (2006) The plant immune system. Nature 444:323–329. https://doi.org/10.1038/nature05286
47. Kiers ET, Duhamel M, Beesetty Y, Mensah JA, Franken O, Verbruggen E, Fellbaum CR, Kowalchuk GA, Hart MM, Bago A, Palmer TM, West SA, Kumar AS, Lakshmanan V, Caplan JL, Powell D, Czymmek KJ, Levia DF, Bais HP (2012) Rhizobacteria Bacillus subtilis restricts foliar pathogen entry through stomata. Plant J: Cell Mol Biol 72(4):694–706. https://doi.org/10.1111/J.1365-313X.2012.05116.X
48. Neal AL, Ahmad S, Gordon-Weeks R, Ton J (2012) Benzoxazinoids in root exudates of maize attract *Pseudomonas putida* to the Rhizosphere. PLoS ONE 7(4):e35498. https://doi.org/10.1371/JOURNAL.PONE.0035498
49. Planchamp C, Glauser G, Mauch-Mani B (2015) Root inoculation with Pseudomonas putida KT2440 induces transcriptional and metabolic changes and systemic resistance in maize plants. Front Plant Sci 5:719. https://doi.org/10.3389/FPLS.2014.00719/ABSTRACT
50. Abdel-Lateif K, Bogusz D, Hocher V (2012) The role of flavonoids in the establishment of plant roots endosymbioses with arbuscular mycorrhiza fungi, rhizobia and Frankia bacteria. Plant Signal Behav 7(6):636. https://doi.org/10.4161/PSB.20039
51. Weller DM, Raaijmakers JM, Mc Spadden Gardener BB, Thomashow LS (2003) Microbial populations responsible for specific soil suppressiveness to plant pathogens. Ann Rev Phytopathol 40(1):309–348. https://doi.org/10.1146/ANNUREV.PHYTO.40.030402.110010
52. Uddin MN, Rahman UU, Khan W, Uddin N, Muhammad M (2018) Effect of trichoderma harzianum on tomato plant growth and its antagonistic activity against *Phythium ultimum* and *Phytopthora capsici*. Egypt J Biol Pest Contr 28(1):1–6. https://doi.org/10.1186/S41938-018-0032-5/FIGURES/8
53. Bordenstein SR, Theis KR (2015) Host biology in light of the microbiome: ten principles of holobionts and hologenomes. PLoS Biol 13(8):e1002226. https://doi.org/10.1371/JOURNAL.PBIO.1002226
54. Rosenberg E, Sharon G, Zilber-Rosenberg I (2009) The hologenome theory of evolution contains Lamarckian aspects within a Darwinian framework. Environ Microbiol 11(12):2959–2962. https://doi.org/10.1111/J.1462-2920.2009.01995.X

55. Rosenberg E, Zilber-Rosenberg I (2016) Microbes drive evolution of animals and plants: the hologenome concept. MBio 7(2):e01395-e1415. https://doi.org/10.1128/mBio.01395-15
56. York LM, Carminati A, Mooney SJ, Ritz K, Bennett MJ (2016) The holistic rhizosphere: integrating zones, processes, and semantics in the soil influenced by roots. J Exp Bot 67(12):3629–3643. https://doi.org/10.1093/JXB/ERW108

# Chapter 7
# Climate Change:- General Overview and Implications on Agriculture and Allied Sectors

**Sheikh Mehraj, A. H. Pandit, Mohammed Tauseef Ali, Asif Mohi Ud Din Rather, I. A. Bisati, H. A. Malik, and Sabiha Ashraf**

**Abstract** Perhaps the greatest problem the world is currently facing is climate change, and the future existence of man depends on how effectively this challenge is currently tackled. Climate change phenomenon has resulted in disasters across the globe. Sustainability of agriculture, habitation and human healthiness depends on how quickly and effectively we are able to tackle this problem. On a global scale, both agriculture and climate are interconnected processes. The projected effects of global warming on agro-climatic conditions, such as temperature and precipitation in particular, moreover on glacier run-off in general, are expected to be significantly increased. These factors determine the biosphere's ability to sustainably generate food for both humans and animals. Crop production would also be affected by rising carbon dioxide levels. The imbalance of climatic factors due to climate change will determine the consequences of climate change on the agriculture and allied sectors. Understanding the global climate change phenomenon, will help us to effectively foresee and modify farming practices to sustain and increase agricultural production. According to recent scientific findings, India will face an adverse effect of global warming. Food security and productivity, fresh water availability, forest biodiversity, fisheries, and other agri allied activities would suffer adversely. Unfortunately, the people that depend on farming, fishing, and living in the forests will be badly impacted through climate change.

---

S. Mehraj (✉) · A. H. Pandit · M. T. Ali · I. A. Bisati
Division of Fruit Science, SKUAST–K, Srinagar, Jammu and Kashmir, India
e-mail: sheikhmehraj10@gmail.com

A. M. U. D. Rather
Division of Vegetable Science, SKUAST–K, Srinagar, Jammu and Kashmir, India

H. A. Malik
Directorate of Extension, SKUAST–K, Srinagar, Jammu and Kashmir, India

S. Ashraf
College of Temperate Sericulture, Sher-e-Kashmir University of Agricultural Sciences and Technology of Kashmir, Srinagar, Jammu and Kashmir, India



J. A. Parray (ed.), *Climate Change and Microbiome Dynamics*, Climate Change Management, https://doi.org/10.1007/978-3-031-21079-2_7

## Introduction

Over a period of time any change in climate which take place whether due to nature or due to anthropogenous is referred to as climate change [1]. On earth, the most essential component influencing patterns of livelihood has been the climate. Today, it is universally accepted that one of the biggest environmental issues of our time is climate change. Additionally, it has caused a serious impact on all over the world and is becoming a local and paticular problem in all countries of the world. However, in order to analyse it properly it created crusade effects in emerging nations in the world viz., India. Mean while, if we analyse the dramatic change in the global warming in J&K (India), it has been seen that lot of alternations take place in rainfall regime over a period of time, not only this degradation of water bodies has been observed at a greater pace, greater human animal conflict in recent years and rising temperature during the flowering period leads to drying of the stigmas which hampers the fruit set becoming a menance to the orchardists in our horticulture state [2]. Growing anthropogenic strains are worsening this situation, and the changing climate is provoking daily human affairs, which are already impacted in different ways. Global warming has resulted in melting of polar ice caps leading to increased water levels of oceans. Global mean temperature over the years is increasing which is mainly depicted in global mean temperature anamoly (Table 7.1). The main causes of this global mean temperature anomly are mainly due to the mushrooming growth of industries and methane gases emissions produced by the ruminants. Global warming has led to onset of earlier spring and late winters, as well as shorter and warmer winters [3, 4]. The vulnerability of economically weaker section of population will be more to climate change and will find it difficult to adapt since they are unable to handle the existing environmental challenges, such as water stress and drought. India must develop numerous measures to deal with the impending hazards of climate change, which are in addition to the already high environmental stress levels. This should include (i) Research to better understand challenges related to climate change; (ii) implementing sustainable development strategies; (iii) Developing the ability of the poor to adapt; and (iv) pursuing a worldwide agreement to reduce the green house gas emissions in a greatest possible way. Any delay in addressing climate change would increase the cost of future initiatives and make them extremely tough to agree upon. It has been found that shortage of food grains in Kashmir valley has reached up to 40%, while as 30% was observed in vegetable production and 69% in case of oil seeds for 6 million population, putting the Kashmir valley under greater threat due to food insecurity [5]. Due to the climate change in Kashmir valley more than 90% of the paddy lands are converted into apple orchard which could lead to the reduction in food grain population may be over 60% in coming decades if we consider the current rate of change into account. Due to erratic rainfall regime in the Kashmir valley at least 11,909 kanals of paddy land have been shifted into orcharding system in most parts of the Kashmir valley viz., Anantnag, Baramulla, Shopian etc. [2]. The change in shifting of land into orcharding system is clearly mentioned in Table 7.2.

**Table 7.1** Global mean temperature anomaly

| Year | Temperature anomalies (°C) |
|---|---|
| 2011 | 0.37 |
| 2012 | 0.42 |
| 2013 | 0.34 |
| 2014 | 0.32 |
| 2015 | 0.49 |
| 2016 | 0.33 |
| 2017 | 0.45 |
| 2018 | 0.48 |
| 2019 | 0.43 |
| 2020 | 0.52 |

**Table 7.2** Shifting of paddy land into orcharding system in most parts of the Kashmir Valley (Figures given in table are approximate)

| Sl. No | District | Total area $km^2$ | Land changed (in kanals*) |
|---|---|---|---|
| 1 | Anantnag | 3,984 | 3700 |
| 2 | Bandipora | 398 | 695 |
| 3 | Baramulla | 4,588 | 1152 |
| 4 | Budgam | 1,371 | 1112 |
| 5 | Kulgam | | 1250 |
| 6 | Pulwama | 1,398 | 2500 |
| 7 | Shopian | | 1500 |

A kanal mostly use as land measuremet in Kashmir valley, equivalent to 505.857 $m^2$ or 0.125 acres

*Source* A report on climate change and its impact in Kashmir [2]

## Causes of Climate Change

Fossil fuel burning.
Deforestation.
Increase in industrialization.
Faulty agriculture activities.
Urbanization.
Wetland destruction.
Land use pattern.

## Impact of Climate Change on Agriculture

The greenhouse effect is a natural phenomenon that significantly affects the climate of the earth. It creates the comparatively warm and affable climate on the surface of the earth that has allows humans and other life forms to flourish. The main factor behind the overall rise in the earth's atmospheric temperature, which led to global warming are the human activities such as indiscriminate cutting of forests which have led to increased levels of greenhouse gases (GHGs) such as methane ($CH_4$), carbon dioxide ($CO_2$), water vapour ($H_2O$), nitrous oxide ($N_2O$), hydrofluorocarbons (HFCs), sulphur hexafluoride (SF6), and perfluorocarbons (PFCs) [5]. It ios expected that the average surface temperature of earth will increase by 1.4–5.8 °C by 2100 AD from 0.74 °C since the late 19th Century with significant regional differences, which leads to increase in sea level and decline in the area covered under snow and glaciers [6]. The net photosynthetic rate will increase with an increase in the atmospheric $CO_2$ concentration. Increase in carbon dioxide level in the atmosphere results in reduction in water loss besides results in reducing the stomatal pores which are main gateways for the water and gaseous exchange [7]. In some crop plants, the reduction in transpiration could be 30%. However, as the response of stomata to $CO_2$ interacts with numerous environmental factors such as light intensity and temperature and plant parameters, it is still very difficult to forecast how the rise in atmospheric $CO_2$ will affect stomata's responsiveness [7]. The primary source of yield losses brought on by simultaneous increases in atmospheric $CO_2$ and temperature is spikelet sterility induced by high temperatures [8]. Stomatal opening decreases under conditions of elevated $CO_2$ which increase resistance to water loss from leaves. When night temperatures are higher than 21 °C, increased $CO_2$ levels may also directly hinder the maintenance of respiration rate. A few hours of exposure to high temperatures can significantly lower pollen viability, which will result in yield loss. Temperatures greater than 35 °C significantly enhance the sterility of spikelets [8] and increased $CO_2$ levels could exacerbate this problem, possibly due to decreased transpirational cooling [8]. Gas emissions and the effects of land use are primarily driven by the agricultural sector. Agriculture consumes a lot of fossil fuels and land, the practices like zero tillage, paddy farming, and livestock raising contribute to greenhouse gas emissions [9]. The process of zero tillage is now prohibited in several countries due to the conversion of sub surface carbon into carbon carbon dioxide. According to IPCC, fossil fuels, land use, and human activity, are the main factors of the rise in greenhouse gases, as it has been observed during the previous 250 years [10]. A wide range of repercussions from climate change will affect agriculture. Crop yield will be reduced. For instance, a rise in temperature from 1–4 $^0$C can result in the reduction of grain output of 0–49% in rice, 5–40% in potatoes, 13–30% in green gram, and 11–36% in soya bean. According to research on the effects of climate change on the of rice productivity in Punjab, rice grain production will decrease by 5.4, 7.4, and 25.1%, with continuous increase of temperatuere respectively. The cooked rice grains produced by plants raised in high $CO_2$ conditions would be firmer than those produced by plants that are being used. But the levels of zinc and iron, which are vital for human nutrition, would be lesser. Additionally, when the temperature and $CO_2$ levels increase simultaneously, the protein content of the grain reduces. Studies have demonstrated that greater $CO_2$ levels result in decreased nitrogen uptake by plants producing crops with decreased nutritional content. This would mostly affect populations in under developed economies who are less able to make up for it by consuming more food, consuming a wider variety of foods, or perhaps taking supplements.

In case of soil temperature it has been ascertained that it affects the rates of release of nutrients and organic matter decomposition. Although nutrient availability will rise at high temperatures in the short term, however organic matter content will decrease significantly, lowering soil fertility in the long run. The quality of produce is affected by high temperatures. Increased temperature can adversely affect basmati grain elongation and aroma as well as test weight and amylase content.

Potential setbacks are currently faced by the dairy industry. Although the ideal temperature for milk production is between 40 and 75 °F, heat stress can result at temperatures as low as 75 °F particularly on humid days, which can cause a 5–20% reduction in milk output [11]. The livestock yields reduced by 10% in U.S. under a 5.0 °C increase in temperature and yield loss for dairy farms in Appalachia, the Delta States, Texas, the Southeast and the Southern Plains was estimated at 1% for a 1.5 °C increase in temperature above normal [12].

It is anticipated that the increased atmospheric temperatures observed in recent decades will result in a more active hydrological cycle which will lead to more intense rainfall events. Degradation of the soil and erosion are more likely to happen. Global warming would also affect the fertility of the soil. Since the proportion of carbon to nitrogen being constant, doubling carbon is likely to indicate a larger storage of nitrogen as nitrates in soils, supplying plants with more fertiliser and improving yields. The option to switch to less expensive fertilisation techniques may arise if the average nitrogen requirements drop.

## Climate Change and Its Consequences on Temperate Fruits.

Climate change has a potential to greatly affect all of agriculture in the same way as agriculture affects climate change. Global warming may have an impact on chilling requirement, risk of frost, flowering time, growing season length, fruit quality and maturity. Increased evapo transpiration, as a result of global warming, will increase the irrigation demand. Pollen viability has been greatly influenced by increased temperatures, which has resulted in flower drop in apricot and peach. The surface temperature of fruit increases due to prolonged exposure to direct sunlight, which result in increased ripening.

The chilling temperature during winters is important for bud formation in temperate fruit crops. Trees produce their vegetative and reproductive buds in the summer, and these buds continue to remain dormant if they have not acquired the required level of chilling temperature. However, because of the continuing global warming, temperate fruits did not receive the necessary amount of chilling, which causes a number of adverse effects such as delayed vegetative growth, decreased fruit set, and decreased fruit quality. Pollen desiccation, reduced viability of pollen grains and ovules, and pollinator mortality are all results of temperature rise [13]. Increase in winter temperature, anticipated in all scenario will result in a very substantial increase in the number of days with temperature above freezing above 5 °C, thus extending and advancing growing season [14]. Some Italian authors have noted a tendency toward an increasing tendency to spring frost, although this is not solely ascribed to climate change, it may also occur due to the proliferation of early flowering cultivars viz., peach or the expansion of growing regions into areas more susceptible to frost (Tables 7.3 and 7.4). Inhibition of respiration and protein synthesis, as well as an increase in protein breakdown and ethylene generation, are often the immediate impacts of increased doses. In case of increase of temperature in temperate fruits following things will happen.

- During the recent years in Kashmir valley (india) increase in temperature due to global warmimg results in advance bud formation which result in the earlier blooming and fruit set. Mean while the higher temperature in the spring cause frost damage to fruit crops (Fig. 7.1).
- Flowering may be delayed as mean temperature increases in winter.

**Table 7.3** Morphological analysis, production and germination in vitro pollen grains of peach trees 'Granda' under distinct environments during the pre-blooming and blooming period [14]

| Treatment/year | 2018 | 2019 |
|---|---|---|
| *Anthers no normal pollen (%)* | | |
| Greenhouse | 8.89 | 88.23 |
| Orchard | 5.84 | 2.22 |
| Average | 7.37 | 45.23 |
| *Anthers with more than 50% abortive pollen grains (%)* | | |
| Greenhouse | 41.11 | 100.00 |
| Orchard | 19.91 | 33.25 |
| Average | 30.51 | 66.63 |
| *Production of pollen grains/anther* | | |
| Greenhouse | 180.00 | 91.67 |
| Orchard | 455.00 | 226.67 |
| Average | 317.50 | 159.17 |
| *Germination at 20 °C (%)* | | |
| Greenhouse | 0.67 | 2.85 |
| Orchard | 41.06 | 4.62 |
| Average | 20.87 | 3.74 |
| *Germination of pollen grains at 25 °C (%)* | | |
| Greenhouse | 5.68 | 4.01 |
| Orchard | 0.00 | 3.65 |
| Average | 2.84 | 3.83 |

**Table 7.4** Fruit set and production in peach variety Granda [14]

| Treatment year | 2004 | 2005 | Average |
|---|---|---|---|
| *Fruit set (%)* | | | |
| Greenhouse | 0.00 | 0.46 | 0.23 |
| Orchard | 2.22 | 5.59 | 3.90 |
| Average | 1.11 | 3.03 | CV = 39.12% |
| *Yield (kg/tree)* | | | |
| Greenhouse | 0.00 | 0.35 | 0.18 |
| Orchard | 9.29 | 28.73 | 19.01 |
| Average | 4.65 | 14.54 | CV = 60.05% |

**Fig. 7.1** Frost damage to apple due to low temperature indicates scarring viz., collapsing of fruits near to calyx

## Potential Consequences of Climate Change on Diseases, Pests and Weeds

Increases in agricultural, forest, and structural insect pests and weeds are likely to be increased by the increase in temperature. Droughts, more frequent storms, higher rainfall, and other extreme weather events are brought on by global warming. All of this will impact plant development and encourage more insects [16, 17]. Warm-weather pests will begin reproducing earlier since winters will be milder and shorter [18]. It is anticipated that as temperatures rise and rainfall increases, the prevalence of many plant diseases, particularly those brought on by fungi, would rise. Plant pathogens overwinter more successfully when the winters are warmer. Many fungal pathogens grow best between 22-28 °C. It has also been observed that increase in plant growth due to increase in temperature also results in host plant densities [19]. In Japan, rice stripe disease is more likely to spread due to erratic climate change [20]. It's possible that global warming has already contributed to the spread and severity of some potato virus diseases in India. The severity of the oak dieback caused by

Phytophthora cinnamomi has been implicated by global warming. Warm, damp soil is favourable to this pathogen. Plant diseases caused by the climate change greatly affect thge most of the food and fruit crops which have direct impact on the human beings [21]. An increase in rainfall due to global warming would prolong the wet seasons and increase atmospheric humidity in some regions. This could facilitate the development of fungal diseases coupled with greater temperatures. Similarly, increased pressure from insects and disease vectors may occur as a result of higher temperatures and humidity.

## Impact of Climate Change on Fisheries

Fresh water fisheries are anticipated to experience short-term impacts from climate change due to changes in nutrient levels, average water temperature, and prolonged dry season and elevated water levels. Such changes will then have an adverse effect on the quality, productivity, output, viability of fish and entrepreneurship development in fisheries sector which will have an adverse impact on the fishing community lead to snatching of their livelihood. According to the IPCC [21], river flow rates during the dry season are anticipated to decrease throughout south Asia and the majority of African river basins, leading to reduced fish production. As glaciers melt and lose their ability to provide predictable and controlled water flows, bigger fluctuations in river flows are projected throughout time. Researchers discovered that the effects of climatic uncertainty on fish productivity have already started to be experienced by lake fisheries.

## Alternative or Cleaner Approaches

### Organic Agriculture

Organic farming produces considerably lower greenhouse gas emissions (GHG) and sequesters carbon in the soil rapidly and efficiently. Global implementation of organic agriculture would result in additional reduction in emissions of approximately 0.6 to 0.7 Gt $CO_2$ through the avoidance of biomass burning ($CH_4$ and $NO_2$ emissions) and the prevention of 0.41 Gt $CO_2$ emitted from the fossil fuel consumption for chemical nitrogen fertilizer production per year [22]. Organic farming eliminates resource and financial constraints in farming, improving the access to local food. As the organic farming does not use expensive external inputs like chemical fertilisers, pesticides, and gasoline, input prices are much lower. Lower expenses eliminate the need for credit and ensuing debt, which reduces financial risk. The cost of external chemical inputs will increase as the price of fossil fuels rises, making reliance on these inputs insecure. Additionally, organic farming lowers risk by diversification of food and income sources, which lowers the risks related to a particular crop failure. In spite of all these potentials the penetration of natural/organic farming is very weak due to the biased Government extension methods. Absence of credible/accessible certification schemes for organic growers prevents them to compete successfully in export markets [23]. This is primarily due to the fact multinational companies are dictating research priorities in food production/processing etc. and hence, there is low priority for locally relevant/self reliant solutions.

In short, organic farming/Agriculture is a farming system which results in maintainting and restoring the ecological balance of whole biosphere. Moreover, organic foods fetches higher prices around 70–80 than the traditional agriculture system [24]. Comparing it with the traditional system, non judicious application of fungicides and pesticides is on peak directely enter the food chain, penetrating into the water bodies, harming the livestock sector and results in depletion of natural ecosystem [25].

## Mitigation and Adaptation Measures

India needs multipronged approach so as to deal with long pending challenges of global warming besides high environmental stress level. The following challenges should comprises of.

- Research to better comprehend concerns related to climate change.
- Implementing sustainable development strategies.
- Improving the adaptive capacity of the impoverished.
- Pursuing a global agreement to cut greenhouse gas emissions at the earliest.
- Understanding the relation between combating climate change and economic development from a longer perspective is necessary. India should not simply concentrate on short-term financial gain from global organisations and procedures related to combating climate change. The government ought to approach it as a major issue with potentially grave socioeconomic and environmental repercussions, in order to minimize the mitigation of climate change on the society and people in genereal long-term solutions need to be sought out.
- Development of new genotypes resistant to increased $CO_2$ concentrations, temperature, and drought.
- Crop diversification.
- It is important to have a well-informed public discussion that includes all the interested parties, including policymakers, experts, environmental non-governmental organizations, industrial groups, mass media, farmers and fishermen's representatives. Given the urgency, the severity, and a variety of implications for various stakeholders, the development of national climate change policies should be broad-based.
- Creating climate impact modules that provide a greater understanding of how agriculture, forestry, and farming are affected by climate change would help to be better prepared at local level.

- The people living on the coastal areas need be shift to safer place and budget for that part need to be discussed and voted.
- Capacity building programs for the rice-fish cultivation needs to adopted through national adaptation program of action on climate change. Dissemination of knowledge regarding the organic cultivation of the crops needs to prioritized.
- Promotion of "best crop-fish farming practices" through farmer's capacity building and networking. Conceptualization and implementation of "National Adaptation program of Action on climate change. Through the judicious application of organic manure, fertilizers, irrigation water, nitrification inhibitors, fertilizer location, and fertilizer scheduling, improvement in the management of rice production can be done.
- Improve the management of the cattle population, particularly the diet of ruminants. By using limited tillage and managing residues, soil organic carbon content can be increased.
- Through improved machinery designs and resource conservation techniques, increase the efficiency of utilization in agriculture.
- Increasing the area under biofuels and agroforestry by altering the land use pattern, without affecting the production of food grains.
- The cost of adaption is considerable. In order to respond urgently to climate change, a new model of development is required. Research funds are needed to develop crop types that can sequester more carbon and produce better biofuels and still being drought, heat, and flood tolerant. In addition, funding are also needed for other industries to adapt.
- Agro forestry, that is the growing of trees along with crops, can assist farmers in coping with some of the adverse effects of climatic change. Cultivation of cover crops and planting of trees along the boundaries of the farm should be done in order to lessen the soil erosion and restoration of soil fertility. Improved fallow practices are also quite promising. Utilizing retained rainwater as effectively as possible through agro forestry techniques may be one of the most efficient ways to increase the systems' ability to adapt to climate changes.

## Potential Research Approaches for Optimizing Yield Increase Under Changing Climatic Scenario

### *Role of Microbes in Mitigating Climate Change*

Climate change results in a significant change in temperature and precipitation causing global heating, increase in sea level, shifting of people to highland areas and tremendous environmental effects. During the recent years a lot of research takes place in mitigating climatic change and it has been found that microbial world could result in more prompt impact. Microbial world have a more important role in mitigating global warming and could result in the reduction in carbon dioxide, methane and other green house gases which is increasing due to indiscriminate cutting of forests. It has been observed that the plant micro climate and plant rhizosphere contains thousand of micro-organisms viz., plant growth promoting bacteria and plant growth promoting fungi. Rhizobacteria plays an important role in fixing atmospheric nitrogen while as mycorrihizae provide phosphate and nitrate to the plant for growth and developmental processes [26]. It has also been observed that certain microbial organisms provide resistance to biotic and abiotic stress factors. It has been found that some mycorrhizal and endophytic fungi and plant growth promoting bacteria offers signicant resistance to the plants against drought, heat, pathogens and certain toxic elements in the rhizosphere [27]. Stomatal closure in plants due to the various droughts shows water loss by increase in the level of Abscisic acid, ethylene and salicyclic acid. Drought tolerance in plants showed many changes mainly in absiscic acid, ethylene and cytokinons to the PGPR. Root morphology in such plants is modified to release endogenous plant harmones by signaling the IAA-induced pathway for root growth which has been found in *A. brasilense* in aerobic conditions [28]. It has been found that Inoculation of plant species with certain bacterium species can increase its tolerance to drought by isolating its drought-responsive gene, ERD15, from A. thaliana when inoculated with *Paenibacillus polymyxa.* Microbes mainly help to improve the plant to an abiotic stress by meticulously alter the plant structure and their physiology [29]. It has been found that microbial electro-synthesis produces important products from the electricity using carbon dioxide and other organic carbon as an input sources. During this process acetate, butyrate, and other commodity chemicals are produced during the reaction subsequently caproate and caprylate are produced which become a source for the building blocks for the various chemical industries. So the efficient harvesting of carbon could lead to microbial carbon reduction [30].

The paradigm shift to combat climate change is to reduce the green house gases by the microbial way. It has been found that microbial world played an important role in optimizing the present concentration of green house gases. The major use of microbial world could solve this of global warming in nearby future [31]. So the microbial system could solve this problem by the use of nutrient cycling in order to reduction of the green house gases and altering the genetic material [32]. In this case the best way to elimination of green house gaseous is to support the mutual existence of microbial communities and biogeochemical cycles. It has been ascertained that the green house gaseous acts as building blocks for the microbial system and formation of their cell structure. In the present world various microbes have been discovered to cope the changing global warming due to the continuous change in climate change. Most dynamic change will come into existence by the research on the DNA sequencing of the microbial and their physiology in order to get advance research on the climate change. So in order to counter the climatic change in the present world more research should take place by knowing all the well known aspects of the microbial biome.

- Development of low chilling stone, pome and nut fruits cultivars [33]

- Cultivation of high-value crops like walnut, peach, apricot and kiwi as a diversification strategy.
- Marker assisted selection and development of transgenics resistant to abiotic and biotic resistance
- Better Weather Forecasting and Crop Insurance Schemes

## Conclusion

It is well recognized that climate influences human affairs in several ways, primarily through its impact on basic amenities of livelihood i.e. food, water and energy resources. However, appropriate measures together with strict laws need enforced at an earliest. In its development policies and plans, the government should put a special thought on concerns related to climate change adaptation. The development, distribution, and adoption of technology among farmers, as well as adequate financial investments, are required to promote climate change adaptation and mitigation. Further, a competent institutional framework is considered necessary for the state's natural resources to be protected, preserved and managed scientifically. Development of a sustainable pathway is considered to be the most efficient way to combat the climate change, besides uses of renewable energy and plantation crops.

## References


1. Climate change IPCC (2022) The supplementary report to the IPCC scientific assessment. Cambridge University Press 44:489–5092021
2. NHB (2021) A report on climate change in Kashmir region and its impact on human resources. pp 201–240
3. Luedeling E, Blanke M, Gebauer J (2021) Climate change effects on winter chill for fruit crops in Germany (Translated from German). Erwerbs-obstbau 51(3):81–94
4. Collins W, Colman R, Haywood J, Manning RR, Mote (2007) The physical science behind climate change. Sci Am 297(2):64–73
5. Talib AH (2007) On the brink-a report on climate change and its impact in Kashmir, Action aid India, Srinagar, Jammu and Kashmir
6. IPCC (2011) Fourth Assessment Report. Intergovernmental Panel on Climate Change Secretariat. Geneva, Switzerland. http://www.ipcc.ch/
7. Kasper LD, Frame DJ, Ackerley D, Aina T, Booth BBB, Christensen C (2020) Broad range of 2050 warming from an observationally constrained large climate model ensemble. Nat Geosci 7:115–2300
8. Rosenzweig C, Parry ML, Fischer G (1995) World food supply. In: Strzepek KM, Smith JB (eds) As climate changes: international impacts and implications. Cambridge University Press, Cambridge, p 27–56
9. Hartland M, Omasa M, Horie T (2020) High temperature-induced spikelet sterility of japonica rice at flowering in relation to air temperature, humidity and wind velocity condition. Jpn J Crop Sci 62: 322–328

10. FAO (2020) Climate Change and global health risk management-technical background document-from the expert consultation held on 28 TO 29 February, 2008. FAO, Rome
11. Anupama M (2014) Climate change and its impact on agriculture. Int J Sci Res Publ 4(4). ISSN 2250–31532020
12. Frumhoff PC, McCarthy JJ, Melillo JM, Moser SC, Wuebbles DJ (2007) Confronting Climate change in the U.S. northeast: science, impacts, and solutions. Synthesis report of the Northeast Climate Impacts Assessment (NECIA). Union of Concerned Scientists (UCS), Cambridge, MA
13. Adams RM, McCarl BA, Segerson K, Rosenzweig C, Bryant KJ, Dixon BL, Conner R, Evenson RE, Ojima D (1998) The economic effects of climate change on U.S. Agriculture 23:450–480
14. Chakraborty S, Newton AC (2011) Climate change, plant diseases and food security: an overview. Plant Pathol 60:1–14
15. Toselli M, Baldi E, Cavani I, Mazzon M, Quartieri M, Sorrenti G, Marzadori C (2019) Soil-plantnitrogen pools in nectarine orchard in response tolong-term compost application. Sci Total Environ Amsterdam 671:10–18
16. Easterling DR, Meehl GA, Parmesan C (2000) Climate extremes, observations, modeling, and impacts. Science 289:2068–2074
17. Stireman JO (2005) Climatic unpredictability and parasitism of caterpillars: implications of global warming. Proc Natl Acad Sci (USA) 102(48):17384–17387
18. Robinson ST, Masters GJ, Hodkinson ID (2021) Role of hervbivours in climage change and global warming. Global Change Biol 8(1):1–16
19. Rayees AH, Dendy SP, Frank EE, Rouse MN, Travers SE (2019) Climate change and its impact on human resources. Econ Affairs 4(1):25–38
20. Yamamura K, Yokozawa M (2002) Prediction of a geographical shift in the prevalence of rice stripe disease transmitted by the small brown planthopper, Laodelphax striatellus (Hemiptera:Delphacidae) under global warming. Appl Entomol Zool 37(1):181–190. [CAB Abstracts]
21. Ferguson PS (2020) Food, climate and carbon dioxide. The global environment and world food production. CRC Press, Boca Raton
22. IPCC (2007b) The physical science basis. In: Solomon, Quiri SD, Manning M, Z.chen M, Marquis KB Averty MT, Miller HL (eds) Contribution of working group Ist to fourth Assesment report of the intergovernmental panel on climate change. Cambridge University Press, Cambridge United kingdom and Newyork, NY, USA, p 9996
23. Kaur K, Toor MS (2015) Organic farming: status and constraints. Indian J Econ Dev 11(1):333–338
24. Mitra S, Devi H (2016) Organic Horticulture in India. Horticulturae 2(17)
25. Ergonul B, Ergonul PG (2015) Consumer motivation for organic food consumption. Emirates J Food Agri 27(5):416–422
26. Wu JY, Sardo V (2010) Sustainable versus organic agriculture. In: Lichtfouse E (ed) Sociology, organic farming, climate change and soil science. Springer, Dordrecht, pp 41–76
27. Geurts R, Lillo A, Bisseling T (2012) Exploiting an ancient signalling machinery to enjoy a nitrogen fixing symbiosis. Curr Opin Plant Biol 15:438–443
28. De Zelicourt A, Al-Yousif M, Hirt H (2013) Rhizosphere microbes as essential partners for plant stress tolerance. Mol Plant 6:242–245
29. Molina-Favero C, Creus CM, Simontacchi M, Puntarulo S, Lamattin L (2008) Aerobic nitric oxide production by *Azospirillum brasilense* Sp245 and its influence on root architecture in tomato. Mol Plant Microbe Interact 21:1001–1009
30. Menon S, Denman KL, Brasseur G, Chidthaison A, Ciais P Cox PM (2007) Couplings between changes in the climate system and biogeochemistry. Medium: ED: Ernest Orlando Lawrence Berkeley National Laboratory, Berkeley
31. Angenent LT, Richter H, Buckel W, Spirito CM, Steinbusch KJ, Plugge CM (2016) Chain elongation with reactor microbiomes: open-culture biotechnology to produce biochemicals. Environ Sci Technol 50:2796–2810
32. Zimmer C (2010) The microbe factor and its role in our climate future

33. Zhou J, Xue K, Xie J, Deng Y, Liyou W (2011) Microbial mediation of carbon-cycle feedbacks to climate warming. Nat Clim Change 1–5
34. Mendelsohn R, Nordhaus WD, Shaw D (2020) The impact of global warming on agriculture: a ricardian analysis. Am Econ Rev 84(4):753–771

# Chapter 8
# Soil Microbial Community and Climate Change Drivers

**Asma Nazir, Madiha Farooq, Bisma Farooq, Shahnaz Anjum, and Shayista Yousuf**

**Abstract** The biogeochemical cycling of macronutrients, micronutrients, and other components necessary for the development of plants and animal life is governed by the soil microbiome. As we focus our research efforts on one of the most serious issues affecting our planet, knowing and anticipating how climate change will affect soil microbiomes and the ecosystem services they provide is a huge challenge and significant potential. Studies predict that factors related to climate change, such as elevated atmospheric [$CO_2$] and heat, will function together to change ecosystem features and processes, influencing species distributions and, presumably, organism interactions. On the other hand, it is harder to forecast how the microbial populations that control ecological processes would respond. In complex ecosystems, organisms interact with thousands of different species, some of which are useful, some of which are poisonous, and some of which have little to no impact. In this chapter, we examine the present level of knowledge about the effects of climate change on soil microorganisms in various climate-sensitive soil ecosystems, as well as prospective approaches that soil microorganisms may be used in to help lessen the detrimental effects of climate change.

A. Nazir
Department of Clinical Biochemistry, University of Kashmir, Hazratbal, Srinagar 190006, J&K, India

M. Farooq
Department of Microbiology, Asian International University, Imphal 795113, Manipur, India

B. Farooq (✉)
Department of Biomedical Engineering, Satyabhama Institute of Science and Technology, Chennai 600119, India
e-mail: Bismafarooq@ese.gov.ae

S. Anjum
Department of Botany, Lovely Professional University, Phagwara 14411, Punjab, India

S. Yousuf
Department of Zoology, Kashmir University, Srinagar 190006, J&K, India

 
J. A. Parray (ed.), *Climate Change and Microbiome Dynamics*, Climate Change Management, https://doi.org/10.1007/978-3-031-21079-2_8

## Introduction

Although soil is one of the planet's most diverse environments, it is also one of the least understood in terms of the identification and ecological functions of the microbiota. A significant amount of the annual $CO_2$ flow to the atmosphere is caused by the activities of heterotrophic soil organisms, which also serve as the greatest repository of organic carbon (C) in the terrestrial biosphere. Global temperatures have risen in step with constantly rising $CO_2$ levels. According to the most recent US national climate assessment [1–4], the climate is expected to continue to change with more unpredictable and intense weather patterns. Since soil microbes play a major part in the cycling of nutrients and soil organic carbon (SOC), they also play a significant part in the production and consumption of greenhouse gases like $CO_2$, $CH_4$, and $N_2O$. However, due to unknown modifications in soil carbon and nitrogen stores, as well as variations in microbial responses between different soil locations, it has been challenging to predict whether soil will act as a source or sink of greenhouse gases under future climate scenarios [5–8]. Therefore, despite the fact that soil microbial ecology is crucial for predicting future climate impacts, integrating it with landscape-scale climate models is still difficult. The fact that soil microbes would mineralize more SOC and significantly increase greenhouse gas ($CO_2$ and $CH_4$) emissions, aggravating warming trends, is one of the main concerns with climate change [9, 10]. This is concerning since the overall amount of soil carbon, including that found in permafrost, is thought to be around 3,300 pentagrams (Pg), which is around five times more than the amount of $CO_2$ present in the atmosphere today [11, 12]. The future growth or decline of this stock of soil carbon is, however, highly unknown according to climate models. Measuring variations in soil respiration has been the main source of empirical data for field studies on climate change. Determining how bulk soil carbon reserves vary with climatic changes is also necessary in order to enhance models of soil carbon-climate feedback [13, 14].

Soil microbes perform the dual functions of mineralizing SOC and stabilizing carbon inputs into organic forms. The net flux of $CH_4$ and $CO_2$ to the atmosphere is controlled by the balance between these two processes. The microbial carbon utilization efficiency is the portion of the carbon substrate that is kept in the microbial biomass as opposed to that which is respired as $CO_2$. Climate change has increased heterotrophic respiration of SOC globally, which has increased atmospheric $CO_2$ inputs [15]. However, higher soil carbon inputs resulting from increased plant growth [16] and autotrophic fixation by soil microbes could offset soil carbon losses to the atmosphere. Additionally, the amount, content, and chemistry of plant litter as well as any pre-existing SOC affect how sensitively SOC decomposes at different temperatures [17]. Thus, even within certain biomes, the local biogeochemical environment has a significant impact on how organisms respond metabolically to climate. In order to improve climate change models, it is imperative to develop a mathematical understanding of the microbial ecology that drives ecosystem carbon use efficiency and the feedback with climate forcing.

## Effects of the Soil Microbiome on the Characteristics of Emerging Ecosystems

The majority of soil microorganisms have developed coping mechanisms to deal with shifting environmental conditions because soil habitats are dynamic systems. The resident microorganisms often adapt, go dormant, or perish when the environment changes. Depending on their genetic and physiological conditions, soil microorganisms respond to environmental stress in various ways [18].

The degree of disruption and the amount of time required to control gene transcription and translation, as well as to amass mutations or new genes through horizontal gene transfer, determine how quickly an organism adapts to change. Quantifying microbial physiological responses, such as drought resistance, dormancy, or reactivation, nonetheless, continues to be a significant challenge in modeling ecological responses to change at the moment [19].

The stability and resistance of the microbial community to future perturbations may change as the community's structure does. The ability of a single species to adapt will be impacted by the interactions between microbial populations in communities as a result of climate change [20]. As a result of differences in how various species react to temperature increases, for instance, their dispersal patterns may shift. It is possible to predict how the soil microbiome will react to various climate change scenarios by focusing on specific functional traits in the soil microbiome, such as the prevalence of fast-growing, opportunistic "r-strategists" as opposed to slow-growing "K-strategists," as well as environmental characteristics [21].

In order to establish a useful baseline for comparison as the climate changes, high-throughput sequencing has proven crucial in exposing the microbial diversity and composition in distinct soil ecosystems. However, it is now understood that compositional data does not always guide function Not every participant in a group, or even every cell within a population, is operational at all times [22]. The complicated interplay of gene regulation primarily controls which genes are expressed and access to resources, controls activity. The soil microbiome's phenotypic response to climate change is impacted by variations in soil moisture, temperature, and local atmospheric chemistry.

Microbial gene expression is induced by the interaction of the heterogeneous genetic potential within the soil microbiome with environmental changes. The metaphenome, which is the microbiome's collective phenotypic output, produces elemental cycling at the ecosystem level [19]. Soil microbiome management in response to climate change and the improvement of climate models depend heavily on our understanding of the factors that link small-scale microbial traits to larger-scale ecosystem responses.

The underlying bacterial-scale mechanisms that regulate environment responses to climate change are currently poorly defined. Instead of reacting to average environmental conditions, soil microorganisms react to sudden microscale variables that set off biochemical pathways, microbial reflexes, and metabolic relationships. Temporal pauses in biogeochemical responses to sudden environmental change are common as soil microorganisms acclimatize. Contrarily, a slow change, like a rise in temperature, gives evolution more time to select for organisms or genotypes that permit endurance to the stress circumstances brought on by the heat. The response of the community is also influenced by its historical background.

## Influence of the Soil Environment on Microbial Responses to Climate Change

It is challenging to generalize the effects of climate change on soil microbiomes across various soil ecosystems due to the vast differences amongst soils in terms of their biotic and abiotic features. There are variations in biogeochemistry within a certain soil class that control the kinds of microorganisms that are present, including pH [23] and salinity [24]. Furthermore, the microbial dwellings and niches [25] that are created in the soil are influenced by its morphology and water content, which has a domino effect on the metabolism of nutrients and carbon. To better understand how species relationships and metabolism are impacted by climate change, it is necessary to study the fine-scale dispersion and interconnectivity of microbial communities in soils [26]. This data is crucial for understanding carbon cycling because how soil bacteria species distribute carbon eventually defines whether or not it persists in soil and how changes in climate alter such processes [27]. It is well recognized that population of microbes communicate and respire $CO_2$, $N_2O$, and $CH_4$ in different soil niches, but the energetics and thermodynamic parameters of the organic carbon electron acceptors that run microbial metabolism are poorly understood in the context of the soil environment. The description of the physiological response surface, or metaphenome, of the microbial communities living in the soils of our planet is the current challenge.

## Effects of Environmental Change

There are many physiological and community responses that soil microorganisms adopt to adapt to the changing environmental conditions brought on by climate change. Due to the varied expected climate change variables across geographic locations, it is impossible to generalize across diverse terrestrial ecosystems, which is why we present some instances to provide context.

## *Raised Carbon Dioxide ($CO_2$)*

Data from a number of $eCO_2$ field studies has been useful in understanding how microbes may come to this impending climate change. Data from a number of $eCO_2$ field studies has been useful in understanding how microbes may come to this impending climate change. In order to contrast prolonged exposure to increased and atmospheric $CO_2$ levels, FACE (free-air $CO_2$ enrichment) experiments been arranged across a variety of ecoregions. The microbiota has changed with $eCO_2$, according to several investigations. Ecosystem-specific responses in addition to typical soil bacterial responses, like with $eCO_2$, acido-bacterial rates increase, found by a one-decade cross-biome investigation [28]. $eCO_2$ led to a shift in archaea and fungus and bacterial strain species in Australian grasslands. Researchers are being diligent to comprehend how ecological characteristics of microbial communities are mirrored by phylogenetic shifts. A foundation for incorporating microbial physiology into ecosystem ecology is provided by a gene-based approach.

By examining the abundance of particular genes in metagenomes, changes in the potential roles played by the soil microbiome under $eCO_2$ have also been identified. By examining the abundance of particular genes in metagenomes, changes in the potential roles played by the soil microbiome under $eCO_2$ have also been identified [29]. In the BioCON grassland experiment, $eCO_2$-stimulated increases in gene families linked to decomposition, nitrogen fixation, and dissimilatory nitrate reduction were observed, while fewer abundances of gene families linked to glutamine formation and anaerobic ammonium oxidation were found. Genes of microbes involved in breakdown, nitrogen fixation, carbon fixation, $CH_4$ metabolism, nitrogen mineralization, and denitrification were all upregulated in arid grasslands exposed to $eCO_2$ [30].

Understanding the changes in gene activities related to the cycling of organic matter in soil (SOM) allows for a better comprehension of how $eCO_2$ affects microorganisms. However, it is still problematic to provide information for globally terrestrial ecosystem models because $eCO_2$ trials have not been conducted widely with duplicate data sets.

The quantification of carbon exchange between the atmosphere and the soil is a key scientific area of plant–microbe connections. Plant biomass, carbon uptake by roots, and soil microbial activity can all be improved by $eCO_2$. An important scientific area for measuring carbon exchange in between environment and the topsoil is plant–microbe relationship. Equivalent $CO_2$ can improve soil microbial activity, carbon sharing to roots, and plant biomass [31–33].

The frequency and pattern of carbon imports to the rhizosphere are influenced by how various species of plants react to elevated $CO_2$. The $eCO_2$-induced rise in rhizodeposition can 'prime' the microbial breakdown of existing SOC [34]. Priming is the process of speeding the degradation of old SOC by introducing new microbiological feedstock, such as production of litter and/or root exudates, both of which could be accelerated by elevated CO. A review consolidating meta-examination and demonstration uncovered that $eCO_2$ at first invigorates photosynthesis and carbon contributions to soil. In any case, over decadal timescales, $eCO_2$ expanded the microbial deterioration of SOM [35, 36]. Anticipating the balance between carbon gathering through mineral affiliation and soil aggregation [37] and sped up decay via priming [38] stays an extra test. This is on the grounds that adjustments of soil carbon stocks are hard to find [39] and the basic science managing SOM deterioration has not been found. The soil may become drier as temperature increases together with a rise in the soil's wetness brought on by elevated $CO_2$ [28]. In the Australian grassland study [28], when $eCO_2$ was linked with warming, there was a decline, even though overall fungal richness expanded under elevated $CO_2$. The supply of water and micronutrients, that affect photosynthesis, microbial breakdown, and the net buildup of carbon sequestration, also affects the indelible effects of elevated $CO_2$ on soil C reserves. Predicting the responses of soil ecosystems' microbiota composition to variations in $CO_2$ necessitates a comprehension of how such changes react with other significant environmental parameters such as temperature, precipitation, and nutrients (such as phosphorus).

## *Elevated Temperature*

The growth rates and outputs of pure microbial cultures are impacted by temperature. The expression of heat shock proteins and alterations in the lipid content of cell membranes, which diminish membrane integrity, are two physiological reactions of microbes to elevated temperature. The growth rates and outputs of pure microbial cultures are impacted by temperature. The expression of heat shock proteins and alterations in the lipid content of cell membranes, which diminish membrane integrity, are two physiological reactions of microbes to elevated temperature. Although technological developments in sequence analysis and functional gene assays have showed colony and functional gene alterations in result of higher temperatures in the fieldwork, evaluating the temperature sensitivity of soil microbes in situ has proven to be more challenging [39, 40]. The biome being examined also influences how the soil microbiome reacts to rising temperatures (for instance, distinguishing between woodland and grassland). For illustration, temperature rise has been demonstrated to have differential effects on soil fungi in various coniferous forest ecosystems, leading to either stimulation [41] or suppression of fungal biomass and activity. These variations are likely caused by variations in soil moisture and/or vegetation at various points [42, 43]. A long-term soil warming experiment was carried out at the Harvard Forest Ecological Research Station Long Term Ecological Research site, wherein soil was thawed by 5 °C above ambient temperature for up to 26 years in order to assess the effects of prolonged soil warming on the soil microbiome of temperate forests [38, 43].

Short-term reductions in microbial biomass and temperature adaptation of soil respiration were implicated for the apparent acclimation of soil respiration [13]. The physiological adaptations of various populations must yet be measured in a field setting. To fill this knowledge vacuum and measure microbial population changes in the field, new isotopic techniques are now available.

The interaction between drought, heat, and plant type ultimately decides how tolerant bacterial communities are to extreme heat. On Wyoming grasslands, the Prairie Heating and $CO_2$ Enrichment (PHACE) experiment investigated the effects of twelve years of elevated $CO_2$ coupled with warming [44]. Under $eCO_2$ itself and in conjunction with warming, genetic variants in the recycling of nitrogen and carbon were amplified. However, heat alone suppressed nitrogen turnover. Variations in precipitation being magnified by the favorable flora community response, which resulted in a rise in biomas [45]. The enhanced plant biomass thereby largely countered the rising carbon loss via respiration, even while warming accelerated both the carbon intake into soil and soil respiration. Collectively, those actions would work to diminish the global warming's positive feedback loop and halt soil C loss. To sum up, whereas most climate analysis shows positive feedback as a result of warming due to increased soil respiration and a decrease in soil storage, there are confounding experimental data that are mostly ecosystem dependent [7, 8, 41].

## References


1. Reidmiller DR, Avery CW, Easterling DR, Kunkel KE, Lewis KL, Maycock TK,Stewart BC (2017) Impacts, risks, and adaptation in the United States: Fourth national climate assessment, volume II (USGCRP, 2018). https://doi.org/10.7930/NCA4.2018
2. Shaban H, Fazeli-Nasab B, Alahyari H, Alizadeh G, Shahpesandi S (2015) An overview of the benefits of compost tea on plant and soil structure. Adv Biores 6(61):154–158. https://doi.org/10.15515/abr.0976-4585.6.1.154158
3. Ali-Soufi M, Shahriari A, Shirmohammadi E, Fazeli-Nasab B (2017) Investigation of biological properties and microorganism identification in susceptible areas to wind erosion in Hamoun wetlands. In: Congress on restoration policies and approaches of Hamoun international wetland Zabol, pp 231–240
4. Ali-Soufi M, Shahriari A, Shirmohammadi E, Fazeli-Nasab B (2017) Seasonal changes biological characteristics of airborne dust in Sistan plain, Eastern Iran. In: International conference on loess research. Gorgan University of Agricultural Sciences and Natural Resources, Gorgan, Iran
5. Friedlingstein P, Cox P, Betts R, Bopp L, von Bloh W, Brovkin V, Cadule P, Doney S, Eby M, Fung I (2006) Climate–carbon cycle feedback analysis: results from the C4MIP model intercomparison. J Clim 19(14):3337–3353. https://doi.org/10.1175/JCLI3800.1
6. Wang K, Peng C, Zhu Q, Zhou X, Wang M, Zhang K, Wang G (2017) Modeling global soil carbon and soil microbial carbon by integrating microbial processes into the ecosystem process model TRIPLEX-GHG. J Adv Model Earth Syst 9(6):2368–2384. https://doi.org/10.1002/2017MS000920
7. Ali-Soufi M, Shahriari A, Shirmohammadi E, Fazeli-Nasab B (2017) Identification and isolation of associated microorganisms with airborne dust loaded over Sistan plain. In: 15th Iranian Soil Science Congress, Isfahan University of Technology, Isfahan, Iran, Congress COI: SSCI15, Article COI: SSCI15_895
8. Ali-Soufi M, Shahriari A, Shirmohammadi E, Fazeli-Nasab B (2019) Investigation of dust microbial community and identification of its dominance species in northern regions of sistan and baluchestan province. J Water Soil Sci (Science and Technology of Agriculture and Natural Resources) 23(1):309–320. https://doi.org/10.29252/jstnar.23.1.23

9. Wieder WR, Bonan GB, Allison SD (2013) Global soil carbon projections are improved by modelling microbial processes. Nat Clim Chang 3(10):909–912. https://doi.org/10.1038/nclimate1951
10. Rasouli H, Popović-Djordjević J, Sayyed RZ, Zarayneh S, Jafari M, Fazeli-Nasab B (2020) Nanoparticles: a new threat to crop plants and soil rhizobia? In: Hayat S, Pichtel J, Faizan M, Fariduddin E (eds) Sustainable agriculture reviews 41: nanotechnology for plant growth and development. Springer International Publishing, Cham, pp 201–214. https://doi.org/10.1007/978-3-030-33996-8_11
11. Tarnocai C, Canadell J, Schuur EA, Kuhry P, Mazhitova G, Zimov S (2009) Soil organic carbon pools in the northern circumpolar permafrost region. Glob Biogeochem Cycles 23(2). https://doi.org/10.1029/2008GB003327
12. Bruhwiler L, Basu S, Butler JH, Chatterjee A, Dlugokencky E, Kenney MA, McComiskey A, Montzka SA, Stanitski D (2021) Observations of greenhouse gases as climate indicators. Clim Change 165(1):1–18. https://doi.org/10.1007/s10584-021-03001-7
13. Bradford MA, Davies CA, Frey SD, Maddox TR, Melillo JM, Mohan JE, Reynolds JF, Treseder KK, Wallenstein MD (2008) Thermal adaptation of soil microbial respiration to elevated temperature. Ecol Lett 11(12):1316–1327. https://doi.org/10.1111/j.1461-0248.2008.01251.x
14. Dhuldhaj UP, Malik N (2022) Global perspective of phosphate soliloquizing microbes and phosphatase for improvement of soil, food and human health. Cell, Mol Biomed Rep 2(3):173–186. https://doi.org/10.55705/cmbr.2022.347523.1048
15. Bond-Lamberty B, Bailey VL, Chen M, Gough CM, Vargas R (2018) Globally rising soil heterotrophic respiration over recent decades. Nature 560(7716):80–83. https://doi.org/10.1038/s41586-018-0358-x
16. Rustad L, Campbell J, Marion G, Norby R, Mitchell M, Hartley A, Cornelissen J, Gurevitch J (2001) A meta-analysis of the response of soil respiration, net nitrogen mineralization, and aboveground plant growth to experimental ecosystem warming. Oecologia 126(4):543–562. https://doi.org/10.1007/s004420000544
17. Nottingham AT, Whitaker J, Turner BL, Salinas N, Zimmermann M, Malhi Y, Meir P (2015) Climate warming and soil carbon in tropical forests: insights from an elevation gradient in the Peruvian Andes. Bioscience 65(9):906–921. https://doi.org/10.1093/biosci/biv109
18. Schimel J, Balser TC, Wallenstein M (2007) Microbial stress-response physiology and its implications for ecosystem function. Ecology 88(6):1386–1394. https://doi.org/10.1890/06-0219
19. Evans SE, Wallenstein MD (2014) Climate change alters ecological strategies of soil bacteria. Ecol Lett 17(2):155–164. https://doi.org/10.1111/ele.12206
20. Berg MP, Kiers ET, Driessen G, Van Der Heijden M, Kooi BW, Kuenen F, Liefting M, Verhoef HA, Ellers J (2010) Adapt or disperse: understanding species persistence in a changing world. Global change Biol 16(2):587–598. https://doi.org/10.1111/j.1365-2486.2009.02014.x
21. De Vries FT, Shade A (2013) Controls on soil microbial community stability under climate change. Front Microbiol 4:265. https://doi.org/10.3389/fmicb.2013.00265
22. Thompson LR, Sanders JG, McDonald D, Amir A, Ladau J, Locey KJ, Prill RJ, Tripathi A, Gibbons SM, Ackermann G (2017) A communal catalogue reveals Earth's multiscale microbial diversity. Nature 551(7681):457–463. https://doi.org/10.1038/nature24621
23. Jansson JK, Hofmockel KS (2018) The soil microbiome—from metagenomics to metaphenomics. Curr Opin Microbiol 43:162–168. https://doi.org/10.1016/j.mib.2018.01.013
24. Fierer N, Jackson RB (2006) The diversity and biogeography of soil bacterial communities. Proc Natl Acad Sci 103(3):626–631. https://doi.org/10.1073/pnas.0507535103
25. Schimel JP, Schaeffer SM (2012) Microbial control over carbon cycling in soil. Front Microbiol 3:348. https://doi.org/10.3389/fmicb.2012.00348
26. Cordero OX, Datta MS (2016) Microbial interactions and community assembly at microscales. Curr Opin Microbiol 31:227–234. https://doi.org/10.1016/j.mib.2016.03.015
27. Dunbar J, Eichorst SA, Gallegos-Graves LV, Silva S, Xie G, Hengartner N, Evans RD, Hungate BA, Jackson RB, Megonigal JP (2012) Common bacterial responses in six ecosystems exposed to 10 years of elevated atmospheric carbon dioxide. Environ Microbiol 14(5):1145–1158. https://doi.org/10.1111/j.1462-2920.2011.02695.x

28. Hayden HL, Mele PM, Bougoure DS, Allan CY, Norng S, Piceno YM, Brodie EL, DeSantis TZ, Andersen GL, Williams AL (2012) Changes in the microbial community structure of bacteria, archaea and fungi in response to elevated $CO_2$ and warming in an Australian native grassland soil. Environ Microbiol 14(12):3081–3096. https://doi.org/10.1111/j.1462-2920.2012.02855.x
29. Allison S (2012) A trait-based approach for modelling microbial litter decomposition. Ecol Lett 15(9):1058–1070. https://doi.org/10.1111/j.1461-0248.2012.01807.x
30. Tu Q, He Z, Wu L, Xue K, Xie G, Chain P, Reich PB, Hobbie SE, Zhou J (2017) Metagenomic reconstruction of nitrogen cycling pathways in a $CO_2$-enriched grassland ecosystem. Soil Biol Biochem 106:99–108. https://doi.org/10.1016/j.soilbio.2016.12.017
31. Yu H, Deng Y, He Z, Van Nostrand JD, Wang S, Jin D, Wang A, Wu L, Wang D, Tai X (2018) Elevated CO2 and warming altered grassland microbial communities in soil top-layers. Front Microbiol 9:1790. https://doi.org/10.3389/fmicb.2018.01790
32. Bréchet LM, Lopez-Sangil L, George C, Birkett AJ, Baxendale C, Castro Trujillo B, Sayer EJ (2018) Distinct responses of soil respiration to experimental litter manipulation in temperate woodland and tropical forest. Ecol Evol 8(7):3787–3796. https://doi.org/10.1002/ece3.3945
33. Abbasi-Moghadam J, Shahriari A, Fazeli-Nasab B (2017) Investigation of bacteria and fungi populations associated with airborne dust during ''wind of 120 days'' blowing in the urban areas of Sistan plain. In: 15th Iranian soil science congress. Isfahan University of Technology, Isfahan, Iran, Congress COI: SSCI15, Article COI: SSCI15_687
34. Qiao N, Schaefer D, Blagodatskaya E, Zou X, Xu X, Kuzyakov Y (2014) Labile carbon retention compensates for $CO_2$ released by priming in forest soils. Glob Change Biol 20(6):1943–1954. https://doi.org/10.1111/gcb.12458
35. Van Groenigen KJ, Qi X, Osenberg CW, Luo Y, Hungate BA (2014) Faster decomposition under increased atmospheric $CO_2$ limits soil carbon storage. Science 344(6183):508–509. https://doi.org/10.1126/science.1249534
36. Drake JE, Gallet-Budynek A, Hofmockel KS, Bernhardt ES, Billings SA, Jackson RB, Johnsen KS, Lichter J, McCarthy HR, McCormack ML (2011) Increases in the flux of carbon belowground stimulate nitrogen uptake and sustain the long-term enhancement of forest productivity under elevated $CO_2$. Ecol Lett 14(4):349–357. https://doi.org/10.1111/j.1461-0248.2011.01593.x
37. Lal R (2004) Soil carbon sequestration impacts on global climate change and food security. Science 304(5677):1623–1627. https://doi.org/10.1126/science.1097396
38. Kuzyakov Y (2010) Priming effects: interactions between living and dead organic matter. Soil Biol Biochem 42(9):1363–1371. https://doi.org/10.1016/j.soilbio.2010.04.003
39. Scharlemann JP, Tanner EV, Hiederer R, Kapos V (2014) Global soil carbon: understanding and managing the largest terrestrial carbon pool. Carbon Manag 5(1):81–91. https://doi.org/10.4155/cmt.13.77
40. Zhang B, Chen S, He X, Liu W, Zhao Q, Zhao L, Tian C (2014) Responses of soil microbial communities to experimental warming in alpine grasslands on the Qinghai-Tibet Plateau. PLoS ONE 9(8):e103859. https://doi.org/10.1371/journal.pone.0103859
41. Heimann M, Reichstein M (2008) Terrestrial ecosystem carbon dynamics and climate feedbacks. Nature 451(7176):289–292. https://doi.org/10.1038/nature06591
42. Zhu X, Liu M, Kou Y, Liu D, Liu Q, Zhang Z, Jiang Z, Yin H (2020) Differential effects of N addition on the stoichiometry of microbes and extracellular enzymes in the rhizosphere and bulk soils of an alpine shrubland. Plant Soil 449(1):285–301. https://doi.org/10.1007/s11104-020-04468-6
43. Yang N, Zhang Y, Li J, Li X, Ruan H, Bhople P, Keiblinger K, Mao L, Liu D (2022) Interaction among soil nutrients, plant diversity and hypogeal fungal trophic guild modifies root-associated fungal diversity in coniferous forests of Chinese Southern Himalayas. Plant Soil 1–14. https://doi.org/10.1007/s11104-022-05646-4

44. Bafana A (2013) Diversity and metabolic potential of culturable root-associated bacteria from origanum vulgare in sub-himalayan region. World J Microbiol Biotechnol 29(1):63–74. https://doi.org/10.1007/s11274-012-1158-3
45. Carol Adair E, Reich PB, Trost JJ, Hobbie SE (2011) Elevated $CO_2$ stimulates grassland soil respiration by increasing carbon inputs rather than by enhancing soil moisture. Global Change Biol 17(12):3546–3563. https://doi.org/10.1111/j.1365-2486.2011.02484.x

# Chapter 9
# Impact of Climate Change on Soil Activity (Nitrifying, Denitrifying) and Other Interactions

**Vishal Hivare, Sonal Kalbande, Rakesh R. Jadhav, and Dattatraya Dalvi**

**Abstract** Though the soil is our motherland, it directly influences quantitative and qualitative crop traits, which determine food security and human health. Unfortunately, it is a complicated environment for microbes, and the anatomy and physiology of microorganisms in soil are immensely complicated. These ambiguities make it difficult to forecast the consequences of climate change on the behavior of soil microorganisms. Drought stress is currently the most severe Impact of climate change and significant, concerning, and dangerous abiotic stresses that cause changes in the soil environment that influence soil organisms such as microbes and plants. It alters the functionality and activity of soil microorganisms in charge of essential ecosystem services and processes. Due to the decrease in microbial activity and production of enzymes (such as oxidoreductases, hydrolases, dehydrogenases, catalase, urease, phosphatases, and glucosidase) and disruption of microbial structure caused by these stress conditions, soil fertility declines, plant productivity falls, and economic loss occurs. To identify more effective strategies for reducing the effects of drought and managing agricultural activities under challenging conditions profitably, a thorough understanding of many factors is needed to address potential approaches like genome editing and molecular analysis (metagenomics, transcriptomics, and metabolomics).

V. Hivare (✉)
International Crops Research Institute for the Semi-Arid Tropics, Patancheru, Telangana, India
e-mail: vishalh2111@gmail.com

S. Kalbande
ICAR-National Institute of Abiotic Stress Management (NIASM), Baramati, MH, India

R. R. Jadhav
Balasaheb Sawant Konkan Krishi Vidyapeeth (Agricultural University), Dapoli, MH, India

D. Dalvi
Vasantrao Naik Marathwada Krishi Vidyapeeth, Parbhani, MH, India



J. A. Parray (ed.), *Climate Change and Microbiome Dynamics*, Climate Change Management, https://doi.org/10.1007/978-3-031-21079-2_9

# Introduction

The most significant threat to human health in the twenty-first century, according to the WHO, is climate change. Modern climate change includes both human-caused global warming and its impact on the Earth's atmospheric circulation. Human activity has caused a 30% increase in the atmospheric concentration of carbon dioxide ($CO_2$), the main greenhouse gas. In addition, plant species' interactions with soil microorganisms are likely to be significantly affected by changes in temperature, ozone, nitrogen deposition, and rainfall patterns [1].

Plant and soil health is essential for all lifestyles on this planet. vegetation displays ecological areas, and flowers reply to climatic variables, including temperature and precipitation. It is likewise nicely understood that plant energy depends on soil traits and fitness and that robust interaction among biota above and below ground govern each domain's functioning [2].

Soil is a wonderful source of medium for plant development and microbial community. Interaction between plants and microbes can be beneficial or harmful based on the climate [3]. Symbiotic or non-symbiotic bacteria and a highly specialized group of fungi are responsible for favourable plant–microbe interactions (mycorrhizal fungi). Beneficial plant-associated bacteria, including those from the genus Azospirillum, the genus Bacillus, the genus Pseudomonas, the genus Rhizobium, the genus Serratia, the genus Stenotrophomonas, and the genus Streptomyces, have been shown to promote plant development and resilience to pathological conditions and abiotic stresses. However, global warming and extreme weather conditions increased $CO_2$ levels and warmth in the atmosphere, hampered microorganism's ability to improve plant development and resistance to infections. It also accelerated the spread and severity of many plant diseases, resulting in the appearance of new lethal mutants and significantly impacting the agricultural system and crop production [4]. Agriculture is regarded as the most sensitive sector to climate change. In the current climate change scenario, utilizing plant–microbe interaction is crucial to increase food production for the population explosion. As individuals, societal action leaders, and researchers with domain expertise, we may work to reverse the current trend.

# Climate Change—A Global Issue

The global development agenda will be influenced and defined by climate resilience attempts to address climate change. However, a climate warming system affects many people's access to necessities, including freshwater, nutrition security, and energy. Climate change and sustainable development are closely related in many ways. Particularly those nations that are least developed and undeveloped will be among those that are most badly impacted and least prepared to handle the anticipated shocks to their social, economic, and environmental systems [5].

The UN Protocol on Climate Change was implemented as part of the "Rio Convention," which was adopted during the Rio Earth Summit in 1992. The political response to climate change on a global scale officially began with this (UNFCCC). The objective of this convention was to prevent "dangerous human interference with the climate system" by outlining a plan for controlling atmospheric greenhouse gas (GHG) concentrations. The COP21/CMP1 Conference of the Parties, which met in Paris, France, in December 2015, adopted the Paris Agreement. This international agreement aims to keep the rise in global temperatures for this century well below 2 degrees Celsius and to support efforts to limit the temperature rise to 1.5 °C above pre-industrial levels.

The Member States reiterate in the 2030 Agenda for Sustainable Development their commitment to halting environmental deterioration and tackling climate change as soon as practicable. The Agenda states that one of the main issues of our day is climate change and claims that it is challenging for all countries to achieve sustainable development due to worries about its negative repercussions. Increasing global temperatures, increasing sea levels, the acidity of the ocean, and other effects of climate change significantly negatively impact coastal regions and low-lying coastal countries, especially those least developed countries and Small Island Developing States. Numerous societies, as well as the planet's biological systems, are in danger of extinction [6].

The World summit on sustainable development Conference's final report, The Future We Want, places a strong emphasis on the immediacy of the global issue of climate change and how it would ultimately influence each nation's capacity to sustain its growth. The study captures the concern of the Member States on the rapidly rising greenhouse gas emissions and the vulnerability of all countries, particularly emerging nations, to the adverse effects of climate change. To execute an acceptable and successful global response to climate change, Member States have asked for the highest level of engagement and cooperation from all nations [7].

## Impact of Climate Change on Plants

The altering environmental conditions affect all living beings within the civilization [8]. Ecological changes impact the terrestrial and worldwide distribution of numerous crops and their yields. Changing climatic circumstances have improved the productivity of plants cultivated in higher latitudes like maize, wheat, and sugar beets while decreasing the productivity of plants grown in numerous lower latitudes like maize and wheat [9]. Numerous studies show that between 1980 and 2008, global wheat and maize yields declined by 5.5% and 3.8%, respectively, compared to their yield forecasts assuming steady climatic circumstances [10].

Numerous climatic conditions are known to impact the growth and productivity of plant systems. Physical characteristics are typically incorporated, such as temperature, rainfall patterns, $CO_2$ levels, changes in agricultural environments, and the adaptability of humanoid groups. Temperature is the most critical aspect of changing environmental conditions because of its apparent nature. Its impacts on the growth of the plant system are only fully comprehended up to the best levels for crop development. Some crops may benefit from the increase in warmth and carbon dioxide levels, but only to a limited extent. For example, crops like wheat and soybeans might benefit from greater $CO_2$ levels when cultivated at appropriate temperatures [11].

Consequently, changing climatic conditions might be advantageous for plant systems, yet, abrupt shifts in environmental factors endanger plant systems. However, the favourable impacts of shifting climatic conditions on plant yields have been predicted to exceed the negative ones until 2030, after which any additional amplification of climatic change will mostly have a negative effect. Consequently, maize, wheat, and rice yields will all suffer in the second half of the twenty-first century, with tropical countries suffering more than temperate ones [12].

## Global Agricultural Ecosystem and Extreme Climate Events

One of the main factors contributing to climate change and the greenhouse effect is the large number of greenhouse gases released by the agricultural sector. Contrarily, climate change considerably impacts agricultural production and risks food security. According to the World Food Programme, people should always have access to an adequate supply of safe and wholesome food to satisfy their dietary demands and food choices. Currently, a food shortage poses the most significant risk to food security. More than 10% of the world's population is underweight even though there is enough food to feed everyone [13]. Climate change is predicted to exacerbate food poverty by increasing food prices and lowering output. The fight against climate change may result in higher food prices. The scarce water supply for food production is strained by drought and increased agricultural water demand. There may be more land competition in areas where the climate is unfavorable for agriculture. Price increases for crops may result from extreme weather phenomena linked to climate change [14].

Agriculture is the industry most at risk from climate change because of its size and susceptibility to weather changes. Changes in temperature and rainfall significantly impact the amount of food that can be cultivated. Temperature, precipitation, and $CO_2$ fertilization affect various crops, locations, and changing things. Warmer temperatures reduce yield, but more rain will likely alleviate this issue [15].

Climate change affects agricultural productivity depending on where you reside and your irrigation type. Extra irrigations may harm the environment, yet they may also increase agricultural productivity [16]. Temperature increases are pretty likely to shorten crop length, reducing agricultural production. Wheat, rice, and maize production are anticipated to fall as it is predicted that temperature will rise by 2 °C in temperate and tropical regions over the next few decades. This indicates that tropical crops are more vulnerable to climate change since they are closer to their high-temperature optimums, making them more susceptible to stress from high temperatures [17].

Insect pests and diseases thrive in warm, moist environments. They all impact how much food we can grow due to factors such as temperature, rainfall, wind speed, and humidity, and their absence could have resulted in an overestimation of the costs of climate change [18]. Due to climate change, droughts are anticipated to worsen in most parts of the world. Drought-affected areas are expected to increase from 15.4 to 44% by 2100. Africa is regarded to be the most vulnerable continent. Because of the dry weather, arid areas are anticipated to lose more than half of their food output by 2050 and more than 90% by 2100 [19].

This year, many people in India may experience temperature surges ranging from 2.33 to 4.78 °C. Climate change would lower food production in many Sub-Saharan African communities by 6–24% during the next few decades. Solomon Islanders are expected to consume more seafood than they produce by 2050 [20]. This is because they are expected to consume more fish than they produce. $CO_2$ levels in the atmosphere should increase agricultural output. During heat waves, $CO_2$ levels will double and stay higher for longer. This could be detrimental to the farming industry. The intensity of climate change's effects on tropical areas of impoverished countries will be dictated by where they are and how hot it is. According to agricultural estimates based on resource and environmental research, wheat and rice yields in northwest India could grow by 28% and 15%, respectively, if $CO_2$ levels rose twice as much as they do currently. Non-leguminous $C_3$ crops grown in high $CO_2$ circumstances have reduced N, Fe, Zn, and S levels, all of which are found in proteins [21]. Weather changes have increased the number of bacteria and enzymes in the soil. There were many more bacteria in the temperature gradient tunnel when the temperature was 4–5 °C higher than in the field, but not as many in the area. This happens when there is a lot of $CO_2$ in the atmosphere. When temperatures hit 29 °C, rice crops develop more quickly, vegetatively and reproductively, and produce more seeds. However, as the temperature rose, the seeds did not set as well as they had previously [22].

## Plants and Microbe Interaction in Response to Climate Change

Plants and a range of taxonomically organized microbial communities are closely related. The microbiome (microbiota and their genomes), composed of bacteria, fungus, protists, nematodes, and viruses, colonizes all exposed plant tissues. The host plant interacts intricately and dynamically with the microbiome in the soil, rhizosphere, roots, and other plant tissues. The environment substantially impacts these interactions, which can improve a plant's resistance to environmental dangers. Despite advances in our consideration of the role of the microbiome in plant development and health, there are still many obstacles to overcome before we can harness microbial connections and features to increase plant flexibility to climate change. External factors, including temperature, moisture content, and nutrient status, can impact the interactions between symbiotic and pathogenic plant microbes. Therefore, it is crucial to understand how climatic conditions affect plant–microbe interactions to anticipate disease outbreaks, develop effective symbioses and biocontrol agents, and create agricultural systems more resilient to climate change [23].

## *Pathogen-Plant Interaction*

Three-way interactions between the environment, the host, and the pathogen, which operate on a scale from resistance to sickness, affect plant health and productivity. The quantity and behavior of pathogens, host–pathogen interplay, and the formation of novel diseases could all be affected by climate change [24]. As global temperatures rise, many plant infections are predicted to spread proportionately more widely [25]. To make matters worse, several commonly employed treatments for diseases don't work well in hot climates [26]. Dryness and high temperatures can weaken ETI (Effector Triggered Immunity) and cause disease in various plant pathobiology [27]. Most research on how climate change affects host-disease interactions has relied on overly simplistic models that only account for one host plant and one pathogen.

In contrast, the interaction and rivalry of the pathobiota and other members of the plant microbiome influence the development of pathogens. In contrast, plants interact in their natural habitat with various potentially harmful microbes [28]. We still don't know how the pathobiota and plant microflora will interact in response to ongoing abiotic stressors.

## *Positive Plant–Microbe Interactions*

Climate change will impact beneficial plant–microbe interactions in a variety of ways. For example, warming might decrease the amount of available photosynthate below ground, restricting the size and diameter of roots [29]. Therefore, it is preferable to use arbuscular mycorrhizal fungus (AMF) species with reduced needs for carbon (C) as they are less prone to colonize roots [30]. Abiotic stresses can have adverse effects on plants. However, some plant microbiome inhabitants have characteristics that mitigate those effects. Extracellular polymeric substances (EPS), which can form hydrophobic biofilms that protect plants from desiccation, are a few examples. Another is the production of 1-aminocyclopropane-1-carboxylate (ACC) deaminase, which enhances stress tolerance by controlling ethylene levels in plants. For instance, a novel mechanism for how heat shock factor A2 (HSFA2) induces thermotolerance in plants methylates heat stress memory genes. It enhances thermotolerance in plants when HSFA2 is produced persistently through the ethylene signaling pathway and the transcription factor EIN3 [31]. It's even conceivable that some bacteria that aid in plant growth may also help plants overcome various challenge [32]. It is likely that multiple microbiome pathways that may be active simultaneously improve plant performance under stress. However, our knowledge of the interconnected molecular pathways that start the series of interactions between plants and the microbiota associated with climate change is insufficient (Table 9.1).

## Nitrifying and Denitrifying Interactions

The consequences of the global shift on belowground nitrogen (N) cycle activities affect plant populations, productivity, and trace gas effluxes. However, few in vivo studies have looked at how different global change components interact to affect nitrification or denitrification.

Over 4 years, the interplay between the nitrifying and denitrifying enzyme activities (NEA and DEA) in an annual grassland ecosystem in response to various aspects of climate change (rising atmospheric $CO_2$ concentration, temperature, precipitation) has studied [33]. To shed insight on the mechanisms behind NEA and DEA's response to environmental change, they looked at the correlations between these activities and soil moisture, microbial biomass C and N, and soil extractable N. Elevated $CO_2$ reduces NEA activity across all examined climate change components and their interactions with other treatments. NEA was unaffected by temperature changes or precipitation. Temperature increase had no discernible impact on DEA.

The duration of climate change affected highland grassland fields, $N_2O$ fluxes and related microbial enzymatic activity, microbial population abundance, and community diversity have been studied [34]. Warming, summer drought, and high $CO_2$ benefitted $N_2O$ fluxes, nitrification, $N_2O$ release through denitrification, and the population size of $N_2O$ reducers and $NH_4$ oxidizers. In situ, $N_2O$ changes were more closely related to microbial population increase in warmer environments than in the control site.

Barnard et al. investigated how NEA and DEA, soil microbial N, and soil organic N responded to increased $CO_2$ in the European grasslands. The study revealed that increasing CO2 had little to no effect on soil extractable [$NH_4^+$] and [$NO_3$], NEA, DEA, and microbial biomass N, DEA, and NEA at some sites. However, it was predicted that DEA and soil [$NO_3$] would decline by 22 and 45% in French grasslands, respectivel [35].

## Alteration in Microbial Distribution

It is generally known that plant communities react to climate changes and that these reactions can change how plants are distributed in space. Several studies have made assumptions about possible alterations in the habitats of numerous plant species under extreme climatic condition [36]. However, there aren't many publications that discuss how allied soil bacteria may alter the host distribution to maintain a good or bad relationship with the host plants. It has been found that plants adapt

**Table 9.1** Types of microbial interactions that can enhance plant uptake of N and related biological processes

| Phylum | Family | N-associated biological process | Specificity | The efficiency of plant N nutrition improves | Intracellular versus extracellular | Specific cellular structure | Bacterial taxa | Plant taxa |
|---|---|---|---|---|---|---|---|---|
| Bacteria | Rhizobia | N fixation | High | High | Intracellular | Nodule | Rhizobium (alpha proteobacteria) Gram-negative | *Fabaceae* |
| | | N fixation | High | High | Intracellular | Nodule | Rhizobium (alpha proteobacteria) Gram-negative | *Parasaponia* spp. |
| | | N fixation | High | High | Intracellular | Nodule | *Frankia* spp. | Actinorhizal plants |
| | | N fixation | Wide range | High | Intracellular/Extracellular | Heterocyst | *Nostoc* spp. | Aquatic plants |
| Fungi | Arbuscular Mycorrhizal Fungi | N uptake stimulation | Wide range | Low/High | Intracellular/Extracellular | Arbuscles | Glomeromycota | Angiosperms |

*Source* Dellagi, A., Quillere, I. & Hirel, B. Beneficial soil-borne bacteria and fungi: A promising way to improve plant nitrogen acquisition. *J. Exp. Bot.* **71**, (2020)

to changing climatic circumstances more quickly than soil-native microbes due to their superior dispersion capabilities. At the level of local communities, there is a shortage of knowledge on microbial dispersal, which only helps to increase worry. Few changes have been caused by scattering in essential microbial functions like a breakdown. However, modifications to plant and microbe dispersion capacities can influence plant establishment, production, and communication within a community, for instance, by changing the input predominance of plant litter [37].

Although it is well known that microbiological species also respond to climate changes, it is usually unknown how quickly or frequently isolated microbiological groups may adapt to climatic changes. Therefore, it is still necessary to answer the questions, such as how much microbiological dispersal restraint matters for ecosystem purposes and how rapidly microbial systems will acclimatize to changing environment. By altering their distribution within the soil systems, the microbial communities that live there may respond to the strain brought on by climate changes. For instance, in search of the ideal thermal range, the higher soil surface temperatures may cause soil bacteria to move deep within the soil profile. This type of microbiota reclassification in soil systems can potentially modify plant–microbe process relations. It is yet unknown to what extent interactions between microorganisms and plants due to direct and indirect effects of climate change may still be necessary for ecosystem functioning. Viral, bacterial, and cyanobacterial members will be more prevalent in future sub-Antarctic zone waters due to shallow mixed layers and rising iron levels. As a result of the region's iron restriction, autotrophic and heterotrophic bacterial and viral populations have declined in the waters of the Polar Frontal Zone. An increase in the number of bacteria in heated plots with higher $CO_2$ proportions has been noticed, but a decrease in bacterial abundance in heated plots with ambient $CO_2$ levels. The relative richness of Acidobacteria and Proteobacteria was affected by variations in rainfall, with Acidobacteria falling and Proteobacteria increasing in wet treatments compared to dry ones [38].

## *Plant–Microbe Communication*

There is a communication pathway between the bacteria and the host plant. Plants release compounds under stress that attracts microorganisms that can boost plant resistance [39]. For instance, actinobacteria are enriched with the genetic ability to transport and utilize glycerol-3-phosphate (G3P) for growth due to glycerol-3-phosphate (G3P) secretion caused by root dryness [40]. Drought decreases the quantity of iron and phytosiderophores available in the rhizosphere, allowing for Actinobacteria enrichment, which may thrive in low iron settings, improving their fitness advantage and capacity to encourage plant development. The host phenotypic plasticity that the plant microbiome also influences can impact plant phenology in a changing climate [41]. For instance, rhizosphere bacteria can regulate the flowering time by modifying the nitrogen (N) cycle and converting the amino acid tryptophan in root exudates to the phytohormone indoleacetic acid [42].

Furthermore, plants communicate with insects, nematodes, and bacteria using volatile organic compounds (VOC). It is suggested that variations in the plant immune system or the host's stress signalling network may be related to variations in the microbiome's makeup caused by drought and warmth that are mediated by root exudates. VOC emissions are increasing due to climate change. To increase plant resistance to climatic stresses, it is essential to comprehend the molecular interactions that abiotic stresses have with metabolites to change the composition and efficiency of the plant microbiome (Fig. 9.1).

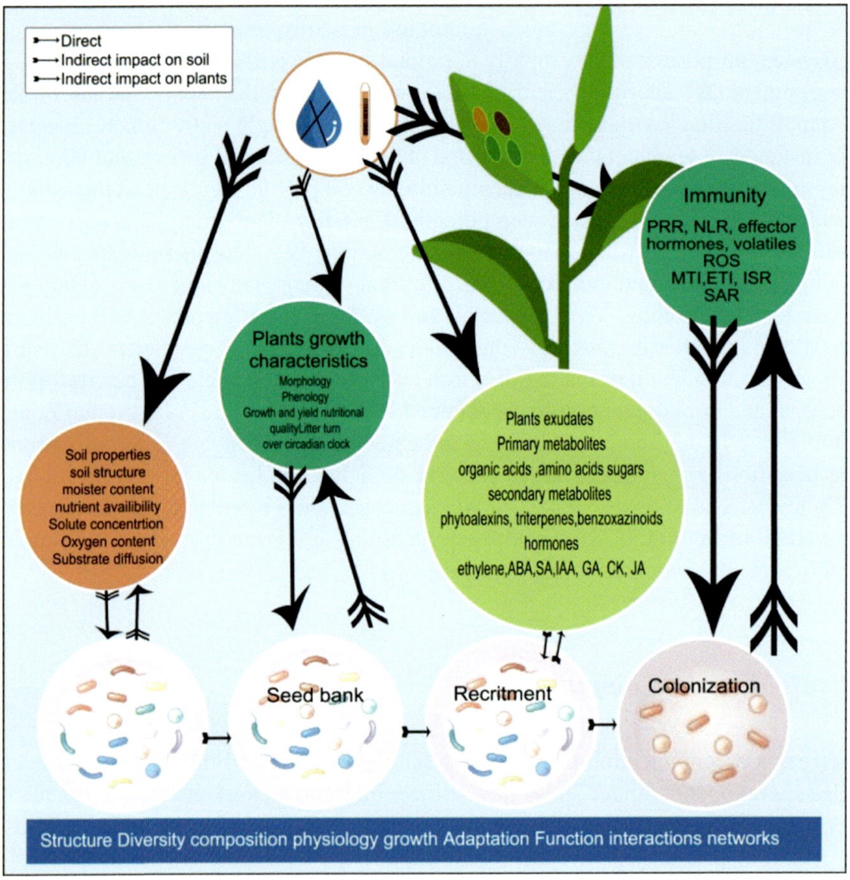


**Fig. 9.1** Impact of climate change on the plant-associated microbiome. *Source* P. T. B. D. B. K. E. B. B. K. S. (2022). *Plant-microbiome interactions under a changing world: responses, consequences and perspectives*. Pankaj Trivedi 1, Bruna D Batista 2, Kathryn E Bazany1, Brajesh K Singh 2 3. https://pubmed.ncbi.nlm.nih.gov/35118660/]

## Climate Change Mitigation and Adaptation Strategies

Farmers' assessments of the severity and threat of climate change serve as the primary drivers of voluntary mitigation. However, the accessibility of crucial information affects the adaption [43]. The number of people who experience water stress will also decrease due to mitigating measures, but those who do will still need adaptation techniques because of the increased stress [44]. Farmers can apply climate-resilient technology by combining conventional and agro-ecological management strategies, such as biodiversification, soil management, and water harvesting. These management strategies result in resilient soils and cropping systems, which boost carbon sequestration, improve soil quality and health, and reduce soil erosion, all of which help ensure food security in the face of climate change [45].

The most successful educational initiatives for raising awareness of climate change for ecological development focus on regional, practical, and local aspects and may be monitored by individual behaviour [46]. The fact that most farmer's favoured adaptations but a tiny percentage favoured GHG reductions highlights the need to focus on programmes with both adaptation and mitigation components. The three main adaptive mitigation strategies are cropping system technologies, resource-conservation technologies, and socioeconomic or policy interventions. Due to a lack of information, small and marginal farmers are less able to adapt to climate change, making them more vulnerable to losses [47]. A lack of management measures and financial repercussions make farmers in African nations particularly susceptible to climate change. Changes in sowing dates are just one agronomic tactic that can be utilized to lessen the consequences of climate change. Simple strategies to cut GHG emissions include alternate rice drying, mid-season drainage, better feeds for cattle, improved N-use efficiency, and soil carbon. The ability of the agroforestry sector to lower atmospheric GHG concentrations and assist small farmers in Kenya in their adaptation to climate change can be advantageous. The use of alternate rice drying, mid-season drainage, better feeds for cattle, improved N-use efficiency, and soil carbon are a few simple ways to lower GHG emissions. Simple adaptation strategies to mitigate climate change's consequences include modifying planting dates and cultivars. The diffusion of technology will significantly impact farmers' responses to climate change. The primary priorities are capacity building, public research assistance, and market integration.

Technologies that maintain soil structure deliver nutrients or water, or both, are most beneficial in reducing climate change. In semi-arid West Africa, it has been demonstrated that Zai, stone bunds, half-moons, and the application of nutrients are appropriate technologies for preserving food production and safeguarding smallholder farmers [48]. In Punjab, Pakistan, studies on climate-smart agriculture practices showed that cotton yield increased with higher returns and more efficient resource utilization. However, the climate is changing, which severely impacts the ability to grow rice and wheat. The Indo-Gangetic plain is particularly vulnerable [49]. Nevertheless, farmers have indicated that they are receptive to utilizing climate-smart agriculture practices that can substitute more profitable farming techniques for traditional ones. The most popular CSA technologies in the western Indo-Gangetic Plains (IGP) are direct sowing, LLL, zero tillage, crop insurance, and irrigation scheduling [50].

In contrast, weather warning services, crop insurance, and laser land levelling (LLL) are most popular in the eastern Indo-Gangetic Plains (IGP). These mitigating strategies have significant potential for flexibility and mitigation. However, they depend on various elements, such as a technology's relevance to the field, public perception, commercial viability, and technical complexity. These techniques perform best when several interventions are employed in conjunction with one another [51].

## Conclusion and Future Perspective

All higher organisms, including those in the plant kingdom, have their origins in the microbial world. Both plants and microbes have developed a few ways to enhance their health. However, plants and microorganisms have developed in specific environments and can only withstand a certain amount of environmental change. In addition to exceeding their tolerance limit, the difference in the climate stresses out microorganisms, reducing both their productivity and the ecological function given to them. Rapid change is constantly testing plants' fitness and operational effectiveness and microbial systems in the world's climatic circumstances. Every conceivable ecological process is recognized to be primarily driven by microbial systems. Extreme weather conditions are known to interfere with these activities, disturbing the functioning of microorganisms. The modification of these processes is also known to interfere with plant productivity, which reducing agricultural output might soon result in a state of food insecurity. Therefore, repairing ecosystem harm brought on by climatic change and further halting these constantly shifting conditions may be practical tools in overcoming this obstacle. Restoration of arable and degraded lands can remove up to 51 gigatons of $CO_2$ from the atmosphere, which can further help increase food production by 17.6 megatons annually. Reducing water use in the agriculture sector without sacrificing agricultural output would also help attain a milestone toward acclimatizing to shifting climatic conditions since agricultural inputs account for 70% of freshwater extractions. Additionally, reducing human intervention and implementing sustainable techniques like afforestation can help limit the effects of climate change.

To conclude this study, we would like to emphasize that despite our focus on how temperature, circadian rhythm, moisture, and nutrients affect plant–microbe interactions, other environmental factors, most notably atmospheric $CO_2$ concentration, have attracted increasing consideration. Furthermore, there are innumerable instances of how the environment affects relationships between animals and microbes. These include (1) the Impact of ultraviolet radiation (UV-R) on the skin microbiome; (2) the disruption of the circadian clock by the gut microbiome; (3) the effects of climate change on the frequency and severity of viral diseases affecting marine animals as well as coral reef bleaching; (4) the role of nutrition in animal immunity. There are probably critical cross-kingdom principles that have not yet been discovered. The study of how climate affects host-microbe interactions in both the plant and animal kingdoms has a more significant impact on our comprehension of how current and future host-microbe interactions in both the plant and animal realms may therefore are influenced by global climatic conditions.

## References

1. Climate change. https://www.who.int/health-topics/climate-change. Accessed 06 Aug 2022
2. Kibblewhite MG, Ritz K, Swift MJ (2008) Soil health in agricultural systems. Philos Trans R Soc Lond B Biol Sci 363(1492). https://doi.org/10.1098/rstb.2007.2178
3. Jacoby R, Peukert M, Succurro A, Koprivova A, Kopriva S (2017) The role of soil microorganisms in plant mineral nutrition—current knowledge and future directions. Front Plant Sci 8. Accessed 06 Aug 2022. [Online]. Available: https://www.frontiersin.org/articles/https://doi.org/10.3389/fpls.2017.01617
4. Velásquez AC, Castroverde CDM, He SY (2018) Plant and pathogen warfare under changing climate conditions. Curr Biol CB 28(10):R619. https://doi.org/10.1016/j.cub.2018.03.054
5. AR4 Climate Change 2007: Synthesis Report—IPCC. https://www.ipcc.ch/report/ar4/syr/. Accessed 06 Aug 2022
6. Martin, "Climate Action," United Nations Sustainable Development. https://www.un.org/sustainabledevelopment/climate-action/. Accessed 06 Aug 2022
7. Report of the World Summit on Sustainable Development, Johannesburg, South Africa, 26 August-4 September 2002. Accessed 06 Aug 2022. [Online]. Available: https://digitallibrary.un.org/record/478154
8. Aristide A (2020) Critical agents of change at earth's surface. Eos. http://eos.org/science-updates/critical-agents-of-change-at-earths-surface. Accessed 07 Aug 2022
9. FAO publications catalogue (2021) FAO, 2021. https://doi.org/10.4060/cb4402en
10. Lobell D, Bänziger M, Magorokosho C, Vivek B (2011) Nonlinear heat effects on African maize as evidenced by historical yield trials. Nat Clim Change 1. https://doi.org/10.1038/nclimate1043
11. Porter JR et al (2022) Food security and food production systems, Report, 2014. Accessed 07 Aug 2022. [Online]. Available: https://cgspace.cgiar.org/handle/10568/68162
12. Challinor AJ, Watson J, Lobell DB, Howden SM, Smith DR, Chhetri N (2014) A meta-analysis of crop yield under climate change and adaptation. Nat Clim Change 4(4):287–291. https://doi.org/10.1038/nclimate2153
13. Food assistance: cash and in-kind | World Food Programme. https://www.wfp.org/food-assistance. Accessed 07 Aug 2022
14. Doney S et al (2014) Chapter 24: Oceans and Marine Resources. Climate Change Impacts in the United States: The Third National Climate Assessmen, U.S. Global Change Research Program. https://doi.org/10.7930/J0RF5RZW
15. Raza A et al (2019) Impact of climate change on crops adaptation and strategies to tackle its outcome: a review. Plants 8:34. https://doi.org/10.3390/plants8020034
16. Chapter 7: Environmental considerations in irrigation development. https://www.fao.org/3/W4347E/w4347e10.htm. Accessed 07 Aug 2022
17. Chapter twenty-four predicted effects of climate change on agriculture: a comparison of temperate and tropical regions. http://www.ciesin.columbia.edu/docs/004-145/004-145.html. Accessed 07 Aug 2022
18. Skendžić S, Zovko M, Živković IP, Lešić V, Lemić D (2021) The impact of climate change on agricultural insect pests. Insects 12(5). https://doi.org/10.3390/insects12050440

19. Guest Article: Droughts Will Change Our World Unless We Act Now | SDG Knowledge Hub | IISD. https://sdg.iisd.org/commentary/guest-articles/droughts-will-change-our-world-unless-we-act-now/. Accessed 07 Aug 2022
20. Dey M, Gosh K, Santos RA, Rosegrant M, Chen O (2016) Economic impact of climate change and climate change adaptation strategies for fisheries sector in Solomon Islands: Implication for food security. Mar Policy 67. https://doi.org/10.1016/j.marpol.2016.01.004
21. Attri SD, Rathore L (2003) Simulation of impact of projected climate change on wheat in India. Int J Climatol 23:693–705. https://doi.org/10.1002/joc.896
22. Chapter 3 : Desertification—Special Report on Climate Change and Land. https://www.ipcc.ch/srccl/chapter/chapter-3/. Accessed 07 Aug 2022
23. Plant–microbiome interactions under a changing world: responses, consequences and perspectives—Trivedi—2022—New Phytologist—Wiley Online Library. https://nph.onlinelibrary.wiley.com/doi/full/https://doi.org/10.1111/nph.18016. Accessed 07 Aug 2022
24. Cohen SP, Leach JE (2020) High temperature-induced plant disease susceptibility: more than the sum of its parts. Curr Opin Plant Biol 56:235–241. https://doi.org/10.1016/j.pbi.2020.02.008
25. Delgado-Baquerizo M et al (2020) Multiple elements of soil biodiversity drive ecosystem functions across biomes. Nat Ecol Evol 4(2):210–220. https://doi.org/10.1038/s41559-019-1084-y
26. Climate change and disease in plant communities | PLOS Biology. https://journals.plos.org/plosbiology/article?id=https://doi.org/10.1371/journal.pbio.3000949. Accessed 07 Aug 2022
27. Cheng YT, Zhang L, He SY (2019) Plant-microbe interactions facing environmental challenge. Cell Host Microbe 26(2):183–192. https://doi.org/10.1016/j.chom.2019.07.009
28. Bartoli C et al (2018) In situ relationships between microbiota and potential pathobiota in Arabidopsis thaliana. https://doi.org/10.1101/261602
29. Warming and elevated ozone induce tradeoffs between fine roots and mycorrhizal fungi and stimulate organic carbon decomposition | Science Advances. https://www.science.org/doi/https://doi.org/10.1126/sciadv.abe9256. Accessed 07 Aug 2022
30. Ma Z et al (2018) Evolutionary history resolves global organization of root functional traits. Nature 555(7694):94–97. https://doi.org/10.1038/nature25783
31. Root endophyte induced plant thermotolerance by constitutive chromatin modification at heat stress memory gene loci | EMBO reports. https://www.embopress.org/doi/full/https://doi.org/10.15252/embr.202051049. Accessed 07 Aug 2022
32. Induction of abiotic stress tolerance in plants by endophytic microbes—PubMed. https://pubmed.ncbi.nlm.nih.gov/29359344/. Accessed 07 Aug 2022
33. Barnard R et al (2006) Several components of global change alter nitrifying and denitrifying activities in an annual grassland. Funct Ecol 20(4):557–564
34. Cantarel A et al (2012) Four years of experimental climate change modifies the microbial drivers of $N_2O$ fluxes in an upland grassland ecosystem. Glob Change Biol 18:2520–2531. https://doi.org/10.1111/j.1365-2486.2012.02692.x
35. Barnard R et al (2004) Atmospheric $CO_2$ elevation has little effect on nitrifying and denitrifying enzyme activity in four European grasslands. Glob Change Biol 10:488–497. https://doi.org/10.1111/j.1529-8817.2003.00746.x
36. Pauli G, Grabherr M, Gottfriedand H (1994) Climate effects on mountain plants. Nature 369(6480). https://doi.org/10.1038/369448a0
37. Rooting theories of plant community ecology in microbial interactions—PubMed. https://pubmed.ncbi.nlm.nih.gov/20557974/. Accessed 07 Aug 2022
38. Evans C et al (2011) Potential climate change impacts on microbial distribution and carbon cycling in the Australian Southern Ocean. Deep-Sea Res Part Ii-Top Stud Oceanogr-DEEP-SEA RES PT II-TOP ST OCE 58:2150–2161. https://doi.org/10.1016/j.dsr2.2011.05.019
39. A genome-wide screen identifies genes in rhizosphere-associated Pseudomonas required to evade plant defenses | bioRxiv. https://doi.org/10.1101/375568v1.full. Accessed 07 Aug 2022
40. Xu J et al (2018) The structure and function of the global citrus rhizosphere microbiome. Nat Commun 9(1):4894. https://doi.org/10.1038/s41467-018-07343-2

41. Dastogeer K, Tumpa F, Sultana A, Akter M, Chakraborty A (2020) Plant microbiome–an account of the factors that shape community composition and diversity. https://doi.org/10.1016/j.cpb.2020.100161
42. Lu T et al (2018) Rhizosphere microorganisms can influence the timing of plant flowering. Microbiome 6(1):231. https://doi.org/10.1186/s40168-018-0615-0
43. Altieri MA, Nicholls CI (2017) The adaptation and mitigation potential of traditional agriculture in a changing climate. Clim Change 140(1):33–45
44. Lal R, Delgado J, Groffman P, Millar N, Dell C, Rotz CA (2011) Management to mitigate and adapt to climate change. J Soil Water Conserv 66:276–285. https://doi.org/10.2489/jswc.66.4.276
45. Del Prado A et al (2014) Synergies between mitigation and adaptation to Climate Change in grassland-based farming systems, pp. 61–74
46. Climate change education | UNESCO. https://www.unesco.org/en/education/sustainable-development/climate-change. Accessed 07 Aug 2022
47. Hatfield J et al (2014) Chapter 6: Agriculture. Climate Change Impacts in the United States: The Third National Climate Assessment, U.S. Global Change Research Program. https://doi.org/10.7930/J02Z13FR.
48. Zakari S, Ouedraogo M, Abasse T, Zougmore R (2019) Farmer's prioritization and adoption of climate-smart agriculture (CSA) Technologies and Practices
49. Singh Y, s Jat H, Jat S (2021) Wheat productivity enhancement through climate smart practices, pp 255–268. https://doi.org/10.1016/B978-0-12-821316-2.00015-7
50. Imran MA, Ali A, Ashfaq M, Hassan S, Culas R, Ma C (2019) Impact of climate smart agriculture (CSA) through sustainable irrigation management on Resource use efficiency: a sustainable production alternative for cotton. Land Use Policy 88:104113. https://doi.org/10.1016/j.landusepol.2019.104113
51. Taneja G, Pal B, Joshi P, Aggarwal PK, Tyagi N (2019) Farmers' preferences for climate-smart agriculture—an assessment in the indo-gangetic plain, pp 91–111. https://doi.org/10.1007/978-981-10-8171-2_5.

# Chapter 10
# Soil Microbial Biochemical Activity and Influence of Climate Change

**Madiha Farooq, Asma Nazir, Shahnaz Anjum, Bisma Farooq, and Shayista Yousuf**

**Abstract** Climate change, particularly temperature rise and increased carbon dioxide ($CO_2$) concentration, is a major source of concern nowadays. Inter-annual climate variability is noticeable and has a big impact on agricultural production. The abundance and activity of beneficial soil microorganisms, which aid in the decomposition of organic matter and the determination of plant nutrient availability, have an impact on soil productivity. It is critical to reduce $CO_2$ and other major greenhouse gas (GHG) emissions by implementing various strategies in land use planning and increasing soil organic matter by employing various techniques that will not only aid in reducing greenhouse gas emissions and mitigating the impact of climate change on the beneficial soil microbial community but will also provide additional benefits to farmers in the form of reduced labour, costs, and grain yields. Changes in land use and human activities have had a substantial impact on gaseous nitrogen (N) losses and the global nitrogen cycle in recent decades, contributing to regional and global atmospheric changes. Microbial activity (nitrifiers and/or denitrifies) and abiotic variables, such as soil temperature, oxygenation, mineral nitrogen, pH, carbon availability, and water content, all influence $N_2O$ emissions. As a result, knowing how microbial and environmental variables interact is crucial for estimating potential $N_2O$ fluxes from soils under climate change.

M. Farooq
Department of Microbiology, Asian International University, Imphal, Manipur 795113, India

A. Nazir
Department of Clinical Biochemistry, University of Kashmir, Hazratbal, Srinagar 190006, India

S. Yousuf
Department of Botany, Lovely Professional University, Phagwara, Punjab 14411, India

S. Anjum (✉)
Department of Biomedical Engineering, Satyabhama Institute of Science and Technology, Chennai 600119, India
e-mail: shinaz.anjum@gmail.com

B. Farooq
Department of Zoology, University of Kashmir, Hazratbal, Srinagar 190006, India

 
J. A. Parray (ed.), *Climate Change and Microbiome Dynamics*, Climate Change Management, https://doi.org/10.1007/978-3-031-21079-2_10

## Introduction

Worldwide changes, for example, warming are straightforwardly adjusting microbial soil breath rates since soil microorganisms, and the cycles they intervene, are temperature delicate. More than 100 years back Svante Arrhenius anticipated that proceeding with the ignition of non-renewable energy sources would prompt a multiplying of carbon dioxide in the environment and related environment warming [1–3]. Despite this advance notice, we are presently confronted with the anticipated multiplying of air carbon dioxide and worldwide temperature increment of 1.3 °C before this century's over if no approach changes are made [3, 4]. Besides, in addition to the fact that we are confronted with climbing worldwide temperatures moving atmospheric conditions, sea fermentation, and the likely loss of numerous species on the planet (Intergovernmental Panel on Climate Change. These elements will all uniquely affect land use, land cover, soil quality, and efficiency. As the environment changes perpetually, it turns out to be additional essential to figure out potential responses from soils to the environmental framework. It's undeniably true that microorganisms, which are related to plants, may animate plant development and improve protection from infection and abiotic stresses. The impacts of environmental change factors, like raised $CO_2$, dry spells, and temperature on valuable plant-microorganism associations are progressively being investigated [5–8]. Organic entities live working together with a huge number of different species, for example, a few helpful and pathogenic species which significantly affect complex networks. Since normal networks are made out of organic entities with altogether different life history characteristics and dispersal capacity, not all of the microbial local areas may answer climatic change factors likewise. Among the various variables connected with environmental change, raised $CO_2$ impacted the overflow of arbuscular and ectomycorrhizal parasites, while the consequences for plant-development-advancing microorganisms and endophytic organisms were more factors. The climb in temperature consequences for gainful plant-related microorganisms was more factor, positive, unbiased, and negative, which were similarly normal and fluctuated significantly with the temperature range. Similarly, plant-development-advancing microorganisms decidedly impacted plants exposed to dry spell pressure. Networks of soil microorganisms (soil microbiomes) assume a significant part in biogeochemical cycles and backing plant development. Here we centre essentially around the jobs that the dirt microbiome plays in cycling soil natural carbon and the effect of environmental change on the dirt carbon cycle. We initially talk about current difficulties in understanding the jobs completed by exceptionally different and heterogeneous soil microbiomes and survey existing information holes in understanding what environmental change will mean for soil carbon cycling by the dirt microbiome. Since soil microbiome dependability is a critical measurement to comprehend as the environment transforms, we examine various parts of steadiness, including obstruction, strength, and practical redundancy [6–8]. We then survey late examination relating to the effect of significant environmental irritations on the dirt microbiome and the capabilities that they do. At long last, we audit new trial philosophies and demonstrate approaches to a work in progress that ought to work with how we might interpret the mind-boggling nature of the dirt microbiome to foresee its future reactions more readily to environmental change. The soil microbiome adds to organic framework prosperity in different ways, including biogeochemical cycling, bioremediation, plant advancement, and fundamental productivity [2–7]. Its work in ozone exhausting substance radiations and mediating soil regular carbon (SOC) is very convincing thinking about future climate assumptions. Natural change and changes in the land the chiefs practices can unfairly impact soil readiness and SOC [9] which subsequently impacts the soil microbiome and its net effect on soil carbon sequestration.

## *Challenges*

Soil environments are profoundly mind-boggling and dependent upon various scene scale bothers that administer whether soil carbon is held or delivered to the air [5–9]. A definitive destiny of SOC is an element of the joined exercises of plants and subterranean organic entities, including soil microorganisms. Although dirt microorganisms are known to help plenty of biogeochemical capabilities connected with carbon cycling [7, 8] by far most of the dirt microbiome stays crude and has generally secretive capabilities. Just a simple part of soil microbial life has been indexed to date, albeit new soil microorganisms [7]. and infections are progressively being found [8]. This absence of information brings about the vulnerability of the commitment of soil microorganisms to SOC cycling and ruins the development of exact prescient models for worldwide carbon transition under environmental change [9]. Thusly, we are continually refining how we might interpret the biochemical capability of the dirt microbiome and the metabolic destiny of SOC.

The absence of data concerning the dirt microbiome metabolic potential makes it especially testing to precisely represent the changes in microbial exercises that happen because of natural change. For instance, plant-determined carbon data sources can prime microbial movement to deteriorate existing SOC at rates higher than model assumptions, bringing about mistakes inside prescient models of carbon motions [10]. To represent this, a reasonable model known as the microbial carbon siphon has been created to characterize how soil microorganisms change and settle soil natural matter [11]. In this model, microbial metabolic exercises for carbon turnover are isolated into two classes: ex vivo adjustment, alluding to the change of plant-determined carbon by extracellular proteins, and in vivo turnover, for intracellular carbon utilized in microbial biomass turnover or stored as dead microbial biomass, alluded to as necromass. The differentiating effects of catabolic exercises that discharge SOC as carbon dioxide ($CO_2$), versus anabolic pathways that produce stable carbon compounds, control net carbon consistency standards. Specifically, microbial carbon sequestration addresses an underrepresented part of soil carbon motion that the microbial carbon siphon model endeavours address [11, 12]. A connected area of vulnerability is the way the kind of plant-determined carbon upgrades microbial SOC stockpiling or on the other hand speeds up SOC decay [12]. For instance, leaf litter and needle litter act as wellsprings of carbon for microbial development in woods soils, yet litter science and pH changes by vegetation type [e.g., among root and foliar litter [13]. or deciduous and coniferous timberland litter [14]. Thus, these biochemical contrasts impact SOC levels through changing decay elements [15]. Likewise, an expanded variety of plant networks builds paces of rhizodeposition, invigorating microbial movement and SOC stockpiling even though dirt ultimately arrives at an immersion point past which they can't store extra carbon [15, 16].

Quiet likewise influences microbial metabolic rates. Many soil microorganisms are fleetingly dynamic, shifting back and forth between lethargic and dynamic states [17]. In any event, during lethargy, some dirt microorganisms are fit for using their energy stores to process SOC and add to soil biomass turnover, but at more slow rates [17–19]. By and by, dynamic individuals from the dirt local area contribute the most to biogeochemical changes, and another worldview is to move examinations from ordered profiles and toward microbiome useful pathways and aggregates [19]. Nonetheless, current sequencing innovations for local area organizations additionally measure torpid microorganisms and, surprisingly, exogenous DNA [20, 21]. and are in this way one-sided against dynamic working individuals from the local area. Refining ways to deal with centre-around capability is consequently expected to help model development through a more exact appraisal of certifiable cycles. Another test is representing the science and actual design of soils themselves, the two of which impact SOC disintegration. Customarily, slow paces of carbon turnover were believed to be owing to actual assurance of carbon particles in micro aggregates or mineral affiliations [22]. or their substance stubbornness to biodegradation [23]. The ongoing worldview develops how mineral affiliations happen, specifically through soil particles' sorption of biopolymers from microbial and plant necromass [24, 25]. for sure, profound soil natural matter is predominantly contained organism-determined items [26]. Also, the spatiotemporal construction of soils is heterogeneous and dynamic, with "problem areas" or "hot minutes" of microbial action [27]. For example, water accessibility is commonly lopsided, so carbon cycling is restricted to regions with adequate water, or to microorganisms equipped for managing to dry up pressure [e.g., through the creation of extracellular polymeric substances to keep a hydrated microenvironment [28]. What's more, different variables impacting SOC mineralization incorporate the presence of anaerobic versus vigorous microsites (anaerobic breath of carbon being less vivaciously ideal than oxygen consuming), accessibility of electron acceptors, and redox status of the dirt [29].

## *Dependability Metrics of Soil Microbiome*

A main pressing issue of environmental change is its effect on soil microbiome steadiness and capability and likewise biological system supportability [30–32]. Meta-examinations have exhibited that in roughly 80% of distributed investigations, soil aggravations evoked quantifiable consequences for microbiome strength [32, 33]. Local area steadiness is normally qualified concerning at least one of three principal measurements: opposition (staying unaltered during unsettling influence), flexibility (recuperation to a steady state), and practical overt repetitiveness (utilitarian profiles are kept up with despite ordered shifts) [32]. In a perfect world, these measurements would be integrated into microbiome aggravation studies, however, limits in examining time and exertion frequently block this chance. Specifically, the level of opposition is much of the time quantifiable during and following an unsettling influence, however, flexibility patterns may just be noticeable years after the fact [34]. As environment aggravations expand in seriousness or recurrence, understanding microbiome response examples will further develop a forecast of future reactions. In this manner, these measurements address a significant thought to consider while planning aggravation tests, and each is checked exhaustively underneath.

## *Obstruction*

Most aggravation studies have zeroed in on opposition as opposed to strength because of its relative simplicity of evaluation. Obstruction is normally estimated as movements in the local area or utilitarian profiles under pressure. For instance, soil water impediment unfavourably influences individuals from the Proteobacteria phylum and increments relative overflows of individuals from Actinobacteria as well as Firmicutes phyla [35]. Through their impacts on phylogenetic profiles, aggravations will thus influence the environment working. For instance, soil drying adjusted the wealth of societies for microorganisms engaged with methane oxidation [36]. while soil warming or raised carbon dioxide ($eCO_2$) impacted smelling salts oxidizing organisms [37]. Anthropogenic nitrogen affidavit (through inordinate manure expansion) can enhance nitrogen-cycling processes, including urea disintegration and tricarboxylate transport [38, 39]. A few natural burdens might frustrate carbon going through diminishing metabolic variety of a local area [40] or by restricting microbial take-up of carbon through diminished dispersion rates [41] For instance, enzymatic action rates, including that of carbon cycling chemicals (beta-glucosidase, aminopeptidase) or other supplement cycling proteins (corrosive phosphatase, arylsulfatase), have been demonstrated to be stifled under a dry spell and following soil consuming [42, 43] As a result, expectations of how stress influences biogeochemical processes for carbon and nitrogen mineralization need to represent microbial reactions.

Microbial life techniques are intently attached to the opposition, specifical proportions of K-to r-chose organic entities. (K-chose microorganisms augment endurance by being slow developing and asset proficient, while r-chose organic entities are energy and asset wasteful yet boost endurance through fast paces of development and proliferation.) In one review, networks with higher proportions of Gram-positive (typically K-chose) to Gram-negative (ordinarily r-chose) microbes were more impervious to $eCO_2$ [44] K-chose living beings are related to more slow development, higher catalyst substrate affinities, and use of additional hard-headed types of carbon [45] qualities attached to pressure obstruction. Conversely, r-chose organic entities are ordinarily more subject to labile carbon compounds for development, for example, those delivered into the rhizosphere through plant root exudates. Since some endemic plant species decline rhizodeposition into the soil under dry spell pressure to keep a carbon supply for their endurance, there is a consumption of labile SOC stocks into the encompassing soil. As the chief excess carbon sources are hard-headed carbon particles, K-tacticians are preferred over r-specialists [35].

Physiological variation is an asset escalated however compelling method for giving pressure obstruction. Some dirt microorganisms have embraced thicker cell walls to endure drying up pressure [35], and additionally layer transformations to endure openness to poisonous metals [34]. Past openness to a pressure condition [34] can prime a local area to oppose future burdens with a comparative method of activity, for instance, through upregulation of as well as an expanded scattering of opposition qualities [32, 46]. Nonetheless, interest in an original opposition instrument frequently has the compromise of losing a past one, and organisms might become helpless to a pressure that they were beforehand impervious to [34]. These patterns have been noticed for various (non-climate change-related) biological unsettling influences: For instance, long-haul copper pressure thwarted the dirt microbiome's ability to answer fluctuating natural circumstances [47]. Essentially, persistently stomped on dryland soils were less ready to answer rewetting than non-stomped-on ones [48]. The safest networks frequently show practical versatility and shift metabolic profiles as a component of ecological circumstances, improving their survivability if a specific speciality is obliterated [33]. Be that as it may, it is not yet clear whether physiological variations and additionally utilitarian pliancy will be boundless enough under environmental change unsettling influences to guarantee the endurance of soil biological systems.

## *Versatility*

The peculiarity of soil microbiome versatility is ostensibly underreported, as studies consolidating a long-enough time course to follow full recuperation are remarkable [32]. In any event, when unequivocally estimated, pre-aggravation profiles might require a very long time to restore [49], and now and again putatively irreversible changes happen [30, 50] these patterns stress the significance of long haul studies consolidating decadal timescales to follow microbial reactions to unsettling influence [51–53]. In a meta-examination of short-and long haul unsettling influences, recuperation was by and large seen in the under portion of the examinations [33]. As aggravations expand in recurrence and term, for example, during environmental change, it is basic to grasp how, if, by any means, microbiomes can recuperate.

Like obstruction, microbiome strength might be evaluated because of order as well as practical profiles. One methodology for estimating strength is through bunching taxa in light of recuperation designs—for instance, taxa that increment under pressure before in this way diminishing during recuperation would frame one group, though taxa that show the contrary pattern would shape another [54]. Flexibility can likewise change by the pace of recuperation. For instance, individuals from the

Planctomycetes, Crenarchaea, and Acidobacteria phyla recuperated quicker after a dirt warming treatment than did Actinobacteria or Verrucomicrobia [55]. Nonetheless, not all individuals from a given phylum answer generally in a similar way. For instance, explicit classes inside the Acidobacteria and Proteobacteria phyla were displayed to vary in their versatilities to dry spell pressure [35]. Particular flexibility patterns by phyla have suggestions for the carbon cycling processes they intervene, as individual taxa have trademark development and carbon absorption designs [56]. For instance, Actinobacteria overflow was adversely connected with carbon mineralization, though Bacteroidetes and Proteobacteria were emphatically related [57]. Subsequently, paces of soil carbon cycling will generally rely on how quick individuals from these phyla recuperate to a given pressure. Also, for practical profiles, versatility relies upon the capability being referred to and the phylogenetic goal that is being analyzed. For instance, nitrification is less tough than denitrification [32, 58], probable since it is intervened by a smaller organization of microorganisms. Thusly, capabilities in light of extensively dispersed proteins by and large have more obstruction but lower flexibility, while those with barely circulated chemicals, like complex polysaccharide debasement, have less opposition yet higher strength [59]. One more disparity between opposition and flexibility is the impact of earlier pressure—past openness to a pressure frequently diminishes paces of versatility to another one, though obstruction is by and large fortified [43].

A few variables add to microbial strength. One is commonness: Highly bountiful as well as broadly scattered life forms are less inclined to be crushed by the pressure. One more technique for strong organisms is to enter lethargy, framing what is known as the microbial seed bank [60]. In the two situations, getting through organisms are better ready to reseed the dirt microbiome upon stress enhancement [33]. Quick ribosome union and more limited age times are favourable characteristics, as they speed up recuperation; in any case, quickly developing taxa (e.g., r-specialists) are frequently exceptionally asset subordinate and accordingly more powerless to push [45]. By and large local area strength is likewise helped by pressure opposition systems, as they might be passed from lenient to vulnerable people using the quality stream to help recuperation [61]. On the other hand, lenient however less charitable living beings might hush up about opposition instruments, developing quickly under a given pressure condition while helpless creatures cease to exist [60]. In outrageous cases, deft people have been displayed to adjust their metabolic pathways to consolidate a generally distressing harmful compound as a carbon/nitrogen source [62]. Indeed, even through and through enmity against other recuperating gatherings might help strength, which was placed as the purpose for expanded survivability for microorganisms compared with parasites after soil warming [63].

## Environmental Change Impacts on the Soil Microbiome

Environmental change-related aggravations can altogether modify soil microbial local area and utilitarian profiles [5]. If dirt carbon or potentially nitrogen cycling are impacted, this can thusly influence environmental change either through certain criticisms to the climate (e.g., ozone harming substance outflows) or negative inputs (e.g., carbon immobilization into microbial or plant biomass) [12]. A better comprehension of how soil microorganisms answer to environmental change will thusly eventually further develop environment models. In any case, environmental change can conjure a few unmistakable bothers or in any event, intensifying aggravations, which can apply to differentiate impacts on the dirt microbiome [5]. Given the vulnerability concerning the transaction between various environmental change factors, ongoing examinations have started to consolidate different elements in the blend [37, 64–67]. Here, we explicitly audit soil microbiome reactions to soil warming and $eCO_2$, and how these variables cooperate straightforwardly and in a roundabout way to impact change in soil local area and utilitarian profiles.

### *Soil Warming*

Current environment models foresee a worldwide temperature climb of generally 3.7 °C by 2100 [68]. Considering that dirt microbial networks are certifiably impacted by warming [5], this addresses an inescapable effect of environmental change on the dirt microbiome. Soil warming is remembered to influence occupant microbial networks in a stepwise design. To start with, natural carbon deterioration rates are improved over a shorter time, expanding microbial biomass. One investigation discovered that the dirt microbial populace size expanded by 40–150% under soil warming [68]. Then, microbial breath has been displayed to decline over the long run as labile carbon is drained [69]. Following quite a while of openness, changes have been seen in microbial physiologies, local area structure, and user profiles, both as microorganisms adjust to warming, and as their digestion movements use the leftover headstrong carbon sources [70]. The subtleties behind these means are illustrated beneath.

Warming has been seen to increment microbiome local area variety and wealth [55, 71, 72], as well as to enhance individuals from the Acidobacteria and Actinobacteria phyla and class Alphaproteobacteria [55, 69, 73]. These ordered movements cross-over with utilitarian profiles: Oligotrophic taxa (i.e., slow-developing microorganisms fit for getting by in supplement unfortunate circumstances, e.g., Actinobacteria) are advanced over copiotrophic taxa (i.e., quickly developing organisms improved for supplement rich conditions, e.g., Bacteroidetes), perhaps as a reaction to changing soil carbon synthesis [74]. For instance, warming medicines enduring 5 to 8 years were displayed to incline toward more stubborn carbon-corrupting taxa from the Actinobacteria or Acidobacteria, despite not many generally quantifiable reactions in local area arrangement [52]. Quantifiable contrasts in utilitarian societies answerable for smelling salts oxidation [37] or diazotrophs [72] have additionally been noticed following soil warming.

Microbial capability can be influenced by warming both straightforwardly (e.g., through speed increase of enzymatic rates) or by implication (invigorating plant development and rhizodeposition and modifying soil properties). For instance, the cycling of phosphorus and sulfur has been demonstrated to be invigorated under warming [70, 75], however, making surmisings for carbon and nitrogen cycling is more troublesome. Warming has been exhibited to raise paces of nitrogen cycling processes, including denitrification, nitrogen obsession, nitrification, and nitrogen mineralization [75], although its accurate impacts rely upon the quality/process under study [70]. For instance, now and again warming stifled specific nitrogen cycling capabilities [65, 72]. One clarification is negative criticism: Warming increments soil inorganic nitrogen and plant nitrogen pool sizes [66], at last, discouraging paces of microbial disintegration and nitrogen cycling [76, 77]. Consequently, it is conceivable that nitrogen cycling can move over the long run as a component of the span/greatness of warming and nitrogen accessibility.

Paradoxically, carbon cycling has been demonstrated to be at first advanced by warming [73, 74] assuming carbon bioavailability is adequate. The temperature optima of extracellular chemicals for carbon corruption are with the end goal that warming can go about as a boost [69]. Over significant stretches of warming, studies have noticed diminished quantities of qualities engaged with labile carbon debasement, with expansions in those for refractory carbon digestion [65, 70, 74] and a higher variety of mindful practical organization [73]. These discoveries might be in some measure part of the way owing to water misfortune from dissipation during warming. At the point when soil dampness is controlled, labile carbon corruption can stay invigorated while debasement of headstrong carbon is unaltered [75]. Carbon cycling shifts likewise fluctuate by soil layer, where natural and mineral skylines have various reactions in sugar corruption potential after decadal timescales of warming [52]. Dissecting soil warming as a solitary element hence addresses a sub-standard methodology, as warming is probably going to be combined with other environmental change factors that likewise impact carbon cycling, consumption of soil dampness as well as $eCO_2$.

## *Raised Carbon Dioxide*

Similarly, as with warming, $eCO_2$ affects the dirt microbiome. For the time being, $eCO_2$ increments breath rates, microbial biomass, and hereditary signs for carbon cycling processes [78]. It additionally animates plant creation and rhizodeposition, thus preparing copiotrophs in the rhizosphere to separate labile and (later) refractory carbon [65, 79, 80]. By the by, ordered patterns for soil microbiomes under $eCO_2$ are in no way, shape or form reliably. One review examining patterns of $eCO_2$ across soil environments observed that the main normal reaction was consumption of Acidobacteria Groups 1 and 2 [81]. Like warming, be that as it may, over a long timescale $eCO_2$ is anticipated to improve for oligotrophs. Following 14 years of $eCO_2$ in a California field, diminishes in copiotrophic (r-chose) Bacteroidetes were noticed, alongside expansions in organisms with lower rRNA duplicate numbers, a typical quality of oligotrophs (K-chose) [82]. Under warming, enhancement of oligotrophic microorganisms is normal, because of diminished soil dampness and consumption of labile carbon. On the other hand, $eCO_2$ is anticipated to invigorate the plant and microbial development, which drains soil nitrogen. Thus, soil carbon cycling is anticipated to decline. To be sure, over longer timescales of $eCO_2$ treatment, there was a detailed stamped decline in soil carbon cycling, with practically no adjustment of carbon corruption [82]. Such circumstances will consequently incline toward more slow developing, asset-effective oligotrophic microorganisms. Under $eCO_2$, enzymatic exercises for phosphorus cycling will generally increment [65, 78, 82, 83], however, nitrogen cycling is more factor. Expansions in plant net essential creation, microbial immobilization of soil nitrogen, and microbial denitrification rates will all drain soil mineral nitrogen [66, 82, 84]. As an outcome, keeping up with soil nitrogen accessibility (and likewise plant/microbial development rates) requires an expansion in relative paces of nitrogen cycling and mineralization. Improved nitrogen cycling under $eCO_2$ has been noticed [37, 53, 78, 79, 85, 86], albeit genuine enzymatic rates are frequently unaltered or decline [82]. This error might be inferable from higher overflows of nitrogen fixers (e.g., Rhizobiales) or smelling salts oxidizers [37, 85], albeit this is certainly not an all-inclusive pattern [44, 77]. Varying outcomes for nitrogen cycling are sporadically seen across $eCO_2$ studies and might be affected by fluctuation in puzzling variables, for example, soil dampness accessibility, nearness to root exudates, soil profundity, and level of nitrogen constraint [44, 79]. What's more, the environment referred to, e.g., agroecosystems may have various outcomes from crude woods [81].

## *Combinatorial and Indirect Effects*

Taking into account any environmental change figure detachment neglects to address the exchange between them that is probably going to affect soils in genuine situations. To represent this information hole, numerous new I nvestigations have integrated multifactorial plans, whether with $eCO_2$ and warming [37, 65–67], $eCO_2$ and raised ozone [79, 85, 87], $eCO_2$ and nitrogen expansion [44], or different mixes. Frequently, varying outcomes are found for blends contrasted with single-factor medicines, featuring the significance of this methodology. For example, in one examination displaying the impacts of warming as well as $eCO_2$ on field soils in a cotton agroecosystem, the blend of warming with $eCO_2$ incited shifts in smelling salts oxidizing microbial networks and expansions in soil nitrification rates, though barely any tremendous impacts were seen for warming alone [37]. Frequently, a blend of irritations brings about one variable constricting the impacts of the other.

Concerning $eCO_2$ and warming, frequently $eCO_2$ checks warming-actuated diminish in soil dampness or elevate plant rhizodeposition to keep up with carbon cycling and heterotrophic breath as carbon is exhausted under warming [65]. Almost certainly, the general significance of the two variables fluctuates by climate. For instance, various patterns may be found in prairies contrasted with woods, or agroecosystems contrasted with icy biomes [5]. For example, in a dryland local area study, warming beat $eCO_2$ [67], while in a prairie concentration the contrary pattern was noticed [65]. In the last option study, the mix of $eCO_2$ and warming had comparable impacts to $eCO_2$ alone—warming diminished signals for carbon cycling, alkali oxidation, and creation, though $eCO_2$ and the blend had the contrary pattern. Prominently, a subset of $eCO_2$-invigorated qualities for nitrogen cycling and carbon corruption were not generally improved under the blend, including qualities for unmanageable carbon debasement [65], which might be a consequence of expanded rhizodeposition of labile carbon blocking the need for such qualities.

A confusing variable for concentrating on unsettling influence reactions in the dirt microbiome is unravelling directly from circuitous impacts. As talked about above, $eCO_2$ by implication influences soil networks through expanded plant rhizodeposition, soil nitrogen limit, and higher soil dampness content ($eCO_2$ actuates plant water protection through diminished stomatal conductance) [65, 83, 88], as well as through root exudate profiles, soil construction, or leaf litter science [85]. On the other hand, warming invigorates plant development however brings down soil dampness through dissipation, and such changes in water accessibility might greatly affect the dirt microbiome than warming alone [55, 89]. In particular, improvement for oligotrophs under warming might be to some extent because of their higher compound substrate affinities addressing a benefit as dispersion diminishes submerged limit [74]. Other confounding variables incorporate treatment length [73, 79], irregularity [83], and soil profundity or skyline [52, 81, 90]. Such errors feature the significance of representing jumbling boundaries during soil annoyance studies.

## Microbial Biochemical Pathways and Climate Change

Albeit the reaction of the dirt microbiome is frequently learned at a significant level, for example, local area-wide ordered shifts, one more significant part of environmental change reaction is how explicit biochemical pathways are impacted. A new report on warmed soils from Arctic and Antarctic conditions tracked down various normal metabolic reactions [84]. For instance, methane creation and digestion of acetic acid derivation and di-and mono-methylamine expanded as temperatures were raised from 1 °C to 30 °C, while diminishes were considered in propionate and acetic acid derivation oxidation to be well as digestion of H2 and formate [91]. Moreover, as temperatures were raised above 7 °C, the rate-restricting step for methane creation moved from propionate oxidation to polysaccharide hydrolysis. Additionally, the drying of Puerto Rican soils expanded signals for carbon digestion catalysts including beta-glucosidase, cellobiohydrolase, N-acetylglucosaminidase, and xylanase [92].

Nonetheless, this impact was diminished through pretreatment of soils with a reenacted dry spell, recommending long haul changes to soil working considering an aggravation might improve impacts of future burdens.

Microbial biochemical pathways are additionally by implication impacted by environmental change influences on plant-microorganism associations. A new report found that organisms take up less plant-inferred carbon under both intensity and dry season pressure [93, 94]. Besides, environmental change might modify plant cover [88] or plant local area profiles, e.g., through plant movement to colder climes [77] or expanded proportions of C4:C3 plant types [75]. Reactions in rhizodeposition under pressure can likewise shift by plant species or cultivar [85]. For instance, wild-type plants were displayed to have higher paces of root exudation under $eCO_2$ than developed assortments [88], as did C4 grasses compared with C3 plants [70, 75]. As a result of changing kind and amount of plant-determined carbon contributions to the dirt, different microbial pathways for carbon take-up and digestion will be invigorated.

## *Climate Change Impacts on Soil Carbon*

An International Soil Carbon Network was as of late settled to distinguish holes in SOC demonstrating [95]. One of the greatest difficulties distinguished was the location of changes in SOC, because of two its spatiotemporal variety across soil biological systems and a deficient comprehension of the cycles overseeing whether SOC is balanced out or decayed. Preferably, models would be gotten from unthinking understandings of SOC elements, however, most are rather founded on reenactments, because of difficulties in acquiring observational information and estimating SOC [96]. Instances of flow research need to remember comprehension of SOC elements for soil (micro)aggregate microenvironments and what preparing means for soil carbon turnover [96]. At last, the joining of robotic bits of knowledge from subatomic information into environment models will better foresee the destiny of soil carbon under environmental change.

One more region that should be tended to is the incorporation of environment-important microbial cycles. Most environment models expect that dirt natural matter deterioration is a first-request rot process between theoretical pools. In 2009 there were 33 SOC models addressed inside the Global Change and Terrestrial Ecosystems Soil Organic Matter Network data set, and 30 of those were multicompartment, process-based models [97], in which rot rates are regularly communicated as a component of carbon focus and a rate steady. Albeit worldwide models consolidate data about soil and environment properties [4], microbial cycles may not be remembered for first-request suspicions [97]. Upon their consideration, notwithstanding, the prescient capacity for SOC destiny under environment is certifiably improved [98]. This has brought about proceeded with the advancement of further developed Earth System Models (ESMs) that incorporate microbial impacts on SOC transition [4, 97], and new models for connecting decay to the size and action of the dirt microbiome [99]. These improvements feature the significance of second-request processes (microbial exercises for SOC change) for anticipating SOC transition either as microbial biomass or as respiratory misfortune to the climate as $CO_2$. Models expect to anticipate what environmental warming will mean for soil-obtained ozone-harming substance discharge from now on, which requires observational judgments of the degree of soil carbon criticisms. Nonetheless, environment forecasts might be founded on obsolete soil models that don't mirror the ongoing logical agreement on soil carbon development and adjustment [100]. For instance, even though SOC is the consequence of net results (breath) and sources of info (carbon obsession) of plant-determined carbon, most observational information has zeroed in on yields alone, neglecting to represent conceivable compensatory impacts like elevated soil carbon development [100]. An equilibrium of carbon results and data sources is caught by ESMs [98], yet isn't yet broadly remembered for worldwide expectations [101].

Conversely, models on SOC motion have started to incorporate parts of the plant-soil biological system, including plant types and mineral communications, which might fluidly affect SOC transition contingent upon explicit conditions. The CORPSE (Carbon, Organisms, Rhizosphere and Protection in the Soil Environment) model incorporates parts of preparing and soil security, which advance soil deterioration and carbon stockpiling, separately [102]. In any case, after getting observational information, they found differentiating results from the two soil warming examinations: At one site (Oak Ridge, Tennessee), carbon adjustment in the dirt surpassed SOC shortfall from preparing under warming, though at a different site (FACE at Duke Forest, North Carolina) the contrary pattern was found, bringing about net SOC deficit [102]. These reproductions showed expanded $CO_2$ levels invigorated preparing to a more noteworthy degree than carbon capacity, which will yield a net worldwide carbon shortfall under environmental change. Different models have consolidated data on plant utilitarian sorts (e.g., C3 versus C4 grasses, broadleaf versus needleleaf) that thus recognize plant soil inputs [97]. As of late another model (MEMS v1.0) was proposed, connecting soil natural matter science with both microbial handling and cooperation with soil minerals, to further develop environment model forecasts [102]. On a connected note, a demonstrating approach has as of late been recommended that considers microbial life methodologies [64]. Even though dirt microbial life methodologies have normally been doled out to two classifications—quickly developing r-planners and more slowly developing, energy-monitoring K-tacticians—the new model parts life techniques into three classifications: Y for development yield, A for asset procurement, and S for stress resistance. Every one of these three classes would address a benefit under an alternate arrangement of ecological circumstances and availabilities, with the end goal that it would be improbable for an organism to have a place with multiple [64, 103–106]. Moreover, as every classification has a particular profile for carbon use, approving this system will assist with foreseeing generally speaking microbial carbon cycling rates and dynamic cycles.

## Conclusions

A new source of inspiration underscored the significance of understanding ecological microorganisms notwithstanding environmental change. Plentiful proof uncovers that dirt microorganisms are impacted by environmental change-related unsettling influences with significant inputs to biological system wellbeing and environment constraining. Under these aggravations, changes in microbial local area creation and work will thus have repercussions for interkingdom collaborations, biogeochemical cycling, and carbon stream, in manners that might compound or weaken environmental change. As we start to completely comprehend key jobs done by microorganisms possessing soil biological systems, this information might be utilized to anticipate what basic metabolic cycles are meant for by ecological change, and might be utilized for alleviation of negative parts of environmental change.

## References

1. Amundson R, Berhe AA, Hopmans JW, Olson C, Sztein AE, Sparks DL (2015) Soil and human security in the 21st century. Science 348(6235):1261071
2. Cavicchioli R, Ripple WJ, Timmis KN, Azam F, Bakken LR et al (2019) Scientists' warning to humanity: microorganisms and climate change. Nat Rev Microbiol 17(9):569–586
3. Lal R (2004) Soil carbon sequestration impacts on global climate change and food security. Science 304(5677)
4. Campbell EE, Paustian K (2015) Current developments in soil organic matter modeling and the expansion of model applications: a review. Environ Res Lett 10(12):1230
5. Jansson JK, Hofmockel KS (2020) Soil microbiomes and climate change. Nat Rev Microbiol 18:35–46
6. Thompson LR, Sanders JG, McDonald D, Amir A, Ladau J et al (2017) A communal catalogue reveals Earth's multiscale microbial diversity. Nature 551:457–463
7. Goel R, Kumar V, Suyal DC, Narayan Soni R (2018) Toward the unculturable microbes for sustainable agricultural production. In: Meena VS (ed) Role of rhizospheric microbes in soil. Springer, Singapore, pp 107–123
8. Schulz F, Alteio L, Goudeau D, Ryan EM, Yu FB et al (2018) Hidden diversity of soil giant viruses. Nat Commun 9(1):488
9. Friedlingstein P, Cox P, Betts R, Bopp L, von Bloh W et al (2006) Climate-carbon cycle feedback analysis: results from the $C^4$MIP Model Intercomparison. J Clim 19(14):3337–3353
10. Kuzyakov Y, Horwath WR, Dorodnikov M, Blagodatskaya E (2019) Review and synthesis of the effects of elevated atmospheric $CO_2$ on soil processes: no changes in pools, but increased fluxes and accelerated cycles. Soil Biol Biochem 128:66–78
11. Liang C, Schimel JP, Jastrow JD (2017) The importance of anabolism in microbial control over soil carbon storage. Nat Microbiol 2(8):17105
12. Sulman BN, Phillips RP, Oishi AC, Shevliakova E, Pacala SW (2014) Microbe-driven turnover offsets mineral-mediated storage of soil carbon under elevated $CO_2$. Nat Clim Change 4(12):1099–1102
13. Harmon ME, Silver WL, Fasth B, Chen H, Burke IC et al (2009) Long-term patterns of mass loss during the decomposition of leaf and fine root litter: an intersite comparison. Glob Change Biol 15(5):1320–1323
14. Qualls RG (2016) Long-term (13 years) decomposition rates of forest floor organic matter on paired coniferous and deciduous watersheds with contrasting temperature regimes. Forests 7(10):231

15. Lange M, Eisenhauer N, Sierra CA, Bessler H, Engels C et al (2015) Plant diversity increases soil microbial activity and soil carbon storage. Nat Commun 6(1):6707
16. Stewart CE, Paustian K, Conant RT, Plante AF, Six J (2007) Soil carbon saturation: concept, evidence and evaluation. Biogeochemistry 86(1):19–31
17. Joergensen RG, Wichern F (2018) Alive and kicking: why dormant soil microorganisms matter. Soil Biol Biochem 116:419–430
18. Jones SE, Lennon JT (2010) Dormancy contributes to the maintenance of microbial diversity. PNAS 107(13):5881–5886
19. Jansson JK, Hofmockel KS (2018) The soil microbiome—from metagenomics to metaphenomics. Curr Opin Microbiol 43:162
20. Graham EB, Knelman JE, Schindlbacher A, Siciliano S, Breulmann M et al (2016) Microbes as engines of ecosystem function: when does community structure enhance predictions of ecosystem processes? Front Microbiol
21. Carini P, Marsden PJ, Leff JW, Morgan EE, Strickland MS, Fierer N (2017) Relic DNA is abundant in soil and obscures estimates of soil microbial diversity. Nat Microbiol 2(3):1624222
22. Balesdent J, Chenu C, Balabane M (2000) Relationship of soil organic matter dynamics to physical protection and tillage. Soil Tillage Res 53(3–4):215–230
23. Marschner B, Brodowski S, Dreves A, Gleixner G, Gude A et al (2008) How relevant is recalcitrance for the stabilization of organic matter in soils? J Plant Nutr Soil Sci 171(1):91–110
24. Cotrufo MF, Wallenstein MD, Boot CM, Denef K, Paul E (2013) The Microbial Efficiency-Matrix Stabilization (MEMS) framework integrates plant litter decomposition with soil organic matter stabilization: do labile plant inputs form stable soil organic matter? Glob Change Biol 19(4):988–995
25. Schmidt MWI, Torn MS, Abiven S, Dittmar T, Guggenberger G et al (2011) Persistence of soil organic matter as an ecosystem property. Nature 478(7367):49–56
26. Erich MS, Plante AF, Fernández JM, Mallory EB, Ohno T (2012) Effects of profile depth and management on the composition of labile and total soil organic matter. Soil Sci Soc Am J 76(2):408–419
27. Bach EM, Hofmockel KS (2016) A time for every season: soil aggregate turnover stimulates decomposition and reduces carbon loss in grasslands managed for bioenergy. GCB Bioenergy 8(3):588–599
28. Vurukonda SSKP, Vardharajula S, Shrivastava M, SkZ A (2016) Enhancement of drought stress tolerance in crops by plant growth promoting rhizobacteria. Microbiol Res 184:13
29. Keiluweit M, Wanzek T, Kleber M, Nico P, Fendorf S (2017) Anaerobic microsites have an unaccounted role in soil carbon stabilization. Nat Commun 8(1):1771
30. Allison SD, Martiny JBH (2008) Resistance, resilience, and redundancy in microbial communities. PNAS 105(Supplement 1):11512–11519
31. Orwin KH, Wardle DA (2005) Plant species composition effects on belowground properties and the resistance and resilience of the soil microflora to a drying disturbance. Plant Soil 278(1–2):205–221
32. Griffiths BS, Philippot L (2013) Insights into the resistance and resilience of the soil microbial community. FEMS Microbiol Rev 37(2):112–129
33. Shade A, Peter H, Allison SD, Baho DL, Berga M et al (2012) Fundamentals of microbial community resistance and resilience. Front Microbiol 3:41
34. Azarbad H, van Gestel C, Niklińska M, Laskowski R, Röling W, van Straalen N (2016) Resilience of soil microbial communities to metals and additional stressors: DNA-based approaches for assessing "stress-on-stress" responses. Int J Mol Sci 17(6):933
35. Naylor D, Coleman- D (2018) Drought stress and root-associated bacterial communities. Front Plant Sci 8:2223
36. van Kruistum H, Bodelier PLE, Ho A, Meima M, Veraart AJ (2018) Resistance and recovery of methane-oxidizing communities depends on stress regime and history; a microcosm study. Front Microbiol 9:1714

37. Nguyen LTT, Broughton K, Osanai Y, Anderson IC, Bange MP et al (2019) Effects of elevated temperature and elevated $CO_2$ on soil nitrification and ammonia-oxidizing microbial communities in field-grown crop. Sci Total Environ 675:81–89
38. Fierer N, Lauber CL, Ramirez KS, Zaneveld J, Bradford MA, Knight R (2012) Comparative metagenomic, phylogenetic and physiological analyses of soil microbial communities across nitrogen gradients. ISME J 6
39. Leff JW, Jones SE, Prober SM, Barberán A, Borer ET et al (2015) Consistent responses of soil microbial communities to elevated nutrient inputs in grasslands across the globe. PNAS 112(35):10967
40. Hueso S, García C, Hernández T (2012) Severe drought conditions modify the microbial community structure, size and activity in amended and unamended soils. Soil Biol Biochem 50:167–173
41. Schaeffer SM, Homyak PM, Boot CM, Roux- D, Schimel JP (2017) Soil carbon and nitrogen dynamics throughout the summer drought in a California annual grassland. Soil Biol Biochem 115:54–62
42. McKew BA, Taylor JD, McGenity TJ, Underwood GJC (2011) Resistance and resilience of benthic biofilm communities from a temperate saltmarsh to desiccation and rewetting. ISME J 5(1):30–41
43. Hinojosa MB, Parra A, Laudicina VA, Moreno JM (2016) Post-fire soil functionality and microbial community structure in a Mediterranean shrubland subjected to experimental drought. Sci Total Environ 573:1178–1189
44. Simonin M, Nunan N, Bloor JMG, Pouteau V, Niboyet A (2017) Short-term responses and resistance of soil microbial community structure to elevated $CO_2$ and N addition in grassland mesocosms. FEMS Microbiol Lett 364(9):fnx077
45. Orwin KH, Dickie IA, Wood JR, Bonner KI, Holdaway RJ (2016) Soil microbial community structure explains the resistance of respiration to a dry-rewet cycle, but not soil functioning under static conditions. Funct Ecol 30(8):1430–1439
46. Jacquiod S, Nunes I, Brejnrod A, Hansen MA, Holm PE et al (2018) Long-term soil metal exposure impaired temporal variation in microbial metatranscriptomes and enriched active phages. Microbiome 6(1):223
47. Jacquiod S, Franqueville L, Cécillon S, Vogel TM, Simonet P (2013) Soil bacterial community shifts after chitin enrichment: an integrative metagenomic approach. PLoS ONE 8(11):e79699
48. Steven B, Belnap J, Kuske CR (2018) Chronic physical disturbance substantially alters the response of biological soil crusts to a wetting pulse, as characterized by metatranscriptomic sequencing. Front Microbiol 9:2382
49. Lourenço KS, Suleiman AKA, Pijl A, vanVeen JA, Cantarella H, Kuramae EE (2018) Resilience of the resident soil microbiome to organic and inorganic amendment disturbances and to temporary bacterial invasion. Microbiome 6(1):142
50. Knelman J, Schmidt S, Garayburu V, Kumar S, Graham E (2019) Multiple, compounding disturbances in a forest ecosystem: fire increases susceptibility of soil edaphic properties, bacterial community structure, and function to change with extreme precipitation event. Soil Syst 3(2):40
51. Melillo JM, Frey SD, DeAngelis KM, Werner WJ, Bernard MJ et al (2017) Long-term pattern and magnitude of soil carbon feedback to the climate system in a warming world. Science 358(6359):101–105
52. Pold G, Billings AF, Blanchard JL, Burkhardt DB, Frey SD et al (2016) Long-term warming alters carbohydrate degradation potential in temperate forest soils. Appl Environ Microbiol 82(22):6518–6530
53. Drake JE, Gallet-Budynek A, Hofmockel KS, Bernhardt ES, Billings SA et al (2011) Increases in the flux of carbon belowground stimulate nitrogen uptake and sustain the long-term enhancement of forest productivity under elevated $CO_2$: C fluxes belowground and long-term FACE productivity. Ecol Lett 14(4):349–357
54. Jiao S, Chen W, Wei G (2019) Resilience and assemblage of soil microbiome in response to chemical contamination combined with plant growth. Appl Environ Microbiol 85(6):e02523-e2618

55. Sheik CS, Beasley WH, Elshahed MS, Zhou X, Luo Y, Krumholz LR (2011) Effect of warming and drought on grassland microbial communities. ISME J 5(10):1692–1700
56. Morrissey EM, Mau RL, Hayer M, Liu X-JA, Schwartz E et al (2019) Evolutionary history constrains microbial traits across environmental variation. Nat Ecol Evol 3(7):1064–1069
57. Fierer N, Bradford MA, Jackson RB (2007) Toward an ecological classification of soil bacteria. Ecology 88(6):1354–1364
58. Wertz S, Degrange V, Prosser JI, Poly F, Commeaux C et al (2007) Decline of soil microbial diversity does not influence the resistance and resilience of key soil microbial functional groups following a model disturbance. Environ Microbiol 9(9):2211–3194
59. Chaer G, Fernandes M, Myrold D, Bottomley P (2009) Comparative resistance and resilience of soil microbial communities and enzyme activities in adjacent native forest and agricultural soils. Microb Ecol 58(2):414–424
60. Ho A, Lüke C, Reim A, Frenzel P (2016) Resilience of (seed bank) aerobic methanotrophs and methanotrophic activity to desiccation and heat stress. Soil Biol Biochem 101:130–138
61. Puglisi E, Hamon R, Vasileiadis S, Coppolecchia D, Trevisan M (2012) Adaptation of soil microorganisms to trace element contamination: a review of mechanisms, methodologies, and consequences for risk assessment and remediation. Crit Rev Environ Sci Technol 42(22):2435–2470
62. Udiković-Kolić N, Devers- M, Petrić I, Hršak D, Martin-Laurent F (2011) Evidence for taxonomic and functional drift of an atrazine-degrading culture in response to high atrazine input. Appl Microbiol Biotechnol 90(4):1547–1554
63. Bárcenas G, Bååth E (2009) Bacterial and fungal growth in soil heated at different temperatures to simulate a range of fire intensities. Soil Biol Biochem 41(12):2517–2526
64. Malik AA, Martiny JBH, Brodie EL, Martiny AC, Treseder KK, Allison SD (2020) Defining trait-based microbial strategies with consequences for soil carbon cycling under climate change. ISME J 14:1–9
65. Yu H, Deng Y, He Z, Van Nostrand JD, Wang S et al (2018) Elevated $CO_2$ and warming altered grassland microbial communities in soil top-layers. Front Microbiol 9:1790
66. Dijkstra FA, Blumenthal D, Morgan JA, Pendall E, Carrillo Y, Follett RF (2010) Contrasting effects of elevated $CO_2$ and warming on nitrogen cycling in a semiarid grassland. New Phytol 187(2):426–437
67. Hu H-W, Macdonald CA, Trivedi P, Anderson IC, Zheng Y et al (2016) Effects of climate warming and elevated $CO_2$ on autotrophic nitrification and nitrifiers in dryland ecosystems. Soil Biol Biochem 92:1–15
68. Solomon S (2007) The intergovernmental panel on climate change. Climate Change 2007: the physical science basis: contribution of working group I to the fourth assessment report of the intergovernmental panel on climate change. United Nations Environment Programme, New York
69. Pold G, Melillo JM, DeAngelis KM (2015) Two decades of warming increases diversity of a potentially lignolytic bacterial community. Front Microbiol 6:480
70. Xue K, Xie J, Zhou A, Liu F, Li D et al (2016) Warming alters expressions of microbial functional genes important to ecosystem functioning. Front Microbiol 7:668
71. Rocca JD, Simonin M, Blaszczak JR, Ernakovich JG, Gibbons SM et al (2019) The microbiome stress project: toward a global meta-analysis of environmental stressors and their effects on microbial communities. Front Microbiol 9:3272
72. Carrell AA, Kolton M, Glass JB, Pelletier DA, Warren MJ et al (2019) Experimental warming alters the community composition, diversity, and $N_2$ fixation activity of peat moss (*Sphagnum fallax*) microbiomes. Glob Change Biol 25:2993–3004
73. DeAngelis KM, Pold G, Topçuoğlu BD, van Diepen LTA, Varney RM et al (2015) Long-term forest soil warming alters microbial communities in temperate forest soils. Front Microbiol 6:104
74. Li Y, Lv W, Jiang L, Zhang L, Wang S et al (2019) Microbial community responses reduce soil carbon loss in Tibetan alpine grasslands under short-term warming. Glob Change Biol 25:3438–3449

75. Zhou J, Xue K, Xie J, Deng Y, Wu L et al (2012) Microbial mediation of carbon-cycle feedbacks to climate warming. Nat Clim Change 2(2):106–110
76. Frey SD, Ollinger S, Nadelhoffer K, Bowden R, Brzostek E et al (2014) Chronic nitrogen additions suppress decomposition and sequester soil carbon in temperate forests. Biogeochemistry 121(2):305–316
77. Lladó S, López- R, Baldrian P (2017) Forest soil bacteria: diversity, involvement in ecosystem processes, and response to global change. Microbiol Mol Biol Rev 81(2):e00063-1678
78. Xiong J, He Z, Shi S, Kent A, Deng Y et al (2015) Elevated $CO_2$ shifts the functional structure and metabolic potentials of soil microbial communities in a $C_4$ agroecosystem. Sci Rep 5(1):9316
79. Cheng L, Booker FL, Burkey KO, Tu C, Shew HD et al (2011) Soil microbial responses to elevated $CO_2$ and $O_3$ in a nitrogen-aggrading agroecosystem. PLoS ONE 6(6):e2137
80. Chen YP, Liu Q, Liu YJ, Jia FA, He XH (2015) Responses of soil microbial activity to cadmium pollution and elevated $CO_2$. Sci Rep 4(1):4287
81. Dunbar J, Eichorst SA, Gallegos-Graves LV, Silva S, Xie G et al (2012) Common bacterial responses in six ecosystems exposed to 10 years of elevated atmospheric carbon dioxide: soil bacterial response in six ecosystems. Environ Microbiol 14(5):1145–1158
82. Yang S, Zheng Q, Yuan M, Shi Z, Chiariello NR et al (2019) Long-term elevated $CO_2$ shifts composition of soil microbial communities in a Californian annual grassland, reducing growth and N utilization potentials. Sci Total Environ 652:1474–1481
83. Ebersberger D, Niklaus PA, Kandeler E (2003) Long term $CO_2$ enrichment stimulates N-mineralisation and enzyme activities in calcareous grassland. Soil Biol Biochem 35(7):965–972
84. Delgado- M, Maestre FT, Escolar C, Gallardo A, Ochoa V et al (2014) Direct and indirect impacts of climate change on microbial and biocrust communities alter the resistance of the N cycle in a semiarid grassland. J Ecol 102(6):1592–1605
85. Wang P, Marsh EL, Ainsworth EA, Leakey ADB, Sheflin AM, Schachtman DP (2017) Shifts in microbial communities in soil, rhizosphere and roots of two major crop systems under elevated $CO_2$ and $O_3$. Sci Rep 7(1):15019
86. Hofmockel KS, GalletBudynek A, McCarthy HR, Currie WS, Jackson RB, Finzi A (2011) Sources of increased N uptake in forest trees growing under elevated $CO_2$: results of a large-scale $^{15}N$ study. Glob Change Biol 17(11):3338–3350
87. Dunbar J, Gallegos-Graves LV, Steven B, Mueller R, Hesse C et al (2014) Surface soil fungal and bacterial communities in aspen stands are resilient to eleven years of elevated $CO_2$ and $O_3$. Soil Biol Biochem 76:227–234
88. Formánek P, Rejšek K, Vranová V (2014) Effect of elevated $CO_2$, $O_3$, and UV radiation on soils. Sci World J 2014:730149
89. Kardol P, Cregger MA, Campany CE, Classen AT (2010) Soil ecosystem functioning under climate change: plant species and community effects. Ecology 91(3):767–781
90. Hayden HL, Mele PM, Bougoure DS, Allan CY, Norng S et al (2012) Changes in the microbial community structure of bacteria, archaea and fungi in response to elevated $CO_2$ and warming in an Australian native grassland soil: climate change effects on microbial communities. Environ Microbiol 14(12):3081–3096
91. Tveit AT, Urich T, Frenzel P, Svenning MM (2015) Metabolic and trophic interactions modulate methane production by Arctic peat microbiota in response to warming. PNAS 112(19):E2507–E2516
92. Bouskill NJ, Wood TE, Baran R, Ye Z, Bowen BP, et al (2016) Belowground response to drought in a tropical forest soil. I. Changes in microbial functional potential and metabolism. Front Microbiol 7:525
93. von Rein I, Gessler A, Premke K, Keitel C, Ulrich A, Kayler ZE (2016) Forest understory plant and soil microbial response to an experimentally induced drought and heat-pulse event: the importance of maintaining the continuum. Glob Change Biol 22(8):2861–2874
94. Pérez Castro S, Cleland EE, Wagner R, Sawad RA, Lipson DA (2019) Soil microbial responses to drought and exotic plants shift carbon metabolism. ISME J 13(7):1776–1787

95. Prommer J, Walker TWN, Wanek W, Braun J, Zezula D et al (2019) Increased microbial growth, biomass, and turnover drive soil organic carbon accumulation at higher plant diversity. Glob Change Biol 26:669
96. Bore EK, Apostel C, Halicki S, Kuzyakov Y, Dippold MA (2017) Soil microorganisms can overcome respiration inhibition by coupling intra- and extracellular metabolism: $^{13}C$ metabolic tracing reveals the mechanisms. ISME J 11(6):1423–1433
97. Harden JW, Hugelius G, Ahlström A, Blankinship JC, Bond- B et al (2018) Networking our science to characterize the state, vulnerabilities, and management opportunities of soil organic matter. Glob Change Biol 24(2):e705–e718
98. Ostle NJ, Smith P, Fisher R, Woodward FI, Fisher JB et al (2009) Integrating plant-soil interactions into global carbon cycle models. J Ecol 97(5):851–863
99. Wieder WR, Bonan GB, Allison SD (2013) Global soil carbon projections are improved by modelling microbial processes. Nat Clim Change 3(10):909–912
100. Todd-Brown KEO, Randerson JT, Post WM, Hoffman FM, Tarnocai C et al (2012) Causes of variation in soil carbon predictions from CMIP5 Earth system models and comparison with observations. Biogeosci Discuss 9(10):14437–14473
101. Allen SC, Jose S, Nair PKR, Brecke BJ, Nair VD, Graetz DA, Ramsey CL (2005) Nitrogen mineralization in a pecan (Carya illinoensis K. Koch) –cotton (Gossypium hirsutum L.) alley cropping system in the southern United States. Biol Fertil Soils 41:28–37
102. Azam F, Gill S, Farooq S, Lodhi A (2004) Effect of $CO_2$ on nitrification and immobilization of NH 4þ 2 N. Biol Fertil Soils 40:427–431
103. Bäckman JSK, Klemedtsson AK, Klemedtsson L, Lindgren PE (2004) Clear-cutting affects the ammonia-oxidizing community differently in limed and non-limed coniferous forest soils. Biol Fertil Soils 40:260–267
104. Baldock JA, Skjemstad JO (2000) Role of the mineral matrix and minerals in protecting natural organic materials against decomposition. Org Geochem 31:697–710
105. Black CA (1957) Soil–plant relationships. Wiley, New York; Bremner JM, McCarty GW (1988) Effects of terpenoids on nitrification in soil. Soil Sci Soc Am J 52:1630–1633
106. Carney KM, Matson PA, Bohannan BJM (2004) Diversity and composition of tropical soil nitrifiers across a plant diversity gradient and among land-use types. Ecol Lett 7:684–694

# Chapter 11
# Climate Change Drivers and Soil Microbe-Plant Interactions

**Sangeeta Singh, Tanmaya Kumar Bhoi, Ifrah Khan, Vipula Vyas, R. Athulya, Atiraj Rathi, and Ipsita Samal**

**Abstract** Climate change is one of the most important global concerns of modern era, with economic, social, scientific, political, moral, and ethical aspects. The soil ecosystem, which encompasses an enormous diversity of microbial life, is critical in this regard because it is a key component of the carbon and nitrogen cycles and is associated in the removal of greenhouse gases in the atmosphere which contribute to climate change. The microbial world is an important component of various biogeochemical cycles, and its role in climate change must be considered. Microbes, on the other hand, are rarely mentioned in climate change discussions. Microbial activity has not been taken into account sufficiently in most climates due to a lack of adequate understanding. Therefore, this book chapter provides an insight into the the intrinsic and extrinsic attributes, direct and indirect mechanism and emerging technologies for understanding of plant–microbe responses to climatic change that confer reason of soil microbial communities to climate extremes.

## Introduction

For more than 12,000 years, Earth's climate remained stable which in turn is vital for human kind's very existence [1]. During the past century, the typical global temperature increased close to a 1.5°F, and in next 100 years, it is expected to rise an additional 0.5°F–8.6°F. This is a critical problem since even little changes in the average global temperature can lead to significant changes in the weather and climate [2]. The microbial communiity is extremely significant for this context because it plays a crucial

S. Singh (✉) · T. K. Bhoi · I. Khan · V. Vyas · A. Rathi
Forest Protection Division, ICFRE-Arid Forest Research Institute (AFRI), Jodhpur, Rajasthan 342005, India
e-mail: singhsangi123@gmail.com

R. Athulya
Forest Protection Division, ICFRE-Institute of Wood Science and Technology (IWST), Bangalore, Karnataka 560003, India

I. Samal
Sri Sri University, Cuttack, Odisha 754 006, India



J. A. Parray (ed.), *Climate Change and Microbiome Dynamics*, Climate Change Management, https://doi.org/10.1007/978-3-031-21079-2_11

role in the nitrogen and carbon cycles and is proportionately involved in the removal and emission of gases that play a partin climate change, such as methane and CO2 [3]. While heterotrophic microorganisms break down organic substances to release greenhouse gases, photosynthetic microbes consume atmospheric carbon dioxide. The net carbon flux is primarily determined by the balance between the two processes, and it varies across different ecosystems based on climatic factors like temperature. As a result, microbial reactions play a critical role in the earth's carbon cycle because they not only lock up large amounts of carbon but also release it, according to [4, 5]. It is important to emphasise that most greenhouse gases, including $CO_2$, $CH_4$, and $N_2O$, are produced by microbes [6]. In this book chapter we have discussed about the various action mechanisms of climate change including the mechanisms affecting the microbial community, alterations in microbial diversity, the physiological alterations, action mechanisms on plants, variations in moisture content, and the various consequences on microorganisms due to change in climate, rising temperatures, altered precipitation, increased $CO_2$ emissions, drought situations and try to elaborate on emerging technologies and better comprehension of plant and microbe responses to variations in climate and their interactions. Respectively the end of the chapter deals with mitigation strategies like mulching, use of organic residues, fertilisers, crop and landscape administration are also taken into account.

## Action Mechanisms of Climate Change

Temperature, precipitation, and changes in length of seasons are all indicators of climate change [7]. Therefore, the major ways in which its mechanism of action is exhibited are changes in temperature and moisture levels.

### *Mechanisms Affecting the Microbes*

Soil microbial populations are affected both directly and indirectly by climate change elements such as increased atmospheric $CO_2$, changing temperature forms, and overall warming [8]. In addition, as a result of multiple components changing abruptly because of climate change, the terrestrial microbial population undergoes complicated alterations [8]. The microbial population, plants, and soil carbon balance may all be notably impacted by such large-scale changes brought on by climate change [9, 10]. Nonetheless, interactivity between different climate alterationelementsare also possibly discerning towards certain soil microorganisms, which can lead to conversions in factions that may ultimately determine the future condition of ecospheres [8].

## *Alterations in the Microbial Variety*

Negative impact like Abiotic stress brought on by climate change can change the variety and functioning of soil microbes [11]. Because different microbial species prefer different temperature scales for activity and growth, an increase in temperature may have an effect on how the microbial population is formed [6, 12]. Swiftness of processing of microorganisms, yield, as well as activity is prompted with an increase in temperature. Therefore, the microbial community shifts in approval of the species with sped up rates of development and better tolerance for higher temperatures [8]. The effects of climate change on two important cyanobacteria, namely *Microcoleus steenstrupii* and *Microcoleus vaginatus*, present in topsoil of arid region of western USA, exemplifies this impact. As global temperatures rise, the former, which is thermotolerant, has been observed to replace the latter and even outcompeteit, which ispsychrotolerant. These microorganisms are essential for preserving the topsoil's microbial community, whose traits are necessary for preventing soil erosion [13]. For the reason thatmicrobial community differ in terms of sensitivity to temperature, physiology, andgrowth rates,it shows that climate change alters both the relative abundance and activity of soil microbial populations. Therefore, as a result it has a direct impact on how these organisms' particular functions are regulated [9, 10]. Warming-related variations in the population of microbes' organisation may also result in a decrease in the amount of substrate that is readily available [14]. In the same context, it shall be noted that both bacteria and fungi abundance is likely to be impacted by global warming [14]. It is noteworthysincecertain microbes control ecological processes like nitrification, denitrification, nitrogen fixation, and methanogenesis. Therefore, changes in their relative abundance have a direct effect on how quickly these processes occur. Although, because a diversity of organisms manage some activities thattake place at a very coarserate (viz., as mineralization of nitrogen), abiotic factors like moisture and temperature have a greater impact on these processes than microbial community makeup [9, 10].

## *Conversions in Physiology*

Rising temperature increases the upkeep of microbes, which leads to escalation in demand of themaintainence of microbial community (respiration per biomass) [15]. As a result, heat increases soil respiration by accelerating soil microbial activity [16, 17]. Changes occurring in the respiration of soil is started because of alterations in the available carbon comparative abundance [18], composition of the microbial community [19], the quantity and quality of plant litter [17] and the availability of substrate [20, 21], which are all associated with temperature elevation. Therefore, it is believed that due to sensitivity to temperature of microbial metabolism and also the activities they engage in, changes all over the globe changes such as temperature increase can directly impact the rates of respiration of soil bacteria [9, 10]. Temperature and moisture levels are firmly connected, and high or low moisture levels may restrict soil respiration [22]. Although, until other factors like moisture and substrate become limited or the conformation/formation of a forest stand is reformed or changed, it is unlikely that the microbial community's makeup will change or that adaptations will occur that indicate a rise in soil respiration [14]. Changes in soil temperature and moisture brought on by differences in precipitation can also affect soil respiration [23]. In this context, enzyme activity should also be taken into account. It is important to note that, as temperatures rise, microbes allocate more nutrients for the development of enzymes (to obtain the additional nutrients needed) [24, 25]. In reality, due to direct and indirect effects on microbial production of enzymes and turnover rates, climatic change causes long-term changes in enzyme pools in addition to minisculechanges in activity of enzymes steered by thermodynamics [26, 27]. Due to their impacts on substrate availability, enzyme efficiency, and microbial efficiency, variations in temperature and moisture have an impact on both the comprehensive and relative rates of production of enzymes. If soil temperature rises, increasing the processing of substrate and the existing enzyme pool becomes available, microorganisms may devote less energy to producing enzymes if biomass of micobes stays constant [28]. It should be recognised that C-degrading enzymes are more temperature sensitive than N-degrading enzymes [29–31]. Substrate temperature sensitivity is a related issue that is influenced by a number of variables including oxygen availability, moisture content and accessibility (surface assimilation and accumulation state) [20]. The relationship between temperature and soil respiration can be understood by looking at substrate usage and microbial development [32]. Additionally, the kind of soil influences soil microbial activity, which may be a relevant role in this case. Due to the characteristics of allophone, it ought to be emphasised that microbiological activity is minimal in soil made of volcanic ash [33]. The fact that microbial biomass turnover, respiration and soil organic matter are all higher in tropical soils than in temperate soils serves as an illustration of the importance of temperature with regard to these processes [34].

## Action Mechanisms on Plants

Plants are significant biotic components that are crucial in this context. By allowing roots to release carbon substrates [35, 36], changing temperature of soilas well asmoisture with the help ofshade and transpiration [37], and changing the quantity of rain that ultimately reaches the soil, they modify rates of soil microbial respiration. Additionally, the type of plant remnants and quality viz., organic matter, that reaches the soil and the respiration of soil, are determined by the constitution of the vegetation. The variation in soil respiration beneath evergreen and deciduous forests serves as one example of this [38]. According to [39], the kind of anthropogenic land use and management and plant cover both affect the nature of organic materials in soils with a comparable geology. This is very significant since the key factors affecting how sensitive soil respiration is to temperature are the availability of temperature-dependent substrate release and rapidly decomposable carbon [32]. Changes in the sensitivity of temperature of organic matter of soil disintegration can result in significant inaccuracies in models of C-cycle [32].

### *Undulation in Moisture*

Changes in moisture, a major variable that significantly affects the patterns of soil respiration in many terrestrial ecosystems, is another way that climate change has an impact on soil ecosystems [40]. Numerous variables that change with the moisture present and amount of water, such as gas diffusion, water movement, solute diffusion, and the motility and survival of microorganisms, have an impact on microbial activity and, consequently, decomposition [22, 41]. Additionally, moisture could reduce activity of microbes in a variety of settings, including soils and saltwater. Less water availability diminishes intracellular water potential, which in turn lessens enzyme activity and hydration [42]. The release and dynamics of $CO_2$ can be significantly impacted by soil moisture [40]. All of this is demonstrated by the observation that in grasslands, temperature and soil moisture are the key regulators of respiration in soil, that in turn controls $CO_2$ response between soil and atmosphere [40].

## Consequences of Climate Change on Microbes

Microbes respond dynamically to both abiotic and biotic stimuli [43], therefore the consequences of change in climate on these microorganisms are evident. In general, soil microbes are extremely active and respond promptly to environmental factors [34]. However, the relevance of each environmental component is regulated by temporal and spatial dimensions [44]. At higher latitudes, the consequences of temperature rise on microbial processes are projected to be most severe [20, 45].

### *Rising Temperature*

By 2100, the average global surface temperature is expected to rise by 1.1 to 6.4 °C, which may have an impact on soil carbon sequestration by potentially accelerating heterotrophic microbial activities [46]. Droughts in the [40] area may become more frequent, intense, and long-lasting as temperatures rise [47]. The structure and activities of soil microbial communities are known to be sensitive to variations in both temperature and water accessibility [48]. Temperature increases hasten microbial breakdown, increasing $CO_2$ released by soil thereby creating a positive feedback loop to climate change [49]. Because of global warming, by 2100, it is anticipated that 25 percent of permafrost might melt resulting in releasing around 100 Petagrams (Pg) of carbon for microbial breakdown [20]. The enormous organic carbon stocks (400 Pg, or 4,000 million tonnes) in these soils are susceptible to higher breakdown rates due to higher melting rates and depths in high-latitude permafrost. Flooding of melted permafrost regions generates anaerobic conditions conducive to methanogenesis breakdown. Increased temperature is closely related to increased soil respiration, and a 2 °C increase in world average temperature is anticipated to increase soil carbon release by 10 Pg, owing mostly to increased microbial activity. The ideal scales of temperature for optimum activity and growth are different for different microbial groups. Rising temperatures can influence the composition of the microflora, which can limit the emission of organic carbon of soil in some circumstances due to the extinction of acclimatised microbiota [50]. Tropical soils emit more CO2 than temperate soils because to higher and longer heat regimes, where the overall rate of disintegration of organic matter is substantially faster due to increased microbial activity. Changes in soil temperature are anticipated to change microbial-operated nitrification and denitrification activities in the environment of soil due to population shifts indenitrifiers and nitrifiers. Changes in the soil microenvironment can induce community changes and changed metabolic reactions in microorganisms engaged in soil nutrient cycle, as well as an increase or decrease in the viability and pathogenicity of soil-mediated pathogenic bacteria such as *Salmonella typhimurium*. As a result of the lower temperature, microbial growth and activities normally reduces in the winter. In general, extremely high temperatures are harmful to many bacteria. Indeed, some organisms may be able to endure such harsh environments by transforming into dormant forms that can withstand high temperatures. Although, such typical periodical/seasonal patterns might differ in individual ecosystems of soil. For example, in arctic soil, microbial density is at its peak in late winter when temperature is reduced [51]. The ideal average temperature for microbe life is about around 20 °C, whereas the upper limit is somewhere near 50 °C [52].

## *Altered Precipitation*

The rate of decomposition of soil organic carbon and another significant regulator of terrestrial microbial community structure is soil moisture, which can be influenced by the IPCC's (Intergovernmental Panel on Climate Change) projected 20 percent increase or decrease in precipitation. Long dry periods may restrict microbial growth and decomposition, having a negative feedback effect on carbon flux in some ecosystems. Carbon dioxide generation is also influenced by the periodic soaking and drying of soil. When dry soils are re wetted, the activities of latent bacteria rises. This adds to increased $CO_2$ evolution during soil rewetting. Soil moisture can have an influence on chemical engineers both directly and indirectly. Soil moisture has a direct impact on bacteria's physiological condition and may impede their ability to breakdown various types of natural substances [53]. The soil moisture values required for optimum microbial activity vary according to type of soil and microbial community diversity [54]. Soil moisture also has an indirect effect on microbial community development, activity, and composition by changing the quality and amount of plant litter formation. These can have an impact on plant–microbe interactions. Since availability of water and temperature are driving forces of N mineralization, denitrification, and microbial activity in dry land soils [55, 56], changing climate will have a significant impact on these processes through its impact on soil water and temperature availability [57, 58].

## *Increased $CO_2$*

Anthropogenic $CO_2$ emissions are to blame for the current rise in atmospheric $CO_2$. Carbon dioxide levels in the atmosphere are rising at a 0.4 percent annual pace and are expected to double by 2100, owing mostly to anthropogenic activities including fossil fuel consumption and land-use changes. An estimated 30–40% of 2o produced by human activities into the atmosphere dissolves in seas, rivers, and reservoirs [59, 60], contributing to ocean acidification. The direct impact of increasing $CO_2$ on above-ground biomass production has indeed been widely researched [61]. It has been demonstrated that increasing above-ground net plant productivity (ANPP) increases C availability below-ground and boosts soil microbial activity [62]. Plants' average growth rate is accelerated by high $_{CO2}$ concentrations, allowing them to store more $CO_2$. Plant development was accompanied by a rise in soil respiration as a result of the increased availability of nutrients for breakdown by producing more $CO_2$ into the atmosphere. Increased $CO_2$ levels have an impact on the root zone's release of pliable sugars, organic acids, and amino acids, which can promote microbial activity. Long-term, it is thought that increased microbial biomass brought on by improved carbon release from roots may cause soil nitrogen to become immobilised, lowering the amount of nitrogen available to plants and creating a feedback loop that restricts further growth in plant development. The improved soil C:N ratio that follows may favour greater fungus diversity and dominance. Fungal cell walls are mostly made of carbon polymers (chitin and melatin), which are significantly more resistant to being destroyed than those found in bacterial membranes and walls (peptidoglycan and phospholipids). This means that fungi are more efficient at assimilating carbon (they store more carbon than they metabolise) than bacteria. As a result, soil respiration rates are often low in fungi-dominated environments, increasing the potential for carbon storage. A rise in atmospheric $CO_2$ may be one of the repercussions of climate change, and it can drastically alter the soil environment by changing the distribution of above and below-ground nutrients. Because $CO_2$ is the basic building block of photosynthesis, a rise in atmospheric $CO_2$ might result in enhanced plant growth. This may lead to an increase in rate of production of litter and a change in molecular structure of litter, which may result in a change in digestibility. Such changes will subsequently have an impact on the type of organic matter accessible to soil microbes [63]. As a result, altered litter generation may alter total carbon supply and N movement between plants and microbes [64]. Furthermore, rising $CO_2$ levels may result in increased root development, which will have a considerable influence on soil structure and serious ramifications for soil biota.

### *Droughts*

As temperatures rise, the intensity and severity of drought episodes in mesic ecosystems are expected to rise as well [65]. Water stress is predicted to have an impact on both microbial and plant populations, by disrupting important nutrient cycles and plant–microbe responses. Drought lowered soil moisture dramatically, generating unfavourable growth circumstances that resulted in a 50–80% fall in microbial population number [66]. Drought stress has been demonstrated to affect both the initiation and functioning of legume Rhizobium symbiosis [67, 68]. According to [69], populations of *Rhizobium leguminosarum* and *Rhizobium japonicum* declined biphasically in drying soils.

## New Developments and Improved Knowledge of Plant–Microbe Response to Climate Change

To understand complex community dynamics and function, studies attempting to understand microbial dynamics have traditionally relied on methods like DGGE (denaturing gel gradient electrophoresis), TRFLP (terminal restriction fragment length polymorphism), PLFA (phospholipid fatty acid analysis), or simply measures of biomass. In general, these methods have shown trends in the make-up of microbial communities [70], but they do not show responses from particular taxa and only offer a scant amount of information regarding functional changes. Researchers are now focusing on microbial interactions with hosts that are more functionally significant and at the highest resolution thanks to the development of new sequencing techniques and the -omics revolution. Researchers can identify changes in microbial communities that will enhance their comprehension of which bacteria are present in an environment and what their potential roles are by employing the methods of meta-genomics, transcriptomics, proteomics, and metabolomics [71, 72]. One tool that can be used to focus on the active microbial community, which is involved in a variety of tasks, is stable isotope probing [73]. When these methods are used more frequently, researchers are faced with a number of difficulties, such as determining which methods produce the most accurate results and how to analyse these enormous datasets in the most precise and pertinent ways. Amplicon sequencing of the 16 s rRNA gene has become popular for determining the makeup of the bacterial community in ecosystems [74]. Although this generates a lot of data at a depth where species accumulation curves are starting to saturate, it has very little to no impact on future functional changes in communities [75]. In order to comprehend the composition of microbial communities as well as their potential for function, some scientists are now using shotgun metagenomics to look at the variety of functional genes that are present in a habitat. The data produced by this method could be used to determine function, but it lacks the depth of amplicon sequencing and might miss rare taxa [76–78]. It is crucial to start sampling microbial communities at a size that is appropriate for the diversity and function of these tiny creatures, especially with the introduction of several new technologies targeted at understanding the dynamics of soil microorganisms.

At such a coarse geographic scale, it could be challenging to detect meaningful diversity patterns about these communities due to the significant variation contained in a soil sample [79]. Microorganisms can interact at the scale of the soil aggregate or at the plant root-soil interface, and there are significant differences between soil aggregates [80]. Future study should take into account the questions regarding diversity and function they are asking and appropriately alter their sampling technique to completely begin understanding how microbes interact with one another and their plant hosts. Beyond the question of what instruments to use to research microbial populations, the problem of how to interpret these significant datasets is a complex one [78]. Today, a variety of software programmes are available to assist with processing and analysis, including qiime [81], mothur [82], and less well-known tools like IMTORNADO [83], which assign taxonomy identity by utilising a variety of different taxonomic databases. The given dataset may produce different results depending on which of these processing approaches is used and which taxonomy is used when accessing the various databases. To enable dataset comparisons between laboratories and research teams, researchers must start contrasting diverse approaches and creating a standard procedure. Researchers must specifically investigate which processing pipeline produces the most pertinent results quickly, which database contains the most up-to-date and accurate taxonomic information for the taxa of interest, and how to standardise analyses across research groups in order to extract the most information from a given dataset. The molecular underpinnings of plant-microbial interactions at the plant root-soil interface, where microorganisms are prevalent and closely interact with plant roots, are also becoming better understood thanks to technological advancements [84]. It is difficult to identify how various soil bacterial subgroups enter the plant root and populate it. We are starting to put together the molecular foundation for these interactions by utilising state-of-the-art sequencing technologies that enable the rapid and affordable sequencing of entire organismal genomes. The genome of the ectomycorrhizal fungus Laccaria exhibits unusual characteristics, such as effector type small-secreted proteins with unknown functions that are only produced in symbiotic tissues, according to studies on the mutualistic relationship between Laccaria and its plant host [85]. Additionally, the plant host Populus has complete D-mannose lectin-like receptor gene deletions, which significantly reduces Laccaria colonisation [86]. By comprehending the molecular underpinnings of these interactions, the microbial population can be controlled to enhance plant and ecosystem level functions. It will also allow researchers to start creating microbial communities that can boost plant growth, carbon allocation, and carbon storage, as well as beginning to forecast which microbes will live in the plant root endosphere.

## Climate Change Effects on Plant–Microbe Interactions

Some plant species are adapting to climate change by moving to higher elevations and latitudes, flowering and leafing out earlier in the growing season, and changing the expression of advantageous features [87–92]. On a smaller scale the arctic has become increasingly shrubby as a result of warming, with woody shrubs replacing grasses and forbs in some parts. This change in the ecosystem's features has led to carbon feedbacks in these systems [93–96]. Soil communities, especially those that are strongly connected with plants, have the potential to speed up or slow down changes in plant communities. Studies by [97–99], for instance, found that microbial communities associated with rootscould have a big impact on phenology, plant survival, and the expression of functional characteristics. All of these characteristics are sensitive to climatic variations. There is currently a lack of knowledge regarding how interactions among plants, the microbial population with which they coexist, and climate change impact ecosystem processes [100, 101]. The carbon balance in the soil, changes in the overwhelming majority of the soil microbial community, and plant growth and establishment may all be adversely affected by climate change for a very long time. In reality, interactions between plants and soil ecosystems, such as plant-soil feedbacks, are among the most important yet poorly understood controllers of soil nitrogen and carbon dynamics. The interactions between plants and soil communities will decide how an ecosystem responds if soil microbial populations shift as a result of climate change, which effects the establishment and growth of plant species. Recent studies have shown that the early responses of the local soil ecosystem might shield plants from drought stress [102]. There is mounting evidence that shifts in microbial diversity may affect the selection of functional characteristics in plants [103]. The indirect impacts of climate on plants and the soil communities that support them can differ greatly from the direct effects of temperature on the majority of the soil community. [43] discovered, for instance, that changes in precipitation had an impact on the soil community and its function in an oldfield in TN (USA), but that the impact of precipitation on the composition and function of the soil community varied depending on the plant the soil was obtained from. To evaluate the influence of climate change on communities and functions, soil samples were collected and homogenised from various parts of the site. These results suggest that the reactions of soil ecosystems to climate change may be cancelled out if the mix of plant communities' changes along with climate change. Most research may not adequately capture these community and functional modifications because soils are collected from many plant species and homogenised together [43]. These interactions may progressively build up in the soil system and alter ecosystem function (like carbon cycling) and trajectory (like plant establishment), given the strong interactions between plants and the soil communities they are linked with;Strong interactions between plants and the soil communities they are linked with may eventually accumulate in the soil system and alter ecosystem function (like carbon cycling) and trajectory (like plant establishment); however, research must be conducted to distinguish these interactions.

## Alleviation Schemes

The same methods that boost productivity and resistance to climate change give favourable co-benefits in terms of agricultural GHG reduction. There are three basic techniques for regulating GHGs in agricultural production: (a) lowering emissions, (b) increasing carbon removal from the atmosphere, and (c) minimizing emissions by using bioenergy or agricultural expansion rather than growth [104]. There is a positive relationship amongst soil organic carbon and crop output; methods that improve fertility of the soil productivity also reduce GHG emissions, especially in places wherein soil degradation is a major concern [105]. Reference [106] distinguishes between actions with high and low mitigation potential, as well as those with high and low food security prospects.

### *Light Soil Sealing/Mulching*

The technique of mulching involves covering the soil's surface to prevent erosion and boost fertility. Mulch is frequently laid down at the beginning of the growing season for crops and can be replaced as necessary. By retaining both heat and moisture, it first helps to warm the soil. Mulch can be created from a variety of substances, such as organic waste products (such as hay, bark, and agricultural residue), manures, wastewater sludge, and rubber or plastic covers.

### *Utilization of Organic Waste (Compost, Manure, and Sludge)*

The amount of organic matter in soil is increased by a variety of carbon-rich wastes, including coffee-berry pulp, sludge, grain and legume straw, animal manure, etc. Before being applied to the field for agricultural reasons, organic leftovers should be given time to degrade. For microbes to grow and flourish, they need both carbon and nitrogen, and the addition of carbon-rich substances makes soil nutrients momentarily immobile.

### *Fertilizers*

Microorganisms become more active when nitrogen is made readily available to them by some inorganic nitrogenous fertilisers in large quantities. As a result, low-quality organic inputs and soil's organic content break down more quickly, leaving less soil carbon behind and the organic matter content of the soil continuing to decline. This causes the soil to become less healthy and its ability to hold water to decline.

### *Crop Administration/Selection of Species of Crop*

The sort of habitat that soil fauna can access depends on the agricultural crop that is chosen. Legumes, for instance, can act as organic fertilisers by boosting soil N levels through a symbiotic relationship with rhizobia. Because crop changes affect the populations of biological regulators, crop rotations can also help to reduce the accumulation of diseases and pests. In order to reduce nitrous oxide emissions, it is essential to employ crop management techniques that encourage N usage efficiency and yield.

### *Landscape Administration/Hedgerows and Grassy Field Margins*

The establishment of bushes and trees or grassland strips next to intensively farmed fields offers soil fauna a permanent habitat, food, and a secure environment. Due to their limited mobility, shrubs, as opposed to grassy field boundaries, are much more advantageous to soil critters, especially bio-controls; soil bacteria will have very little spread into the fields. This is important since 10% of the soil-dwelling species found on farms are only found in field edges.

## Microbial Communities and Mitigation Strategies

### *Managing Microbial Communities and Reducing $CO_2$ Release*

Around 2,000 Pg of organic carbon may be found in soils, which is double the quantity in the atmosphere and three times the amount in plants [46, 107]. It has been suggested that land use may be adjusted to sequester an additional 1 Pg of carbon every year in soil since different land types, such as woodlands, pastures, and agricultural land, have varying capacities to store carbon [107, 108].

### *Using Microbial Community Management to Lower Methane Emissions*

Worldwide, methane memissions are perhaps more directly regulated by microbes than carbon dioxide ($CO_2$) emissions. Microbial methanogenesis, it's a process which is performed by a variety of anaerobic archaea in seas, termite guts, wetlands, etc., accounts for the majority of natural emissions of methane ranging approximately up to 250 million tonnes methane per year. However, emissions from human activity, majorly fossil fuel extraction and landfills, outnumbers the natural sources.

## Conclusion

It is admirable that microbes play a role in regulating the amount of greenhouse gases in the atmosphere, but the scientific community still needs to fully comprehend and value this contribution. Given the reported unpredictability, it is obvious that knowing the immediate and long-term impacts of climate change on these bacteria, as well as their associated short- and long-term feedbacks, would aid in our comprehension of the potential contributions of these microbes. If used appropriately, microbes have the potential to be an important natural resource for reducing climate change. It might become a big problem rapidly if not managed carefully. It is imperative that we research this topic thoroughly and comprehend the underlying mechanics and then effectively apply what we learn to the formulation of solutions.

## Future Perspectives

According to projections on the World Meteorological Organization Website, the average global surface air temperature could rise from 1.4 °C and 5.8 °C by the year 2100, and predictions state that a 2 °C rise in global temperature would result in an increase in the release of soil carbon of 106 kg (i.e., 10 petagrams) of CO2 and some other greenhouse gases [62, 109, 110]. This could set off a chain reaction that would cause the temperature to rise even more and the surroundings to alter. Climate change is predicted to result in more precipitation throughout the winter months in northern medium and high latitudes as well as Antarctica. Instead of being spread out over multiple mild occurrences, larger amounts of rainwateris more probable to be discharged withiin a few extremely large outbreaks (World Meteorological Organization Website). As a result, various ecological factors in terrestrial and aquatic environments are anticipated to alter, which will have a significant effect on microbes. There are several models that forecast how such environmental changes may affect bacteria [111, 112]. Recent modelling methods and research, however, have shown that soil warming over a long period of timedepicts a larger greater than initially believed positive feedback between atmospheric soil organic matter release and climate warming [113]. Terrestrial ecosystems in the region of arctic are predicted to be especially hard hit by the issue. Consequently, the Arctic has been emphasised as a crucial area for identifying climate change [114]. But there are few mechanically determined models that forecast how soils will respond to climate change [115]. Separate ecosystems are probably going to react to the problem in different ways. For instance, it has been predicted that in reaction to climate change, European forest soils will behave as $CO_2$ sinks, on the other handsoils in the agricultural areacould lose organic matter and subsequently release $CO_2$ [116, 117].

## References


1. NASA (2015) http://climate.nasa.gov/solutions/adaptation-mitigation/. Accessed 15 Dec 2015
2. US EPA (2015) Climate change: basic information. http://www3.epa.gov/climatechange/basics/. Accessed 15 Dec 2015
3. Microbiology online (2015) Microbes and climate change. http://www.microbiologyonline.org.uk/aboutmicrobiology/microbesandclimatechange. Accessed 15 Dec 2015
4. Weiman S (2015) Microbes help to drive global carbon cycling and climate change. Microb Mag 10(6):233–238. https://doi.org/10.1128/microbe.10.233.1
5. Zimmer C (2010) The microbe factor and its role in our climate future. http://e360.yale.edu/feature/the_microbe_factor_and_its_role_in_our_climate_future/2279/
6. Singh BK, Bardgett RD, Smith P, Reay DS (2010) Microorganisms and climate change: terrestrial feedbacks and mitigation options. Nat Rev Microbiol 8(11):779–790. https://doi.org/10.1038/nrmicro2439,PMID20948551
7. Smith P, Fang C, Dawson JJC, Moncrieff JB (2008) Impact of global warming on soil organic carbon. Adv Agron 97:1–43. https://doi.org/10.1016/S0065-2113(07)00001-6
8. Castro HF, Classen AT, Austin EE, Norby RJ, Schadt CW (2010) Soil microbial community responses to multiple experimental climate change drivers. Appl Environ Microbiol 76(4):999–1007. https://doi.org/10.1128/AEM.02874-09,PMID20023089
9. Classen AT, Sundqvist MK, Henning JA, Newman GS, Moore JAM, Cregger MA et al (2015) Direct and indirect effects of climate change on soil microbial and soil microbial plant interactions: what lies ahead? Ecosphere 6(8):1–21. https://doi.org/10.1890/ES15-00217.1
10. Classen AT, Sundqvist MK, Henning JA, Newman GS, Moore JAM, Cregger MA et al (2015) Direct and indirect effects of climate change on soil microbial and soil microbial-plant interactions: what lies ahead? Ecosphere 6(8):130. https://doi.org/10.1890/ES15-00217.1
11. Shade A, Peter H, Allison SD, Baho DL, Berga M, Bürgmann H et al (2012) Fundamentals of microbial community resistance and resilience. Front Microbiol 3:417. https://doi.org/10.3389/fmicb.2012.00417,PMID23267351
12. Fierer N, Schimel JPA (2003) A Proposed mechanism for the pulse in carbon dioxide production commonly observed following the rapid rewetting of a dry soil. Soil Sci Soc Am J 67(3):798–805. https://doi.org/10.2136/sssaj2003.7980
13. DiGregorio BE (2015) Climate change affecting microbes in North America soils. American society for microbiology. https://www.microbemagazine.org/index.php?option=com_contentandview=articleandid=6497:climatechangeaffectingmicrobesinnorthamericasoils. Accessed 15 Dec 2015

14. Schindlbacher A, Rodler A, Kuffner M, Kitzler B, Sessitsch A, Zechmeister-Boltenstern S (2011) Experimental warming effects on the microbial community of a temperate mountain forest soil. Soil Biol Biochem 43(7):1417–1425. https://doi.org/10.1016/j.soilbio.2011.03.005,PMID21760644
15. Anderson JPE, Domsch KH (2010) A physiological method for the quantitative measurement of microbial biomass in soil. Soil Biol Biochem 2010:215–21
16. Wu Z, Dijkstra P, Koch GW, Peñuelas J, Hungate BA (2011) Responses of terrestrial ecosystems to temperature and precipitation change: a meta-analysis of experimental manipulation. Glob Change Biol 17(2):927–942. https://doi.org/10.1111/j.1365-2486.2010.02302.x
17. Rustad L, Campbell J, Marion G, Norby R, Mitchell M, Hartley A et al (2001) A meta-analysis of the response of soil respiration, net nitrogen mineralization, and aboveground plant growth to experimental ecosystem warming. Oecologia 126(4):543–62. https://doi.org/10.1007/s004420000544, PMID 28547240
18. Fierer N, Craine JM, McLauchlan K, Schimel JP (2005) Litter quality and the temperature sensitivity of decomposition. Ecology 86(2):320–326. https://doi.org/10.1890/04-1254
19. Balser TC, McMahon KD, Bart D, Bronson D, Coyle DR, Craig N et al (2006) Bridging the gap between micro- and macro-scale perspectives on the role of microbial communities in global change ecology. Plant Soil 289(1–2):59–70. https://doi.org/10.1007/s11104-006-9104-5
20. Davidson EA, Janssens IA, Luo Y (2006) On the variability of respiration in terrestrial ecosystems: moving beyond Q(10). Glob Change Biol 12(2):154–164. https://doi.org/10.1111/j.1365-2486.2005.01065.x
21. Eliasson PE, McMurtrie RE, Pepper DA, Stromgren M, Linder S, Agren GI (2005) The response of heterotrophic $CO_2$ flux to soil warming. Glob Change Biol 11(1):167–181. https://doi.org/10.1111/j.1365-2486.2004.00878.x
22. Luo Y, Zhou X (2006) Soil respiration and the environment. Academic Press, London
23. Aanderud ZT, Jones SE, Schoolmaster DR Jr, Fierer L (2013) Sensitivity of soil respiration and microbial communities to altered snowfall. Soil Biol Biochem 57:217–227
24. Wang G, Post WM, Mayes MA (2013) Development of microbialenzyme-mediated decomposition model parameters through steady-state and dynamic analyses. Ecol Appl 23(1):255–272. https://doi.org/10.1890/12-0681.1, PMID 23495650
25. Wang G, Post WM (2012) A theoretical reassessment of microbial maintenance and implications for microbial ecology modeling. FEMS Microbiol Ecol 81(3):610-617. https://doi.org/10.1111/j.1574-6941.2012.01389.x, PMID 22500928
26. Trasar-Cepeda C, Gil-Sotres F, Leirós MC (2007) Thermodynamic parameters of enzymes in grassland soils from Galicia, NW Spain. Soil Biol Biochem 39(1):311–319. https://doi.org/10.1016/j.soilbio.2006.08.002
27. Schimel J, Balser TC, Wallenstein M (2007) Microbial stressresponse physiology and its implications for ecosystem function. Ecology 88(6):1386–1394. https://doi.org/10.1890/06-0219,PMID17601131
28. Allison SD, Vitousek PM (2005) Responses of extracellular enzymes to simple and complex nutrient inputs. Soil BiolBiochem 37:937–944
29. Stone MM, Weiss MS, Goodale CL, Adams MB, Fernandez IJ, German DP et al (2012) Temperature sensitivity of soil enzyme kinetic sunder N-fertilization in two temperate forests. Glob Change Biol 18(3):1173–1184. https://doi.org/10.1111/j.1365-2486.2011.02545.x
30. Wallenstein MD, McMahon SK, Schimel JP (2009) Seasonal variation in enzyme activities and temperature sensitivities in Arctic tundra soils. Glob Change Biol 15(7):1631–1639. https://doi.org/10.1111/j.1365-2486.2008.01819.x
31. Wallenstein MD, Haddix ML, Lee DD, Conant RT, Paul EA (2012) A litter-slurry technique elucidates the key role of enzyme production and microbial dynamics in temperature sensitivity of organic matter decomposition. Soil Biol Biochem 47:18–26. https://doi.org/10.1016/j.soilbio.2011.12.009
32. Larionova AA, Yevdokimov IV, Bykhovets SS (2007) Temperature response of soil respiration is dependent on concentration of readily decomposable C. Biogeosciences 4(6):1073–1081. https://doi.org/10.5194/bg-4-1073-2007

33. Joa J, Moon K, Chun S, Kyung-San C, Hae-Nam HH (2010) Effect of temperature on soil microbial biomass, enzyme activities and PLFA content during incubation period of soil treated with organic materials. In: Proceedings of the 19th world congress of soil science, soil solutions for a changing world, Brisbane, Australia. Published on DVD
34. Joergensen RG (2010) Organic matter and micro-organisms in tropical soils. In: Dion P (ed) Soil biology and agriculture in the tropics. Springer, Berlin pp 17–44
35. Cardon ZG, Gage DJ (2006) Resource exchange in the rhizosphere: molecular tools and the microbial perspective. Annu Rev Ecol Evol Syst 37(1):459–488. https://doi.org/10.1146/annurev.ecolsys.37.091305.110207
36. Scott-Denton LE, Rosenstiel TN, Monson RK (2006) Differential controls by climate and substrate over the heterotrophic and rhizospheric components of soil respiration. Glob Change Biol 12(2):205–216. https://doi.org/10.1111/j.1365-2486.2005.01064.x
37. Lauenroth WK, Bradford JB (2006) Ecohydrology and the partitioning AET between transpiration and evaporation in a semiarid steppe. Ecosystems 9(5):756–767. https://doi.org/10.1007/s10021-006-0063-8
38. Rey A, Jarvis P (2006) Modelling the effect of temperature on carbon mineralization rates across a network of European forest sites (FORCAST). Glob Change Biol 12(10):1894–1908. https://doi.org/10.1111/j.1365-2486.2006.01230.x
39. Guntiñas ME, Gil-Sotres F, Leirós MC, Trasar-Cepeda C (2013) Sensitivity of soil respiration to moisture and temperature. J Soil Sci Plant Nutr 13(2):445–461
40. Aanderud ZT, Schoolmaster DR Jr, Lennon (2011) Plants mediate the sensitivity of soil respiration to rainfall variability. Ecosystems 14:156
41. Rodrigo A, Recous S, Neel C, Mary B (1997) Modelling temperature and moisture effects on C-N transformations in soils: comparison of nine models. Ecol Modell 102(2–3):325–339. https://doi.org/10.1016/S0304-3800(97)00067-7
42. Stark JM, Firestone MK (1995) Mechanisms for soil moisture effects on activity of nitrifying bacteria. Appl Environ Microbiol 61(1):218–221
43. Kardol P, Cregger MA, Campany CE, Classen AT (2010) Soil ecosystem functioning under climate change: plant species and community effects. Ecology 91(3):767–781. https://doi.org/10.1890/09-0135.1, PMID 20426335
44. Savage K, Davidson EA, Richardson AD, Hollinger DY (2009) Three scales of temporal resolutionfrom automated soil respiration measurements. Agric For Meteorol 149(11):2012–2021. https://doi.org/10.1016/j.agrformet.2009.07.008
45. The Core Writing Team (2007) Climate change 2007: synthesis report contribution of working groups I, II and III to the fourth assessment. Report of the intergovernmental panel on climate change. IPCC, Geneva, Sweden
46. IPCC (2007) The physical science basis. Clim change.
47. Global climate change impacts in the United States (GCCI) (2009). In: Karl TR, Melillo JM, Peterson TC (eds) Cambridge University Press, New York
48. Hartel PG (2005) Soil abiotic environmental factors. Sylvia, 2nd edn. Principles and applications of soil microbiology, pp 41–51
49. Allison SD, Wallenstein MD, Bradford MA (2010) Soil carbon response to warming dependent on microbial physiology. Nat Geosci 3:336–340
50. Li WKW, Dickie PM (1987) Temperature characteristics of photosynthetic and heterotrophic activities: seasonal variation in temperate microbial plankton. Appl Environ Microbiol 53(10):2282–2295. https://doi.org/10.1128/aem.53.10.2282-2295.1987, PMID 16347449
51. Schadt CW, Martin AP, Lipson DA, Schmidt SK (2003) Seasonal dynamics of previously unknown fungal lineages in tundra soils. Science 301(5638):1359–1361. https://doi.org/10.1126/science.1086940, PMID 12958355
52. Vannier G (1994) The thermo biological limits of some freezing intolerant insects -the super cooling and thermo stupor points. Acta Oecol 15:31–42
53. Harris RF (1980) Effect of water potential on microbial growth and activity. In: Water potential relat soil microbiol water potential, pp 23–95. https://doi.org/10.2136/sssaspecpub9.c2

54. Prado AGS, Airoldi C (1999) The influence of moisture on microbial activity of soils. Thermochim Acta 332(1):71–74. https://doi.org/10.1016/S0040-6031(99)00062-3
55. Gallardo A, Merino J (1998) Soil nitrogen dynamics in response to carbon increase in a Mediterranean shrubland of SW Spain. Soil Biol Biochem 30(10–11):1349–1358. https://doi.org/10.1016/S0038-0717(97)00265-4
56. Gallardo A, Schlesinger WH (1992) Carbon and nitrogen limitations of soil microbial biomass in desert ecosystems. Biogeochemistry 18(1):1–17. https://doi.org/10.1007/BF00000423
57. Schlesinger WH, Bernhardt ES (2013) Biogeochemistry: an analysis of global change. Academic Press, San Diego
58. Robertson GP, Groffman P (2007) Soil microbiology, biochemistry, ecology. Springer, New York
59. Feely RA, Sabine CL, Lee K, Berelson W, Kleypas J, Fabry VJ et al (2004) Impact of anthropogenic $CO_2$ on the $CaCO_3$ system in the Oceans. Science 305(5682):362–366. https://doi.org/10.1126/science.1097329,PMID15256664
60. Millero FJ (1995) Thermodynamics of the carbon dioxide system in the oceans. Geochimicaet Cosmochim Acta 59(4):661–677. https://doi.org/10.1016/0016-7037(94)00354-O
61. Pan Y, Melillo JM, McGuire AD, Kicklighter DW, Pitelka LF, Hibbard K et al (1998) Modeled responses of terrestrial ecosystems to elevated atmospheric $CO_2$: a comparison of simulations by the biogeochemistry models of the vegetation/ecosystem modeling and analysis project (VEMAP). Oecologia114(3):389–404. https://doi.org/10.1007/s004420050462, PMID 28307783
62. Pendall E, Bridgham S, Hanson PJ, Hungate B, Kicklighter DW, Johnson DW et al (2004) Below-ground process responses to elevated $CO_2$ and temperature: a discussion of observations, measurement methods and models. New Phytol 162(2):311–322. https://doi.org/10.1111/j.1469-8137.2004.01053.x
63. Zak DR, Pregitzer KS, Curtis PS, Holmes WE (2000) Atmospheric $CO_2$ and the composition and function of soil microbial communities. Ecol Appl
64. Berntson GM, Bazzaz FA (1997) Nitrogen cycling in microcosms of yellow birch exposed to elevated $CO_2$: simultaneous positive and negative below-ground feedbacks. Glob Change Biol 3(3):247–258. https://doi.org/10.1046/j.1365-2486.1997.00070.x
65. Knapp AK, Beier C, Briske DD, Classen AT, Luo Y, Reichstein M, Smith MD, Smith SD, Bell JE, Fay PA, Heisler JL (2008) Consequences of more extreme precipitation regimes for terrestrial ecosystems. Bioscience 58(9):811–821
66. Sheik CS, Beasley WH, Elshahed MS, Zhou X, Luo Y, Krumholz LR (2011) Effect of warming and drought on grassland microbial communities. Int Soc Microgr Ecol 5(10):1692–1700. https://doi.org/10.1038/ismej.2011.32,PMID21451582
67. Sprent JI (1971) Effects of water stress on nitrogen fixation in root nodules. Plant Soil (Special Volume):225–228
68. Kirda C, Danso SKA, Zapata F (1989) Temporal water stress effects on nodulation, nitrogen accumulation and growth of soybean. Plant Soil 120(1):49–55. https://doi.org/10.1007/BF02370289
69. Pena-Cabriales JJ, Alexander M (1979) Survival of Rhizobium in soils undergoing drying. Soil Sci Soc Am J 43(5):962–966. https://doi.org/10.2136/sssaj1979.03615995004300050030x
70. Gray SB, Classen AT, Kardol P, Yermakov Z, Mille RM (2011) Multiple climate change factors interact to alter soil microbial community structure in an old-field ecosystem. Soil Sci Soc Am J 75(6):2217–2226. https://doi.org/10.2136/sssaj2011.0135
71. Castro HF, Classen AT, Austin EE, Crawford KM, Schadt CW (2012) Development and validation of a citrate synthase directed quantitative PCR marker for soil bacterial communities. Appl Soil Ecol 61:69–75. https://doi.org/10.1016/j.apsoil.2012.05.007
72. Muller EE, Glaab E, May P, Vlassis N, Wilmes P (2013) Condensing the omics fog of microbial communities. Trends Microbiol 21(7):325–333. https://doi.org/10.1016/j.tim.2013.04.009,PMID23764387
73. Mau RL, Liu CM, Aziz M, Schwartz E, Dijkstra P, Marks JC et al (2015) Linking soil bacterial biodiversity and soil carbon stability. ISME J9(6):1477–1480. https://doi.org/10.1038/ismej.2014.205,PMID25350158

74. Sanschagrin S, Yergeau E (2014) Next-generation sequencing of 16S ribosomal RNA gene amplicons. J Vis Exp 90(90):e51709–e51709. https://doi.org/10.3791/51709,PMID25226019
75. Fierer N, Leff JW, Adams BJ, Nielsen UN, Bates ST, Lauber CL et al (2012) Cross-biome metagenomic analyses of soil microbial communities and their functional attributes. Proc Natl Acad Sci USA 109(52):21390–21395. https://doi.org/10.1073/pnas.1215210110,PMID23236140
76. Shade A, Jones SE, Caporaso JG, Handelsman J, Knight R, Fierer N et al (2014) Conditionally rare taxa disproportionately contribute to temporal changes in microbial diversity. MBio5(4):e01371-14. https://doi.org/10.1128/mBio.01371-14, PMID 25028427
77. Lynch MD, Neufeld JD (2015) Ecology and exploration of the rare biosphere. Nat Rev Microbiol 13(4):217–229. https://doi.org/10.1038/nrmicro3400,PMID25730701
78. Zhou J, He Z, Yang Y, Deng Y, Tringe SG, Alvarez-Cohen L (2015) High-throughput metagenomic technologies for complex microbial community analysis: open and closed formats. MBio6(1):e02288–02214. https://doi.org/10.1128/mBio.02288-14, PMID 25626903
79. Ranjard L, Lejon DPH, Mougel C, Schehrer L, Merdinoglu D, Chaussod R (2003) Sampling strategy in molecular microbial ecology: influence of soil sample size on DNA fingerprinting analysis of fungal and bacterial communities. Environ Microbiol 5(11):1111–1120. https://doi.org/10.1046/j.1462-2920.2003.00521.x,PMID14641591
80. Lombard N, Prestat E, van Elsas JD, Simonet P (2011) Soil-specific limitations for access and analysis of soil microbial communities by metagenomics. FEMS Microbiol Ecol 78(1):31–49. https://doi.org/10.1111/j.1574-6941.2011.01140.x,PMID21631545
81. Caporaso JG, Kuczynski J, Stombaugh J, Bittinger K, Bushman FD, Costello EK et al (2010) QIIME allows analysis of high-throughput community sequencing data. Nat Methods 7(5):335–336. https://doi.org/10.1038/nmeth.f.303,PMID20383131
82. Schloss PD, Westcott SL, Ryabin T, Hall JR, Hartmann M, Hollister EB et al (2009) Introducing Mothur: opensource, platform-independent, community-supported software for describing and comparing microbial communities. Appl Environ Microbiol 75(23):7537–7541. https://doi.org/10.1128/AEM.01541-09,PMID19801464
83. Jeraldo P, Kalari K, Chen XF, Bhavsar J, Mangalam A, White B et al (2014) IM-Tornado: a tool for comparison of 16S reads from paired-end libraries. PLoS ONE 9(12):e114804. https://doi.org/10.1371/journal.pone.0114804,PMID25506826
84. Hol WHG, Bezemer TM, Biere A (2013) Getting the ecology into interactions between plants and the plant growth-promoting bacterium *Pseudomonas fluorescens*. Front Plant Sci 4:81. https://doi.org/10.3389/fpls.2013.00081,PMID23596447
85. Martin F, Aerts A, Ahrén D, Brun A, Danchin EG, Duchaussoy F et al (2008) The genome of Laccaria bicolor provides insights into mycorrhizal symbiosis. Nature 452(7183):88–92. https://doi.org/10.1038/nature06556,PMID18322534
86. Labbé J, Jorge V, Kohler A, Vion P, Marçais B, Bastien C et al (2011) Identification of quantitative trait loci affecting ectomycorrhizal symbiosis in an interspecific F1 poplar cross and differential expression of genes in ectomycorrhizas of the two parents: populus deltoides and Populus trichocarpa. Tree Genet Genomes 7(3):617–627. https://doi.org/10.1007/s11295-010-0361-3
87. Grabherr G, Gottfried M, Paull H (1994) Climate effects on mountain plants. Nature 369(6480):448. https://doi.org/10.1038/369448a0, PMID 23320303
88. Walther GR, Post E, Convey P, Menzel A, Parmesan C, Beebee TJC et al (2002) Ecological responses to recent climate change. Nature 416(6879):389–395. https://doi.org/10.1038/416389a,PMID11919621
89. Parmesan C, Yohe G (2003) A globally coherent fingerprint of climate change impacts across natural systems. Nature 421(6918):37–42. https://doi.org/10.1038/nature01286, PMID 12511946.
90. Verheijen L, Aerts R, Brovkin V, Cavender-Bares J, Cornelissen J, Kattge J et al (2015) Inclusion of ecologically based trait variation in plant functional types reduces the projected land carbon sink in an earth system model. Glob Change Biol. https://doi.org/10.1111/gcb.12871

91. Hudson JMG, Henry GHR, Cornwell WK (2011) Taller and larger: shifts in Arctic tundra leaf traits after 16 years of experimental warming. Glob Change Biol 17(2):1013–1021. https://doi.org/10.1111/j.1365-2486.2010.02294.x
92. Walker MD, Wahren CH, Hollister RD, Henry GH, Ahlquist LE, Alatalo JM et al (2006) Plant community responses to experimental warming across the tundra biome. Proc Natl Acad Sci USA 103(5):1342–1346. https://doi.org/10.1073/pnas.0503198103,PMID16428292
93. Sturm M, Racine C, Tape K (2001) Climate change—increasing shrub abundance in the Arctic. Nature 411(6837):546–547. https://doi.org/10.1038/35079180,PMID11385559
94. Hinzman LD, Bettez ND, Bolton WR, Chapin FS, Dyurgerov MB, Fastie CL et al (2005) Evidence and implications of recent climate change in northern Alaska and other arctic regions. Clim Change 72(3):251–298. https://doi.org/10.1007/s10584-005-5352-2
95. Lawrence DM, Swenson SC (2011) Permafrost response to increasing Arctic shrub abundance depends on the relative influence of shrubs on local soil cooling versus large-scale climate warming. Environ Res Lett 6(4):8. https://doi.org/10.1088/1748-9326/6/4/045504
96. Pearson RG, Phillips SJ, Loranty MM, Beck PSA, Damoulas T, Knight SJ et al (2013) Shifts in Arctic vegetation and associated feedbacks under climate change. Nat Clim Change 3(7):673–677. https://doi.org/10.1038/nclimate1858
97. van der Heijden MGA, Klironomos JN, Ursic M, Moutoglis P, Streitwolf-Engel R, Boller T et al (1998) Mycorrhizal fungal diversity determines plant biodiversity, ecosystem variability and productivity. Nature 396(6706):69–72. https://doi.org/10.1038/23932
98. Friesen ML, Porter SS, Stark SC, von Wettberg EJ, Sachs JL, Martinez-Romero E (2011) Microbially mediated plant functional traits. Annu Rev Ecol Evol Syst 42(1):23–46. https://doi.org/10.1146/annurev-ecolsys-102710-145039
99. Wagner MR, Lundberg DS, Coleman-Derr D, Tringe SG, Dangl JL, Mitchell-Olds T (2014) Natural soil microbes alter flowering phenology and the intensity of selection on flowering time in a wild Arabidopsis relative. Ecol Lett 17(6):717–726. https://doi.org/10.1111/ele.12276,PMID24698177
100. Fischer DG, Chapman SK, Classen AT, Gehring CA, Grady KC, Schweitzer JA et al (2014) Marschner review: plant genetic effects on soils under climate change. Plant Soil 37:91–19
101. Mohan JE, Cowden CC, Baas P, Dawadi A, Frankson PT, Helmick K et al (2014) Mycorrhizal fungi mediation of terrestrial ecosystem responses to global change: mini-review. Fungal Ecol 10:3–19. https://doi.org/10.1016/j.funeco.2014.01.005
102. Lau JA, Lennon JT (2012) Rapid responses of soil microorganisms improve plant fitness in novel environments. Proc Natl Acad Sci USA 109(35):14058–14062. https://doi.org/10.1073/pnas.1202319109,PMID22891306
103. Lau JA, Lennon JT (2011) Evolutionary ecology of plant-microbe interactions: soil microbial structure alters selection on plant traits. New Phytol 192(1):215–224. https://doi.org/10.1111/j.1469-8137.2011.03790.x,PMID21658184
104. Smith P, Martino D, Cai Z, Gwary D, Janzen H, Kumar P et al (2007) Policy and technological constraints to implementation of greenhouse gas mitigation options in agriculture. Agric Ecosyst Environ 118(1–4):6–28. https://doi.org/10.1016/j.agee.2006.06.006
105. Lal R (2004) Soil carbon sequestration impacts on global climate change and food security. Science 304(5677):1623–1627. https://doi.org/10.1126/science.1097396,PMID15192216
106. FAO (2009) Food Security and agricultural mitigation in developing countries: options for capturing synergies. Rom
107. Smith P, Smith P (2004) Soils as carbon sinks: the global context. Soil Use Manag 20(2):212–218. https://doi.org/10.1079/SUM2004233
108. Houghton RA (2007) Balancing the global carbon budget. Annu Rev Earth Planet Sci 35(1):313–347. https://doi.org/10.1146/annurev.earth.35.031306.140057
109. Luo Y, Wan S, Hui D, Wallace LL (2001) Acclimation of soil respiration to warming in a tall grass prairie. Nature 413(6856):622–625. https://doi.org/10.1038/35098065,PMID11675783
110. Raich JW, Potter CS (1995) Global patterns of carbon dioxide emission from soils. Global Biogeochem Cycles 9(1):23–36. https://doi.org/10.1029/94GB02723

111. Wilson PDG, Brocklehurst TF, Arino S, Thuault D, Jakobsen M, Lange M et al (2002) Modelling microbial growth in structured foods: towards a unified approach. Int J Food Microbiol 73(2–3):275–289. https://doi.org/10.1016/s0168-1605(01)00660-2,PMID11934035
112. Dens EJ, Van Impe JF (2001) On the need for another type of predictive model in structured foods. Int J Food Microbiol 64(3):247–260. https://doi.org/10.1016/s0168-1605(00)00472-4,PMID11294347
113. Kirschbaum MUF (2004)Soil respiration under prolonged soil warming: are rate reductions caused by acclimation of substrate loss? Glob Change Biol 10(11):1870–1877. https://doi.org/10.1111/j.1365-2486.2004.00852.x
114. Ruess L, Michelsen A, Schmidt IK, Jonasson S (1999) Simulated climate change affecting microorganisms, nematode density and biodiversity in subarctic soils. Plant Soil 212(1):63–73. https://doi.org/10.1023/A:1004567816355
115. Liski J, Palosuo T, Peltoniemi M, Sievänen R (2005) Carbon and decomposition model Yasso for forest soils. Ecol Modell 189(1–2):168–182. https://doi.org/10.1016/j.ecolmodel.2005.03.005
116. Kirkby KJ, Smart SM, Black HIJ, Bunce RGH, Corney PM, Smithers RJ (2005) Long term ecological change in British woodland (1971–2001). In: English nature research report, vol 653. Peterborough: English Nature
117. Sleutel S, De Neve S, Hofman G (2003) Estimates of carbon stock changes in Belgian cropland. Soil Use Manag 19(2):166–171. https://doi.org/10.1079/SUM2003187

# Chapter 12
# Climate Changing Impact on Microbes and Their Interactions with Plants: An Overview

**Niraj Singh and Pranjal Pratim Das**

**Abstract** Global warming and climate change are the burning issues that affect all domains of life on earth directly or indirectly. In the context of microorganisms, it is now well known that diverse communities of microbes are associated with plants and play a crucial role in plant health by stimulating plant growth, enhancing resistance to diseases, biotic and abiotic stresses. Climate change is an emerging threat to disrupt plant–microbe interactions network at local to global scales. Interactions between plants and their associated microbes have critical influences on population dynamics, community composition, plant ecosystem, and on evolutionary processes. In the recent past, several researchers have highlighted that the plant microbiome has an important role in maintaining soil nutrient balance which is easily available for plants and it also provides strength to plants under stress conditions. In this review, we have highlighted recent research works related to climate influence on plant–microbe interactions and the mechanisms by which environmental factors create an impact on diverse plant-associated microbes, symbiotic associations, and plant-microbiota interactions. This review has indicated that presently, there is a great need for in-depth research in this area to increase an accurate understanding of climate change's impact on plant–microbe interactions in nature.

## Introduction

Climate is defined as the long-term weather conditions of a place that includes humidity, temperature, atmospheric pressure, wind, precipitation, etc. A climate is a complex system that has developed due to the interactions of multiple factors, such as water bodies, oceans, ponds, the earth's environment, atmosphere, glaciers

N. Singh (✉)
Department of Microbiology, The Assam Royal Global University, Guwahati, Assam 781035, India
e-mail: neerajsingh710@gmail.com; n.singh@rgu.ac

P. P. Das
Department of Biotechnology, Darrang College (Affiliated to Gauhati University), Gauhati, Assam 784001, India

 
J. A. Parray (ed.), *Climate Change and Microbiome Dynamics*, Climate Change Management, https://doi.org/10.1007/978-3-031-21079-2_12

as well as different forms of life or organisms. Climate change has been considered a major emerging threat to all domains of life on this planet. As the most diverse and abundant organisms, microorganisms and their associated activities are greatly affected by a changing climate. Microorganisms have a diverse community, colonizing soil, plants, and animals in aquatic and terrestrial conditions [5, 29, 33]. In the global warming era, the study to understand the impact of the resilient climate on microfauna and microflora is very limited and it requires more attention on it. Being ubiquitous, microorganisms are affected in all different environments such as marine, terrestrial, and agricultural ecosystems by climate change [7]. Due to variations in temperature, rainfall patterns, and biotic and abiotic stress conditions such as drought, salinity, ozone stress, pathogens, climate durability, and seasonal abnormalities have created a great impact on the structural and bio-diversity of microbial communities associated with plants. Plants harbor diverse types of microbiomes, such as endosphere-if the microbes are present inside termed and episphere-if present outside of the tissue and/or plant [39]. Soil is also considered as one of the best repositories of microbial population, whereas plants interact with this diverse microbial community and other entities present in nature. The microbial communities are classified as a rhizosphere, endosphere, and phyllosphere in the environment [78]. Plants and microbial populations encompass to form a "halobiont". Generally, halobionts are defined as the association of a host with various microbial species around them and together they form an ecological unit. Microbes interact with plants at diverse locations and do ecological functions in the both above-ground and below-ground environments. Microbes also act as major drivers of the different element cycles, such are nitrogen, carbon, and phosphorus. Over and above, microbes also play a potential role in maintaining resistance to the climatic change impact and against other types of stress responses. Earlier in the case of microbes, most of the studies were focused on their role as pathogens while at present, advances in the high-throughput sequencing and molecular techniques have helped us to improve our understanding of the beneficial roles of microbial communities for hosts and ecosystems [5, 29, 33]. Network of plant–microbe interactions involve a diverse variety of microbial populations from a number of different kingdoms [64, 85]. Plant-associated microbes are further defined by species of host plants and parts of plants, such as leaf, stem, root, and tissue location [3, 41]. Beneficial impacts of plant–microbe interactions like plant health have been demonstrated in number of research findings, such as root-associated microbiota play an important role in plant growth promotion and also enhance resistance against biotic and abiotic stresses [50, 85]. The leaf-associated microbial community has also been shown to be involved in fitness and growth promotion of the host [18, 66], resilience to abiotic stresses [41], and plant protection against pathogens and disease [31]. Further, positive correlations have been found between the plant-associated diverse microbial community and ecosystem's productivity [42], and it also has been observed that the decrease in microbial diversity was correlated with pathogen infection and disease propagation [40]. These observations and experimental data of plant-associated microbes have been strongly indicating the importance of an accurate understanding of the molecular mechanisms of host-microbe interactions to drive the better adaptation of plants to climate change and the global warming effect. While climate change is emerging as a big threat but conversations on its crucial link with plant-associated microbes are still rare outside of the microbiologist and allied science community. It has already been well reported that climate change severely affects the crop yield of many agriculturally important crops around worldwide [40, 76, 92] and it is expected to intensify the negative impact of climate variability on crop yield in the coming decades. Therefore, some innovative approaches are urgently required to minimize the influence of climate change on plants. In order to tackle climate change and plant health issues, it is very much important to learn how a changing climate will affect the microbes and their relationships with plants and their environment.

## Impact of Global Warming and Drought

Due to climate change, different forms of life across the earth are currently experiencing the rising temperatures [46]. O'Brien and Lindow have reported that due to rising temperature, bacterial colonizing pathways are affected in plant leaf surfaces [56]. A few recently published articles also have shown the influence of rising temperatures on plant–microbe interactions, [15, 82], immune system of host plants, [12] and soil microbes [26]. Numbers of studies have shown that temperature plays as one of the main drivers to develop and maintain the community composition of soil microbes [23], phyllosphere fungal [16], and ectomycorrhizal fungi [74]. The impact of intra-annual changing of temperature also have been studied via seasonality, which is observed to be a key determinant to maintain the composition of the microbial community in soil, such as in the rhizosphere [24] and the phyllosphere [25, 34, 58, 63]. Apart from temperature, abiotic factors also vary with seasonality and these variations also have a crucial role in the shaping of microbial communities. For example, in case of phyllosphere fungal assemblage, the number of days of frost in spring was found to be one of the key factors for dissimilarities in assemblage of phyllosphere fungal [16]. Peñuelas et al. have found that the bacterial and fungal phyllosphere communities were higher under spring and winter in comparison to Mediterranean summer [58]. Other studies also have reported that the root associated microbial communities are extremely variable during the growing season but clear predictable patterns in community composition still to be detected [21, 61]. Furthermore, Grady et al. observed variation of core leaf bacterial and archaeal communities in the early, mid, and late phases of switch grass growing season [25]. These observations have indicated that seasonality also influence the assembly of microbial community. Soil community composition of bacterial and archaeal has been found to vary significantly between semi-arid, arid and Mediterranean climates, which is indicating the availability of water acting as a key component in shaping communities composition in ecosystems [4]. Further, dry climatic condition influences the fungal community of soil, such as it increases the fungal diversity and total abundance in soil [28, 35]. While it reduces bacterial diversity, the soil with a history of water stress have shown lower bacterial diversity and abundance [44]. Generally, warm climate condition increases bacterial diversity and bacterial abundance used to enhanced under normal precipitation patterns, while drought along with warming condition significantly decreases the bacterial abundance [68]. However, the combined impact of drought and warming on microbial growth and diversity determination still have to be investigated for proper understanding. It is also reported that drought-adapted microbiota are observed for plants subjected to subsequent water stress [35, 44]. It is also now well known that some endophytes can improve host drought resilience, such as *Lolium* sp. and its endophyte *Epichloe* [47]. It was also found that functionality of leaf and root microbiota is affected by drought condition [30]. As global change in climates accelerating, these research findings and observation have highlighted the need to enhance our understanding about the impact of climate variation on microbial community diversity/abundance and its role for the maintenance of ecosystem productivity to tackle the prolonged warming and drought.

## Climate Variation Impact on Plant Microbiomes Assemblage

Plant microbiome assembly is a complex ecological process of continuous coevolution over millions of years that is governed by a number of factors. Plants system attracts the desired soil microbe's community to colonize and develop the plant microbiomes. The seed-associated microbiome facilitates the germination of seed and plant growth. Last few decades, host microbiome is gaining more attention from concerned researchers. Host microbiome is defined as the microbial community present in a particular species, irrespective of environmental conditions, seasons, and management, and plays a crucial role in the host's functions [72, 77]. Microbes are very sensitive to temperature and moisture to perform their physiological and metabolic function and therefore, climate variability act as an important factor that affect the assembly of the plant microbiome directly or indirectly.

Due to rapid fluctuation in the environmental conditions, the direct influence of climate change is likely to be more pronounced on the microbial communities covering the outer surface (phyllosphere) in comparison to those microbes in the internal plant tissue environments *i.e.,* the endosphere community [77]. The soil microbiome is directly influenced by climate, while the rhizosphere microbiome is not only impacted by the external climate but also indirectly influenced by the plant host responses such as root exudation, plant physiology variation, morphology, and immune response. In the recent past, many researchers suggested that plant-associated microbiome always give a consistent response to climate change [55, 83, 88]. Under reduced moisture conditions and drought, it was observed that a number of plant species selectively recruit gram-positive bacteria (due to thicker cell walls) to enhance tolerant against desiccation, while it reduces the gram-negative bacterial population in the root region and rhizosphere [55, 89]. The exact understanding of impacts of climate change on plant–microbiome assembly is a big challenge due to the interconnection of multiple factors and complex interaction processes.

Under climatic stress, plant–microbiome interactions are modulated by chemical communications. For example, plants have developed an exudation-mediated 'cry for help' mechanism for the recruitment of a stress-relieving microbiome when plant is exposed to stressful environmental conditions [49]. In the current scenario, there is a very limited knowledge available regarding the indirect influence of climate change on microbiome assembly in the host plants. Over and above, plants also have developed a mechanism to incorporate desired microbes and it acts as a multi-layered microbial management system for the most favourable microbes for incorporation into the plant tissues and to distinguish friend from foe [27, 75]. In the plant's first immune layer, where pattern recognition receptors recognize the microbe-associated molecular patterns, such as bacterial flagellin or fungal chitin, it induces microbe-associated molecular patterns' triggered immunity (MTI) while in the case of the second immune layer, pathogen effectors are recognized by nucleotide-binding leucine-rich repeat (NLR) resulting to the plants' effector-triggered immunity (ETI). Changing climate, warming, and drought have altered the plant immunity and the shape of the plant–microbiome. It is reported that warming can affect both an increase [11] and decrease [32] in MTI and to suppress ETI in plants [12, 19]. Suppression of ETI disrupts the host-mediated microbial colonization network which may cause dysbiosis in endosphere microbial communities living inside the plant tissues. Further, under the suitable environmental condition of mechanism of effector-triggered immunity suppression, plants reduce their immune response to the colonization of beneficial microbes and these microbes coordinate with the host to provide stress relief. During rapidly fluctuating surrounding environments, plants also modulate immunity through dynamic changes in hormonal pathways, such as salicylic acid (SA), abscisic acid (ABA), jasmonic acid (JA), and ethylene [48]. During warm and drought climates, salicylic acid production decreases and it is involved, via interaction with some other plant hormones such as jasmonic acid, in the assembly of both epiphytic and endophytic microbial communities [45]. Drought-induced production of abscisic acid, antagonistically it acts to salicylic-mediated immune signalling, changes in the different classes of defense metabolites and it allocation or distribution, plant hormones, and signalling molecules under climate change play a role in plant microbiome assembly.

## Climate Changing Impact on Plant–Microbe Interactions

At present, climate changing is increasing globally and it is a prevalent phenomenon affecting our food security worldwide. Climate change has resulted in the increased concentrations of $CO_2$ in the atmosphere, temperature elevation, and has changed the patterns of rainfall across the various parts of the globe (Fig. 12.1). The progression of climatic aberration may lead to several abiotic stresses and pathogen attack, which is detrimental to plants and crops. Besides, the changing climatic patterns disturb the hydrological cycle and availability of water which may also influence on agricultural

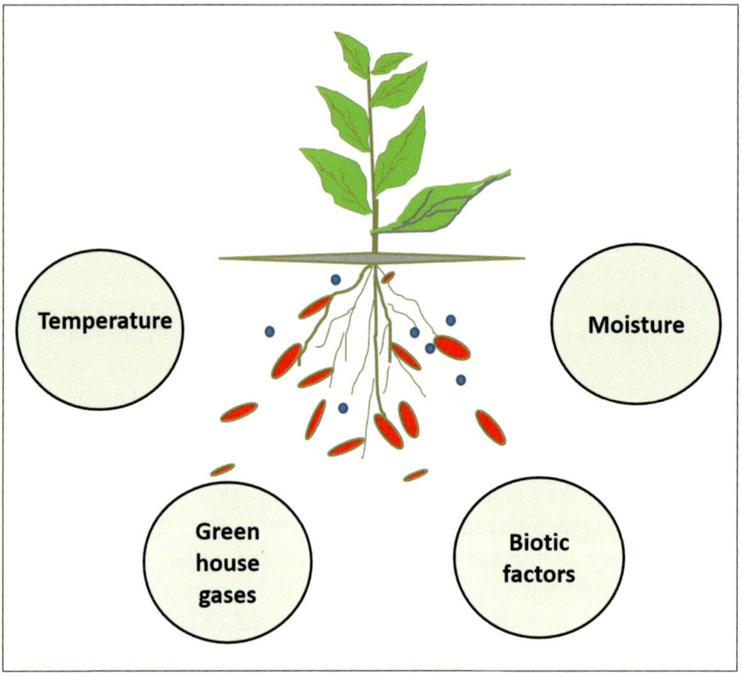


**Fig. 12.1** Climate change and plant-microbe interactions

production [36]. The variable climatic conditions also have affected the structure, function, assemblage, and interactions of microbes with host and non-host plants [17].

## *Plant–Microbiome Communication*

A system of communication exists between the host plant and microbiomes. Under stress environment, generally, plants exude some metabolites to recruit selective microorganisms to enhance plant resilience [44]. For example, in the drought-affected area, plants induce the secretion of glycerol-3-phosphate (G3P) in the roots enriching actinobacteria with the genetic potential to transport and utilize G3P for growth [89]. Drought induces a reduction in the iron and phytosiderophore availability in the rhizosphere and facilitates the actinobacteria population, which can adapt under low iron conditions and increase both the fitness and plant growth promotion ability [89]. The microbiome, associated with plants, also influences the host's phenotypic plasticity, which can impact plant phenology in a changing climate [17]. It is reported that microorganisms present in the rhizosphere may modulate the timing of flowering through the nitrogen (N) cycle and by the conversion of tryptophan in root exudates to the phytohormones-indoleacetic acid [49]. Moreover, plants also use volatile organic compounds to attract microbes, insects, and nematodes [87]. Rising of climate temperature increases the volatile compound emissions and it is also hypothesized that root exudate-mediated variation in microbial community composition may influence the changes in the immune responses of host plants, or signalling within the host. Deciphering the molecular mechanism of abiotic stresses influence the reshaping the microbes' composition and function of the plant microbiome is very much required to understand for developing strategies to enhance plant resilience under climate stresses.

## *Beneficial Plant–Microbe Interactions*

Climate change also has the diverse type of effects on beneficial plant–microbe interactions [12]. A warming climate decreases the photosynthate allocation in the underground part of the plants which affects the development of roots in diameter as well as in length [60]. Consequently, root colonization of arbuscular mycorrhizal fungi (AMF) is reduced or AMF species with lower carbon (C) requirements get more favoured [8, 51]. Few members of the plant-associate microbial community have traits that alleviate the impact of abiotic stresses on host plants [77, 79] (Table 12.1). For example, 1-aminocyclopropane-1-carboxylate (ACC) deaminase increases the stress tolerance by regulating the level of plant ethylene, and extracellular polymeric substances (EPS) developed the hydrophobic biofilms that help plants from desiccation. Plant hormone enhances level to stimulate the plant growth and induces accumulation of osmolytes and/or detoxifies reactive oxygen species. It also influence nutrient and water uptake by increasing root surface area and modulating the plant's epigenetic regulation that help in the adaptation to new environmental conditions. For example, Enterobacter sp. SA-87 induces thermotolerance by developing a novel mechanism in which heat shock factor A2 (HSFA2) constitutively expressed via ethylene signalling pathway and transcription factor EIN3 and these processes enhance the thermotolerance in plants [69]. In some plants, it also reported that growth-promoting bacteria help the plants to cope with multiple stresses [9, 43]. Researchers have validated that under stress conditions, improved plant performance is the result of the number of interactions in the plant microbiome that provide support against abiotic and biotic stresses. The scientific community still has a very limited understanding of the intertwined molecular mechanisms and the interactions between host plants and their microbiota under today's climate change. Identifying and understanding these mechanisms and the factors that influence them will help in the development of some noval approaches to neutralize the impacts of climate change on plant health.

**Table 12.1** Plant–microbe interaction under adverse climatic conditions

| Sl No | Microbes | Plant species | Abiotic stress | References |
|---|---|---|---|---|
| 1 | *S. meliloti* | *M. sativa* | Drought tolerance | [54] |
| 2 | Azotobacter | Maize | Drought stress | [71] |
| 3 | Salep gum and *Spirulina platensis* | Maize | Cd toxicity | [67] |
| 4 | *Achromobacter xylosoxidans* | Mustard green | Cu toxicity | [59] |
| 5 | *Glucoacenatobacter diazotrophicus* | Sugarcane | Drought | [81] |
| 6 | *Pseudomonas* sp. and *Bacillus* sp. | Spinach | Cd, Pb, Zn toxicity tolerance | [70] |
| 7 | *Bacillus aryabhatti* | Soybean | Heat stress tolerance | [57] |
| 8 | PGPB | Sorghum | Cr stress & heat stress tolerance | [52] |
| 9 | *Rhizobium* sp. | Sunflower | Drought | [2] |
| 10 | Microalgae-cyanobacteria | Tomato | Salt stress | [53] |
| 11 | *Burkholderia* sp. | Tomato | Cd toxicity tolerance | [20] |
| 12 | Cyanobacteria | Arabidopsis | Heat stress | [13] |
| 13 | *Serratia* sp. | Chickpea | Nutrient stress tolerance | [91] |
| 14 | *Bacillus subtilis* and *Paenibacillus illinoinensis* | Pepper | Drought tolerance | [84] |
| 15 | *Pseudomonas frederiksbergensis* OS261 | Red pepper | Salt stress | [10] |
| 16 | *Varivorax paradoxus* 5C-2 | Pea | Salinity tolerance | [86] |
| 17 | *Pseudomonas vancouverensis* | Tomato | Chilling stress tolerance | [73] |
| 18 | *Bacillus subtilis*, *Arthrobacter* species | Wheat | Salinity stress | [80] |
| 19 | *Cellulosimicrobium cellulans* | Chili | Chromium toxicity tolerance | [80] |

## *Pathogen–Plant Interactions*

A tripartite environment–host–pathogen interaction regulates the plant's health and productivity from resistance to different diseases. Climate change and associated factors alter the pathogen behaviour by changing the host–pathogen interactions and they also enhance the emergence of new pathogens and disease conditions [14].

Simultaneously, pathogens also can adopt different strategies of infection by modifying their virulence system that potentially leading to the breakdown of R gene-mediated plant resistance. It is already reported that warming and drought conditions can break down effector-triggered immunity (ETI) and promote disease in plant system [12]. Climate change studies on host–pathogen interactions have generally used a simplified model system which composed of a single pathogen interaction with a single host plant. However, in nature, plants interact with a large number of pathogenic microbes (pathobiota) [6] wherein the pathogen establishment and disease state depend on the competition between the pathogens and members of the plant-associated microbiome. Currently, there is lack of proper understanding of the interaction between pathogenic microbes and plant microbiota under exposure to long-term abiotic stresses.

## Hormonal Crosstalk with Plant–Microbe Interactions Under Changing Climatic Conditions

Plant hormones are organic substances that stimulate plant physiological processes. Phytohormones act as key regulators of plant growth and development and plant response to the surrounding environmental conditions. Climate change causes various stresses in plants, such as drought, salinity, heavy metals toxicity, incidence of pathogen attacks, and different diseases. Phytohormones are important regulators which provide a defence system to plants under abiotic and biotic stress conditions. Studies have shown that phytohormones improve plant growth and metabolic process under stress. ABA and auxin play key regulators role in abiotic stress tolerance [1]. The adverse impact of Pb on sunflower plant was mitigated by using the low auxin concentration with increased root growth. It is reported that seeds priming with auxin alleviate in many plant species that helps under abiotic stress [62]. Microbial communities associated with plants play key role for stress tolerance under changing climatic perturbations. Besides, the microbial colony associated with the plants influenced the plant hormone [22]. Microbes, associated in root, stimulate mitigation of osmotic stress and salt stress by the production of phytohormones [90]. PGPR also provides protection to the plant under stress by inducing phytohormone signalling as well as activating the defence responses [37, 38, 70]. Thus, it is now well established that hormones have a positive role against stress in changing climatic scenarios. However, to identify the phytohormone modulation in plant system by the microbial population, under stress response, require further in-depth study. Over and above, identification of transcription factors and receptors are needed to understand gene expression levels after the application of microbial phytohormone. Hormonal signalling and crosstalk mechanism of plant associated microbe like PGPR and plant growth promoting fungus in nutrient acquisition requires deeper attention. The role of biotechnological approaches in plant–microbe hormonal crosstalk, under stress conditions, warrants a thorough investigation.

## Conclusion and Future Prospects

Plant-associated microbial communities and their dynamics are well-known and concerned researchers in this aspect is still working toward a better understanding of the interaction networks of the microbe and its assemblages in plant systems. Above and below the ground, detail understanding on the several factors, that influence the plants-associated microbes, is still lacking. Recruitment number of desired microbes by plants near their root is called the rhizosphere that later enters inside the root system. Subsequently, various signal molecules coordinate the gathering of the plant microbiomes of the rhizosphere and phyllosphere. Molecular mechanisms linked with microbiome assembly, composition, and diversity in their function will provide tremendous scope for future research. Climate change is a global concern and it is enhancing climatic adversity as well as adversely affecting plant and microbial growth. The negative impact of climate change on microbial structure and their functioning in ecological niches is also a matter of concern. It is very much necessary to understand to what extent manipulation of plant-associated microbial composition could be done to enhance crop yield through sustainable agriculture that could maintain the environment in an eco-friendly manner. To explore in-depth knowledge about the plant–microbe interaction and host specificity, there is an urgent need for an advanced level of integrated innovative molecular approaches, such as meta-omics, ecological models, and bioinformatics, which may confirm the interlink of the correlation between plant-associated microbial community and environmental factors under climate changes.

## References

1. Ahmad I, Maathuis FJ (2014) Cellular and tissue distribution of potassium: physiological relevance, mechanisms and regulation. J Plant Physiol 171:708–714
2. Alami Y, Achouak W, Marol C, Heulin T (2000) Rhizosphere soil aggregation and plant growth promotion of sunflowers by an exopolysaccharide-producing Rhizobium sp. strain isolated from sunflower roots. Appl Environ Microbiol 66:3393–3398
3. Andrews JH, Harris RF (2000) The ecology and biogeography of microorganisms on plant surfaces. Annu Rev Phytopathol 38:145
4. Angel R, Soares MIM, Ungar ED, Gillor O (2010) Biogeography of soil archaea and bacteria along a steep precipitation gradient. ISME J 4:553–563
5. Banerjee S, Schlaeppi K, van der Heijden MG (2018) Keystone taxa as drivers of microbiome structure and functioning. Nat Rev Microbiol 16:567–576
6. Bartoli C et al (2018) In situ relationships between microbiota and potential pathobiota in Arabidopsis thaliana. ISME J 12:2024–2038
7. Bastiaansen R, Doelman A, Eppinga MB, Rietkerk M (2020) The effect of climate change on the resilience of ecosystems with adaptive spatial pattern formation. Ecol Lett 23:414–429
8. Bergmann J et al (2020) The fungal collaboration gradient dominates the root economics space in plants. Sci Adv 6:eaba3756
9. Bokhari A et al (2019) Bioprospecting desert plant Bacillus endophytic strains for their potential to enhance plant stress tolerance. Sci Rep 9:1–13

10. Chatterjee P, Samaddar S, Anandham R, Kang Y, Kim K, Selvakumar G, Sa T (2017) Beneficial soil bacterium Pseudomonas frederiksbergensis OS261 augments salt tolerance and promotes red pepper plant growth. Front Plant Sci 8:705
11. Cheng C, Gao X, Feng B, Sheen J, Shan L, He P (2013) Plant immune response to pathogens differs with changing temperatures Nat Commun 4:1–9
12. Cheng YT, Zhang L, He SY (2019) Plant-microbe interactions facing environmental challenge. Cell Host Microbe 26:183–192
13. Chua A, Sherwood OL, Fitzhenry L, Ng CK-Y, McCabe PF, Daly CT (2020) Cyanobacteria-derived proline increases stress tolerance in Arabidopsis thaliana root hairs by suppressing programmed cell death. Front Plant Sci 11:490075
14. Cohen SP, Leach JE (2020) High temperature-induced plant disease susceptibility: more than the sum of its parts. Curr Opin Plant Biol 56:235–241
15. Compant S, Van Der Heijden MG, Sessitsch A (2010) Climate change effects on beneficial plant–microorganism interactions. FEMS Microbiol Ecol 73:197–214
16. Cordier T, Robin C, Capdevielle X, Fabreguettes O, Desprez-Loustau ML, Vacher C (2012) The composition of phyllosphere fungal assemblages of E uropean beech (F agus sylvatica) varies significantly along an elevation gradient. New Phytol 196:510–519
17. Dastogeer KM, Tumpa FH, Sultana A, Akter MA, Chakraborty A (2020) Plant microbiome–an account of the factors that shape community composition and diversity. Curr Plant Biol 23:100161
18. Davison J (1988) Plant beneficial bacteria. Bio/technology 6:282–286
19. Desaint H, Aoun N, Deslandes L, Vailleau F, Roux F, Berthomé R (2021) Fight hard or die trying: when plants face pathogens under heat stress. New Phytol 229:712–734
20. Dourado MN et al (2013) Burkholderia sp. SCMS54 reduces cadmium toxicity and promotes growth in tomato. Ann Appl Biol 163:494–507
21. Edwards JA et al (2018) Compositional shifts in root-associated bacterial and archaeal microbiota track the plant life cycle in field-grown rice. PLoS Biol 16:e2003862
22. Foo E, Plett JM, Lopez-Raez JA, Reid D (2019) The Role of plant hormones in plant-microbe symbioses, vol 10. Frontiers Media SA
23. Frindte K, Pape R, Werner K, Löffler J, Knief C (2019) Temperature and soil moisture control microbial community composition in an arctic–alpine ecosystem along elevational and micro-topographic gradients. ISME J 13:2031–2043
24. Gomes T, Pereira JA, Benhadi J, Lino-Neto T, Baptista P (2018) Endophytic and epiphytic phyllosphere fungal communities are shaped by different environmental factors in a Mediterranean ecosystem. Microb Ecol 76:668–679
25. Grady KL, Sorensen JW, Stopnisek N, Guittar J, Shade A (2019) Assembly and seasonality of core phyllosphere microbiota on perennial biofuel crops. Nat Commun 10:1–10
26. Guerra CA, Delgado-Baquerizo M, Duarte E, Marigliano O, Görgen C, Maestre FT, Eisenhauer N (2021) Global projections of the soil microbiome in the Anthropocene. Glob Ecol Biogeogr 30:987–999
27. Hacquard S, Spaepen S, Garrido-Oter R, Schulze-Lefert P (2017) Interplay between innate immunity and the plant microbiota. Annu Rev Phytopathol 55:565–589
28. Hawkes CV, Kivlin SN, Rocca JD, Huguet V, Thomsen MA, Suttle KB (2011) Fungal community responses to precipitation. Glob Change Biol 17:1637–1645
29. Huttenhower C, Gevers D, Knight R, Abubucker S, Badger JH, Chinwalla AT et al (2012) Structure, function and diversity of the healthy human microbiome. Nature 486:207–214
30. Ibekwe AM et al (2020) Functional relationships between aboveground and belowground spinach (Spinacia oleracea L., cv. Racoon) microbiomes impacted by salinity and drought. Sci Total Environ 717:137207
31. Innerebner G, Knief C, Vorholt JA (2011) Protection of Arabidopsis thaliana against leaf-pathogenic Pseudomonas syringae by Sphingomonas strains in a controlled model system. Appl Environ Microbiol 77:3202–3210
32. Janda M, Lamparová L, Zubíková A, Burketová L, Martinec J, Krčková Z (2019) Temporary heat stress suppresses PAMP-triggered immunity and resistance to bacteria in Arabidopsis thaliana. Mol Plant Pathol 20:1005–1012

33. Jansson JK, Hofmockel KS (2020) Soil microbiomes and climate change. Nat Rev Microbiol 18:35–46
34. Jumpponen A, Jones K (2010) Seasonally dynamic fungal communities in the Quercus macrocarpa phyllosphere differ between urban and nonurban environments. New Phytol 186:496–513
35. Kaisermann A, de Vries FT, Griffiths RI, Bardgett RD (2017) Legacy effects of drought on plant–soil feedbacks and plant–plant interactions. New Phytol 215:1413–1424
36. Kashyap A (2018) India-Canada relations: environment and climate change. Indian Foreign Aff J 13:37–43
37. Kashyap AS et al (2021) Screening and biocontrol potential of rhizobacteria native to gangetic plains and hilly regions to induce systemic resistance and promote plant growth in chilli against bacterial wilt disease. Plants 10:2125
38. Khan N, Bano A, Ali S, Babar M (2020) Crosstalk amongst phytohormones from planta and PGPR under biotic and abiotic stresses. Plant Growth Regul 90:189–203
39. Khondoker NA, Uddin FJ, Sarker MAR, Rahman A (2020) Influence of nitrogen and phosphorus level for the performance of French Bean (Phaseolus Vulgaris L.). Acta Sci Malays (ASM) 4:34–38
40. Koskella B, Meaden S, Crowther WJ, Leimu R, Metcalf CJE (2017) A signature of tree health? Shifts in the microbiome and the ecological drivers of horse chestnut bleeding canker disease? New Phytol 215:737–746
41. Laforest-Lapointe I, Messier C, Kembel SW (2016) Host species identity, site and time drive temperate tree phyllosphere bacterial community structure. Microbiome 4:1–10
42. Laforest-Lapointe I, Paquette A, Messier C, Kembel SW (2017) Leaf bacterial diversity mediates plant diversity and ecosystem function relationships. Nature 546:145–147
43. Lata R, Chowdhury S, Gond SK, White JF Jr (2018) Induction of abiotic stress tolerance in plants by endophytic microbes. Lett Appl Microbiol 66:268–276
44. Lau JA, Lennon JT (2012) Rapid responses of soil microorganisms improve plant fitness in novel environments. Proc Natl Acad Sci 109:14058–14062
45. Lebeis SL et al (2015) Salicylic acid modulates colonization of the root microbiome by specific bacterial taxa. Science 349:860–864
46. Lenssen NJ, Schmidt GA, Hansen JE, Menne MJ, Persin A, Ruedy R, Zyss D (2019) Improvements in the GISTEMP uncertainty model. J Geophys Res: Atmos 124:6307–6326
47. Li F, Deng J, Nzabanita C, Li Y, Duan T (2019) Growth and physiological responses of perennial ryegrass to an AMF and an Epichloë endophyte under different soil water contents. Symbiosis 79:151–161
48. Li N, Euring D, Cha JY, Lin Z, Lu M, Huang L-J, Kim WY (2021) Plant hormone-mediated regulation of heat tolerance in response to global climate change Frontiers in Plant. Science 11:627969
49. Liu H, Brettell LE, Qiu Z, Singh BK (2020) Microbiome-mediated stress resistance in plants. Trends Plant Sci 25:733–743
50. Lugtenberg B, Kamilova F (2009) Plant-growth-promoting rhizobacteria. Annu Rev Microbiol 63:541–556
51. Ma Z et al (2018) Evolutionary history resolves global organization of root functional traits. Nature 555:94–97
52. Masuda S, Otomo S, Maruo C, Nishimura O (2018) Contribution of dissolved $N_2O$ in total $N_2O$ emission from sewage treatment plant. Chemosphere 212:821–827
53. Mutale-joan C et al (2021) Microalgae-cyanobacteria–based biostimulant effect on salinity tolerance mechanisms, nutrient uptake, and tomato plant growth under salt stress. J Appl Phycol 33:3779–3795
54. Naya L, Ladrera R, Ramos J, González EM, Arrese-Igor C, Minchin FR, Becana M (2007) The response of carbon metabolism and antioxidant defenses of alfalfa nodules to drought stress and to the subsequent recovery of plants. Plant Physiol 144:1104–1114
55. Naylor D, DeGraaf S, Purdom E, Coleman-Derr D (2017) Drought and host selection influence bacterial community dynamics in the grass root microbiome. ISME J 11:2691–2704

56. O'Brien RD, Lindow SE (1988) Effect of plant species and environmental conditions on ice nucleation activity of Pseudomonas syringae on leaves. Appl Environ Microbiol 54:2281–2286
57. Park Y-G et al. (2017) Bacillus aryabhattai SRB02 tolerates oxidative and nitrosative stress and promotes the growth of soybean by modulating the production of phytohormones. PLoS One 12:e0173203
58. Penuelas J, Rico L, Ogaya R, Jump A, Terradas J (2012) Summer season and long-term drought increase the richness of bacteria and fungi in the foliar phyllosphere of Quercus ilex in a mixed Mediterranean forest. Plant Biol 14:565–575
59. Pickering CM, Rossi S, Barros A (2011) Assessing the impacts of mountain biking and hiking on subalpine grassland in Australia using an experimental protocol. J Environ Manag 92:3049–3057
60. Qiu Y et al (2021) Warming and elevated ozone induce tradeoffs between fine roots and mycorrhizal fungi and stimulate organic carbon decomposition. Sci Adv 7:eabe9256
61. Redford AJ, Fierer N (2009) Bacterial succession on the leaf surface: a novel system for studying successional dynamics. Microb Ecol 58:189–198
62. Rhaman MS, Imran S, Rauf F, Khatun M, Baskin CC, Murata Y, Hasanuzzaman M (2020) Seed priming with phytohormones: an effective approach for the mitigation of abiotic stress. Plants 10:37
63. Rico L, Ogaya R, Terradas J, Peñuelas J (2014) Community structures of N 2-fixing bacteria associated with the phyllosphere of a H olm oak forest and their response to drought. Plant Biol 16:586–593
64. Sapp M, Ploch S, Fiore-Donno AM, Bonkowski M, Rose LE (2018) Protists are an integral part of the Arabidopsis thaliana microbiome. Environ Microbiol 20:30–43
65. Sarhan TZ, Ismael SF (2014) Effect of low temperature and seaweed extracts on flowering and yield of two cucumber cultivars (Cucumis sativus L.). Int J Agric Food Res 3
66. Schauer S, Kutschera U (2011) A novel growth-promoting microbe, Methylobacterium funariae sp. nov., isolated from the leaf surface of a common moss. Plant Signal Behav 6:510–515
67. Seifikalhor M, Hassani SB, Aliniaeifard S (2020) Seed priming by cyanobacteria (Spirulina platensis) and salep gum enhances tolerance of maize plant against c admium toxicity. J Plant Growth Regul 39:1009–1021
68. Sheik CS, Beasley WH, Elshahed MS, Zhou X, Luo Y, Krumholz LR (2011) Effect of warming and drought on grassland microbial communities. ISME J 5:1692–1700
69. Shekhawat K et al (2021) Root endophyte induced plant thermotolerance by constitutive chromatin modification at heat stress memory gene loci. EMBO Rep 22:e51049
70. Shilev S, Babrikova I, Babrikov T (2020) Consortium of plant growth-promoting bacteria improves spinach (Spinacea oleracea L.) growth under heavy metal stress conditions. J Chem Technol Biotechnol 95:932–939
71. Shirinbayan S, Khosravi H, Malakouti MJ (2019) Alleviation of drought stress in maize (Zea mays) by inoculation with Azotobacter strains isolated from semi-arid regions. Appl Soil Ecol 133:138–145
72. Singh BK, Trivedi P, Egidi E, Macdonald CA, Delgado-Baquerizo M (2020) Crop microbiome and sustainable agriculture. Nat Rev Microbiol 18:601–602
73. Subramanian P, Mageswari A, Kim K, Lee Y, Sa T (2015) Psychrotolerant endophytic Pseudomonas sp. strains OB155 and OS261 induced chilling resistance in tomato plants (Solanum lycopersicum Mill.) by activation of their antioxidant capacity. Mol Plant-Microbe Interact 28:1073–1081
74. Tedersoo L et al (2012) Towards global patterns in the diversity and community structure of ectomycorrhizal fungi. Mol Ecol 21:4160–4170
75. Teixeira PJP, Colaianni NR, Fitzpatrick CR, Dangl JL (2019) Beyond pathogens: microbiota interactions with the plant immune system. Curr Opin Microbiol 49:7–17
76. Tito R, Vasconcelos HL, Feeley KJ (2018) Global climate change increases risk of crop yield losses and food insecurity in the tropical Andes. Glob Change Biol 24:e592–e602
77. Trivedi P, Leach JE, Tringe SG, Sa T, Singh BK (2020) Plant–microbiome interactions: from community assembly to plant health. Nat Rev Microbiol 18:607–621

78. Trivedi P, Leach JE, Tringe SG, Sa T, Singh BK (2021) Author Correction: Plant–microbiome interactions: from community assembly to plant health. a 19:72–72
79. Trivedi P, Mattupalli C, Eversole K, Leach JE (2021) Enabling sustainable agriculture through understanding and enhancement of microbiomes. New Phytol 230:2129–2147
80. Upadhyay SK, Singh JS, Saxena AK, Singh DP (2012) Impact of PGPR inoculation on growth and antioxidant status of wheat under saline conditions. Plant Biol 14:605–611
81. Vargas L et al (2014) Drought tolerance conferred to sugarcane by association with Gluconacetobacter diazotrophicus: a transcriptomic view of hormone pathways. PLoS ONE 9:e114744
82. Velásquez AC, Castroverde CDM, He SY (2018) Plant–pathogen warfare under changing climate conditions. Curr Biol 28:R619-R634
83. Vescio R, Malacrinò A, Bennett AE, Sorgonà A (2021) Single and combined abiotic stressors affect maize rhizosphere bacterial microbiota. Rhizosphere 17:100318
84. Vigani G et al (2019) Root bacterial endophytes confer drought resistance and enhance expression and activity of a vacuolar H+-pumping pyrophosphatase in pepper plants. Environ Microbiol 21:3212–3228
85. Vorholt J (2012) Microbial life in the phyllosphere. Nat Publ Gr 10:828–840
86. Wang Q, Dodd IC, Belimov AA, Jiang F (2016) Rhizosphere bacteria containing 1-aminocyclopropane-1-carboxylate deaminase increase growth and photosynthesis of pea plants under salt stress by limiting Na+ accumulation. Funct Plant Biol 43:161–172
87. Weisskopf L, Schulz S, Garbeva P (2021) Microbial volatile organic compounds in intra-kingdom and inter-kingdom interactions. Nat Rev Microbiol 19:391–404
88. Wipf HM-L, Bùi T-N, Coleman-Derr D (2021) Distinguishing between the impacts of heat and drought stress on the root microbiome of Sorghum bicolor. Phytobiomes J 5:166–176
89. Xu L et al (2018) Drought delays development of the sorghum root microbiome and enriches for monoderm bacteria. Proc Natl Acad Sci 115:E4284–E4293
90. Yandigeri MS et al (2012) Drought-tolerant endophytic actinobacteria promote growth of wheat (Triticum aestivum) under water stress conditions. Plant Growth Regul 68:411–420
91. Zaheer A, Mirza BS, Mclean JE, Yasmin S, Shah TM, Malik KA, Mirza MS (2016) Association of plant growth-promoting Serratia spp. with the root nodules of chickpea. Res Microbiol 167:510–520
92. Zhao C et al (2017) Temperature increase reduces global yields of major crops in four independent estimates. Proc Natl Acad Sci 114:9326–9331

# Chapter 13
# Soil Salinity and Climate Change: Microbiome-Based Strategies for Mitigation of Salt Stress to Sustainable Agriculture

**Manisha Phour and Satyavir S. Sindhu**

**Abstract** Global climate change, environmental stresses, intensification of cropping practices, changed precipitation cycles, depleted water resources and reduction in soil fertility are the major constraints limiting crop productivity. Among various environmental (abiotic) stresses, soil salinity is one of the serious climate change impact, which affects about 20 and 33% of the total cultivated and irrigated agricultural lands, respectively. In recent years, soil salinization of agricultural land, along with water and environmental pollution; have emerged as significant threats to worldwide food security and agricultural sustainability. Salt stress results from excessive accumulation of salts in the soil that significantly affects soil fertility, stability, biodiversity, and consequently affects crop productivity. These problems necessitated the search of sustainable and eco-friendly agri-technologies to ameliorate the adverse effects of salt stress on plant growth and crop yield. In this context, some microorganisms inhabiting either the plant rhizosphere in extreme environments, or within halophytic plant roots, also possessing other plant growth-promoting traits, showed enormous potential in enhancing the adaptation ability of stressed plants to salinity stress conditions. These plant-associated beneficial microbes play key role in salt stress mitigation by producing osmoprotectants, antioxidants, ACC deaminase enzyme, hormones, exopolysaccharides, organic acids, nitric oxide and siderophores along with increased nutrient availability. Subsequent inoculations of crop plants with such salt–tolerant plant growth–promoting bacteria (PGPB) were found to increase the plant growth and crop yield of different plants grown in saline soils. This review briefly summarizes the different biochemical and molecular mechanisms employed by rhizospheric microbial communities for alleviation of salinity stress. Further, in-depth knowledge related to beneficial interactions of salt-tolerant microbes with the native crop plants is needed to facilitate plant growth and crop productivity under saline agro-ecosystems.

---

M. Phour
University Institute of Biotechnology, Chandigarh University, Mohali, India

S. S. Sindhu (✉)
Department of Microbiology, CCS Haryana Agricultural University, Hisar 125004, India
e-mail: sindhuss@hau.ac.in



J. A. Parray (ed.), *Climate Change and Microbiome Dynamics*, Climate Change Management, https://doi.org/10.1007/978-3-031-21079-2_13

## Introduction

Increasing crop production to fulfill world food demand is a key agricultural challenge for sustaining 70% of food sources in order to feed 9 billion people by 2050 [1]. Changing agro-climatic factors, using integrated management techniques, as well as current intensive cropping systems are the limiting constraints for increasing crop yield in agricultural systems [2, 3]. Climate change, declining water sources, soil salinization, water pollution and limited availability of cultivated land are the other major constraints to twenty-first century agriculture [4–7]. Moreover, crop yield is hampered by high winds, dryness, soil salinity, high temperatures, and flooding. Among all these constraints, soil salinity is a worst environmental stress that reduces area of productive land, plant growth, crop yield as well as quality of agri-produce [8–10]. In addition, farmers use excessive amount of nitrogenous and phosphatic fertilizers in intensive farming system for increasing food production [11, 12]. The injudicious use of chemical fertilizers in modern agriculture has further degraded soil and water quality, rendering soils biologically inert and often excessively saline, and it has even polluted surface and ground water [13]. It is estimated that between 20 and 33% of the world's agricultural lands have been damaged as a result of soil salinity, which has led to losses of $27.3 billion worldwide [3, 14].

Due to increasing problem of soil salinity, alternative strategies are needed to sustain agriculture production in salt-stressed soil and to increase crop yield in an eco-friendly and sustainable manner [15, 16]. The major strategies include plant genetic engineering, conventional breeding, and the use of salt-tolerant plant growth promoting rhizobacteria (ST-PGPR) as bio-inoculants in order to alleviate deleterious effects of high salt stress on plant growth and development [17–19]. In addition, increased salinity levels have also been reported to adversely affect microbial population and their plant-growth-promoting (PGP) properties [20]. These observations suggested the isolation and utilization of salt-tolerant plant-growth-promoting bacteria (PGPR) to protect crops from salinization and climate change. Therefore, different laboratories worldwide are currently involved in screening of salt-tolerant microorganisms obtained from different habitats and agroclimatic zones, and from various plant parts and regions i.e., phyllosphere, rhizosphere, and endorhizosphere, for their tolerance to high salt concentrations to cope up with high soil salinity levels. These halo-tolerant microorganisms are subsequently tested for mitigation of salinity stress on plants, for increasing nutrient uptake [21] and to enhance plant growth [22, 23]. Thus, application of selected salt-tolerant microbes in the form of bio-enhancers/bioprotectants may lead to increased survivability of crop plants under extreme saline conditions through alteration in various physiological, biochemical, and molecular pathways, resulting in enhancement of crop productivity [24–26].

Several microorganisms belonging to different genera, such as *Azotobacter*, *Azospirillum*, *Bacillus*, *Burkholderia*, *Enterobacter*, *Enterococcus*, *Pantoea*, *Paenibacillus*, *Pseudomonas*, *Rhizobium*, *Serratia*, *Stenotrophomonas* and *Variovorax* have been found to induce stress tolerance in different crops and positively influence plant growth under adverse saline conditions [19, 26–30]. For instance, salt-tolerant indigenous species of *Pseudomonas*, *Agrobacterium*, *Klebsiella*, *Bacillus*, and *Ochrobactrum* isolated from the halophytic plant, *Arthrocnemum indicum*, showed tolerance to 4–8% NaCl and improved productivity of groundnut in saline soil over uninoculated control plants [31]. Many salt-tolerant strains of *Bacillus* also possessed other plant growth promoting (PGP) traits along with high tolerance to excess of salt (4% NaCl) [32]. Some of the bacterial isolates showed salt tolerance even upto 10% NaCl along with excellent PGP attributes including solubilization of P, K and Zn, and production of indole-3-acetic acid (IAA), cell wall degrading enzymes, exopolysaccharides, biofilm, antibiotics, and siderophores [33–35].

Salt-tolerant bacteria employ several direct and indirect mechanisms to survive and proliferate under salt-stressed conditions in soil, and subsequently contribute towards amelioration of salt stress and stimulation of plant growth resulting into increased crop yield. Some of these salt-tolerant bacteria are currently being developed as biofertilizers; as a cost-effective environmental-friendly agri-technology to increase food production [36–39]. This chapter summarizes the characterization of salt-tolerant microbes and discusses various mechanisms involved in amelioration of salt stress. The use of salt-tolerant PGPR as bio-inoculants to improve crop production under salt stress conditions is also documented. Information provided in this chapter will help in understanding of plant-microbe interactions under saline environments to improve saline soil-based agriculture.

## Climate Change and Soil Salinization

Agriculture is the most vulnerable sector that is often exposed to plethora of climate changes. Global warming, changes in precipitation patterns and recent abrupt changes in climatic conditions has increased incidence of abiotic and biotic stresses [6, 40]. The exposure of plants to stressed environments has been accounted for as major cause for stagnation of productivity in agriculture and horticulture crops [40, 41]. Recent climate changes accompanied by altered precipitation cycles and depleted water resources are further expected to exacerbate crop stresses [42]. Several abiotic stresses such as extreme temperatures (heat stress, cold and frost), drought, flooding, soil salinity and nutrient stress have been found to adversely affect crop cultivation, plant development and production of cereal as well as legume crops under field conditions [43, 44]. Besides, intensive utilization of agricultural lands for growth of exhaustive crops has further declined soil fertility and environmental degradation.

Inter-seasonal climatic variability is a major concern among abiotic stress factors, which is normally reflected from year-to-year fluctuations in crop yields. The abiotic stresses, for example, extreme temperatures, dry season, flooding, salinity, metal stress and nutrient stress are the results of climate change and global warming, which causes alteration in precipitation patterns [6]. Abiotic stresses also cause land degradation, which make soil nutrient deficient and more stress prone [43]. Abiotic stresses are blended and associated with each other. For instance, increase or reduction in rainfall, rise or fall in temperature brings dry spell stress. Dry spell stress at last gives rise to salinity stress, which causes alkalization of soil. The nutrients stay inaccessible to the plants developed in alkaline saline soil and it leads to nutrient-deprived circumstances or nutrient stress. Humidity in environment is another climatic variability. In moist regions, pace of precipitation is high and soil leaching decreases soil pH because of decrease of basic cations. Hence, decline in soil pH results in acidification stress. Because of acidification stress, nutrient become inaccessible to plants and further leads to nutrient stress in the soil. Accordingly, abiotic stresses appear to be interconnected with each other and function as a chain because of variations in climatic environments [44].

The probability of occurrence of extreme climatic changes has increased in the last couple of decades and has reshaped the Earth's ecosystems [43, 45]. Climate change has accelerated tenfold in the last century and green house gas (GHG) emissions have caused a rise of 0.9 °C in average temperature in the nineteenth century. Warming could reach 1.5 °C by 2050 due to deforestation, GHG emissions, and pollution of soil, water, and air. The enormous temperature rise has exacerbated droughts, food shortages, unexpected precipitation, and heat waves. On the other hand, farmers lack the appropriate management technologies to sustain agricultural productivity under forced abiotic stress conditions, which adversely influence plant growth and yield [43]. The climate change has also far-reaching effects on survival and functioning of beneficial microorganisms and climate-smart agricultural practices, which is vital to food supply and the global economy [45]. Climate change models have anticipated that warmer temperatures and increase in the frequency and term of dry spells during twenty-first century will have net negative consequences for productivity of agricultural and horticultural crops. Natural disaster damages have topped $200 billion annually since 2016, and 95% of these losses are due to climate-related weather events like cyclones, floods, and droughts. The world's population is predicted to top 9 billion by 2050, straining agricultural areas, which are already impacted by climate change. Thus, rapid climate change has threatened global food security due to its adverse effects on crop productivity [43].

## Global Distribution of Saline Soils

Human activities have disrupted the natural hydrological equilibrium in many agro-climatic regions since the beginning of industrialization. These operations disrupt the natural distribution of salt in various landscapes and deteriorated the natural and agricultural environments. Soil salinization is a major threat to global food supply with changes in climatic conditions [46]. Poor drainage, brackish water irrigation, and long-term agricultural irrigation increase the salinity in soil [47]. The primary salinization area is less than one billion acres, where as secondary salinization has covered an area of 77 million hectares (with 58% occurring in irrigated areas and 20% of all irrigated lands) [48, 49]. About 5.2 billion hectares of agricultural land worldwide are salt-affected and are unsuitable for crop cultivation [50]. Low rainfall, erosion of native rocks, excessively surface evaporation, use of inorganic fertiliser, irrigation with salty water, and unsustainable farming techniques all lead to soil salinization [51, 52]. By 2050, half of all arable land may be salt-affected. More than $7 \times 10^6$ hectares of soil in India are salt-affected [53, 54].

Excessive accumulation of salts in the soil limits uptake of plant nutrients and water absorption, thereby disrupting plant growth and development processes [55]. Excess calcium, magnesium, sodium, sulphate, and chloride ions limit plant development by causing soil salinization. Farmland salinization is increasing by 0.3–1.5 million hectares per year, resulting in agricultural production losses of more than 20%. The salinization of arable land will have an impact on agricultural revenue and economic development, along with global food supply; and crop productivity losses may cost about 12–27.3 billion dollars per year [14, 56]. Chemical or physical methods used for salt extraction from salt affected soils may contribute towards restoration of saline soils [14] (Fig. 13.1). For example, lime and gypsum are two chemical neutralizers [57], whereas, leaching, scraping and flushing are physical methods for salinity management [58]. In addition, crops that are tolerant to salt, such as barley, rice, wheat, mung bean, cotton, and canola, are being developed [59]. Only a small number of salt-tolerating genes have been investigated for their potential to enhance crop production in both normal and saline soil [60]. It is common to increase agricultural output by employing environment-friendly methods and upgrading irrigated land. Biotic and abiotic factors have an effect on the current agricultural system, making it more efficient and sustainable is a major challenge for agriculture scientists [61].

Recently, use of salt-tolerant plant growth-promoting rhizobacteria as biofertilizers has emerged as novel agri-technology for improving soil health and crop yield under salt stress conditions [7, 44, 62–65]. These salt-tolerant rhizobacteria produce osmo-regulators, antioxidants, exopolysaccharides, ACC deaminase, nitric oxide, phytohormones, siderophores and transporter proteins, which act as promising bio-enhancer for increasing crop productivity and phytopathogen resistance, thereby sustaining soil health under salt stress conditions [3, 18, 39, 56, 64].

## Salinity Stress and Impact on Plants and Microbes

Soil salinity has emerged as a major environmental issue due to disastrous consequences of salt deposition in soils and its detrimental influence on agriculture production [4, 6, 14]. Plants acquire an array of protective genetic and metabolic mechanisms during the course of evolution to combat adverse environmental fluctuations and stresses. Many a times, the burden of abiotic stresses is reduced with the contribution of associated microbes. Various studies on plant-microbe interactions established that salinity has profound effect on the survival and activity of soil-inhabiting microorganisms as well as on the growth of plants.

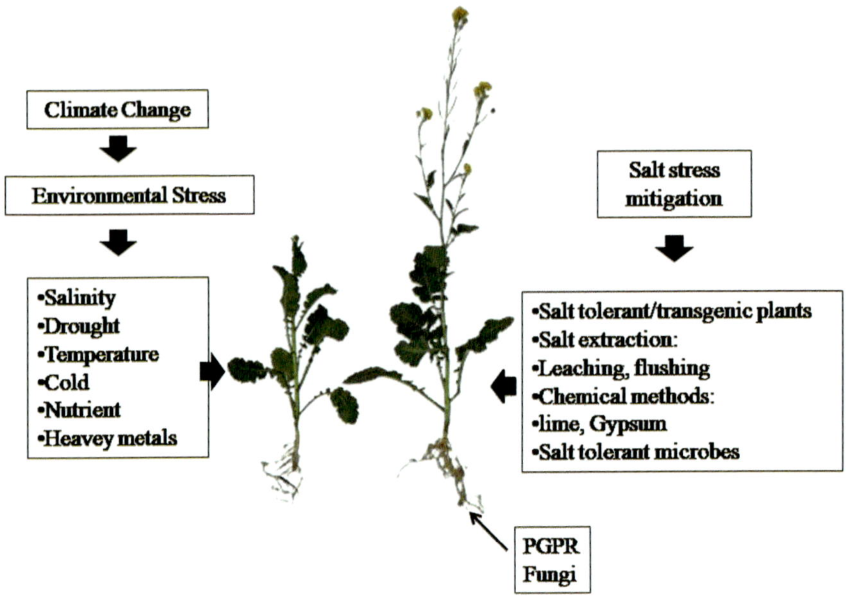


**Fig. 13.1** Environmental stresses induced by climate change and mechanisms involved in salt stress mitigation

## *Effect of Salinity Stress on Plants*

Presence of excess salt in soil is detrimental to plant health. Many stages of plant development, from germination through maturity, are known to be influenced by salinity. However, agricultural crops respond differentially to salinized soil conditions. Usually, salinity reduces the agricultural output of most cereals, legumes, forages, and horticultural crops. In addition, salinity also alters the ecological balance and physicochemical properties of the soil. Salt stress leads in low agricultural yield, significant soil erosion, and limited socio-economic returns [66]. Additionally, salt stress has an effect on the morphological, biochemical, and physiological processes of the plant. These processes include germination of seeds, plant health, photosynthetic activity, protein synthesis, lipid metabolism, water holding capacity and absorption of nutrients [67–69]. For instance, the accumulation of sodium ions in leaf laminae hindered flowering in chickpea (*Cicer arietinum* L.) plants [70]. The buildup of sodium ions in plant tissues leads to the formation of different reactive oxygen species (ROS), which impede photosynthesis [71]. ROS are known to damage DNA and further induce lipid peroxidation, protein oxidation, enzyme inactivation, and chlorophyll degradation [72].

Under these saline conditions, plants use the salt overload sensitive (SOS) pathway, which is an essential protective mechanism used for sodium ion extrusion, potassium/sodium ion levels retention, and ion homeostasis [3, 73]. The negative effects on the SOS system under salt stress include a reduction in germination of seeds, leaf area, and pigmentation; an increase in defoliation and senescence; and a reduction in reproductive capability. In addition, salt stress causes ion toxicity, restricts water transfer from soil solutions, limits nutrient absorption, and causes osmotic and oxidative stress. Thus, it affects the overall plant health [56, 74–77]. Additionally, salt stress suppresses the plant growth and development, including enzyme activity [78], DNA, RNA, and protein synthesis, and mitosis during the reproductive stage of the plant [19, 79]. Salinity also impairs reproductive development in plants by retarding sporogenesis and stamen filament elongation, triggering ovule abortion and fertilized embryo senescence, and promoting programmed cell death in plant tissues. The survival and development of plants are monitored to determine their resistance to salt stress because they include the up- and down-regulation of physiological systems, such as osmotic balance [80]. Failure to attain equilibrium between these systems results in cell dehydration, loss of turgidity, and ultimately plant death [76]. Some studies have linked salt stress to stunted plant development and symbiosis in field peas, causing a decrease in biomass and production [81]. Even nutrient-rich weeds, such as *Portulaca oleracea* L., are significantly affected by salt, as seen by alterations in physiology and root architecture, as well as decreases in biomass and yield [82]. Thus, salt stress is hazardous at various stages of cereal and legume cropping systems, producing 15–100% loss in legumes and endangering food security [3, 83, 84].

The drastic effect of salt stress can be observed in terms of crop yield losses. The primary effects related to crop yield can be in terms of germination which either decreases or sometimes ceases under extreme saline conditions. Ali Khan et al. [85] showed that under saline conditions growth, yield, and biomass of pearl millet is adversely affected in terms of germination percentage, plant height, leaf area, total biomass and grain yield plant$^{-1}$. Impact of salinity on pea was also found to adversely affect growth, yield and biomass [81]. Farooq et al. [83] also reviewed the effects of salt stress on grain legumes, and they described that in different legumes salinity may reduce crop yield by 12–100%. Salt tolerance of black cumin (*Nigella sativa* L.) and its effect on seed emergence and germination, and yield were studied by Faravani et al. [86]. They showed that an increase in salinity level from 0.3 to 9 dS m$^{-1}$ reduced the average seed and biological yield. Similarly, the effect of different levels of salinity on a weed plant *Portulaca oleracea* showed a reduction in biomass and yield, changes in physiological attributes, and alteration in stem and root structure [87]. Salinity has thus a wide level of impacts on seed germination, plant growth and crop yield of different crops.

It was observed that chickpea crop is extremely susceptible to salt stress and salinity is main restrictive factor bringing about low production. Salinity additionally brought about poor plant growth, low nitrogen fixing ability, reduced nodule numbers and decreased percentage of tissue nitrogen in arid and semiarid regions, in this manner, bringing about 8–10% losses in chickpea yield [17, 88]. To distinguish tolerant genotypes from sensitive genotypes of chickpea, a concentration of 40 mmol L$^{-1}$ NaCl was accounted for as optimum level of salinity [89]. Reduction in salinity levels was found to cause excellent recovery with substantial new shoot growth. The critical point of salinity level for seed yield reduction of chickpea was reported as low as 3 dS $m^{-1}$ in field soils [90]. Rhizobial isolates also showed different growth rate at higher NaCl concentrations. Maximum growth rate was seen at 1% (w/v) NaCl and minimum growth rate was seen at 4% (w/v) NaCl [91]. Only 11.1% of isolates tolerated 5% NaCl concentration [91, 92].

## *Effects of Salinity on Soil Microorganisms*

In dry and semiarid locations, where precipitation is scarce and often insufficient to eliminate salts from the plant root zone, soil salinity is a significant constraint on agricultural output [93, 94]. Both microorganisms and plants are negatively impacted by high salt concentrations [95]. However, the metabolic burden imposed by these stress tolerance systems might be deleterious to sensitive bacteria, reducing the activity of the cells that survive the stress [96–98]. Various reports on naturally saline soils have indicated that salinity has negative effects on microbial soil communities and their activity [95, 99–101]. The impact is usually more prominent in the rhizosphere due to enhanced consumption of water absorption by the plants as a result of transpiration. Accordingly, osmoadaptation necessitates a significant amount of energy [102, 103].

Omar et al. [104] reported that higher salt concentration upto 5% significantly decreased the entire microbiota. Other biotic and abiotic stresses (including soil salinity) have been reported to affect rhizosphere microbial composition, biodiversity, microbial metabolic activity and functioning, agricultural residue decomposition and nutrient availability, soil health and plant development [19, 76, 105, 106]. There is genetic variation in salt tolerance among rhizobia, which can have a substantial impact on the productivity of legume crops. The capacity for growth and survival of different chickpea rhizobial strains in salt conditions varies greatly [107, 108]. It also has been found that rapid rhizobia growth is associated with greater salt tolerance. Changes in cell shape and size or abnormalities in the pattern of extracellular polysaccharides (EPS) and lipopolysaccharides (LPS) have been seen in rhizobia that have been exposed to salt stress [108–110]. The symbiosis is more vulnerable to salt stress than free-living rhizobia because legume plants are more sensitive to salinity stress in general. Many strains of *Rhizobium* spp. have had their inoculum viability, nodulation, and nitrogen-fixing abilities reported to be negatively impacted by salt stress [109].

Only 33% of bacterial isolates were able to survive in solutions containing more than 8% NaCl (w/v), and of those, only 19% displayed PGP characteristics at these concentrations, as reported by Upadhyay et al. [111]. Isolate SU8 had the highest proline content and synthesis, with 2.73 and 11.95 g mg protein at 0% and 10% NaCl (w/v), respectively. The synthesis of reducing sugars (RS) and total soluble sugars (TSS) in rhizobacterial isolates was inversely related to the concentration of salt (NaCl), which had the potential to lower salinity levels and promote the development of agricultural crops grown in salty conditions. All of the rhizobacterial-isolated strains were able to grow up to a concentration of 4% NaCl, but their capacity to tolerate salt decreased with rising salt concentrations. The experiment involved screening of 40 rhizobacterial isolates for different concentrations of sodium chloride, ranging from 2 to 8% [20]. Garg and Sharma [112] identified and tested 10 rhizobia from *Trigonella foenumgraecum* for stress tolerance. To evaluate the growth of the isolates, a yeast mannitol medium with a wide pH range (4–10) and varying NaCl concentrations (0.05–5%) was used. Increasing salt concentrations inhibited the development of *Rhizobium* strains. Shultana et al. [113] also isolated salt tolerant rhizospheric bacteria from rice roots grown in saline conditions (0.41–17.64 dS $m^{-1}$). Salt tolerant rhizobacterial isolates were grown on Tryptic Soy Agar media plates with different NaCl concentration (0, 0.5, 1, 1.5, 2.0M) to check their salt tolerant capacity. Five highly salt tolerant strains were found to grow upto 2.0M NaCl, however increasing salt concentrations inhibited the growth of isolated rhizobacteria.

## Mechanisms of Salinity Stress Tolerance in Microbes and Plants

The rhizosphere is the most favourable environment for microbial populations [114]. Numerous microorganisms, such as bacteria, fungi, actinomycetes, and archaea, populate the rhizosphere of different plants. These soil or rhizosphere-inhabiting bacteria influence the ecosystem function, plant health, and pollutant degradation [115, 116]. These microbial communities act as a catalyst for the transformation, decomposition, and recycling of soil nutrients and organic matter in the soil. Thus, microbial populations have been found to affect crop development both directly and indirectly. Some of these soil- or rhizosphere-inhabiting microorganisms have acquired the ability to survive high salt (NaCl) concentrations. These salt-tolerant microbes have the potential to boost productivity of both grains and legumes in arid and semi-arid regions for sustainable agriculture [117]. It has been demonstrated that various bacterial genera such as *Klebsiella, Streptomyces, Pseudomonas, Agrobacterium, Bacillus, Enterobacter, Stenotrophomonas, Rhizobium* and *Ochromobacter* enhance grain and legume production in saline circumstances [31, 118, 119]. As salinity increased in the rhizosphere, it affects root exudation, microbial population and degradation of organic materials [120]. A negative correlation was observed between EC of soil and microbial biomass, indicating that soil salinity has a negative effect on microbial biomass [121]. In similar studies, Nelson and Mele [122] observed that NaCl has an indirect influence on rhizospheric microbial diversity through root exudates and plant quantity/quality, as well as a direct effect via microbial toxicity. Under salt soil conditions, molecular signaling among microorganisms and plants play a substantial effect on their rhizospheric microbiota [123].

When rhizospheric bacteria are exposed to a highly osmosis conditions, fast turgor loss and dehydration occur to compensate for the outflow of water. Ionic instability is caused by elevated ion concentrations, which maintains $K^+$ osmotic balance, activates osmotic responses, and up-regulates genes involved in adaptation, metabolism, defensive, and amino acid transport pathways in the cytoplasm. Furthermore, the production of organic solutes increases intracellular osmotic strength, which helps to stabilise biomolecules under salt stress conditions [124, 125]. The second mechanism of salt tolerant rhizospheric bacteria is exopolysaccharide (EPS) production, which alters membrane compositions such as periplasmic glucans, proteins, fatty acids, shorter peptidoglycans, and interpeptide bridges, and capsule content for accelerated water retention, regulating carbon source usage in microbial cells, and protecting microbiota from osmotic stress [126–128]. Flexible surface appendages surrounding the microbial cell also act as a protective barrier at low electrolyte concentrations, decreasing osmotic stress and minimizing the damaging effects of ionic strength changes [129].

Various microorganisms, inhabiting the phyllosphere, rhizosphere, and endorhizosphere, have been found to help plants in adaptation during salt stress by absorbtion of nutrients from soil leading to improvement in plant growth and development [21]. Besides this, metabolic activity and functioning of microbial enzymes under salt stress may improve seed germination, root architecture, chlorophyll content, biomass, and disease resistance. In brief, salt mitigation strategies include direct and indirect mechanisms leading to promotion of plant growth and increases in crop yield in saline soils (Fig. 13.2). Direct mechanisms include enhanced accumulation of osmoprotectants such as glycine, betaine, trehalose, and proline [130, 131], upregulating production of antioxidant enzymes, such as SOD, CAT, APX, and GR, to provide protection against oxidative stress [72, 132], maintaining high $K^+/Na^+$ ratio (ion homeostasis) and regulating the expression of ion transporters to protect against ion toxicity [72, 133, 134], lowering of stress-induced hormone (ethylene) level with expression of ACC deaminase activity [37, 135], synthesizing of exopolysaccharides and biofilm formation to reduce $Na^+$ ion accumulation in plant roots by binding to excessive $Na^+$ ions and preventing their translocation to plant leaves via xylems [132, 136], and maintaining high levels of photosynthetic activity and stomatal conductance [137]. Other indirect mechanisms employed for salt stress amelioration by PGPR include enhancing nutrient availability and uptake, siderophore production for iron uptake, phosphate solubilization [136, 138], modulating plant growth hormones for root and shoot development, and by conferring disease resistance through inducing systemic tolerance, production of organic acids and nitric oxide [139], and secretion of extracellular polymeric substances for increased soil aggregation to improve plant growth under salt stress [76, 140–142].

Under salinity stress, *Pseudomonas* sp. and *Novosphingobium* sp. from citrus and *Distichlis spicata* rhizobacterial strains reduced salicylic acid (SA), abscisic acid (ABA), and ethylene, as well as root proline and chloride accumulation and photosystem II activity [143]. He et al. [144] identified a novel salt-tolerant *Pseudomonas* sp. in the rhizosphere of the desert shrub *Haloxylon ammodendron*, which caused perennial ryegrass to become salt-tolerant. Proteomic, genomics and transcriptomics studies characterized various transcription factors, gene expression, protein expression and microbial interactions in plant cells and microbes in response to salt tolerance [145]. For instance, *Burkholderia phytofirmans* strain induced long-term metabolic and transcriptional changes in *Arabidopsis thaliana* involving expression

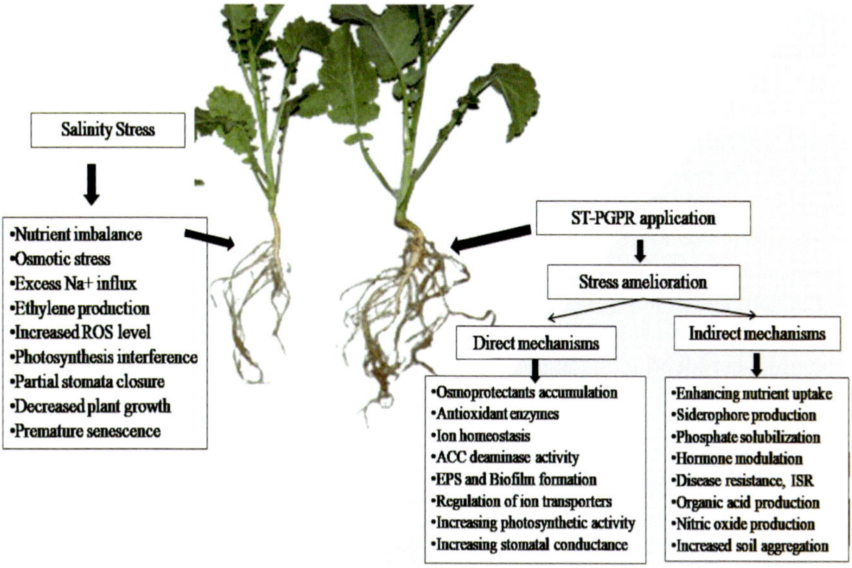


**Fig. 13.2** Rhizobacteria-mediated adaptive responses of plants to salinity stress to promote plant growth

of genes related to ROS scavenging (APX2), lipoxygenase-2 reduction, and detoxifying (Glyoxalase 7) under salt stress [146]. Some of the salt-tolerating PGPR strains regulated the expression of dehydroascorbate reductase, catalase (CAT), ascorbate peroxidase (APX), and glutathione reductase (GR) genes under salt stress conditions [147]. Functional metagenomics was used to find numerous salt-tolerant genes in PGPR and some of the salt-tolerant PGPR strains alleviated salt stress along with biological control of phytopathogens [3, 84, 148–150]. Several PGPR strains synthesized phytohormones [e.g., IAA, cytokinins (CK), and gibberellins] as well as having ACC deaminase activity [64, 119, 135, 151].

## *Production and Accumulation of Osmoprotectants*

Stressed plants produce osmoprotectants like quaternary ammonium compounds such as betaine, proline, polyamines, glycine and other amino acids that improve water intake and reduce water losses, and dilute the concentrations of toxic ion [152, 153]. Various plant growth promoting strains have been characterized, which possess the capacity to tolerate osmotic stress from $K^+$ ions and osmolytes in the cytoplasm [154, 155]. At 2.5 mol $L^{-1}$ NaCl, upregulation of the *pro*A, *pro*H, and *pro*J genes was observed in salt tolerant PGPR strains [156]. During salt stress, *Halomonas* sp.

SBS 10 and *Azospirillum* spp. were found to accumulate proline, ectoine, glycine, betaine and trehalose, making maize plants resistant to salt stress [157, 158]. Inoculation of salt-stressed tomato plants with *Pseudomonas extremorientalis* TSAU20 reported to have increased proline levels [159]. Similarly, increase in glycine and betaine levels conferring salt tolerance under osmotic stress, was observed in rice and sugarcane, when inoculated with *B. pumilus*, *Pseudomonas pseudoalcaligenes*, and *Enterobacter* sp. EN-21 [160, 161]. In wheat crop, inoculation with *Dietzia natronolimnaea* STR1 exhibited strong antioxidants activity and accumulation of proline under salt stress conditions [162]. Inoculating salt-stressed *Acacia gerrardii* with *B. subtilis* strain was reported to enhance proline levels and enhanced salt tolerance by maintaining water balance [150]. Trehalose, an osmoprotectant sugar, was found to confer salinity resistance, and many genes involved in biosynthesis of trehalose were identified in halo-tolerant PGPR strains [3, 163–165].

## *Antioxidant Enzyme Activity*

Salt-stressed plants produce different reactive oxygen species (ROS) that damage various proteins, lipids, and DNA [166]. The level of ROS-scavenging enzymes such as superoxide dismutase (SOD), APX and CAT was reduced on exposure of plants to abiotic stress i.e., salt and drought, and increased the lipid peroxidation [84, 167]. A wide range of antioxidant enzymes, such as superoxide dismutases, nitrate reductase (NR), catalase (CAT), peroxidase (POD) and glutathione reductase (GR) were produced by salt-tolerant PGPR strains under salinity stress [3, 145, 168]. Interestingly, inoculation of chickpea plants with *Azospirillum lipoferum* strain FK1 caused enhanced expression of the anti-oxidant genes and also improved nutrient absorption, non-enzymatic metabolites and flavonoids leading to improvement in symbiotic efficiency [169]. Wheat plants co-inoculated with *Azospirillum brasilense* DSM1690 and *Pseudomonas fluorescens* Ms-01 showed higher POD levels than uninoculated control plants [170]. Salt-tolerant *Bacillus cereus* strain Pb25 enhanced the level of antioxidant enzymes catalase and peroxidase in mungbean (*Vigna radiata*), when grown at 9 dS $m^{-1}$ saline conditions [72]. After PGPR inoculation, salt-stressed plants may stimulate the expression of antioxidant enzyme-related genes, resulting in enhanced ROS-scavenging enzyme activity [171].

During salinity stress, peroxidation of lipids has been reported to increase malondialdehyde (MDA) concentration, indicating damage to structural integrity of cell membrane and inoculation with salt tolerant PGPR strains reduced MDA accumulation and thus, helped plants to combat salinity stress [172]. Similarly, the decreased levels of MDA were observed in rice seedlings after inoculation with *Enterobacter* sp. P23 even during salt stress [119]. Inoculation of PGPR strains viz. *Serratia* sp. SL-12 in wheat [118], and *Klebsiella* sp. IG3 in oat plants [173], were found to reduce MDA level. Thus, inoculation of plants with PGPR was found to increase biomass and nutrient efficiency in stressed plants by altering the level of antioxidants and stomatal conductance [174]. Therefore, use of salt-tolerant rhizobacteria as bio-inoculants causes enhancement of plant growth under salinity stress conditions through modulation of osmoprotectants levels, upregulation of the stress-related genes, and by enhanced production of enzymatic and non-enzymatic antioxidants in stressed plants.

## *Reduced Uptake of Salt Ions by Microbes and Plants*

Another strategy for PGPR tolerance to high salt concentration is minimization of salt absorption by ion affinity transporter control, root structure alteration via broad rhizosheath, and cation trapping in exopolysaccharides. Microbes maintain ion homeostasis by boosting the $K^+/Na^+$ ratio and decreasing $Na^+$ and $Cl^-$ in the shoot and leaves, respectively. Salt stress changes the expression of genes such as KT1, NHX2, SOS1, and HKT1, and these molecular alterations result in salt tolerance [146]. Niu et al. [175] reported that *Pucciniella tenuiflora* infected with *Bacillus subtilis* GB30 caused lower $Na^+$ buildup, which was corroborated by the down-regulation of ptHKT2 and up-regulation of ptHKT1 and ptSOS1 genes in roots exposed to high NaCl concentrations (200 mmol $L^{-1}$).

Volatile organic compounds (VOCs) have been reported to play crucial role in many cases of PGPR interaction with plants especially antibiosis and biocontrol of plant pathogens, and regulation of auxins [176, 177]. During salt stress, VOCs down-regulate the expression of high affinity $K^+$ transporter (HKT1), but it is upregulated in shoots, which results in lower accumulation of $Na^+$ inside the plant [178].

## *ACC Deaminase Activity and Lowering of Ethylene Formation*

Ethylene, a stress hormone, is synthesized under stressed conditions and affects a number of metabolic activities within plants [136]. In addition, plants release 1-aminocyclopropane-1-carboxylic acid (ACC) in root exudates, which is converted to the stress hormone ethylene ($C_2H_4$) by the enzyme ACC oxidase. Ethylene has been demonstrated to play fundamental roles in root branching, root hair formation, nodule development and for amelioration of biotic as well as biotic stresses [33]. On the other hand, many plant growth-promoting bacteria possess the enzyme ACC deaminase; which scavenges the exuded ACC and thereby down-regulates ethylene production by cleaving ACC into $\alpha$-ketobutyrate and ammonia [135, 179–182]. Low levels of ethylene acts in plant defence against different abiotic stresses [183], but excessive levels of ethylene can cause ethylene stress, which slows growth and development in plants [184, 185]. Under stress conditions, plants produce ethylene, which subsequently affects legume nodule formation [186, 187]. Under salt stress, PGPR can convert ACC into ammonia and $\alpha$-ketobutyrate, providing nitrogen to the plants [33, 76]. Rhizobacteria with ACC deaminase activity were reported to reduce salt stress and enhanced plant growth of tomato and rice [188].

ACC deaminase activity has been reported in various salt–tolerant bacterial genera belonging to *Arthrobacter*, *Acidovorax, Bacillus, Brevibacterium, Enterobacter, Exiguobacterium, Gracilibacillus, Klebsiella, Methylobacterium, Planococcus, Pseudomonas, Rhizobium, Salinicoccus, Stenotrophomonas, Variovorax* and *Virgibacillus* [189]. Inoculation of ACC deaminase-containing halotolerant bacteria was found to ameliorate salt stress in plants and improved crop productivity under salinity stress [151, 190–192]. For instance, inoculation of salt-tolerant ACC deaminase activity containing *Enterobacter cloacae* strain KBPD improved nodulation and symbiotic efficiency in *Vigna radiata* at 50, 100, and 150 mmol $L^{-1}$ NaCl concentrations [64]. Similarly, Tiwari et al. [193] found that ACC deaminase-producing salt tolerant PGPR strains improved plant cell biochemical characteristics such as bio-compatible solute formation, membrane permeability, stability, and photosynthetic pigment production under salt and drought stress. Ali et al. [194] reported that inoculation with endophytc strains i.e., *Pseudomonas migulae* and *Pseudomonas fluorescens* containing ACC deaminase activity improved physiological indices in plants under stress conditions.

In another study, inoculation of oat (*Avena sativa*) with *Klebsiella* sp. strain IG 3 enhanced shoot and root lengths, plant biomass, and relative water contents under NaCl stress (100 mmol $L^{-1}$) [173]. The concomitant higher expression of *acds* genes (encoding ACC deaminase) and *ipdc* genes (encoding IAA) was observed under stress conditions. Expression of ACC deaminase in ST-PGPR strains was demonstrated to enhance the infection thread persistence during nodulation in legume crops, which is adversely affected by ethylene under salt stress conditions [187]. Shaharoona et al. [195] reported that the coinoculation of an ACC deaminase-possessing PGPR and *Bradyrhizobium* in mungbean (*Vigna radiata* L.) improved nodulation and other symbiotic traits by reducing ethylene as compared with the single *Bradyrhizobium* treatment. The ACC deaminase-producing halo-tolerant bacterial strains *Brevibacterium iodinum* RS16, *Zimmermannella alba* RS11, and *Bacillus licheniformis* RS56 have been reported to reduce the secondary ethylene peak in red pepper plants at 150 mmol $L^{-1}$ NaCl [196]. The inoculation of lentils with ACC deaminase-producing PGPR led to higher plant growth, nodulation, and grain yield under oxidative stress conditions [197]. *Arthrobacter* sp. and *Bacillus* sp. producing IAA and ACC deaminase increased proline content under salt stress in sweet pepper and chickpea [198, 199]. Chandra et al. [200] reported that three ACC deaminase-producing bacterial strains viz. *Pseudomonas palleroniana* DPB16, *Pseudomonas* sp. UW4, and *V. paradoxus* RAA3, enhanced growth, nutrient uptake, osmolyte production, antioxidant enzyme activities, and grain yield of wheat under salt and drought stress conditions in contrast to the uninoculated control treatment. Thus, various inoculation studies in different crops suggested that $ACC^{+}$ bacteria could be used as an eco-friendly inoculant to improve growth of salinity-sensitive crop plants [29, 192, 201].

Hussein et al. [202] evaluated eight yeast strains i.e., *Yarrowia lipolytica* YEAST-1, *Candida diddensiae* YEAST-2, *Trichosporon gamsii* YEAST-5, *T. ovoides* YEAST-6, *Y. lipolytica* YEAST-16, *C. subhashii* YEAST-17, *Saccharomyces cerevisiae* YEAST-30, and *S. cerevisiae* YEAST-34) for plant growth-promotion (PGP) traits, biofilm formation, seed germination and for alleviation of salinity stress in wheat (*Triticum aestivum* L.). *Y. lipolytica* YEAST-1 strain was found to enhance the plumule length of *T. aestivum* seedling by more than 4.0, 3.0, and 2.0 cm at salinity stress of 50, 100, and 200 mM NaCl, respectively, after 96 h of treatment. Highest expression of amino-cyclopropane-1-carboxylate deaminase (ACCD) genes was observed in *S. cerevisiae* YEAST-34, at 5 mM ACC. Inoculation of *Y. lipolytica* YEAST-1 enhanced the radicle length of *T. aestivum* seedling significantly by 0.8 cm at 50 mM NaCl, 0.7 at 100 mM NaCl, and 0.06 cm at 200 mM NaCl stress.

## *Exopolysaccharide Production and Biofilm Formation*

Salt-tolerant PGPR strains were found to form exopolysaccharides (EPSs), which promote biofilm formation and root colonization leading to better plant-microbe interactions. Root colonization by exoploysaccharide producing salt tolerant rhizospheric strains improves uptake of nutrients (i.e., potassium and phosphorus), disease resistance, plant development and growth [203]. EPSs improve soil particle aggregation, promote cation exchange, water and nutrient retention, environmental changes, and root colonization [204, 205]. Inoculation of *Bacillus subtilis* in salt-stressed *Arabidopsis* plants suppressed the upregulation of HKT1 (high-affinity potassium ion transporters), prevented excessive $Na^+$ ion absorption by plant tissues and sustained ion homeostasis [132]. Similarly, salt resistance in oilseeds crops such as *Brassica napus* increased $K^+$ retention and decreased $K^+$ ion-permeable channel by activating $H^+$-ATPase activity and maintaining a negative membrane potential [206]. Increased plasma membrane sodium/hydrogen ions or potassium/sodium ions exchange activity also increased ROS-mediated $Na^+$ extrusion from plant roots [206]. Microorganisms and host plants, such as *Triticum aestivum*, *Brassica* sp., and *Hordeum vulgare*, were discovered to have a close link with salt tolerance [206–208].

Bacterization with salt-tolerant EPS-producing rhizobacteria was found to improve germination of seeds [203]. The development of biofilm, which was helped along by the synthesis of EPS, contributed to an increase in PGPR's resilience to both abiotic or biotic stresses [209]. EPS-producing *Enterobacter* sp. P23 reduced $Na^+$ ion concentration in rice seedlings by binding excess $Na^+$ ions [119]. Similarly, co-inoculation of salt-tolerant *Halomonas variabilis* HT1 and *Planococcus rifietoensis* RT4 at 200 mmol $L^{-1}$ NaCl concentration was found to increase plant growth and soil aggregation by EPS, and biofilm development in chickpea [141]. Treatment with *Enterobacter* sp. MN17 and *Bacillus* sp. MN54 was reported to improve plant water condition and growth of *Chenopodium quinoa* at 400 mmol $L^{-1}$ NaCl irrigation conditions [210]. Salt-tolerant EPS and biofilm-producing *Marinobacter lipolyticus* strain SM19, and *B. subtilis* sub sp. *inaquosorum* alleviated the deleterious effects of drought and salinity stress in *Triticum aestivum* [211]. Recently, Chu et al. [212] demonstrated the essential role of EPS-producing halo-tolerant *Pseudomonas* PS01 in the regulation of the *LOX2* gene related to salt stress tolerance in *Arabidopsis thaliana*.

## *Siderophore Production*

Iron is the fourth most prevalent metal in the Earth and it also acts as a cofactor in 140 enzymes in plants. It is generally present as $Fe^{3+}$ (ferric), insoluble (OH) hydroxides, and oxyhydroxides O (OH) under abundant $O_2$ availability [213]. Soil- or rhizosphere-inhabiting microorganisms produce low-molecular-weight, iron chelators termed as siderophores [214]. Plants assimilate iron from bacteria-produced siderophores either via ligand exchange, direct absorption of siderophore-Fe complexes, or iron uptake [215, 216]. Numbers of studies have demonstrated that inoculations with siderophore-producing rhizobacteria enhance plant development through increased siderophore-mediated Fe-uptake [213]. Crowley and Kraemer [217] found a siderophore-mediated iron transport system in oat plants. They concluded that siderophores synthesized by rhizospheric bacteria transport the iron to oat plants, which has capabilities for absorbing Fe-siderophore complexes even when there is a scarcity of iron in the soil. In a similar manner, the Fe-pyoverdine complex that was produced by *Pseudomonas fluorescens* C7 was absorbed by *Arabidopsis thaliana* plants, which resulted in an increase of iron within the plant tissues as well as an improvement in plant development [218].

Sadeghi et al. [219] reported that siderophore production in *Streptomyces* increase wheat growth under saline conditions. Tank and Saraf [220] also found PGPR promotes growth of tomato plants grown under 2% NaCl conditions; and PGPR were demonstrated to solubilize insoluble P and produced siderophores. Similarly, bacterial strains viz. *Halobacillus* SL3 and *Bacillus halodenitrificans* PU62 were found to produce siderophores in saline conditions [23]. Siderophore-producing *Pseudomonas* sp. GRP-3 improved iron nutrition in *Vigna radiata* by reducing chlorosis and increasing chlorophyll content. Siderophore-producing rhizobacteria increased plant height and improved nitrogen uptake [221]. Rajkumar et al. [213] also found siderophore-producing *Ensifer meliloti* strains that suppressed groundnut charcoal rot. Siderophore-producing salt-tolerant *Bacillus aryabhattai* MS3 strain was isolated from the root area of salt-prone rice fields and highest siderophore production was observed, which estimated at 60% and 43% under non-saline and saline (200 mM NaCl) conditions, respectively [222]. The expression of the *ent*D gene (involved in the biosynthesis of siderophore) was evidenced irrespective of saline states. The salt-tolerant *Bacillus aryabhattai* MS3 strain may enhance plant growth in saline soils with iron limitation.

## *Phosphate Solubilization*

Phosphorous is the second most important vital macronutrient for plant growth and crop production [223, 224]. It plays a key role in the development of the root stem, the formation of seeds and flowers, nitrogen fixation, and disease resistance. Phosphorous exists in a bound and insoluble form with calcium in neutral soils [as tricalcium phosphate $(Ca_3PO_4)^2$], or with iron and aluminium in acidic soils [as aluminium phosphate $(Al_3PO_4)$ or iron phosphate $(Fe_3PO_4)$ in soil [225]. Thus, the concentration of soluble or inorganic available phosphorus i.e., orthophosphate is very low in the soil [226]. Phosphate-solubilizing microorganisms (PSMs) possess the capability to transform insoluble form of phosphate into inorganic utilizable form mainly through organic acids production [227–230]. Various soil microbes including *Azotobacter, Bacillus, Bradyrhizobium, Pseudomonas, Erwinia* sp., *Flavobacterium* sp., *Micrococcus* sp., *Corynebacterium, Nostoc, Serratia phosphaticum, Calothrix brauna, Burkholderia, Sarcina* sp., *Scytonema* and *Advenella kashmirensis* have been found to solubilize phosphorous in soil [231–234].

Salt stress in soil has been reported to affect population of phosphorus solublizing microbes and their P-solubilization capability [235]. Alkaline soils containing high level of calcium-phosphate were found to increase P-solubilization [236]. High salt-tolerating rhizobacteria may solubilize phosphate, Zn and K, fix nitrogen, and produce ACC deaminase activity and phytohormones, making more nutrients available to plants even under abiotic (drought and salt) stress conditions [84]. For instance, *Alteromonas* sp. and *Pseudomonas* sp. solubilize phosphate at 2 mol $L^{-1}$ NaCl concentrations [237]. Some bacteria mineralize P by producing phosphatases and phytases enzymes. Mahdi et al. [238] reported that phosphate-solubilizing bacteria produce phosphatase enzyme, which releases P from minerals by replacing phosphate cations with $H^+$. Potassium (K) is another most important nutrient for growth of plants after nitrogen and phosphorous [239]. Therefore, inoculation of halotolerant K solubilizing bacterial strains i.e., *Acinetobacter pitti* strain L1/4, *A. pitti* strain L3/3, *Rhizobium pusense* strain L3/4, *Caprivadus oxalaticus* strain L4/12 and *Ochrobactrum ciceri* strain L5/1 caused significant increases in the shoot length, fresh weight, dry weight and chlorophyll contents of rice plants under saline conditions [240].

## *Production of Phytohormones*

Auxins (indole acetic acid; IAA), cytokinins, gibberellins (GA), ethylene, and abscisic acid (ABA) constitute up the five major groups of phytohormones [241–244]. These plant and bacterial hormones, known as phytohormones, regulate a wide variety of physiological processes, including as cell division, development, gene expression, and stress responses, as well as the rate and form of root and shoot growth [245–247]. Phytohormones have been shown to improve a plant's nutrient availability, water absorption capacity, and resistance to salt stress by increasing root hair length and root surface area [248–252]. The capability of plants to acclimatize to salinity stress depends on their interaction with beneficial potent microbes that have the ability to produce IAA, CK, and gibberellic acids (GAs) [145, 253]. Therefore, attempts are being made to identify PGPR strains that can help plants to overcome and mitigate salt stress by producing phytohormones. For instance, auxin-producing salt-tolerant *Leclercia adecarboxylata* strain MO1 enhanced carbohydrate synthesis, chlorophyll fluorescence, *ipdc* gene expression, and organic acid production in tomato [254]. IAA-producing PGPR strains were demonstrated to enhance ACC deaminase activity via a signalling cascade that hydrolyzed ACC into ammonia and α-ketobutyrate [33], allowing the plant to proliferate even under salt stress by lowering ethylene levels.

Application of *Enterobacter* sp. found to enhance seed germination (48%) of rice at 150 mmol $L^{-1}$ salt concentration [119]. *Bacillus amyloliquefaciens* SQR9 strain improved maize seedling development, antioxidant enzyme activities (CAT, POD, and GR), total sugar content, and $K^+/Na^+$ ratio, under salt stress conditions. PGPR-inoculated plants retain $K^+$ ions to minimise $Na^+$ toxicity under salt stress [255]. *Streptomyces* lowers salt stress in wheat by producing auxins, according to Sadeghi et al. [219]. Auxins and GAs were found to lower the inhibitory effects of salt's on wheat seedlings [23]. *Enterobacter* sp. EJ01 obtained from halophyte *Dianthus* increased salt tolerance (200 mmol $L^{-1}$) in tomato plants by boosting desiccation, embryogenesis, proline biosynthesis, and stress-inducing and priming activities [256]. *Ensifer meliloti* genetically modified for enhanced production of IAA conferred the ability to tolerate 0.3 mol $L^{-1}$ salt in *Medicago truncatula* [257]. Zahir et al. [258] found that inoculating a beneficial rhizospheric microbiome increased mungbean (*Vigna radiata*) growth and yield in saline environments via better IAA production. Thus, PGPR's phytohormone synthesis is an exploitable trait; more research is needed to use these rhizosphere bacteria to lessen salinity's effects. *Pantoea agglomerans* strain lma2 can produce 161 g $mL^{-1}$ IAA at 200 mmol $L^{-1}$ of NaCl, making it a potential PGPR under salt stress [259]. Numan et al. [78] showed extensive IAA production in durum wheat with osmotolerant PGPR *Azospirillum brasilense* NH at high salt concentrations, underlining IAA's role in salt tolerance. *Micrococcus luteus* also increase maize growth by producing IAA [78]. As potential auxin makers, many rhizobia and rhizobacteria strains also found to withstand salt and osmotic stress in mung bean [260, 261]. Kuzmina et al. [234] reported that IAA production and phosphate-mobilizing activity of *Advenella kashmirensis* strain IB-K1 showed plant growth-promoting effects on wheat seedlings. Additionally, the presowing treatment of wheat (*Triticum durum* Desf.) seeds with *A. kashmirensis* strain IB-K1 effectively relieved the deleterious effect of salt stress on plant growth under moderate salinization level of cultured soil, which ultimately resulted in higher plant output.

Gibberellin-producing bacterial isolates, such as *Azospirillum* sp., *Bacillus pumilus*, *Bacillus licheniformis*, and *P. fluorescens*, were reported by Bottini et al. [262]. Salinity stress reduces GA synthesis in plants, while PGPR inoculation increases endogenous GA [263], inducing salinity tolerance and preventing tissue damage [264]. For instance, increased endogenous GA levels in *Promicromonospora* sp. SE188, *Burkholderia cepacia* SE4, and *Acinetobacter calcoaceticus* SE370 improved cucumber plant growth under salt stress, with increased proteins and antioxidant enzymes, and decreased sugars and ribonuclease [84, 265]. Attia et al. [266] showed that seed priming with gibberellic acid (GA3, 3 μM) partially attenuated the salt stress effect and efficiently reduced polyamines (PA; putrescine, spermidine and spermine) levels in salt-stressed seedlings of fennel (*Foeniculum vulgare* Mill.) as compared to the control. Organ and treatment-specific reduction in peroxidase and catalase activities were observed. In a similar manner, the responses of PA genes to salinity were found to be varied. In hypocotyls and cotyledons (H+C), up-regulation was observed for SPMS1, ODC1, and ADC1, whereas down-regulation was shown for SAMDC1 in the radicle.

Another phytohormone abscisic acid (ABA) is produced by salt-tolerant strains of *Achromobacter xylosoxidans*, *B. licheniformis*, *Proteus mirabilis*, *P. fluorescens*, *Stenotrophomonas maltophilia*, and *Bacillus megaterium* produce [3, 243]. Recent reports suggested that ABA-mediated signalling increases salt tolerance in different crops. For instance, inoculation of *Dietzia natronolimnaea* STR1 and *Bacillus amyloliquefaciens* RWL-1 in wheat and rice altered auxin and ABA signalling cascades, resulting in increase of salinity tolerance [162, 267]. The mechanism involved in lowering the inhibitory effect of salt on plant development by abscisic acid is through increasing $K^+$ and $Ca^{2+}$ ions, reducing sugar and proline in the root, and neutralizing $Na^+$ and chloride ($Cl^-$) ions concentrations [268, 269]. Patel and Saraf [270] also identified salt-tolerant *Pseudomonas putida*, *Pseudomonas stutzeri*, and *Stenotrophomonas maltophilia* in *Coleus* rhizospheres with elevated CK, gibberellins, and IAA level under salt stress conditions.

Cytokinins (CK) are involved in tissue differentiation and cell proliferation function, and act as master regulators in mitigating salinity stress in plants [271]. Many salt-tolerant species of *Bacillus*, *Arthrobacter*, *Pseudomonas*, *Azospirillum*, *Azotobacter*, *Arthrobacter*, *Halomonas*, and *Stenotrophomonas* were reported to produce cytokinins [272]. Increased cytokinin production decreased ethylene, reducing leaf senescence in cereals and legumes, hence boosting plant growth [273, 274]. Sita and Kumar [275] provided a more in-depth explanation of the function of gamma-aminobutyric acid (GABA) in the resistance of legumes to abiotic stress. Another phytohormone 5-aminolevulinic acid (ALA) has recently received wide applications due to its potential use as herbicide, insecticide, antimicrobial, alleviation of abiotic stress and plant growth stimulator under different stress conditions [276]. Growth rate of root and shoot, and leaf water relations of canola (*Brassica napus*) plants were improved by ALA application under different NaCl (100, 200 mM) concentrations [277]. Bacterial inoculation of mustard plants with ALA producing and salt tolerant (8% NaCl) isolate JMM15 showed 190.89% (at 0 dS $m^{-1}$ EC), 123.18% (at 8 dS $m^{-1}$ EC) and 230.86% (at 12 dS $m^{-1}$ EC) increase in shoot dry weight at 80 days of growth under controlled greenhouse conditions [10].

## *Organic Acids Role in Amelioration of Salt Stress*

One of the most severe abiotic stressors that plants can experience is salinity stress, which can cause disruptions in their physiological, biochemical, and metabolic processes. The application of natural metabolites to the plant is a viable technique for mitigating the deleterious effects of stresses on plants. It has been observed that the use of salicylic acid (SA) has tremendous agronomic potential in terms of enhancing the stress response of a variety of agronomically valuable crops, such as barley, maize, sunflower, wheat, bean, strawberry, and chamomile, amongst others [278]. Under salt stress conditions, SA application has been reported to provide several beneficial effects for plants i.e., the mitigation or reduction of photosynthetic pigments and photosynthetic performance, preservation of membrane integrity, stimulation of ABA and proline accumulation, reduction in lipid peroxidation and membrane permeability, lowering $Na^+$ content and higher $K^+$ concentration, etc. [278]. Treatment of wheat seedlings with sinapic acid, caffeic acid, ferulic acid, and p-coumaric acid, in addition to SA, resulted in enhanced growth of the plants despite the presence of salt stress [279]. Caffeic acid protected cucumber from chilling stress [280], and application with ellagic acid expedited the germination and seedling growth of chickpea under osmotic stress conditions. In addition, the treatment with vanillic acid lowered the deleterious effects of salt stress in tomato plants [280–282]. It has been also observed that all of these phenolic acids enhance the antioxidant capacity of plants by improving the activity of antioxidant enzymes and the accumulation of nonenzymatic antioxidants.

When comparing three different *Brassica* crops (kale, white cabbage, and Chinese cabbage), Lini'c et al. [283] found a positive correlation between phenolic acid levels and salinity tolerance, with kale being the most tolerant, white cabbage being moderately tolerant, and Chinese cabbage being the most sensitive species. Salicylic acid (SA) and ferulic acid (FA) were applied topically to plants and their effects on Chinese cabbage (*Brassica rapa* L. ssp. *pekinensis* (Lour.) Hanelt cv. Cantonner Witkrop) that had been exposed to short-term salt stress (150 mM NaCl, 72 h) were assessed [284]. Under salt stress conditions, rise in SA and proline concentration was reported whereas a decline in phenolic compounds, antioxidant activity, and photosynthetic performance (particularly owing to the degradation of PSI function) was observed.

Both proline and SA levels dropped when salt-stressed plants were treated with SA and FA (10 mM). Interestingly, in FA and SA treatments, the content of polyphenolic chemicals, notably FA, sinapic acid (SiA), kaempferol (KAE), and quercetin (QUE), enhanced in salt-stressed plants. As a result, there was an increase in antioxidant activities and a rise in photosynthetic efficiency. When comparing FA and SA, the latter was found to have a more beneficial alleviating impact on salt stress. Gholamnia et al. [285] also examined the effects of three different salt levels and two different temperatures on peppermint (*Mentha piperita* L.) by comparing the expression profiles of genes encoding proteins involved in the rosmarinic acid production pathway and various physiological responses. The upregulation of C4H and HPPR genes indicates the functions of these genes in defence mechanisms as well as the impacts of phenolic chemicals on oxidative stress inhibition.

## *Nitric Oxide Production and Mitigation of Salt Stress*

Nitric oxide (NO) is a gaseous and highly reactive nitrogen species, which is produced under normal as well as environmental stress conditions in living cells. NO has been reported to regulate various developmental processes during plant growth such as seed germination, root growth, stomata closure, flowering, stress response, and cell death [286, 287]. NO also modulates production of reactive oxygen species (ROS) in plants after exposure to various abiotic stresses and subsequently, activate defence mechanisms through enhanced production of antioxidants [288, 289]. Production of nitric oxide also leads to altered gene expression and activation of various redox regulated genes encoding antioxidant enzymes such as glutathione peroxidase (GPx), superoxide dismutase (SOD), glutathione reductase (GR), ascorbate peroxidase (APX) and chloramphenicol acetyltransferase (CAT), and may result in suppression of lipid peroxidation or malondialdehyde (MDA) synthesis [290–292]. The exogenous application of NO enhanced the production of ascorbate, glutathione, total phenolic content, proline, and flavonoids in NaCl-treated spinach [293] and tomato plants [294]. In addition, NO acts as an endogenous modulator of several plant hormones resulting in the inhibition of the induced programmed cell death and aid in stomatal function in *Arabidopsis*, wheat and pea plants [295, 296].

Salt stress normally has a negative effect on ion homeostasis and osmotic balance of plant cells [94]. Intracellular ion imbalance inhibits soil nutrient uptake leading to nutrition deficiencies. Furthermore, salt stress provokes membrane disintegration, loss of metabolic function ion leakage, DNA defragmentation, and subsequent cell death [297]. Plants have evolved various protective mechanisms to ameliorate the negative effects of salt stress. NO mediated mitigation of stress and the underlying mechanisms have been extensively studied using different approaches [298–301]. For instance, regulation of $Na^+/K^+$ ratio and $H^+$-ATPase of the plasma membrane is caused by NO, which confers salt tolerance in axenically grown cucumber plant [302]. Similarly, NO was demonstrated to activate the synthesis of $H^+$-ATPase in maize seedlings, resulting in production of $H^+$ gradient, which force the exchange of $Na^+/H^+$ and causes homeostasis of $Na^+$ and $K^+$ [303]. In another similar study, Zhao et al. [290] showed that NO acted as a signal in salt resistance in the calluses from two ecotypes of reed and induced the expression of the plasma membrane $H^+$-ATPase, which provided protection by making a balance in $K^+$:$Na^+$ ion ratio. NO mediated protection against salt stress in vivo has been shown in *Arabidopsis Atnoa1* mutants with impaired endogenous NO levels as these plants show enhanced sensitivity to salt stress, as well as reduced survival rates compared to wild type plants. Additionally, these mutants exhibited a greater $Na^+/K^+$ ratio in the shoots than the wild type plants [304, 305].

Besides this, NO donor sodium nitroprusside (SNP) has been found to alleviate osmotic stress tolerance and enhances seedling growth under salt stress in several plant species including rice, lupin, and cucumber [306, 307]. Increases in dry weights have been reported in maize, and seashore mallow seedlings after NO application under salt stress [303, 308]. In addition, the release of the nanoparticle known as chitosan nanoparticles (CS NPs) by NO treatment has been found to mitigate the toxic effects of salinity in maize plants [309]. The induction of polyamines is known to be closely associated with NO production or exogenous application of NO donors [308]. Therefore, NO-polyamine interaction may cause adaptive responses in plants for stress tolerance [310]. The increased levels of $H_2O_2$ in soybean due to the long-term salinity stress treatment, were reduced to the basal levels with exogenously-applied NO donor [311]. Adamu et al. [312] observed that treatment of salt-susceptible rice seedlings with SNP (NO donor) under salt stress caused a significant increase in the expression of *OsHIPP38*, *OsGR1*, and *OsP5CS2* and provided a resistant response to salt stress. On the other hand, untreated control plants (lacking NO donor treatment) succumbed to salt-stress. Furthermore, SNP-treated plants produced more plant biomass under salt stress conditions.

## Inoculation Effects of Salt-Tolerant Bacteria in Improving Plant Growth of Different Crops

The deleterious effects of salinity have been observed on plant growth and yield in various crops including mungbean, soybean, groundnut, pigeon pea, common bean, chickpea, groundnut, maize, tomato and cucumber. The main problem in the agriculture sector is to find an alternate solution for salt-stressed soil that ensures agricultural sustainability while increasing yield production in an environment friendly manner [15, 16].

The capacity of halo-tolerant rhizobacteria to deal with high soil salinity problem is well acknowledged and bacterization with salt-tolerant rhizobacterial strains has been found to mitigate the deleterious effects of salts on plants [9, 313–317]. Thus, use of salt-tolerant microbes as bio-enhancers/bioprotectants not only increases agricultural yield but also ensures plant survival in extreme salty conditions via physiological, biochemical, and molecular routes [24]. Besides salt tolerance, other PGP traits of salt-tolerant rhizobacteria contributes towards improvement of plant growth and increases in crop yield of different crops including cereals, legumes, oil seeds, and vegetables (Table 13.1). Thus, development of microbial consortia consisting of different bacteria or bacteria with mycorrhizal fungi has emerged as another feasible approach for improved amelioration of plant abiotic and biotic stresses [26, 318–320].

When pepper plants were inoculated with *Bacillus* sp. TW4, they showed a decrease in osmotic stress, which is often seen in the form of salt (and/or drought) stress. Under the influence of abiotic stress, the expression of genes associated with ethylene metabolism was found to be suppressed in these pepper plant [198]. *Bacillus* sp. TW4 exhibited ACC deaminase activity, which may be associated to the lowered expression of these genes. It has also been found that salt stress also affects nodulation during *Phaseolus-Rhizobium* interactions. However, in contrast to application of *Rhizobium* strain, *Azospirillum* inoculation of salt-stressed plants resulted in a longer exudation of plant flavonoids, suggesting an upregulation of flavonoid genes [313]. In barley seedlings, inoculation with *Azospirillum* seemed to alleviate NaCl stress, exhibiting the response to salt stress [321]. Salinity reduced the dry mass of the roots and shoots of lettuce plants compared to the control plants growing in non-saline environments [322]. At both medium and high salt conditions, the plants inoculated with *Pseudomonas mendocina* exhibited significantly higher shoot biomass than the control plants. Reduced chlorosis, necrosis, and drying were also seen in salt-stressed Mt-RD64 plants in comparison to salt-stressed Mt-1021 plants [322]. The antioxidant enzymes such as superoxide dismutase, ascorbic peroxidase, glutathione reductase, and proline oxidase were also associated to mitigate the salt stress.

Misra et al. [323] revealed the occurrence of most prominent group of ACC deaminase-producing salt tolerant *Bacillus* sp., which caused salt stress mitigation and improved grain yield of rice in different agro-ecological zones. Similarly, inoculation with *Pseudomonas* strain 002 [314] and *Staphylococcus sciuri* strain SAT-17 [324], which were able to tolerate 75 and 150 mmol $L^{-1}$ NaCl, respectively, were found to improve plant growth and biomass under salinity treatments. The inoculation with saline-adapted *Azospirillum* strains was found to improve grain productivity in wheat [22]. Nadeem et al. [325] documented significantly improved plant height, root length, chlorophyll content, and grain yield in maize under salt stress conditions using ACC deaminase-producing PGPR. Similarly, significant stimulation of growth and seed germination was observed in cotton under saline conditions with the inoculation of *P. putida* strain RS 198 [21]. Likewise, Upadhyay et al. [111] demonstrated that combined inoculation of *B. subtilis* and *Arthrobacter* sp. was found to mitigate soil salinity effects in wheat and caused improvement in plant biomass, total soluble sugars, and proline content. Inoculation of *Halobacillus* sp. and *B. halodenitrificans* also enhanced the growth parameters of wheat in salt-affected soils as compared with the uninoculated control at 320 mmol $L^{-1}$ NaCl [23]. In similar studies, inoculation of wheat (*Triticum aestivum* L.) var. WH157 with salinity-tolerant *Azotobacter* strains i.e., ST3, ST6, ST9, ST17 and ST24 caused significant increase in total nitrogen, biomass and grain yield in earthern pots containing saline soil under pot house conditions [326]. Maximum increase in plant growth parameters were obtained after inoculation with *Azotobacter* strain ST24 at fertilization dose of 120 kg N $ha^{-1}$ and its inoculation resulted in attaining 89.9 cm plant height, 6.1 g seed yield, 12.0 g shoot dry weight and 0.7% total nitrogen.

Significant increases in seed germination and enhancement in plant growth have been reported by several workers due to osmoprotectant accumulation, modulation of gene expression associated with salt stress, and by induction of antioxidative enzymes against the ROS pathway [119, 327, 328]. Recently, Damodaran et al. [329] demonstrated enhanced grain yield in rice and wheat by using *Lysinibacillus* sp. that mitigated the harmful effects caused by high salt stress. Similarly, bacterization of soybean with *Bacillus firmus* SW5 resulted in significant improvement in nutrient uptake, photosynthesis, gas exchange, flavonoid and phenolic contents, osmoprotectants, and antioxidant enzymes under salt stress conditions [330]. Treatment of sunflower (*Helianthus annus*) with fluorescent *Pseudomonas* was found to positively affect plant biomass in salt stress conditions [331] whereas other bacterial genera belonging to species of *Pseudomonas*, *Ochrobactrum*, *Agrobacterium*, and *Klebsiella* induced salt tolerance in groundnut [31]. Similarly, inoculation of a *Pseudomonas* strain isolated from halophilic grass *Distichlis spicata* was observed to improve the growth of different crops under salt stress [332].

**Table 13.1** PGPRs conferring salt tolerance in plants

| PGPR strains | Crop | PGPR attributes | References |
|---|---|---|---|
| *Aeromonas* sp. | Wheat (*Triticum aestivum*) | EPS production | [80] |
| *Acinetobacter johnsonii* | Maize (*Zea mays* L.) | Enzymatic activities, nutrient uptake and antioxidant defence | [335] |
| *Azotobacter chroococcum* | Maize | Improved $K^+/Na^+$ ratio, polyphenol content and proline | [336] |
| *Bacillus amyloliquefaciens* | Rice (*Oryza sativa*) | Betaine, sucrose and trehalose | [327] |
| *Bacillus amyloliquefaciens* | Rice | Proline content | [337] |
| *Glutamicibacter* sp | Rice | Production of ACC deaminase | [338] |
| *Micrococcus* sp. | *Arabidopsis thaliana* and rice | Production of IAA and siderophore | [339] |
| *Klebsiella oxytoca* and *Bacillus* sp. | Cotton seeds | Antioxidative enzymes and photosynthetic pigment | [340] |
| *Klebsiella* sp. | Oat (*Avena sativa*) | Proline content, malondialdehyde content, antioxidant enzymes | [173] |
| *Curtobacterium*sp. | Barley (*Hordeum vulgare* L.), soybean (*Glycine max* L.) | Production of proline and IAA | [341] |
| *Bacillus baekryungensis* DPM17 | okra (*Abelmoschus esculentus*) | Phosphate solubilization, nitrogen fixation, production of ammonia, IAA and gibberellins | [342] |
| *Arthrobacter woluwensis* AK1 | Soybean (*Glycine max* L.) | Production of IAA and ABA | [343] |
| *Mesorhizobium* sp. | Chick pea (*Cicer arietinum*) | ACC deaminase activity | [344] |
| *Bacillus licheniformis*, *Pseudomonas plecoglossicida* | Sunflower | Production of IAA, biofilm formation, phosphate solubilization, and ACC deaminase activity | [345] |

(continued)

**Table 13.1** (continued)

| PGPR strains | Crop | PGPR attributes | References |
|---|---|---|---|
| *Bacillus marisflavi* sp., *Bacillus cereus* | *Pisum sativum* | Production of ACC deaminase | [346] |
| *Orchbactrum* sp | Groundnut (*Arachis hypogaea* L.) | Production of IAA and ACC deaminase | [347] |
| *Pseudomonas* sp. | Tomato | Production of IAA, ACC deminase and EPS | [164] |
| *Pantoea* sp. | Mungbean (*Vigna radiata* L.) | ACC deaminase activity | [348] |
| *Tsukamurella tyrosinosolvens, Burkholderia pyrrocinia* | Peanuts | Increased catalase, superoxide dismutase and peroxidase activities | [349] |
| *Streptomyces* sp. and *Bacillus* sp. | Ice-plant (*Mesembryanthemum crystallinum* L.) | IAA, phosphorus solubilization, ACC deaminase, siderophore production | [350] |

Saravanakumar and Samiyappan [333] showed that ACC deaminase-producing *P. fluorescens* strain TDK-1 significantly enhanced the growth of groundnut seedlings under salt stress conditions as compared with inoculation of strain lacking ACC deaminase activity. Inoculation of wheat with *Chryseobacterium gleum* sp. strain SUK possessing ACC deaminase activity showed significant stimulation of plant growth and enhancement in grain yield under salt stress conditions [64]. In another experiment, combined application of rhizobia and ACC deaminase-producing *Pseudomonas* on mungbean (*Vigna radiata*) showed superior growth, nodulation, and yield under salt stress conditions [261]. Similarly, coinoculation of soybean with salt-tolerating *P. putida* TSAU1 and ACC deaminase-producing *Bacillus japonicum* USDA 110 improved plant growth, macro- and micro-nutrient acquisitions, and seed protein content by modulating root architecture under salt stress conditions [159]. The combined inoculation of *Variovorax paradoxus* 5C-2 and *Mesorhizobium loti* strains possessing ACC deaminase activity had additive and synergistic effects on nodulation, root growth, and uptake of elements (e.g., N, P, Mg, Ca, Na, and Zn) in *Lotus ornithopodioides* and *L. edulis* [334]. Separate inoculation with the two bacterial strains viz. *Rhizobium* sp. LSMR-32 and *Enterococcus mundtii* LSMRS-3, possessing multifunctional growth promoting traits, ameliorated salinity stress effects and increased seed germination, grain yield, plant height, biomass, chlorophyll content, and nutrient uptake compared to uninoculated plants under salt stress conditions [319]. Inoculation with both the strains increased nodule number, nodule biomass, and leghaemoglobin amount in spring mungbean along with increase in soil phosphatase and dehydrogenase levels.

Ullah et al. [351] inoculated wheat cv. Inqlab-91 seeds with cultures of *Pseudomonas mendocina* Khsr2, *Pseudomonas putida* Khsr4, *Pseudomonas stutzeri* Khsr3 and *Azotobacter vinelandii* Khsr1. The applied PGPR strains significantly improved the transfer of K, Ca, Mg and Zn from soil to plant shoots and reduced the transfer of Cr in inoculated plants over that of uninoculated soil. The maximum $K^+/Na^+$ ratio of rhizosphere soil and wheat leaves was recorded in *Pseudomonas putida* Khsr4 inoculation. The applied PGPR helped in selective uptake of K over Na and enhanced transfer of nutrients resulting in higher yield. Yield of ridge sown plot was 3.59% higher than drill sown plot, and 10.87% higher than broadcast sown plot respectively. Oliveira Lopes et al. [352] reported that synergistic interactions between five different rhizobia (*B. elkanii* BR 2003, *B. pachyrhizi* BR 3262, *B. yuanmingense* BR 3267, *B. paxllaeri* BR 10,398, and *B. icense* BR 10,399) and *Azospirillum baldaniorum* strain (Sp245), alone or in combination, attenuated the deleterious effects of salt stress (75 mM NaCl) on lima bean. Plants coinoculated with rhizobia and *A. baldaniorum* showed the highest value for root length, plant biomass (shoot, root, and nodules), number of nodules, and photosynthetic pigments. Coinoculated plants under salt stress showed a minor increase in sodium and the highest potassium content values, and nitrogen fixation efficiency than plants inoculated with rhizobia.

Three isolates e.g., E-2, T-2, and T-1 (identified as *Klebsiella* sp. strain BAB-6433, *Citrobacter freundii* strain R2A5, and *Citrobacter* sp. DY1981, respectively) showed salt (NaCl) tolerance at concentrations of 7%, 6%, and 6%, respectively [353]. Inoculation of these salt-tolerant isolates significantly improved plant growth of paddy plants in a hydroponic study, ensuring nutrient availability to the plants grown under a nutrient (nitrate or phosphate) deprived growth matrix. Naseri et al. [354] reported that highest saline stress, 10 dS $m^{-1}$, reduced shoot and root dry weight and root volume of tomato up to 51.3, 41.5, and 51.8%, respectively. In addition, it also increased stomatal resistance and proline content 2.01- and 3.66-folds and decreased $K^+/Na^+$ ratio 4.16-folds, respectively. Inoculation of *Bacillus megaterium* P2 on tomato plants was found to modulate salt tolerance mechanisms, improved plant growth factors, soil biological indicators and also balanced $K^+/Na^+$ uptake even at 10 dS $m^{-1}$ salinity level. However, the efficiency of strains was dependent on the magnitude of salt stress. In similar studies, Gritli et al. [26] evaluated the effect of different microbial inocula consisting of nodule-forming and nitrogen-fixing *Rhizobium laguerreae* and arbuscular mycorrhizal fungus (AMF) *Rhizophagus irregularis*, along with various plant growth-promoting bacteria (PGPB) including *Bacillus subtilus*, *Bacillus simplex* and *Bacillus megaterium* on alleviating salt stress in *Lathyrus cicera* under pot trial studies. Exposure of plants to salinity (100 mM NaCl) significantly reduced growth of *L. cicera*. On the other hand, inoculation with different inocula enhanced plant growth and markedly promoted various biochemical traits, and resulted in mitigating deleterious effects of salinity stress on *L. cicera*. Coinoculation also upregulated the expression of two marker genes (*LcHKT1* and *LcNHX7*) related to salinity tolerance.

## Genetic Engineering of Plants and Microbes for Efficient Alleviation of Salinity Stress

In response to a wide range of environmental challenges, plants have evolved a wide range of strategies for modulating their rhizosphere. A deeper knowledge of the interkingdom signaling and biological processes occurring between microbes and plants may provide insights as to how the rhizosphere might be controlled to enhance plant health and production [355–358]. In the long term, rhizosphere engineering might lessen our need for herbicides and pesticides by substituting beneficient microbiota, biostimulants, or transgenic plants for agrochemicals [359]. Rhizosphere engineering is possible through the appropriate selection of crop species and cultivars, by application of stress-tolerant microbes as soil/seed treatments [360, 361]. Microorganisms can be developed to enhance nutritional availability in addition to resistance to abiotic or biotic stresses, inhibition of harmful bacteria, or that can support the survival of beneficial microorganisms. Crops can be chosen by breeders to have beneficial attributes, or beneficial microorganisms can be developed [5, 59, 362, 363]. The development of genetic techniques and the growing field of metagenomics will speed up research on the rhizosphere's microbial diversity, and rhizosphere engineering will lead to efficient modification of microorganisms for ecologically sustainable farming practices [315, 364, 365].

Various genetic engineering techniques and molecular biology approaches are being employed recently to improve the beneficial traits in plants and microorganisms to improve soil health resulting into increased plant growth and crop yield [361, 366]. In addition, identification of novel effective microbial inoculants, detecting particular bacterial gene sequences, analyzing population density with copy number of particular functional genes and the persistence of microbial inoculants in soil is a never-ending process to achieve desirable impacts on crop productivity [367–370]. The genetic diversity of rhizobacterial isolates is shown by DNA finger printing [371]. For instance, two efficient bacterial isolates i.e., *Bacillus cereus* (P31) and *Achromobacter xylosoxidans* (P35) were identified by 16S rDNA analysis out of seven bacterial strains isolated from surface-sterilized sweet potato roots and these strains were recommended to decrease chemical fertilizer consumption in sustainable agriculture [372]. *Enterobacter* spp. exhibiting PGP features and isolated from maize roots was phylogenetically described using the MicroSeqTM 16S rDNA technology, and it showed the closest similarity (99.4%) with *Enterobacter asburiae* [373]. *Bacillus, Delftia, Methylobacterium, Microbacterium, Paenibacillus, Staphylococcus,* and *Stenotrophomonas* were identified in common bean based on 16S rDNA sequences [374]. The inoculation of *Dianthus caryophyllus* roots with *Klebsiella* SGM 81 having *ipd*C gene significantly altered plant development in both laboratory and field environments, and caused an increase in root hair formation suggesting increased synthesis of auxins [375]. The presence of the *acd*S gene was detected in nine strains using PCR amplification and *Microbacterium* sp. ECI-12A showed the highest ACC deaminase activity (539.1 nmol α-ketobutyrate $mg^{-1}$

protein $h^{-1}$) [19]. Amplification of the *pqq* gene (involved in phosphate solubilization) revealed similarities between the indigenous and previously sequenced *Bacillus licheniformis* strains in this gene and its surrounding regions [376].

Multiple strategies are utilized by halotolerant PGPR in order to overcome the effects of salinity stress. In saline agroecosystems, salt-tolerant rhizobacteria boost plant performance under abiotic stress, which leads to higher crop output [377]. There is still a paucity of knowledge on the salt tolerance mechanisms of halotolerant PGPR. This lack of knowledge includes bacterial genes and proteogenomics in osmotolerance as well as plant-microbial interactions in saline soil. In spite of this, numerous studies on salt-resistant rhizobacteria have been carried out in the last ten years in order to investigate the molecular processes of gene expression when salt is present in the environment [378]. Ma et al. [379] have proposed that understanding the regulation networks of salt-tolerant rhizobacteria during abiotic stress could be a critical way of combating such stressors and promoting global food production in an environmentally acceptable manner. This method might be used to develop either specific microbes or beneficent microbial consortium to boost plant development in a variety of soil conditions. Thus, plant/soil-optimized microorganisms may be employed as inoculum for various crops in various soils. Various reports indicated that crop-specific soil microbiomes improve plant-microbe interactions over time [380].

Recently, functional metagenomics provided a magnificent way of identification of various genes responsible for salt resistance in microorganisms. Liu et al. [381] carried out whole genome analysis of a halotolerant PGPR *Klebsiella* sp. D5A and it revealed the presence of salt tolerance genes with a wide range of pH adaptability and PGP traits including phosphate solubilization, IAA biosynthesis, acetoin, and 2,3-butanediol synthesis, siderophore production, and $N_2$ fixation. The salt-stress induced damage in citrus plants was reduced by treatment with *Pseudomonas putida* and *Novosphingobium* sp., which resulted in lowering the level of abscisic acid (ABA) and salicylic acid (SA), reducing the efficiency of photosystem II (Fv/Fm), increasing accumulation of IAA in the leaf and inhibiting accumulation of root chloride and proline during salt stress [382]. A salt-tolerant *Enterobacter* sp. UPMR18 strain containing ACC deaminase showed plant growth-promoting effects through induction of reactive oxygen species scavenging enzymes including superoxide dismutase (SOD), ascorbate peroxidase (APX) and catalase (CAT) and upregulating to ROS pathway genes [143]. In similar studies, a novel salt-tolerant bacterial strain *Pseudomonas* sp. M30-35 was obtained from the rhizosphere of succulent xerohalophyte shrub *Haloxylon ammodendron*, which showed salt and drought tolerance capabilities. *Pseudomonas* sp. M30-35 was found to contain 34 genes possessing homology with certain genes associated with PGP traits and abiotic stress tolerance [144]. *Bacillus safensis* VK strain showed salt tolerance up to 14% NaCl and pH ranging from 4 to 8 [383]. Several genes were characterized by genomic studies of *B. safensis* strain, which were associated with functioning of PGP traits under conditions of high salt concentrations, drought, heavy metals, and polyaromatic hydrocarbons contamination. Sapre et al. [173] isolated *Klebsiella* sp. IG 3 from the rhizosphere of wheat and it showed salt tolerance up to 20%. This strain positively modulated the expression profile of *rbc*L (codes for the ribulose-1,5-bisphosphate carboxylase/oxygenase RuBisCo) and WRKY1 (transcription factor dealing with plants reaction to biotic stress) genes under salt-stress conditions.

An integrated strategy that included already identified genetic variants, using diversified and new sources to produce novel variations. Moreover, instead of focusing on a single attribute or characterstics during breeding, it may be more productive to look for combinations of characteristics (Table 13.2). A variety of genes involved in various pathways that increase plant tolerance to abiotic stresses have been used in the development of transgenic plants in recent years. Genes encoding different enzymes involved in promoting tolerance to multiple abiotic stresses through modifications in membrane phospholipids, production of osmoprotectants, and late embryogenesis proteins can be introduced into cereal or legume plants using single-gene transformation [384]. In legumes, mass screening is being used to identify salt-tolerant germplasm for enhancement of legume genotype. Sehrawat et al. [385] assessed 117 mungbean genotypes for salt tolerance and observed significant diversity in their efficiency under salt treatment, and classifying them as highly tolerant, moderately tolerant, sensitive, and extremely susceptible genotypes. Characteristics such as germination and seedling growth, proline content, photosynthetic efficiency, osmoregulation, crop yield, nodule formation, and ion homeostasis were used to screen genotypes for salt stress resistance.

Various reports on the salt tolerant transgenic plants have shown that activating a stress-response signal transduction pathway is an effective and potential method for increasing plant tolerance to biotic stresses [406–408]. Co-activation of various stress-response pathways, with either synergistic or antagonistic effects, may emerge from simultaneous exposure of a plant to multiple abiotic stress conditions. To deal with abiotic stresses, numerous distinct stress hormones, including ethylene, jasmonic acid, and abscisic acid or reactive oxygen species activation, receptors and signaling complexes, and networks of transcription factors and mitogen-activated protein kinase (MAPK) cascades are likely to communicate with one another. It was recently discovered that ethylene plays a fundamental role in the response of *Arabidopsis* to heat and osmotic stress. It was also observed that the expression of the transcriptional co-activator MBF1c in *Arabidopsis* enhances the tolerance of transgenic plants to these stresses by activating the ethylene-response signal transduction pathway [409].

*ERF1* genes in various species have been frequently reported to participate in abiotic and biotic stress responses. The overexpression of *ERF1* gene in *Arabidopsis* enhanced the defense of transgenic plants against *P. cucumerina* [410], as well as their resistance against drought and salt stress [411]. The overexpression of *ERF1* gene in wheat strengthened the responses of the transgenic plants to pathogen stress and several abiotic stresses [412]. In *Arabidopsis*, *AtERF1* gene played a positive role in salt, drought, and heat stress tolerance by regulating stress-specific gene, and by integrating jasmonic acid, ethylene, and abscisic acid signals [413]. Overexpression of the pepper *CaERF5* gene in tobacco plants enhanced the resistance to *Ralstonia solanacearum* infection under the influence of salicylic acid, methyl jasmonate, and

**Table 13.2** Transgenic plants having improved salt tolerance

| Crops | Transferred gene | Observations | References |
|---|---|---|---|
| Wheat | Mt1 D | Turgor maintenance | [386] |
| Brassica | SOS1 | Plasma membrane $Na^{+}/K^{+}$ antiporter | [387] |
| | h-type Trx proteins, AtTrx-*h*2 | Improved antioxidant enzyme activity | [388] |
| Tomato | BADH1 | Improves salt tolerance; accumulation of betaine | [389] |
| | SIMYB 102 | Salt tolerance by regulating $Na^{+}$-$K^{+}$ homeostasis and ROS balance | [390] |
| *Arabidopsis thaliana* | JcDREB | Transcription factor | [391] |
| Soybean | WRKY11 | Improves salt tolerance | [392] |
| Chickpea | P5CS | Synthesis and accumulation of proline | [393] |
| Mungbean | codA | Improve abiotic stress tolerance | [394] |
| | VrWRKY | Enhance abiotic stress tolerance | [395] |
| Common bean | Asr1, Asr2 | ABA signaling pathway | [396] |
| Alfalfa | CsALDH12A1 | Improves salt tolerance | [397] |
| | GmDREB1 | Conferred salt tolerance | [398] |
| | IbOr | Increased tolerance to multiple abiotic stresses | [399] |
| Faba bean | PR10a | Synthesis and accumulation of osmolytes | [400] |
| Populus | *OsCYP714D1* | Improved the salt tolerance | [401] |
| Pigeon pea | OsRuvB | Improve salt tolerance through increases in chlorophyll content, relative water content, peroxidase and catalase activity | [402] |
| Peanut | AhWRKY75 | Increased antioxidant activity | [403] |
| Potato | StCYS1 | High proline and chlorophyll content | [404] |
| Birch | BpERF1.1 | Improved tolerance to cold, salt and drought stress | [405] |

ethylene [410]. In similar studies, overexpression of the soybean *GmERF3* gene, an AP2/ERF type transcription factor improved the tolerance of transgenic tobacco against drought, salinity, and even mosaic disease [411]. Zhang et al. [405] over-expressed *BpERF1.1* gene in birch (*Betula platyphylla* Suk.) using *Agrobacterium*-mediated infection method and obtained 11 transgenic lines with improved tolerance against multiple abiotic stresses. RNA-seq analysis identified 689 differentially expressed genes (DEGs) in the transgenic birch compared with WT, including 228 up-regulated genes and 461 down-regulated genes. Gene ontology enrichment analysis showed that among these DEGs, 273 genes were involved in various plant biological processes, and 83% of them were involved in cellular processes, metabolic processes, biological regulation and response to stimulus (11%). Thus, *BpERF*1.1 gene was found to improve the tolerance and resistance of birch against cold, salt and drought stress, probably by interconnecting with other genes involved in plant response to abiotic stresses.

## Conclusions and Future Perspectives

Extensive studies have been carried out to analyze various environmental factors, which affect soil fertility and cause agricultural yield losses due to salt stress [6, 45, 414]. The study of ecological and evolutionary responses to salt stress in agroecosystems could benefit from the identification and examination of significant local microorganisms that are found in salty environments [415]. It is impossible to exaggerate the significance of using metagenomic, proteogenomic, and metabolomic approaches in order to harness and discover new PGPR, as well as specific metabolites and upregulated gene expression for the salt tolerance [145]. Given the effects of climate change, screening of sufficient salt-resistant PGPR strains is needed that may provide tolerance to abiotic stresses in order to maintain crop quality [416–418]. For developing novel and effective bio-enhancers, bioinoculants, and bio-protectants, characterization of essential metabolites, such as osmoprotectants, anti-oxidant enzymes, biosurfactants, phytohormone precursors and nutrients are needed. In agriculture, microbial consortia have become increasingly popular that may provide tolerance not only to abiotic stress, but also give resistance against phytopathogens [419].

Abiotic stresses are one of the most serious barriers to agricultural production on a global scale. Salt-tolerant microorganisms that are associated with rhizoplane, rhizosphere, and endophytic bacteria can play an important role in conferring abiotic stress resistance to plants. Currently, a lot of efforts are being made to improve the field efficacy of ACC deaminase-producing halo-tolerant bacteria. For instance, significant efforts are invested in development of improved biofertilizer formulations and bioinoculants to resist salt stress in wheat and cucumber such as chitosan-immobilized aggregated *Methylobacterium oryzae* strain CBMB20 [420], super absorbent polymer [421], and *Paenibacillus beijingensis* BJ-18 and *Bacillus* sp. L-56 [422]. Inoculations of effective salt-tolerant bioinoculants will assist in the mitigation of the adverse effects of climate change and help in enhancing crop productivity in salt-stressed soils contributing to an expanded global food supply for ever-growing global population. These salt-tolerant biofertilizers will provide phytohormones and nutrients, lower ethylene levels, induce novel plant genes to accelerate osmolyte accumulation, increase $K^+$ concentration, reduce $Na^+$ uptake, and ultimately maintaining a high $K^+$ ions. Numerous plant species have demonstrated salt tolerance as a result of bacterization with PGPR.

The production of stress-tolerant cultivars through conventional breeding and genetic engineering is essential, but the process is time-consuming and expensive. In comparison, the utilization of microorganisms to alleviate the negative effects of abiotic stresses is less expensive, friendlier to the environment, and requires less time. To maximize the benefits of microbial inoculants and enhance plant development and tolerance to a variety of biotic and abiotic stressors, new strategies will be developed once it is understood how the various microbial populations and plant systems are connected to one another. In the future, more in-depth research focusing on the gene expression level and multi-functional PGP features of salt-tolerating rhizobacteria needs to be carried out in order to build tailor-made bioformulations that may mitigate the effects of salinity stress under changing climate conditions and may boost plant growth under abiotic stresses in saline soil [423, 424].

## References


1. Global Agricultural Productivity Report (GAP Report). Global Harvest Initiative, Washington, https://globalagriculturalproductivity.org/wpcontent/uploads/2019/01/GHI_2018-GAP-Report_FINAL-10.03.pdf
2. Tilman D, Cassman KC, Matson PA, Naylor R, Polasky S (2002) Agricultural sustainability and intensive production practices. Nature 418:671–677
3. Egamberdieva D, Wirth S, Bellingrath-Kimura SD, Mishra J, Arora NK (2019) Salt-tolerant plant growth promoting rhizobacteria for enhancing crop productivity of saline soils. Front Microbiol 10:2791
4. Shahbaz M, Ashraf M (2013) Improving salinity tolerance in cereals. Crit Rev Plant Sci 32:237–249
5. Patel MK, Kumar M, Li W, Luo Y, Burritt DJ, Alkan N, Tran LSP (2020) Enhancing salt tolerance of plants: from metabolic reprogramming to exogenous chemical treatments and molecular approaches. Cells 9:2492
6. Corwin DL (2021) Climate change impacts on soil salinity in agricultural areas. Eur J Soil Sci 72:842–862
7. Bakka K, Gopika PV, Sreelakshmi H, Challabathula D (2022) Halotolerant plant growth promoting rhizobacteria: a futuristic direction to salt stress tolerance. In: Roy S, Mathur P, Chakraborty AP, Saha SP (eds) Plant stress: challenges and management in the new decade. Advances in science, technology & innovation. Springer, Cham
8. Foyer CH, Rasool B, Davey JW, Hancock RD (2016) Cross-tolerance to biotic and abiotic stresses in plants: a focus on resistance to aphid infestation. J Exp Bot 67:2025–2037
9. Nadeem SM, Ahmad M, Naveed M, Imran M, Zahir ZA, Crowley DE (2016) Relationship between *in vitro* characterization and comparative efficacy of plant growth-promoting rhizobacteria for improving cucumber salt tolerance. Arch Microbiol 198:379–387
10. Phour M, Sindhu SS (2020) Amelioration of salinity stress and growth stimulation of mustard (*Brassica juncea* L.) by salt-tolerant *Pseudomonas* species. Appl Soil Ecol 149:103518
11. Schultz RC, Colletti JP, Faltonson RR (1995) Agroforestry opportunities for the United States of America. Agrofor Syst 31:117–142

12. Huang R, McGrath SP, Hirsch PR, Clark IM, Storkey J, Wu L, Zhou J, Liang Y (2019) Plant–microbe networks in soil are weakened by century-long use of inorganic fertilizers. Microbiol Biotechnol 12:1464–1475
13. Socolow RH (1999) Nitrogen management and the future of food: lessons from the management of energy and carbon. Proc Natl Acad Sci USA 96:6001–6008
14. Qadir M, Quillérou E, Nangia V, Murtaza G, Singh M, Thomas RJ, Drechsel P, Noble AD (2014) Economics of salt-induced land degradation and restoration. Nat Resour Forum 38:282–295
15. Gepstein S, Glick BR (2013) Strategies to ameliorate abiotic stress-induced plant senescence. Plant Mol Biol 82:623–633
16. Hamilton CE, Bever JD, Labbé J, Yang XH, Yin HF (2016) Mitigating climate change through managing constructed-microbial communities in agriculture. Agric Ecosyst Environ 216:304–308
17. Flowers TJ, Galal HK, Bromham L (2010) Evolution of halophytes: multiple origins of salt tolerance in land plants. Funct Plant Biol 37:604–612
18. Dimkpa C, Weinand T, Asch F (2009) Plant-rhizobacteria interactions alleviate abiotic stress conditions. Plant Cell Environ 32:1682–1694
19. Shrivastava P, Kumar R (2015) Soil salinity: a serious environmental issue and plant growth promoting bacteria as one of the tools for its alleviation. Saudi J. Biol. Sci. 22:123–131
20. Gauri Singh AK, Bamania M (2012) Characterization of *Mesorhizobium* sp. isolated from root nodules of *Cicer arietinum*. Intern J Agril Sci Res 2:142–154
21. Yao LX, Wu ZS, Zheng YY, Kaleem I, Li C (2010) Growth promotion and protection against salt stress by *Pseudomonas putida* Rs-198 on cotton. Eur J Soil Biol 46:49–54
22. Nia SH, Zarea MJ, Rejali F, Varma A (2012) Yield and yield components of wheat as affected by salinity and inoculation with *Azospirillum* strains from saline or non-saline soil. J Saudi Soc Agric Sci 11:113–121
23. Ramadoss D, Lakkineni VK, Bose P, Ali S, Annapurna K (2013) Mitigation of salt stress in wheat seedlings by halotolerant bacteria isolated from saline habitats. Springer Plus 2:6
24. Palacios OA, Bashan Y, de-Bashan LE (2014) Proven and potential involvement of vitamins in interactions of plants with plant growth-promoting bacteria—an overview. Biol Fertil Soils 50:415–432
25. Sindhu SS, Sharma R (2020) Plant growth promoting rhizobacteria (PGPR): a sustainable approach for managing soil fertility and crop productivity. In: Malik DK, Rathi M, Kumar R, Bhatia D (eds) Microbes for humankind and application. Daya Publishing House, New Delhi, pp 97–130
26. Gritli T, Boubakri H, Essahibi A, Hsouna J, Ilahi H, Didier R, Mnsari B (2022) Salt stress mitigation in *Lathyrus cicera* by combining different microbial inocula. Physiol Mol Biol Plants 28:1191–1206
27. Rajput L, Imran A, Mubeen F, Hafeez FY (2013) Salt-tolerant PGPR strain *Planococcus rifietoensis* promotes the growth and yield of wheat (*Triticum aestivum* L.) cultivated in saline soil. Pak J Bot 45:1955–1962
28. Gupta G, Parihar SS, Ahirwar NK, Snehi SK, Singh V (2015) Plant growth promoting rhizobacteria (PGPR): current and future prospects for development of sustainable agriculture. J Microb Biochem Technol 7:96–102
29. Zhou C, Zhu L, Xie Y, Li FY, Xiao X, Ma ZY, Wang JF (2017) *Bacillus licheniformis* SA03 confers increased saline-alkaline tolerance in *Chrysanthemum* plants by induction of abscisic acid accumulation. Front Plant Sci 8:1143
30. Zhang SY, Fan C, Wang YX, Xia YS, Xiao W, Cui XL (2018) Salt-tolerant and plant-growth-promoting bacteria isolated from high-yield paddy soil. Can J Microbiol 64:968–978
31. Sharma S, Kulkarni J, Jha B (2016) Halotolerant rhizobacteria promote growth and enhance salinity tolerance in peanut. Front Microbiol 7:1600
32. Sharma A, Singh P, Kumar S, Kashyap PL, Srivastava AK, Chakdar H, Singh RN, Kaushik R, Saxena AK, Sharma AK (2015) Deciphering diversity of salt-tolerant bacilli from saline soils of Eastern Indo- Gangetic Plains of India. Geomicrobiol J 32:170–180

33. Glick BR (2014) Bacteria with ACC deaminase can promote plant growth and help to feed the world. Microbiol Res 169:30–39
34. Ramesh A, Sharma SK, Sharma MP, Yadav N, Joshi OP (2014) Inoculation of zinc solubilizing *Bacillus aryabhattai* strains for improved growth, mobilization and biofortification of zinc in soybean and wheat cultivated in Vertisols of central India. Appl Soil Ecol 73:87–96
35. Amaresan N, Kumar K, Madhuri K, Usharani GK (2016) Isolation and characterization of salt tolerant plant growth promoting rhizobacteria from plants grown in tsunami affected regions of Andaman and Nicobar Islands. Geomicrobiol J 33:942–947
36. Zaidi A, Ahmad E, Khan MS, Saif S, Rizvi A (2015) Role of plant growth promoting rhizobacteria in sustainable production of vegetables: current perspective. Sci Hortic Amst 193:231–239
37. Du Jardin P (2015) Plant biostimulants: definition, concept, main categories and regulation. Sci Hortic Amst 196:3–14
38. Sharma R, Dahiya A, Sindhu SS (2019) Harnessing proficient rhizobacteria to minimize the use of agrochemicals. Intern J Curr Microbiol Appl Sci 7(10):3186–3197
39. Kumar S, Diksha Sindhu SS, Kumar R (2022) Biofertilizers: an ecofriendly technology for nutrient recycling and environmental sustainability. Curr Res Microbial Sci 3:100094
40. Barrios S, Ouattara B, Strobl E (2008) Impact of climatic change on agricultural production: is it different for Africa? Food Policy 33(4):287–298
41. Kalra N, Suneja P, Mendiratta N, Gupta N (2013) Simulating impact of climate change and its variability on growth and yield of crops. Clim Change Environ Sustain 1(1):11–19
42. Bazany KE, Wang JT, Delgado-Baquerizo M, Singh BK, Trivedi P (2022) Water deficit affects inter-kingdom microbial connections in plant rhizosphere. Environ Microbiol 24(8):3722–3734
43. Clair SB, Lynch JP (2010) The opening of Pandora's box: climate change impacts on soil fertility and crop nutrition in developing countries. Plant Soil 335:101–115
44. Grover M, Ali SZ, Sandhya V, Rasul A, Venkateswarlu B (2011) Role of microorganisms in adaptation of agriculture crops to abiotic stresses. World J Microbiol Biotechnol 27(5):1231–1240
45. Ajala OA, Ajibade FO, Oluwadipe OR, Nwogwu NA, Adelodun B, Guadie A, Ajibade TF, Lasisi KH, Adewumi JR (2022) Microbial impact on climate-smart agricultural practices. In: Kumar A, Singh J, Luiz Fernando Romanholo Ferreira LFR (eds) Microbiome under changing climate. Woodhead Publishing, pp 203–236
46. Turral H, Burke J, Faurès JM (2011) Climate change, water and food security. Food and Agriculture Organization of the United Nations, Rome
47. Rengasamy P (2010) Soil processes affecting crop production in salt-affected soils. Funct Plant Biol 37:613–620
48. FAO (2015) Status of the World's Soil Resources (SWSR)—Main Report. Rome: Food and Agriculture Organization of the United Nations
49. Abbaspour K, Ashraf Vaghefi S (2019) Harmonized world soil database in SWAT format. Pangaea, Los Angeles, CA
50. Ahmed MZ, Gul B, Khan MA, Watanabe KN (2016) "1—characterization and function of sodium exchanger genes in *Aeluropus lagopoides* under NaCl stress. In: Khan MA, Ozturk M, Gul B, Ahmed MZ (eds) Halophytes for food security in dry lands. Academic Press, San Diego, CA, pp 1–16
51. Parihar P, Singh S, Singh R, Singh VP, Prasad SM (2015) Effect of salinity stress on plants and its tolerance strategies: a review. Environ Sci Pollut Res 22:4056–4075
52. Arora NK, Fatima T, Mishra I, Verma M, Mishra J, Mishra V (2018) Environmental sustainability: challenges and viable solutions. Environ Sust 1:309–340
53. Patel BB, Dave RS (2011) Studies on infiltration of saline-alkali soils of several parts of Mehsana and Patan districts of North Gujarat. J Appl Technol Environ Sani 1:87–92
54. Naik RS (2014) Impact of soil salinity on agriculture in the Lift Irrigation Commands of Shirol Taluka. PhD Dissertation, Shivaji University

55. Munns R (2002) Comparative physiology of salt and water stress. Plant Cell Environ 25:239–250
56. Porcel R, Aroca R, Ruiz-Lozano JM (2012) Salinity stress alleviation using arbuscular mycorrhizal fungi. Rev Agron Sustain Develop 32:181–200
57. Keren R (2005) Salt-affected soils, reclamation. In: Hillel D (ed) Encyclopaedia of soils in the environment. Elsevier, Oxford, pp 454–461
58. Ayyam V, Palanivel S, Chandrakasan S (2019) Approaches in land degradation management for productivity enhancement. In: Ayyam V, Palanivel S, Chandrakasan S (eds) Coastal ecosystems of the tropics—adaptive management. Springer, Singapore, pp 463–491
59. Fita A, Rodríguez-Burruezo A, Boscaiu M, Prohens J, Vicente O (2015) Breeding and domesticating crops adapted to drought and salinity: a new paradigm for increasing food production. Front Plant Sci 6:978
60. Morton MJL, Awlia M, Al-Tamimi N, Saade S, Pailles Y, Negrão S, Tester M (2019) Salt stress under the scalpel-dissecting the genetics of salt tolerance. Plant J 97:148–163
61. Vaishnav A, Kumari S, Jain S, Varma A, Tuteja N, Choudhary DK (2016) PGPR-mediated expression of salt tolerance gene in soybean through volatiles under sodium nitroprusside. J Basic Microbiol 56:1274–1288
62. Berg G, Martinez JL (2015) Friends or foes: can we make a distinction between beneficial and harmful strains of the *Stenotrophomonas maltophilia* complex? Front Microbiol 6:241
63. Egamberdieva D, Jabborova D, Hashem A (2015) *Pseudomonas* induces salinity tolerance in cotton (*Gossypium hirsutum*) and resistance to *Fusarium* root rot through the modulation of indole-3-acetic acid. Saud J Biol Sci 22:773–779
64. Bhise KK, Bhagwat PK, Dandge PB (2017) : Synergistic effect of *Chryseobacterium gleum* sp. SUK with ACC deaminase activity in alleviation of salt stress and plant growth promotion in *Triticum aestivum* L. 3Biotech 7:1–13
65. Zhu Z, Zhang H, Leng J, Niu H, Chen X, Liu D, Chen Y, Gao N, Ying H (2020) Isolation and characterization of plant growth-promoting rhizobacteria and their effects on the growth of *Medicago sativa* L. under salinity conditions. Antonie Van Leeuwenhoek 113:1263–1278
66. Hu Y, Schmidhalter U (2002) Limitation of salt stress to plant growth. In: Hock B, Elstner CF (eds) Plant Toxico Marcel Dekker Inc., New York, pp 91–224
67. Singh KN, Chatrath R (2001) Salinity tolerance. In: Reynolds MP, OrtizMonasterio JI, McNab A (eds) Application of physiology in wheat breeding. CIMMYT, Mexico, pp 101–110
68. Parida AK, Das AB (2005) Salt tolerance and salinity effects on plants: a review. Ecotoxicol Environ Saf 60:324–349
69. Akbarimoghaddam H, Galavi M, Ghanbari A, Panjehkeh N (2011) Salinity effects on seed germination and seedling growth of bread wheat cultivars. Trakia J Sci 9:43–50
70. Pushpavalli R, Quealy J, Colmer TD, Turner NC, Siddique KHM, Rao MV, Vadez V (2016) Salt stress delayed flowering and reduced reproductive success of chickpea (*Cicer arietinum* L.), a response associated with $Na^+$ accumulation in leaves. J Agron Crop Sci 202:125–138
71. Ahmad P, Nabi G, Ashraf M (2011) Cadmium-induced oxidative damage in mustard [*Brassica juncea* (L.) Czern. & Coss.] plants can be alleviated by salicylic acid. South Afric J Bot 77:36–44
72. Islam F, Yasmeen T, Arif MS, Ali S, Ali B, Hameed S, Zhou WJ (2016) Plant growth promoting bacteria confer salt tolerance in *Vigna radiata* by upregulating antioxidant defense and biological soil fertility. Plant Growth Regul 80:23–36
73. Ji HT, Pardo JM, Batelli G, Van Oosten MJ, Bressan RA, Li X (2013) The salt overly sensitive (SOS) pathway: established and emerging roles. Mol Plant 6:275–286
74. Bano A, Fatima M (2009) Salt tolerance in *Zea mays* (L.) following inoculation with *Rhizobium* and *Pseudomonas*. Biol Fertil Soils 45:405–413
75. Kim JS, Mele PM, Crowley DE (2013) Application of PCR primer sets for detection of *Pseudomonas* sp. functional genes in the plant rhizosphere. J Agric Chem Environ 1:8–15
76. Paul D, Lade H (2014) Plant-growth-promoting rhizobacteria to improve crop growth in saline soils: a review. Agron Sustain Dev 34:737–752

77. Ryu JY, Lee HJ, Seo PJ, Jung JH, Ahn JH, Park CM (2014) The *Arabidopsis* floral repressor BFT delays flowering by competing with FT for FD binding under high salinity. Mol Plant 7:377–387
78. Numan M, Bashir S, Khan Y, Mumtaz R, Shinwari ZK, Khan AL, Khan A, Al-Harrasi A (2018) Plant growth promoting bacteria as an alternative strategy for salt tolerance in plants. Microbiol Res 209:21–32
79. Javid MG, Sorooshzadeh A, Moradi F, Sanavy SAMM, Allahdadi I (2011) The role of phytohormones in alleviating salt stress in crop plants. Aust J Crop Sci 5:726–734
80. Ashraf M, Hasnain S, Berge O, Mahmood T (2004) Inoculating wheat seedling with exopolysaccharide-producing bacteria restricts. Biol Fertil Soils 40(3):157–162
81. Wolde G, Adamu C (2018) Impact of salinity on seed germination and biomass yields of field pea (*Pisum sativum* L.). Asian J Sci Tech 09:7565–7569
82. Alam A, Juraimi AS, Rafii MY, Hamid AA, Aslani F, Alam MZ (2015) Effects of salinity and salinity-induced augmented bioactive compounds in purslane (*Portulaca oleracea* L.) for possible economical use. Food Chem 169:439–447
83. Farooq M, Gogoi N, Hussain M, Barthakur S, Paul S, Bharadwaj N, Migdadi HM, Alghamdi SS, Siddique KHM (2017) Effects, tolerance mechanisms and management of salt stress in grain legumes. Plant Physiol Biochem 118:199–217
84. Bhise KK, Dandge PB (2019) Mitigation of salinity stress in plants using plant growth promoting bacteria. Symbiosis 79:191–204
85. Ali Khan M, Shahid Shaukat S, Shahzad A, Arif H (2012) Growth and yield responses of pearl millet (*Pennisetum glaucum* L.) irrigated with treated effluent from waste stabilization ponds. Pak J Bot 46:1011–1018
86. Faravani M, Emami SD, Gholami BA, Faravani A (2013) The effect of salinity on germination, emergence, seed yield and biomass of black cumin. J Agric Sci 58:41–49
87. Amirul Alam M, Juraimi AS, Rafii MY, Hamid AA, Aslani F, Alam MZ (2015) Effects of salinity and salinity-induced augmented bioactive compounds in purslane (*Portulaca oleracea* L.) for possible economical use. Food Chem 169:439–447
88. Garg N, Singla R (2009) Variability in response of chickpea cultivars to short-term salinity in terms of water retention capacity, membrane permeability and osmo-protection. Turkish J Agric Forage 33:57–63
89. Samineni KHM, Siddique PM, Gaur TD, Colmer LS (2011) Salt sensitivity of vegetative and reproductive stages in chickpea podding is particular sensitive stage. Environ Exp Bot 71:260–268
90. Rao DLN, Giller KE, Yeo AR, Flowers TJ (2002) Effects of salinity and sodicity upon nodulation and nitrogen fixation in chickpea. Ann Bot 89:563–570
91. Wadhwa Z, Srivastava V, Rani R, Tanvi Makkar K, Jangra S (2017) Isolation and characterization of *Rhizobium* from chickpea (*Cicer arietinum*). Int J Curr Microbiol Appl Sci 6(11):2880–2893
92. Sehrawat A, Khandelwal A, Sindhu SS (2018) Characterization of *Mesorhizobium* strains for salt tolerance and wilt control: their potential for plant growth promotion of chickpea (*Cicer arietinum* L.). Legume Res 43(1):146–150
93. Tester N, Davenport R (2003) $Na^+$ tolerance and $Na^+$ transport in higher plants. Ann Bot 91:1–25
94. Munns R, Tester M (2008) Mechanisms of salinity tolerance. Annu Rev Plant Biol 59:651–681
95. Zahran HH (1997) Diversity, adaptation and activity of the bacterial flora in saline environments. Biol Fertil Soils 25:211–223
96. Schimel J, Balser TC, Wallenstein M (2007) Microbial stress response physiology and its implications for ecosystem function. Ecology 88:1386–1394
97. Ibekwe AM, Poss JA, Grattan SR, Grieve CM, Suarez D (2010) Bacterial diversity in cucumber (*Cucumis sativus*) rhizosphere in response to salinity, soil pH and boron. Soil Biol Biochem 42:567–575
98. Chowdhury N, Marschner P, Burns RG (2011) Soil microbial activity and community composition: impact of changes in matric and osmotic potential. Siol Biol Biochem 43:1229–1236

99. Batra L, Manna MC (1997) Dehydrogenase activity and microbial biomass carbon in salt affected soils of semi arid regions. Arid Soil Res Rehbit 3:293–303
100. Rietz DN, Haynes RJ (2003) Effects of irrigation induced salinity and sodicity on soil microbial activity. Soil Biol Biochem 35:845–854
101. Sardinha M, Muller T, Schmeisky H, Joergensen RG (2003) Microbial performance in soils along a salinity gradient under acidic conditions. Appl Soil Ecol 23:237–244
102. Oren A (2002) Molecular ecology of extremely halophilic archaea and bacteria. FEMS Microbiol Ecol 39:1–7
103. Jiang H, Dong H, Yu B, Liu X, Li Y, Ji S, Zhang CL (2007) Microbial response to salinity change in Lake Chaka, a hypersaline lake on Tibetan plateau. Environ Microbiol 9:2603–2621
104. Omar SA, Abdel-Sater MA, Khallil AM, Abd-Alla MH (1994) Growth and enzyme activities of fungi and bacteria in soil salinized with sodium chloride. Folia Microbiol 39:23–28
105. Rütting T, Aronsson H, Delin S (2018) Efficient use of nitrogen in agriculture. Nutr Cycl Agroecosyst 110:1–5
106. Schirawski J, Perlin MH (2018) Plant-microbe interaction 2017—the good, the bad and the diverse. Int J Mol Sci 19:1374
107. Zurayk R, Adlan M, Baalbaki R, Saxena MC (1998) Interactive effects of salinity and biological nitrogen fixation on chick-pea (*Cicer arietinum* L.) growth. J Agron Crop Sci 180:249–258
108. Slattery JF, Conventry DR, Slattery WJ (2001) Rhizobial ecology as affected by the soil environment. Aust J Exp Agric 41:289–298
109. Zahran HH (1999) *Rhizobium*-Legume symbiosis and nitrogen fixation under severe conditions and in an arid climate. Microbiol Mol Biol Res 3:968–989
110. Vanderlinde EM, Harrison JJ, Muszynski A, Carlson RW, Turner RJ, Yost CK (2010) Identification of a novel ABC transporter required for desiccation tolerance and biofilm formation in *Rhizobium leguminosarum* bv. *viciae* 3841. FEMS Microbiol Ecol 71:327–340
111. Upadhyay SK, Singh JS, Saxena AK, Singh DP (2012) Impact of PGPR inoculation on growth and antioxidant status of wheat under saline conditions. Plant Biol 14:605–611
112. Garg A, Sharma M (2013) Study of stress tolerant forms of rhizobia isolated from *Trigonella foenumgraecum*. in semi arid region of Rajasthan. Microbiology 2:2277–8179
113. Shultana R, Kee Zuan AT, Yusop MR, Saud HM (2020) Characterization of salt-tolerant plant growth-promoting rhizobacteria and the effect on growth and yield of saline-affected rice. PLoS ONE 15(9):e0238537
114. Bais HP, Weir TL, Perry LG, Gilroy S, Vivanco JM (2006) The role of root exudates in rhizosphere interactions with plants and other organisms. Annu Rev Plant Biol 57:233–266
115. Nie M, Zhang XD, Wang JQ, Jiang LF, Yang J, Quan ZX, Cui XH, Fang CM, Li B (2009) Rhizosphere effects on soil bacterial abundance and diversity in the Yellow River Deltaic ecosystem as influenced by petroleum contamination and soil salinization. Soil Biol Biochem 41:2535–2542
116. Wenzel WW (2009) Rhizosphere processes and management in plant assisted bioremediation (phytoremediation) of soils. Plant Soil 321:385–408
117. Niu XG, Song LC, Xiao YN, Ge WD (2018) Drought-tolerant plant growth-promoting rhizobacteria associated with foxtail millet in a semiarid agroecosystem and their potential in alleviating drought stress. Front Microbiol 8:2580
118. Singh RP, Jha PN (2016) Alleviation of salinity-induced damage on wheat plant by an ACC deaminase-producing halophilic bacterium *Serratia* sp. SL-12 isolated from a salt lake. Symbiosis 69:101–111
119. Sarkar A, Ghosh PK, Pramanik K, Mitra S, Soren T, Pandey S, Mondal MH, Maiti TK (2018) A halotolerant *Enterobacter* sp. displaying ACC deaminase activity promotes rice seedling growth under salt stress. Res Microbiol 169:20–32
120. Ondrasek G, Rengel Z, Romic D, Savic R (2010) Environmental salinisation processes in agro-ecosystem of Neretva River estuary. Novenytermeles 59:223–226
121. Li YQ, Zhao HL, Yi XY, Zuo XA, Chen YP (2006) Dynamics of carbon and nitrogen storages in plant-soil system during desertification process in Horqin sandy land. Environ Sci 27:635–640

122. Nelson DR, Mele PM (2007) Subtle changes in rhizosphere microbial community structure in response to increased boron and sodium chloride concentrations. Soil Biol Biochem 39:340–351
123. Kapoor R (2015) Bacterial diversity of salt tolerant nitrogen fixers around the salt mines of Himachal Pradesh. PhD Dissertation, Chaudhary Sarwan Kumar Himachal Pradesh Krishi Vishvavidyalaya
124. Miller KJ, Wood JM (1996) Osmoadaptation by rhizosphere bacteria. Ann Rev Microbiol 50:101–136
125. Shabala S, Pottosin I (2014) Regulation of potassium transport in plants under hostile conditions: implications for abiotic and biotic stress tolerance. Physiol Plant 151:257–279
126. Klein W, Weber MHW, Marahiel MA (1999) Cold shock response of *Bacillus subtilis*: Isoleucine- dependent switch in the fatty acid branching pattern for membrane adaptation to low temperatures. J Bacteriol 181:5341–5349
127. Piuri M, Sanchez-Rivas C, Ruzal SM (2005) Cell wall modifications during osmotic stress in *Lactobacillus casei*. J Appl Microbiol 98:84–95
128. Sandhya V, Ali SZ, Grover M, Reddy G, Venkateswarlu B (2009) Alleviation of drought stress effects in sunflower seedlings by the exopolysaccharides producing *Pseudomonas putida* strain GAP-P45. Biol Fertil Soils 46:17–26
129. Francius G, Polyakov P, Merlin J, Abe Y, Ghigo JM, Merlin C, Beloin C, Duval JF (2011) Bacterial surface appendages strongly impact nanomechanical and electrokinetic properties of *Escherichia coli* cells subjected to osmotic stress. PLoS ONE 6:e20066
130. Creus CM, Sueldo RJ, Barassi CA (2004) Water relations and yield in *Azospirillum*-inoculated wheat exposed to drought in the field. Can J Bot 82:273–281
131. Arora NK, Tewari S, Singh S, Lal N, Maheshwari DK (2012) PGPR for protection of plant health under saline conditions. In: Maheshwari DK (ed) Bacteria in agrobiology: stress management. Springer, Heidelberg, Berlin, pp 239–258
132. Qin Y, Druzhinina IS, Pan XY, Yuan ZL (2016) Microbially mediated plant salt tolerance and microbiome-based solutions for saline agriculture. Biotechnol Adv 34:1245–1259
133. Pérez-Alfocea F, Albacete A, Ghanem ME, Dodd IC (2010) Hormonal regulation of source-sink relations to maintain crop productivity under salinity: a case study of root-to-shoot signaling in tomato. Funct Plant Biol 37:592–603
134. Etesami H, Beattie GA (2018) Mining halophytes for plant growth promoting halotolerant bacteria to enhance the salinity tolerance of non-halophytic crops. Front Microbiol 9:148
135. Glick BR, Cheng ZY, Czarny J, Duan J (2007) Promotion of plant growth by ACC deaminase-producing soil bacteria. Eur J Plant Pathol 119:329–339
136. Etesami H, Beattie GA (2017) Plant-microbe interactions in adaptation of agricultural crops to abiotic stress conditions. In: Kumar V, Kumar M, Sharma S, Prasad R (eds) Probiotics and plant health. Springer, Singapore, pp 163–200
137. Del Amor FM, Cuadra-Crespo P (2012) Plant growth-promoting bacteria as a tool to improve salinity tolerance in sweet pepper. Funct Plant Biol 39:82–90
138. Nagpal S, Sharma P, Kumawat KC (2019) Assessment of native single and dual inoculants of *Mesorhizobium* sp. and endophytic rhizobacteria for plant growth promotion in chickpea. Agric Res J 56:746–751
139. Nabi RBS, Tayade R, Hussain A, Kulkarni KP, Imran QM, Mun BG, Yun BW (2019) Nitric oxide regulates plant responses to drought, salinity, and heavy metal stress. Environ Expert Bot 161:120–133
140. Yang J, Kloepper JW, Ryu CM (2009) Rhizosphere bacteria help plants tolerate abiotic stress. Trends Plant Sci 14:1–4
141. Qurashi AW, Sabri AN (2012) Bacterial exopolysaccharide and biofilm formation stimulate chickpea growth and soil aggregation under salt stress. Braz J Microbiol 43:1183–1191
142. Nadeem SM, Zahir ZA, Naveed M, Nawaz S (2013) Mitigation of salinity-induced negative impact on the growth and yield of wheat by plant growth-promoting rhizobacteria in naturally saline conditions. Ann Microbiol 63:225–232

143. Habib SH, Kausar H, Saud HM (2016) Plant growth-promoting rhizobacteria enhance salinity stress tolerance in okra through ROS-scavenging enzymes. BioMed Res Int 6284547
144. He AL, Niu SQ, Zhao Q, Li YS, Gou JY, Gao HJ, Suo SZ, Zhang JL (2018) Induced salt tolerance of perennial ryegrass by a novel bacterium strain from the rhizosphere of a desert shrub *Haloxylon ammodendron*. Int J Mol Sci 19:469
145. Bhise KK, Dandge PB (2019) Alleviation of salinity stress in rice plant by encapsulated salt tolerant plant growth promoting bacteria *Pantoea agglomerans* strain KL and its root colonization ability. Arch Agron Soil Sci 65:1955–1968
146. Pinedo I, Ledger T, Greve M, Poupin MJ (2015) *Burkholderia phytofirmans* PsJN induces long-term metabolic and transcriptional changes involved in *Arabidopsis thaliana* salt tolerance. Front Plant Sci 6:466
147. Habib SH, Kausar H, Saud HM, Ismail MR, Othman R (2016) Molecular characterization of stress tolerant plant growth promoting rhizobacteria (PGPR) for growth enhancement of rice. Int J Agric Biol 18:184–191
148. Egamberdieva D, Jabborova D, Berg G (2016) Synergistic interactions between *Bradyrhizobium japonicum* and the endophyte *Stenotrophomonas rhizophila* and their effects on growth, and nodulation of soybean under salt stress. Plant Soil 405:35–45
149. Egamberdieva D, Li L, Lindström K, Räsänen LA (2016) A synergistic interaction between salt-tolerant *Pseudomonas* and *Mesorhizobium* strains improves growth and symbiotic performance of liquorice (*Glycyrrhiza uralensis* fish.) under salt stress. Appl Microbiol Biotechnol 100:2829–2841
150. Hashem A, Abd_Allah EF, Alqarawi AA, Al-Huqail AA, Wirth S, Egamberdieva D (2016) The interaction between arbuscular mycorrhizal fungi and endophytic bacteria enhances plant growth of *Acacia gerrardii* under salt stress. Front Microbiol 7:1089
151. Gontia-Mishra I, Sapre S, Kachare S, Tiwari S (2017) Molecular diversity of 1-aminocyclopropane-1-carboxylate (ACC) deaminase producing PGPR from wheat (*Triticum aestivum* L.) rhizosphere. Plant Soil 414:213–227
152. Sandhya V, Ali SZ, Grover M, Reddy G, Venkateswarlu B (2010) Effect of plant growth promoting *Pseudomonas* spp. on compatible solutes, antioxidant status and plant growth of maize under drought stress. Plant Growth Regul 62:21–30
153. Ashraf M, Shahbaz M, Ali Q (2013) Drought-induced modulation in growth and mineral nutrients in canola (*Brassica napus* L.). Pak J Bot 45:93–98
154. Mishra J, Fatima T, Arora NK (2018) Role of secondary metabolites from plant growth-promoting rhizobacteria in combating salinity stress. In: Egamberdieva D, Ahmad P (eds) Plant microbiome: stress response. Springer, Singapore, pp 127–163
155. Bremer E, Krämer R (2019) Responses of microorganisms to osmotic stress. Annu Rev Microbiol 73:313–334
156. Saum SH, Müller V (2007) Salinity-dependent switching of osmolyte strategies in a moderately halophilic bacterium: glutamate induces proline biosynthesis in *Halobacillus halophilus*. J Bacteriol 189:6968–6975
157. Rodríguez-Salazar J, Suárez R, Caballero-Mellado J, Iturriaga G (2009) Trehalose accumulation in *Azospirillum brasilense* improves drought tolerance and biomass in maize plants. FEMS Microbiol Lett 296:52–59
158. Kushwaha B, Jadhav I, Verma HN, Geethadevi A, Parashar D, Jadhav K (2019) Betaine accumulation suppresses the de-novo synthesis of ectoine at a low osmotic concentration in *Halomonas* sp. SBS 10, a bacterium with broad salinity tolerance. Mol Biol Rep 46:4779–4786
159. Egamberdieva D, Wirth S, Jabborova D, Räsänen LA, Liao H (2017) Coordination between *Bradyrhizobium* and *Pseudomonas* alleviates salt stress in soybean through altering root system architecture. J Plant Interact 12:100–107
160. Jha Y, Subramanian RB, Patel S (2011) Combination of endophytic and rhizospheric plant growth promoting rhizobacteria in *Oryza sativa* shows higher accumulation of osmoprotectant against saline stress. Acta Physiol Plantarum 33(3):797–802
161. Kruasuwan W, Thamchaipenet A (2018) 1-Aminocyclopropane-1-carboxylate (ACC) deaminase-producing endophytic diazotrophic *Enterobacter* sp. EN-21 modulates salt–stress response in sugarcane. J Plant Growth Regul 37(3):849–858

162. Bharti N, Pandey SS, Barnawal D, Patel VK, Kalra A (2016) Plant growth promoting rhizobacteria *Dietzia natronolimnaea* modulates the expression of stress responsive genes providing protection of wheat from salinity stress. Sci Rep 6:34768
163. Qin S, Feng WW, Zhang YJ, Wang TT, Xiong YW, Xing K (2018) Diversity of bacterial microbiota of coastal halophyte *Limonium sinense* and amelioration of salinity stress damage by symbiotic plant growth promoting actinobacterium *Glutamicibacter halophytocola* KLBMP 5180. Appl Environ Microbiol 84:e01533–18
164. Orozco-Mosqueda M, Duan J, DiBernardo M, Zetter E, Campos-Garcia J, Glick BR, Santoyo G (2019) The production of ACC deaminase and trehalose by the plant growth promoting bacterium *Pseudomonas* sp. UW4 synergistically protect tomato plants against salt stress. Front Microbiol 10:1392
165. Shim JS, Seo JS, Seo JS, Kim Y, Koo Y, Do Choi Y, Jung C (2019) Heterologous expression of bacterial trehalose biosynthetic genes enhances trehalose accumulation in potato plants without adverse growth effects. Plant Biotechnol. Rep. 13:409–418
166. Das K, Roychoudhury A (2014) Reactive oxygen species (ROS) and response of antioxidants as ROS-scavengers during environmental stress in plants. Front Environ Sci 2:53
167. Halo BA, Khan AL, Waqas M, Al-Harrasi A, Hussain J, Ali L, Adnan M, Lee IJ (2015) Endophytic bacteria (*Sphingomonas* sp. LK11) and gibberellin can improve *Solanum lycopersicum* growth and oxidative stress under salinity. J Plant Interact 10:117–125
168. Ansari FA, Ahmad I, Pichtel J (2019) Growth stimulation and alleviation of salinity stress to wheat by the biofilm forming *Bacillus pumilus* strain FAB10. Appl Soil Ecol 143:45–54
169. El-Esawi MA, Al-Ghamdi AA, Ali HM, Alayafi AA (2019) *Azospirillum lipoferum* FK1 confers improved salt tolerance in chickpea (*Cicer arietinum* L.) by modulating osmolytes, antioxidant machinery and stress-related genes expression. Environ Exp Bot 159:55–65
170. Kadmiri IM, Chaouqui L, Azaroual SE, Sijilmassi B, Yaakoubi K, Wahb I (2018) Phosphate-solubilizing and auxin-producing rhizobacteria promote plant growth under saline conditions. Arab J Sci Eng 43:3403–3415
171. Gururani MA, Upadhyaya CP, Baskar V, Venkatesh J, Nookaraju A, Park SW (2013) Plant growth-promoting rhizobacteria enhance abiotic stress tolerance in *Solanum tuberosum* through inducing changes in the expression of ROS-scavenging enzymes and improved photosynthetic performance. J Plant Growth Regul 32:245–258
172. Wu ZS, Peng YJ, Guo LN, Li C (2014) Root colonization of encapsulated *Klebsiella oxytoca* Rs-5 on cotton plants and its promoting growth performance under salinity stress. Eur J Soil Biol 60:81–87
173. Sapre S, Gontia-Mishra I, Tiwari S (2018) *Klebsiella* sp. confers enhanced tolerance to salinity and plant growth promotion in oat seedlings (*Avena sativa*). Microbiol Res 206:25–32
174. Hidri R, Barea JM, Mahmoud OMB, Abdelly C, Azcon AR (2016) Impact of microbial inoculation on biomass accumulation by Sulla carnosa provenances, and in regulating nutrition, physiological and antioxidant activities of this species under non-saline and saline conditions. J Plant Physiol 201:28–41
175. Niu SQ, Li HR, Paré PW, Aziz M, Wang SM, Shi HZ, Li J, Han QQ, Guo SQ, Li J, Guo Q, Ma Q, Zhang JL (2016) Induced growth promotion and higher salt tolerance in the halophyte grass *Puccinellia tenuiflora* by beneficial rhizobacteria. Plant Soil 407:217–230
176. Vespermann A, Kai M, Piechulla B (2007) Rhizobacterial volatiles affect the growth of fungi and *Arabidopsis thaliana*. Appl Environ Microbiol 73(17):5639–5641
177. Zhang H, Kim MS, Krishnamachari V, Payton P, Sun Y, Grimson M, Farag MA, Ryu CM, Allen R, Melo IS (2007) Rhizobacterial volatile emissions regulate auxin homeostasis and cell expansion in *Arabidopsis*. Planta 226(4):839
178. Zhang H, Kim MS, Sun Y, Dowd SE, Shi H, Paré PW (2008) Soil bacteria confer plant salt tolerance by tissue-specific regulation of the sodium transporter HKT1. Mol Plant Microbe Interact 21(6):737–744
179. Nadeem SM, Zahir ZA, Naveed M, Asghar HN, Arshad M (2010) Rhizobacteria capable of producing ACC deaminase may mitigate salt stress in wheat. Soil Sci Soc Am J 74:533–542

180. Khandelwal A, Sindhu SS (2013) ACC deaminase containing rhizobacteria enhance nodulation and plant growth in clusterbean (*Cyamopsis tetragonoloba* L.). J Microbiol Res 3:117–123
181. Bouffaud ML, Renoud S, Dubost A, Moenne-Loccoz Y, Muller D (2018) 1-aminocyclopropane-1- carboxylate deaminase producers associated to maize and other *Poaceae* species. Microbiome 6:114
182. Orozco-Mosqueda MC, Glick BR, Santoyo G (2020) ACC deaminase in plant growth-promoting bacteria (PGPB): An efficient mechanism to counter salt in crops. Microbio Res 235:126439
183. Dodd IC, Zinovkina NY, Safronova VI, Belimov AA (2010) Rhizobacterial mediation of plant hormone status. Ann Appl Biol 157:361–379
184. Van Loon LC, Rep M, Pieterse CMJ (2006) Significance of inducible defense-related proteins in infected plants. Ann Rev Phytopathol 44:135–162
185. Yoon GM, Kieber JJ (2013) 14-3-3 regulates 1-aminocyclopropane-1- carboxylate synthase protein turnover in *Arabidopsis*. Plant Cell 25:1016–1028
186. Babalola OO (2010) Beneficial bacteria of agricultural importance. Biotechnol Lett 32:1559–1570
187. Nascimento FX, Brígido C, Glick BR, Rossi MJ (2016) The role of rhizobial ACC deaminase in the nodulation process of leguminous plants. Int J Agron 1369472
188. Bal HB, Nayak L, Das S, Adhya TK (2013) Isolation of ACC deaminase producing PGPR from rice rhizosphere and evaluating their plant growth promoting activity under salt stress. Plant Soil 366:93–105
189. Aslam F, Ali B (2018) Halotolerant bacterial diversity associated with *Suaeda fruticosa* (L.) Forssk. improved growth of maize under salinity stress. Agronomy 8:131
190. Singh RP, Jha P, Jha PN (2015) The plant-growth-promoting bacterium Klebsiella sp. SBP-8 confers induced systemic tolerance in wheat (*Triticum aestivum*) under salt stress. J Plant Physiol 184:57–67
191. Bacilio M, Moreno M, Bashan Y (2016) Mitigation of negative effects of progressive soil salinity gradients by application of humic acids and inoculation with *Pseudomonas stutzeri* in a salt-tolerant and a saltsusceptible pepper. Appl Soil Ecol 107:394–404
192. Etesami H, Glick BR (2020) Halotolerant plant growth-promoting bacteria: Prospects for alleviating salinity stress in plants. Environ. Expt. Botany 178:104–124
193. Tiwari G, Duraivadivel P, Sharma S, Hariprasad P (2018) 1-Aminocyclopropane-1-carboxylic acid deaminase producing beneficial rhizobacteria ameliorate the biomass characters of *Panicum maximum* Jacq. by mitigating drought and salt stress. Sci Rep 8:17513
194. Ali S, Charles TC, Glick BR (2014) Amelioration of high salinity stress damage by plant growth-promoting bacterial endophytes that contain ACC deaminase. Plant Physiol Biochem 80:160–167
195. Shaharoona B, Arshad M, Zahir ZA (2006) Effect of plant growth promoting rhizobacteria containing ACC-deaminase on maize (*Zea mays* L.) growth under axenic conditions and on nodulation in mung bean (*Vigna radiata* L.). Letters Appl Microbiol 42:155–159
196. Siddikee MA, Glick BR, Chauhan PS, Jong Yim W, Sa T (2011) Enhancement of growth and salt tolerance of red pepper seedlings (*Capsicum annuum* L.) by regulating stress ethylene synthesis with halotolerant bacteria containing 1-aminocyclopropane-1-carboxylic acid deaminase activity. Plant Physiol Biochem 49(4):427–34
197. Zafar-ul-Hye M, Ahmad M, Shahzad SM (2013) Short communication synergistic effect of rhizobia and plant growth promoting rhizobacteria on the growth and nodulation of lentil seedlings under axenic conditions. Soil Environ 32:79–86
198. Sziderics AH, Rasche F, Trognitz F, Sessitsch A, Wilhelm E (2007) Bacterial endophytes contribute to abiotic stress adaptation in pepper plants (*Capsicum annum* L.). Can J Microbiol 53:1195–1202
199. Panwar M, Tewari R, Gulati A, Nayyar H (2016) Indigenous salt-tolerant rhizobacterium *Pantoea dispersa* (PSB3) reduces sodium uptake and mitigates the effects of salt stress on growth and yield of chickpea. Acta Physiol Plant 38:278

200. Chandra D, Srivastava R, Gupta VVSR, Franco CMM, Sharma AK (2019) Evaluation of ACC-deaminase-producing rhizobacteria to alleviate water-stress impacts in wheat (*Triticum aestivum* L.) plants. Can J Microbiol 65:387–403
201. Mayak S, Tirosh T, Glick BR (2004) Plant growth-promoting bacteria confer resistance in tomato plants to salt stress. Plant Physiol Biochem 42:565–572
202. Hussein KA, Tohamy TA, El-Maraghy SS (2022) Amino-cyclopropane-1-carboxylate deaminase (ACCD) producing yeasts improved salinity tolerance of *Triticum aestivum* L. Rhizosphere 23:100548
203. Tewari S, Arora NK (2014) Multifunctional exopolysaccharides from *Pseudomonas aeruginosa* PF23 involved in plant growth stimulation, biocontrol and stress amelioration in sunflower under saline conditions. Curr Microbiol 69:484–494
204. Upadhyay SK, Singh DP (2015) Effect of salt-tolerant plant growth promoting rhizobacteria on wheat plants and soil health in a saline environment. Plant Biol 17:288–293
205. Kasim WA, Gaafar RM, Abou-Ali RM, Omar MN, Hewait HM (2016) Effect of biofilm forming plant growth promoting rhizobacteria on salinity tolerance in barley. Ann Agric Sci 61:217–227
206. Chakraborty K, Bose J, Shabala L, Shabala S (2016) Difference in root $K^+$ retention ability and reduced sensitivity of $K^+$-permeable channels to reactive oxygen species confer differential salt tolerance in three *Brassica* species. J Exp Bot 67:4611–4625
207. Cuin TA, Zhou M, Parsons D, Shabala S (2012) Genetic behaviour of physiological traits conferring cytosolic $K^+/Na^+$ homeostasis in wheat. Plant Biol 14:438–446
208. Wu HH, Zhu M, Shabala L, Zhou MX, Shabala S (2015) $K^+$ retention in leaf mesophyll, an overlooked component of salinity tolerance mechanism, a case study for barley. J Integr Plant Biol 57:171–185
209. Sandhya V, Ali SZ (2015) The production of exopolysaccharide by *Pseudomonas putida* GAP-P45 under various abiotic stress conditions and its role in soil aggregation. Microbiology 84:512–519
210. Yang AZ, Akhtar SS, Iqbal S, Amjad M, Naveed M, Zahir ZA, Jacobsen SE (2016) Enhancing salt tolerance in quinoa by halo-tolerant bacterial inoculation. Func Plant Biol 43:632–642
211. Atouei MT, Pourbabaee AA, Shorafa M (2019) Alleviation of salinity stress on some growth parameters of wheat by exopolysaccharide-producing bacteria. Iran J Sci Technol Trans 43:2725–2733
212. Chu TN, Tran BTH, Van Bui L, Hoang MTT (2019) Plant growth-promoting rhizobacterium *Pseudomonas* PS01 induces salt tolerance in *Arabidopsis thaliana*. BMC Res Notes 12:11
213. Rajkumar M, Ae N, Prasad MNV, Freitas H (2010) Potential of siderophore producing bacteria for improving heavy metal phytoextraction. Trends Biotechnol 28:142–149
214. Crowley D (2000) Function of siderophores in the plant rhizosphere. In: The Rhizosphere. CRC Press, pp 239–278
215. Schmidt W (1999) Mechanisms and regulation of reduction-based iron uptake in plants. New Phytol 141(1):1–26
216. Sahu GK, Sindhu SS (2011) Disease control and plant growth promotion of green gram by siderophore producing *Pseudomonas* sp. Res J Microbiol 6:735–749
217. Crowley DE, Kraemer SM (2007) Function of siderophores in the plant rhizosphere. In: Pinton R et al (eds) The rhizosphere. CRC Press, Biochemistry and organic substances at the soil-plant interface, pp 73–109
218. Vansuyt G, Robin A, Briat JF, Curie C, Lemanceau P (2007) Iron acquisition from Fe-pyoverdine by *Arabidopsis thaliana*. Mol Plant Microbe Interact 20(4):441–447
219. Sadeghi A, Karimi E, Dahaji PA, Javid MG, Dalvand Y, Askari H (2012) Plant growth promoting activity of an auxin and siderophore producing isolate of *Streptomyces* under saline soil conditions. World J Microbiol Biotechnol 28:1503–1509
220. Tank N, Saraf M (2010) Salinity-resistant plant growth promoting rhizobacteria ameliorates sodium chloride stress on tomato plants. J Plant Interact 5:51–58
221. Shen XM, Hu HB, Peng HS, Wang W, Zhang XH (2013) Comparative genomic analysis of four representative plant growth-promoting rhizobacteria in *Pseudomonas*. BMC Genomics 14:271

222. Sultana S, Alam S, Karim MM (2021) Screening of siderophore-producing salt-tolerant rhizobacteria suitable for supporting plant growth in saline soils with iron limitation. J Agric Food Res 4:100150
223. Soetan K, Olaiya C, Oyewole O (2010) The importance of mineral elements for humans, domestic animals and plants—a review. Afr J Food Sci 4:200–222
224. Bamagoos AA, Alharby HF, Belal EE, Khalaf AEA, Abdelfattah MA, Rady MM, Ali EF, Mersal GAM (2021) Phosphate-solubilizing bacteria as a panacea to alleviate stress effects of high soil $CaCO_3$ content in *Phaseolus vulgaris* with special reference to P-releasing enzymes. Sustainability 13(13):7063
225. Miller JJ, Chanasyk DS, Curtis TW, Olson BM (2011) Phosphorus and nitrogen in runoff after phosphorus-or nitrogen-based manure applications. J Environ Qual 40(3):949–958
226. Wang D, Lv S, Jiang P, Li Y (2017) Roles, regulation, and agricultural application of plant phosphate transporters. Front Plant Sci 8:817
227. Rodriguez H, Fraga R (1999) Phosphate solubilizing bacteria and their role in plant growth promotion. Biotechnol Adv 17:319–339
228. Bhattacharyya PN, Jha DK (2012) Plant growth-promoting rhizobacteria (PGPR): emergence in agriculture. World J Microbiol Biotechnol 28:1327–1350
229. Oteino N, Lally RD, Kiwanuka S, Lloyd A, Ryan D, Germaine KJ (2015) Plant growth promotion induced by phosphate solubilizing endophytic *Pseudomonas* isolates. Front Microbiol 6:745
230. Saboor A, Ali MA, Husain S, Tahir MS, Irfan M, Bilal M, Baig KS, Datta R, Ahmed N, Danish S, Glick BR (2021) Regulation of phosphorus and zinc uptake in relation to arbuscular mycorrhizal fungi for better maize growth. Agronomy 11:2322
231. Chabot R, Beauchamp CJ, Kloepper JW, Antoun H (1998) Effect of phosphorus on root colonization and growth promotion of maize by bioluminescent mutants of phosphate-solubilizing *Rhizobium leguminosarum* biovar phaseoli. Soil Biol Biochem 30:1615–1618
232. Sindhu SS, Phour M, Choudhary SR, Chaudhary D (2014) Phosphorus cycling: prospects of using rhizosphere microorganisms for improving phosphorus nutrition of plants. In: Geomicrobiology and biogeochemistry. Springer, Berlin, Heidelberg, pp 199–237
233. Dahiya A, Kumar R, Sindhu SS (2021) Microbial endophytes mediated phosphorus solubilization: Sustainable approach to improve soil fertility and plant growth. In: Maheshwari DK (ed) Endophytes: Mineral nutrients management in the series 'sustainable development and biodiversity'. Springer Nature, Gewerbestrasse, Switzerland, pp 35–75
234. Kuzmina LY, Gilvanova EA, Galimzyanova NF (2022) Characterization of the novel plant growth-stimulating strain *Advenella kashmirensis* IB-K1 and evaluation of its efficiency in saline soil. Microbiology 91:173–183
235. Sanjotha P, Mahantesh P, Patil CS (2011) Isolation and screening of efficiency of phosphate solubilizing microbes. Int J Microbiol Res 3:56–58
236. Ahemad M, Kibret M (2014) Mechanisms and applications of plant growth promoting rhizobacteria: current perspective. J King Saud Univ Sci 26:1–20
237. Srinivasan R, Yandigeri MS, Kashyap S, Alagawadi AR (2012) Effect of salt on survival and P-solubilization potential of phosphate solubilizing microorganisms from salt affected soils. Saudi J Biol Sci 19:427–434
238. Mahdi SS, Hassan GI, Hussain A, Rasool FU (2011) Phosphorus availability issue—its fixation and role of phosphate solubilizing bacteria in phosphate solubilization. Res J Agric Sci 2:174–179
239. Patel VK, Vikram B, Sikarwar PS, Sengupta J (2021) Effect of different levels of nitrogen and phosphorus on growth and yield of spinach (*Spinacea oleraceu* L.) cv. all green. J Pharmacogn Phytochem 10(1):2229–2231
240. Ashfaq M, Hassan HM, Ghazali AH, Ahmad M (2020) Halotolerant potassium solubilizing plant growth promoting rhizobacteria may improve potassium availability under saline conditions. Environ Monit Assess 192:697
241. Cassán F, Vanderleyden J, Spaepen S (2014) Physiological and agronomical aspects of phytohormone production by model plant-growth-promoting rhizobacteria (PGPR) belonging to the genus *Azospirillum*. J Plant Growth Regul 33:440–459

242. Duca D, Lorv J, Patten CL, Rose D, Glick BR (2014) Indole-3-acetic acid in plant-microbe interactions. Antonie Van Leeuwenh 106:85–125
243. Salomon MV, Bottini R, de Souza Filho GA, Cohen AC, Moreno D, Gil M, Piccoli P (2014) Bacteria isolated from roots and rhizosphere of *Vitis vinifera* retard water losses, induce abscisic acid accumulation and synthesis of defense-related terpenes in *in vitro* cultured grapevine. Physiol Plant 151:359–374
244. Sarmiento-López LG, López-Meyer M, Maldonado-Mendoza IE, Quiroz-Figueroa FR, Sepúlveda-Jiménez G, Rodríguez-Monroy M (2022) Production of indole-3-acetic acid by *Bacillus circulans* E9 in a low-cost medium in a bioreactor. J Biosci Bioeng 134(1):21–28
245. Jangu OP, Sindhu SS (2011) Differential response of inoculation with indole acetic acid producing *Pseudomonas* sp. in green gram (*Vigna radiata* L.) and black gram (*Vigna mungo* L.). Microbiol J 1:159–173
246. Khan N, Bano A, Ali S, Babar MA (2020) Crosstalk amongst phytohormones from plants and PGPR under biotic and abiotic stresses. Plant Growth Regul 90:189–203
247. Eichmann R, Richards L, Scha¨fer P (2021) Hormones as go-betweens in plant microbiome assembly. Plant J 105:518–541
248. Egamberdieva D, Kucharova Z (2009) Selection for root colonising bacteria stimulating wheat growth in saline soils. Biol Fertil Soils 45:563–571
249. Iqbal M, Ashraf M (2013) Gibberellic acid mediated induction of salt tolerance in wheat plants: growth, ionic partitioning, photosynthesis, yield and hormonal homeostasis. Environ Exp Bot 86:76–85
250. Alqarawi A, Hashem A, Abd_Allah E, Alshahrani T, Huqail A (2014) Effect of salinity on moisture content, pigment system, and lipid composition in *Ephedra alata* Decne. Acta Biol Hung 65:61–71
251. Patel RR, Patel DD, Thakor P, Patel B, Thakkar VR (2015) Alleviation of salt stress in germination of *Vigna radiata* L. by two halotolerant Bacilli sp. isolated from saline habitats of Gujarat. Plant Growth Regul 76:51–60
252. Tsegaye Z, Assefa F, Beyene D (2017) Properties and application of plant growth promoting rhizobacteria. Int J Curr Trend Pharmacobiol Med Sci 2(1):30–43
253. Berg G, Alavi M, Schmidt CS, Zachow C, Egamberdieva D, Kamilova F, Lugtenberg BJJ (2013) Biocontrol and osmoprotection for plants under salinated conditions. In: de Bruijn FJ (ed) Molecular microbial ecology of the rhizosphere. John Wiley & Sons, New York, pp 561–573
254. Kang SM, Shahzad R, Bilal S, Khan AL, Park YG, Lee KE, Asaf S, Khan MA, Lee IJ (2019) Indole-3-acetic acid and ACC deaminase producing *Leclercia adecarboxylata* MO1 improves *Solanum lycopersicum* L. growth and salinity stress tolerance by endogenous secondary metabolites regulation. BMC Microbiol 19:80
255. Wang QY, Dodd IC, Belimov AA, Jiang F (2016) Rhizosphere bacteria containing 1-aminocyclopropane-1-carboxylate deaminase increase growth and photosynthesis of pea plants under salt stress by limiting $Na^+$ accumulation. Funct Plant Biol 43:161–172
256. Kim K, Jang YJ, Lee SM, Oh BT, Chae JC, Lee KJ (2014) Alleviation of salt stress by *Enterobacter* sp. EJ01 in tomato and *Arabidopsis* is accompanied by up-regulation of conserved salinity responsive factors in plants. Mol Cells 37:109–117
257. Forni C, Duca D, Glick BR (2017) Mechanisms of plant response to salt and drought stress and their alteration by rhizobacteria. Plant Soil 410:335–356
258. Zahir ZA, Shah MK, Naveed M, Akhter MJ (2010) Substrate-dependent auxin production by *Rhizobium phaseoli* improves the growth and yield of *Vigna radiata* L. under salt stress conditions. J Microbiol Biotechnol 20:1288–1294
259. Silini-Cherif H, Silini A, Ghoul M, Yadav S (2012) Isolation and characterization of plant growth promoting traits of a rhizobacteria: *Pantoea agglomerans* lma2. Pak J Biol Sci 15:267–276
260. Ahmad M, Zahir ZA, Nazli F, Akram F, Arshad M, Khalid M (2013) Effectiveness of halo-tolerant, auxin producing *Pseudomonas* and *Rhizobium* strains to improve osmotic stress tolerance in mung bean (*Vigna radiata* L.). Braz J Microbiol 44:1341–1348

261. Ahmad M, Zahir ZA, Khalid M, Nazli F, Arshad M (2013) Efficacy of *Rhizobium* and *Pseudomonas* strains to improve physiology, ionic balance and quality of mung bean under salt-affected conditions on farmer's fields. Plant Physiol Biochem 63:170–176
262. Bottini R, Cassán F, Piccoli P (2004) Gibberellin production by bacteria and its involvement in plant growth promotion and yield increase. Appl Microbiol Biotechnol 65:497–503
263. Shahzad R, Waqas M, Khan AL, Asaf S, Khan MA, Kang SM, Yun BW, Lee IJ (2016) Seed-borne endophytic *Bacillus amyloliquefaciens* RWL-1 produces gibberellins and regulates endogenous phytohormones of *Oryza sativa*. Plant Physiol Biochem 106:236–243
264. Kang SM, Khan AL, You YH, Kim JG, Kamran M, Lee IJ (2014) Gibberellin production by newly isolated strain *Leifsonia soli* SE134 and its potential to promote plant growth. J Microbiol Biotechnol 24:106–112
265. Kang SM, Khan AL, Waqas M, You YH, Kim JH, Kim JG, Hamayun M, Lee IJ (2014) Plant growth-promoting rhizobacteria reduce adverse effects of salinity and osmotic stress by regulating phytohormones and antioxidants in *Cucumis sativus*. J Plant Interact 9:673–682
266. Attia H, Alamer K, Algethami B (2022) Gibberellic acid interacts with salt stress on germination, growth and polyamine gene expression in fennel (*Foeniculum vulgare* Mill.) seedlings. Physiol Mol Biol Plants 28:607–622
267. Shahzad R, Khan AL, Bilal S, Waqas M, Kang SM, Lee IJ (2017) Inoculation of abscisic acid-producing endophytic bacteria enhances salinity stress tolerance in *Oryza sativa*. Environ Exp Bot 136:68–77
268. Egamberdieva D (2011) Survival of *Pseudomonas extremorientalis* TSAU20 and *P. chlororaphis* TSAU13 in the rhizosphere of common bean (*Phaseolus vulgaris*) under saline conditions. Plant Soil Environ 57:122–127
269. Jha Y, Subramanian RB (2013) Paddy plants inoculated with PGPR show better growth physiology and nutrient content under saline conditions. Chil J Agric Res 73:213–219
270. Patel T, Saraf M (2017) Biosynthesis of phytohormones from novel rhizobacterial isolates and their in vitro plant growth-promoting efficacy. J Plant Interact 12:480–487
271. Fahad S, Hussain S, Bano A, Saud S, Hassan S, Shan D, Khan FA, Khan F, Chen YT, Wu C, Tabassum MA, Chun MX, Afzal M, Jan A, Jan MT, Huang JL (2015) Potential role of phytohormones and plant growth-promoting rhizobacteria in abiotic stresses: consequences for changing environment. Environ Sci Pollut Res 22:4907–4921
272. Parray JA, Jan S, Kamili AN, Qadri RA, Egamberdieva D, Ahmad P (2016) Current perspectives on plant growth-promoting rhizobacteria. J Plant Growth Regul 35:877–902
273. Pallai R, Hynes RK, Verma B, Nelson LM (2012) Phytohormone production and colonization of canola (*Brassica napus* L.) roots by *Pseudomonas fluorescens* 6-8 under gnotobiotic conditions. Can J Microbiol 58:170–178
274. Kapoor R, Kaur M (2016) Cytokinins production by fluorescent *Pseudomonas* isolated from rhizospheric soils of Malus and Pyrus. Afric J Microbiol Res 10:1274–1279
275. Sita K, Kumar V (2020) Role of gamma amino butyric acid (GABA) against abiotic stress tolerance in legumes: a review. Plant Physiol Rep 25:654–663
276. Phour M, Ghai A, Rose G, Dhull N, Sindhu SS (2018) Role of aminolevulininc acid in stress adaptation and crop productivity. Int J Curr Microbiol Appl Sci 7(5):1516–1524
277. Naeem MS, Rasheed M, Liu D, Jin ZL (2011) 5-Aminolevulinic acid ameliorates salinity-induced metabolic, water-related and biochemical changes in *Brassica napus* L. Acta Physiol Plant 33:517–528
278. Miura K, Tada Y (2014) Regulation of water, salinity, and cold stress responses by salicylic acid. Front Plant Sci 5:4
279. Kaur H, Bhardwaj RD, Grewal SK (2017) Mitigation of salinity-induced oxidative damage in wheat (*Triticum aestivum* L.) seedlings by exogenous application of phenolic acids. Acta Physiol Plant 39:221–236
280. Wan YY, Zhang Y, Zhang L, Zhou ZQ, Li X, Shi Q, Wang XJ, Bai JG (2015) Caffeic acid protects cucumber against chilling stress by regulating antioxidant enzyme activity and proline and soluble sugar contents. Acta Physiol Plant 37:1706–1715

281. El-Soud WA, Hegab MM, Elgawad HA, Zinta G, Asard H (2013) Ability of ellagic acid to alleviate osmotic stress on chickpea seedlings. Plant Physiol Biochem 71:173–183
282. Parvin K, Nahar K, Hasanuzzaman M, Bhuyan MB, Mohsin SM, Fujita M (2020) Exogenous vanillic acid enhances salt tolerance of tomato: Insight into plant antioxidant defense and glyoxalase systems. Plant Physiol Biochem 150:109–120
283. Lini´c I, Šamec D, Grúz J, Vujˇci´c Bok V, Strnad M, Salopek-Sondi B (2019) Involvement of phenolic acids in short-term adaptation to salinity stress is species specific among *Brassicaceae*. Plants 8:155
284. Lini´c I, Mlinari´c S, Brkljaˇci´c L, Pavlovi´c I, Smolko A, Salopek-Sondi B (2021) Ferulic acid and salicylic acid foliar treatments reduce short-term salt stress in Chinese cabbage by increasing phenolic compounds accumulation and photosynthetic performance. Plants 10:2346
285. Gholamnia A, Mosleh Arani A, Sodaeizadeh H (2022) Expression profiling of rosmarinic acid biosynthetic genes and some physiological responses from *Mentha piperita* L. under salinity and heat stress. Physiol Mol Biol Plants 28:545–557
286. Lamattina L, García-Mata C, Graziano M, Pagnussat G (2003) Nitric oxide: the versatility of an extensive signal molecule. Annu Rev Plant Biol 54(1):109–136
287. Besson-Bard A, Pugin A, Wendehenne D (2008) New insights into nitric oxide signaling in plants. Annu Rev Plant Biol 59:21–39
288. Lipton SA, Choi YB, Pan ZH, Lei SZ, Chen HSV, Sucher NJ, Loscalzo J, Singel DJ, Stamler JS (1993) A redox-based mechanism for the neuroprotective and neurodestructive effects of nitric oxide and related nitroso-compounds. Nature 364:626–632
289. Zottini M, Formentin E, Scattolin M, Carimi F, Lo Schiavo F, Terzi M (2002) Nitric oxide affects plant mitochondrial functionality *in vivo*. FEBS Lett 515:75–78
290. Zhao L, Zhang F, Guo J, Yang Y, Li B, Zhang L (2004) Nitric oxide functions as a signal in salt resistance in the calluses from two ecotypes of reed. Plant Physiol 134(2):849–857
291. Shi Q, Ding F, Wang X, Wei M (2007) Exogenous nitric oxide protects cucumber roots against oxidative stress induced by salt stress. Plant Physiol Biochem 45:542–550
292. Nalousi AM, Ahmadiyan S, Hatamzadeh A, Ghasemnezhad M (2012) Protective role of exogenous nitric oxide against oxidative stress induced by salt stress in bell-pepper (*Capsicum annum* L.). Am-Eurasian J Agric Environ Sci 12(8):1085–1090
293. Du S, Liu Y, Zhang P, Liu H, Zhang X, Zhang R (2015) Atmospheric application of trace amounts of nitric oxide enhances tolerance to salt stress and improves nutritional quality in spinach (*Spinacia oleracea* L.). Food Chem 173:905–911
294. Hayat S, Yadav S, Wani AS, Irfan M, Alyemini MN, Ahmad A (2012) Impact of sodium nitroprusside on nitrate reductase, proline content, and antioxidant system in tomato under salinity stress. Hortic Environ Biotechnol 53:362–367
295. Leshem YY, Haramaty E (1996) The characterization and contrasting effects of the nitric oxide free radical in vegetative stress and senescence of *Pisum sativum* Linn. Foliage. J Plant Physiol 148:258–263
296. Bright J, Desikan R, Hancock JT, Weir IS, Neill SJ (2006) ABA-induced NO generation and stomatal closure in *Arabidopsis* are dependent on $H_2O_2$ synthesis. Plant J 45:113–122
297. Hasegawa PM, Bressan RA, Zhu JK, Bohnert HJ (2000) Plant cellular and molecular responses to high salinity. Annu Rev Plant Physiol 51(1):463–499
298. Sung CH, Hong JK (2010) Sodium nitroprusside mediates seedling development and attenuation of oxidative stresses in Chinese cabbage. Plant Biotechnol Rep 4(4):243–251
299. Spoel SH, Loake GJ (2011) Redox-based protein modifications: the missing link in plant immune signalling. Curr Opin Plant Biol 14:358–364
300. Begara-Morales JC, Sanchez-Calvo B, Chaki M, Valderrama R, Mata-Perez C, Lopez-Jaramillo J, Padilla MN, Carreras A, Corpas FJ, Barroso JB (2014) Dual regulation of cytosolic ascorbate peroxidase (APX) by tyrosine nitration and S-nitrosylation. J Exp Bot 65:527–538
301. Ahmad P, Abdel Latef AA, Hashem A, Abd Allah EF, Gucel S, Tran LSP (2016) Nitric oxide mitigates salt stress by regulating levels of osmolytes and antioxidant enzymes in chickpea. Front Plant Sci 7:1–11

302. Wang Y, Brown H, Crowley D, Szaniszlo P (1993) Evidence for direct utilization of a siderophore, ferrioxamine B, in axenically grown cucumber plant. Cell Environ 16(5):579–585
303. Zhang Y, Wang L, Liu Y, Zhang Q, Wei Q, Zhang W (2006) Nitric oxide enhances salt tolerance in maize seedlings through increasing activities of proton-pump and $Na^+/H^+$ antiport in the tonoplast. Planta 224(3):545–555
304. Guo FQ (2003) Identification of a plant nitric oxide synthase gene involved in hormonal signaling. Science 302:100–103
305. Zhao MG, Tian QY, Zhang WH (2007) Nitric oxide synthase-dependent nitric oxide production is associated with salt tolerance in *Arabidopsis*. Plant Physiol 144:206–217
306. Fan H, Guo S, Jiao Y, Zhang R, Li J (2007) Effects of exogenous nitric oxide on growth, active oxygen species metabolism, and photosynthetic characteristics in cucumber seedlings under NaCl stress. Front Agric China 308–314
307. Barakat A, Staton M, Cheng CH, Park J, Yassin NBM, Ficklin S, Yeh CC, Hebard F, Baier K, Powell W, Schuster SC, Wheeler N, Abbott A, Carlson JE, Sederoff R (2012) Chestnut resistance to the blight disease: insights from transcriptome analysis. BMC Plant Biol 12:38
308. Guo Y, Tian Z, Yan D, Zhang J, Qin P (2009) Effects of nitric oxide on salt stress tolerance. Life Sci. J. 6:67–75
309. Oliveira HC, Gomes BCR, Pelegrino MT, Seabra AB (2016) Nitric oxide-releasing chitosan nanoparticles alleviate the effects of salt stress in maize plants. Biol Chem 61:10–19
310. Fan HF, Du CX, Guo SR (2013) Nitric oxide enhances salt tolerance in cucumber seedlings by regulating free polyamine content. Environ Exp Bot 86:52–59
311. Egbichi I, Keyster M, Ludidi N (2014) Effect of exogenous application of nitric oxide on salt stress responses of soybean. South Afr J Bot 90:131–136
312. Adamu T, Mun BG, Lee SU, Hussain A, Yun BW (2018) Exogenously applied nitric oxide enhances salt tolerance in rice (*Oryza sativa* L.) at seedling stage. Agronomy 8:276
313. Ghorai S, Pal KK, Dey R (2015) Alleviation of salinity stress in groundnut by application of PGPR. Int Res J Eng Technol 2:742–750
314. Zerrouk IZ, Benchabane M, Khelifi L, Yokawa K, Ludwig-Müller J, Baluska F (2016) A *Pseudomonas* strain isolated from date palm rhizospheres improves root growth and promotes root formation in maize exposed to salt and aluminum stress. J Plant Physiol 191:111–119
315. Meena KK, Sorty AM, Bitla UM, Chaudhary K, Gupta P, Pareek A, Singh DP, Prabha R, Sahu PK, Gupta VK, Singh HB, Krishanani KK, Minhas PS (2017) Abiotic stress responses and microbe-mediated mitigation in plants: the comic strategies. Front Plant Sci 8:172
316. Yasin NA, Khan WU, Ahmad SR, Ali A, Ahmad A, Akram W (2018) Imperative roles of halotolerant plant growth-promoting rhizobacteria and kinetin in improving salt tolerance and growth of black gram (*Phaseolus mungo*). Environ Sci Pollut Res 25:4491–4505
317. Jatan R, Chauhan PS, Lata C (2019) *Pseudomonas putida* modulates the expression of *miRNAs* and their target genes in response to drought and salt stresses in chickpea (*Cicer arietinum* L.). Genomics 111:509–519
318. Pandey P, Maheshwari DK (2007) Two-species microbial consortium for growth promotion of *Cajanus cajan*. Curr Sci 92:1137–1142
319. Kumawat KC, Sharma P, Nagpal S, Gupta RK, Sirari A, Nair RM, Bindumadhava H, Singh S (2021) Dual microbial inoculation, a game changer–bacterial biostimulants with multifunctional growth promoting traits to mitigate salinity stress in spring mungbean. Front Microbiol 11:600576
320. Santoyo G, Gamalero E, Glick BR (2021) Mycorrhizal-bacterial amelioration of plant abiotic and biotic stress. Front Sustain Food Syst 5:672881
321. Zawoznik MS, Ameneiros M, Benavides MP, Vázquez S, Groppa MD (2011) Response to saline stress and aquaporin expression in *Azospirillum*-inoculated barley seedlings. Appl Microbiol Biotechnol 90:1389–1393
322. Bianco C, Defez R (2009) *Medicago truncatula* improves salt tolerance when nodulated by an indole-3-acetic acid-overproducind *Sinorhizobium meliloti* strain. J Exp Bio 4:763–767

323. Misra S, Dixit VK, Khan MH, Mishra SK, Dviwedi G, Yadav S, Lehri A, Chauhan PS (2017) Exploitation of agro-climatic environment for selection of 1-aminocyclopropane-1-carboxylic acid (ACC) deaminase producing salt tolerant indigenous plant growth promoting rhizobacteria. Microbiol Res 205:25–34
324. Akram MS, Shahid M, Tariq M, Azeem M, Javed MT, Saleem S, Riaz S (2016) Deciphering *Staphylococcus sciuri* SAT-17 mediated antioxidative defense mechanisms and growth modulations in salt stressed maize (*Zea mays* L.). Front Microbiol 7:867
325. Nadeem SM, Zahir ZA, Naveed M, Arshad M (2007) Preliminary investigations on inducing salt tolerance in maize through inoculation with rhizobacteria containing ACC deaminase activity. Can J Microbiol 53:1141–1149
326. Chaudhary D, Narula N, Sindhu SS, Behl RK (2013) Plant growth stimulation of wheat (*Triticum aestivum* L.) by inoculation of salinity tolerant *Azotobacter* strains. Physiol Mol Biol Pants 19:515–519
327. Nautiyal CS, Srivastava S, Chauhan PS, Seem K, Mishra A, Sopory SK (2013) Plant growth-promoting bacteria *Bacillus amyloliquefaciens* NBRISN13 modulates gene expression profile of leaf and rhizosphere community in rice during salt stress. Plant Physiol Biochem 66:1–9
328. Rima FS, Biswas S, Sarker PK, Islam R, Seraj ZI (2018) Bacteria endemic to saline coastal belt and their ability to mitigate the effects of salt stress on rice growth and yields. Ann Microbiol 68:525–535
329. Damodaran T, Mishra VK, Jha SK, Pankaj U, Gupta G, Gopal R (2019) Identification of rhizosphere bacterial diversity with promising salt tolerance, PGP traits and their exploitation for seed germination enhancement in sodic soil. Agric Res 8:36–43
330. El-Esawi MA, Alaraidh IA, Alsahli AA, Alamri SA, Ali HM, Alayafi AA (2018) *Bacillus firmus* (SW5) augments salt tolerance in soybean (*Glycine max* L.) by modulating root system architecture, antioxidant defense systems and stress-responsive genes expression. Plant Physiol Biochem 132:375–384
331. Tewari S, Arora NK (2016) Fluorescent *Pseudomonas* sp. PF17 as an efficient plant growth regulator and biocontrol agent for sunflower crop under saline conditions. Symbiosis 68:99–108
332. Palacio-Rodríguez R, Coria-Arellano JL, López-Bucio J, Sánchez-Salas J, Muro-Pérez G, Castañeda-Gaytán G Sáenz-Mata J et al. (2017) Halophilic rhizobacteria from *Distichlis spicata* promote growth and improve salt tolerance in heterologous plant hosts. Symbiosis 73:179–189
333. Saravanakumar D, Samiyappan R (2007) ACC deaminase from *Pseudomonas fluorescens* mediated saline resistance in groundnut (*Arachis hypogea*) plants. J Appl Microbiol 102:1283–1292
334. Safronova VI, Piluzza G, Zinovkina NY, Kimeklis AK, Belimov AA, Bullitta S (2012) Relationships between pasture legumes, rhizobacteria and nodule bacteria in heavy metal polluted mine waste of SW Sardinia. Symbiosis 58:149–159
335. Shabaan M, Asghar HN, Zahir ZA, Zhang X, Sardar MF, Li H (2022) Salt-tolerant PGPR confer salt tolerance to maize through enhanced soil biological health, enzymatic activities, nutrient uptake and antioxidant defense. Front Microbiol 13:901865
336. Rojas-Tapias D, Moreno-Galván A, Pardo-Díaz S, Obando M, Rivera D, Bonilla R (2012) Effect of inoculation with plant growth-promoting bacteria (PGPB) on amelioration of saline stress in maize (*Zea mays*). Appl Soil Ecol 61:264–272
337. Chauhan PS, Lata C, Tiwari S, Chauhan AS, Mishra SK, Agrawal L, Chakrabarty D, Nautiyal CS (2019) Transcriptional alterations reveal *Bacillus amyloliquefaciens*-rice cooperation under salt stress. Sci Rep 9(1):1–13
338. Ji J, Yuan D, Jin C, Wang G, Li X, Guan C (2020) Enhancement of growth and salt tolerance of rice seedlings (*Oryza sativa* L.) by regulating ethylene production with a novel halotolerant PGPR strain *Glutamicibacter* sp. YD01 containing ACC deaminase activity. Acta Physiol Plant 42(4):1–17
339. Sukweenadhi J, Kim YJ, Choi ES, Koh SC, Lee SW, Kim YJ, Yang DC (2015) *Paenibacillus yonginensis* DCY84 T induces changes in *Arabidopsis thaliana* gene expression against aluminum, drought, and salt stress. Microbiol Res 172:7–15

340. Wu Z, Yao L, Kaleem I, Li C (2012) Application efficacy of biological seed coating agent from combination of PGPR on cotton in the field. In: Zhu E, Sambath S (eds) Information technology and agricultural engineering. Advances in intelligent and soft computing, vol 134. Berlin, Springer
341. Cardinale M, Ratering S, Suarez C, Montoya AMZ, Geissler-Plaum R, Schnell S (2015) Paradox of plant growth promotion potential of rhizobacteria and their actual promotion effect on growth of barley (*Hordeum vulgare* L.) under salt stress. Microbiol Res 181:22–32
342. AlAli HA, Khalifa A, Almalki M (2022) Plant growth-promoting bacterium from non-agricultural soil improves okra plant growth. Agriculture 12:873
343. Khan MA, Ullah I, Waqas M, Hamayun M, Khan AL, Asaf S, Kang SM, Kim KM, Jan R, Lee IJ (2019) Halo-tolerant rhizospheric *Arthrobacter woluwensis* AK1 mitigates salt stress and induces physio-hormonal changes and expression of GmST1 and GmLAX3 in soybean. Symbiosis 77(1):9–21
344. Chaudhary D, Sindhu SS (2015) Inducing salinity tolerance in chickpea (*Cicer arietinum* L.) by inoculation of 1-aminocyclopropane-1-carboxylic acid deaminase-containing *Mesorhizobium* strains. Afr J Microbiol Res 9: 117–124
345. Yasmeen T, Ahmad A, Arif MS, Mubin M, Rehman K, Shahzad SM, Iqbal S, Rizwan M, Ali S, Alyemeni MN, Wijaya L (2020) Biofilm forming rhizobacteria enhance growth and salt tolerance in sunflower plants by stimulating antioxidant enzymes activity. Plant Physiol Biochem 156:242–256
346. Gupta A, Bano A, Rai S, Kumar M, Ali J, Sharma S, Pathak N (2021) ACC deaminase producing plant growth promoting rhizobacteria enhance salinity stress tolerance in *Pisum sativum*. 3 Biotech 11(12):1–17
347. Paulucci NS, Gallarato LA, Reguera YB, Vicario JC, Cesari AB, de Lema MBG, Dardanelli MS (2015) *Arachis hypogaea* PGPR isolated from Argentine soil modifies its lipids components in response to temperature and salinity. Microbiol Res 173:1–9
348. Panwar M, Tewari R, Nayyar H (2016) Native halo-tolerant plant growth promoting rhizobacteria *Enterococcus* and *Pantoea* sp. improve seed yield of mungbean (*Vigna radiata* L.) under soil salinity by reducing sodium uptake and stress injury. Physiol Mol Biol Plants 22(4):445–459
349. Xu Y, Li Y, Long C, Han L (2022) Alleviation of salt stress and promotion of growth in peanut by *Tsukamurella tyrosinosolvens* and *Burkholderia pyrrocinia*. Biologia 1–11
350. Mahmood A, Amaya R, Turgay OC, Yaprak AE, Taniguchi T, Kataoka R (2019) High salt tolerant plant growth promoting rhizobacteria from the common ice-plant *Mesembryanthemum crystallinum* L. Rhizosphere 9:10–17
351. Ullah S, Bano A, Ullah A, Shahid MA, Khan N (2022) A comparative study of plant growth promoting rhizobacteria (PGPR) and sowing methods on nutrient availability in wheat and rhizosphere soil under salinity stress. Rhizosphere 23:100571
352. Oliveira Lopes ÁL, Setubal IS, da Costa Neto VP, Zilli JE, Rodrigues AC, Bonifacio A (2022) Synergism of *Bradyrhizobium* and *Azospirillum baldaniorum* improves growth and symbiotic performance in lima bean under salinity by positive modulations in leaf nitrogen compounds. Appl Soil Ecol 180:104603
353. Kumari M, Swarupa P, Kumar A (2022) Validation and evaluation of plant growth promoting potential of rhizobacteria towards paddy plants. J Pure Appl Microbiol 16(2):1209–1225
354. Naseri S, Agha ABA, Sharifi R, Bahraminejad S (2022) Rhizobacteria modify soil biological indices and induce tolerance to osmotic stress in tomato depending on the salinity level and bacteria species. Braz J Microbiol 53(3):1473–1481
355. Sindhu SS, Malik DK, Dadarwal KR (2003) Enhancing the potential of biological nitrogen fixation by genetic manipulations of diazotrophic bacteria for sustainable agriculture. In: Singh RP, Jaiwal PK (eds) Plant genetic engineering, vol 1, Applications and limitations. Sci Tech Publ LCC Houston, USA, pp 199–228
356. Chaparro JM, Sheflin AM, Mentor DK, Vivanco JM (2012) Manipulating the soil microbiome to increase soil health and plant fertility. Biol Fertil Soils 48:489–499

357. Sindhu SS, Sehrawat A, Sharma R, Dahiya A, Khandelwal A (2018) Belowground microbial cross-talk and rhizosphere biology. In: Singh DP (ed) Plant-microbe interactions in agro-ecological perspectives, vol 1, Fundamental mechanisms, methods and functions. Springer Nature, Singapore Pte Ltd, pp 695–752
358. Phour M, Sindhu SS (2022) Mitigating abiotic stress: microbiome engineering for improving agricultural production and environmental sustainability. Planta 256(5):1–34
359. Orozco-Mosqueda MC, Fadiji AE, Babalola OO, Glick BR, Santoyo G (2022) Rhizobiome engineering: unveiling complex rhizosphere interactions to enhance plant growth and health. Microbiol Res 263:127–137
360. Singh RP, Manchanda G, Singh RN, Srivastava AK, Dubey RC (2016) Selection of alkalotolerant and symbiotically efficient chickpea nodulating rhizobia from North-West Indo Gangetic plains. J Basic Microbiol 56:14–25
361. Kaul S, Choudhary M, Gupta S, Dhar MK (2021) Engineering host microbiome for crop improvement and sustainable agriculture. Front Microbiol 12:635917
362. Gopal M, Gupta A (2016) Microbiome selection could spur next generation breeding strategies. Front Microbiol 7:1971
363. Sindhu SS, Sehrawat A, Sharma R, Dahiya A (2016) Biopesticides: use of rhizosphere bacteria for biological control of plant pathogens. Defence Life Sci J 1:135–214
364. Uno Y, Furihata T, Abe H, Yoshida R, Shinozaki K, Yamaguchi-Shinozaki K (2000) *Arabidopsis* basic leucine zipper transcription factors involved in an abscisic acid-dependent signal transduction pathway under drought and high-salinity conditions. Proc Natl Acad Sci USA 97(21):11632–11637
365. Zorner P, Farmer S, Alibek K (2018) Quantifying crop rhizosphere microbiome ecology: the next frontier in enhancing the commercial utility of agricultural microbes. Ind Biotechnol (New Rochelle NY) 14(3):116–119
366. Qiu Z, Egidi E, Liu H, Kaur S, Singh BK (2019) New frontiers in agriculture productivity: optimised microbial inoculants and *in situ* microbiome engineering. Biotechnol Adv 37(6):107371
367. Kim J, Geng RS, Gallenstein RA, Somers DE (2013) The F-box protein ZEITLUPE controls stability and nucleocytoplasmic partitioning of *Gigantea*. Development 140:4060–4069
368. Mueller UG, Sachs JL (2015) Engineering microbiomes to improve plant and animal health. Trends Microbiol 23:606–617
369. Ke J, Wang B, Yoshikuni Y (2021) Microbiome engineering: Synthetic biology of plant-associated microbiomes in sustainable agriculture. Trends Biotechnol 39:244–261
370. Shah AN, Tanveer M, Abbas A, Fahad S, Baloch MS, Ahmad MI, Saud S, Song Y (2021) Targeting salt stress coping mechanisms for stress tolerance in *Brassica*: a research perspective. Plant Physiol Biochem 158:53–64
371. Muresu R, Polone E, Sulas L, Baldan B, Tondello A, Delogu G, Cappuccinelli P, Alberghini S, Benhizia Y, Benhizia H, Benguedouar A, Mori B, Calamassi R, Dazzo FB, Squartini A (2008) Coexistence of predominantly nonculturable rhizobia with diverse, endophytic bacterial taxa within nodules of wild legumes. FEMS Microbiol Ecol 63:383–400
372. Dawwam GE, Elbeltagy A, Emara HM, Abbas IH, Hassan MM (2013) Beneficial effect of plant growth promoting bacteria isolated from the roots of potato plant. Ann Agric Sci 58:195–201
373. Ogbo F, Okonkwo J (2012) Some characteristics of a plant growth promoting *Enterobacter* sp. isolated from the roots of maize. Adv Microbiol 2:368–374
374. de Oliveira Costa LE, de Queiroz MV, Borges AC, de Moraes CA, de Araújo EF (2012) Isolation and characterization of endophytic bacteria isolated from the leaves of the common bean (*Phaseolus vulgaris*). Braz J Microbiol 43:1562–1575
375. Gang S, Saraf M, Waite CJ, Buck M, Schumacher J (2018) Mutualism between *Klebsiella* SGM 81 and *Dianthus caryophyllus* in modulating root plasticity and rhizospheric bacterial density. Plant Soil 424:273–288
376. Abdel-Salam MS, Ibrahim SA, Abd-El-Halim MM, Moharam HFA (2013) Cloning of indigenous *Bacillus licheniformis* pyrroloquinoline quinone gene and its role in enhancement of phosphate solubilization in *Escherichia coli*. Int J Chem Tech Res 5:2236–2242

377. Pan J, Peng F, Xue X, You QG, Zhang WJ, Wang T, Huang CH (2019) The growth promotion of two salt-tolerant plant groups with PGPR inoculation: a meta-analysis. Sustainability 11:378
378. Zhong NQ, Han LB, Wu XM, Wang LL, Wang F, Ma YH, Xia GX (2012) Ectopic expression of a bacterium NhaD-type $Na^+/H^+$ antiporter leads to increased tolerance to combined salt/alkali stresses. J Integr Plant Biol 54:412–421
379. Ma Y, Vosátka M, Freitas H (2019) Beneficial microbes alleviate climatic stresses in plants. Front Plant Sci 10:595
380. Berendsen RL, Pieterse CM, Bakker PA (2012) The rhizosphere microbiome and plant health. Trends Plant Sci 17(8):478–486
381. Liu WX, Wang QL, Hou JY, Tu C, Luo YM, Christie P (2016) Whole genome analysis of halotolerant and alkalotolerant plant growth-promoting rhizobacterium *Klebsiella* sp. D5A. Sci Rep 6:26710
382. Vives-Peris V, Gomez-Cadenas A, Perez-Clemente RM (2018) Salt stress alleviation in citrus plants by plant growth-promoting rhizobacteria *Pseudomonas putida* and *Novosphingobium* sp. Plant Cell Rep 37:1557–1569
383. Kothari VV, Kothari RK, Kothari CR, Bhatt VD, Nathani NM, Koringa PG (2013) Genome sequence of salt-tolerant *Bacillus safensis* strain VK, isolated from saline desert area of Gujarat India. Genome Announc 1:e671–e713
384. Kishor PBK, Hong Z, Miao CH, Hu CAA, Verma DPS (1995) Overexpression of A1-pyrroline-5-carboxylate synthetase increases proline production and confers osmotolerance in transgenic plants. Plant Physiol 108:1387–1394
385. Sehrawat N, Bhat KV, Kaga A, Tomooka N, Yadav M, Jaiwal PK (2014) Development of new gene-specific markers associated with salt tolerance for mungbean (*Vigna radiata* L.Wilczek). Spanish J Agric Res 12:732–741
386. Abebe T, Guenzi AC, Martin B, Cushman JC (2003) Tolerance of mannitol-accumulating transgenic wheat to water stress and salinity. Plant Physiol 131(4):1748–1755
387. Chakraborty K, Sairam RK, Bhattacharya R (2012) Differential expression of salt overly sensitive pathway genes determines salinity stress tolerance in *Brassica* genotypes. Plant Physiol Biochem 51:90–101
388. Ji MG, Park HJ, Cha JY, Kim JA, Shin GI, Jeong SY, Lee ES, Yun DJ, Lee SY, Kim WY (2020) Expression of *Arabidopsis thaliana* thioredoxin-h2 in *Brassica napus* enhances antioxidant defenses and improves salt tolerance. Plant Physiol Biochem 147:313–321
389. Zhang HX, Hodson JN, Williams JP, Blumwald E (2001) Engineering salt-tolerant *Brassica* plants, characterization of yield and seed oil quality in transgenic plants with increased vacuolar sodium accumulation. Proc Natl Acad Sci 98(22):12832–12836
390. Zhang X, Chen L, Shi Q, Ren Z (2020) SlMYB102, an R2R3-type MYB gene, confers salt tolerance in transgenic tomato. Plant Sci 291:110356
391. Tang M, Liu X, Deng H, Shen S (2011) Over-expression of JcDREB, a putative AP2/EREBP domain-containing transcription factor gene in woody biodiesel plant *Jatropha curcas*, enhances salt and freezing tolerance in transgenic *Arabidopsis thaliana*. Plant Sci 181(6):623–631
392. Wang L, Zhu J, Li X, Wang S, Wu J (2018) Salt and drought stress and ABA responses related to bZIP genes from *V. radiata* and *V. angularis*. Gene 651:152–160
393. Kumar K, Ghanti S, Sujata KG, Vijay Kumar BM, Nataraja Karba N, Reddy JK, Rao SM, Kishor KPB (2011) Heterologous expression of P5CS gene in chickpea enhances salt tolerance without affecting yield. Biol Plant 55:634–640
394. Baloda A, Madanpotra S, Aiwal PK (2017) Transformation of mung bean plants for salt and drought tolerance by introducing a gene for an osmoprotectant glycine betaine. J Plant Stress Physiol 3:5–11
395. Srivastava R, Kumar S, Kobayashi Y, Kusunoki K, Tripathi P, Kobayashi Y, Koyama H, Sahoo L (2018) Comparative genome-wide analysis of WRKY transcription factors in two Asian legume crops: Adzuki bean and mung bean. Sci Rep 8:169–171
396. Cortes AJ, This D, Carolina C, Madrinan S, Blair MW (2012) Nucleotide diversity patterns at the drought-related DREB2 encoding genes in wild and cultivated common bean (*Phaseolus vulgaris* L.). Theor Appl Genet 125:1069–1085

397. Duan Z, Zhang D, Zhang J, Di H, Wu F, Hu X, Meng X, Luo K, Zhang J, Wang Y (2015) Co-transforming bar and CsALDH genes enhanced resistance to herbicide and drought and salt stress in transgenic alfalfa (*Medicago sativa* L.). Front Plant Sci 6:1–9
398. Jin T, Chang Q, Li W (2010) Stress-inducible expression of GmDREB1 conferred salt tolerance in transgenic alfalfa. Plant Cell Tissue Organ Cult 100:219–227
399. Wang Z, Ke Q, Kim MD, Kim SH, Ji CY (2015) Transgenic alfalfa plants expressing the sweet potato orange gene exhibit enhanced abiotic stress tolerance. PLoS ONE 10:1–17
400. Hanafy MS, El-Banna A, Schumacher HM, Jacobsen HJ, Hassan FS (2013) Enhanced tolerance to drought and salt stresses in transgenic faba bean (*Vicia faba* L.) plants by heterologous expression of the PR10a gene from potato. Plant Cell Rep 32:663–674
401. Gao H, Huang H, Lu K, Wang C, Liu X, Song Z, Zhou H, Yang L, Li B, Yu C, Zhang H (2021) OsCYP714D1 improves plant growth and salt tolerance through regulating gibberellin and ion homeostasis in transgenic poplar. Plant Physiol Biochem 168:447–456
402. Singh R, Sharma S, Kharb P, Saifi S, Tuteja N (2020) OsRuvB transgene induces salt tolerance in pigeon pea. J Plant Interact 15(1):17–26
403. Zhu H, Jiang Y, Guo Y, Huang J, Zhou M, Tang Y, Sui J, Wang J, Qiao L (2021) A novel salt inducible WRKY transcription factor gene, AhWRKY75, confers salt tolerance in transgenic peanut. Plant Physiol Biochem 160:175–183
404. Liu MM, Li YL, Li GC, Dong TT, Liu SY, Pei LI, Wang QG (2020) Overexpression of StCYS1 gene enhances tolerance to salt stress in the transgenic potato (*Solanum tuberosum* L.) plant. J Integr Agric 19(9):2239–2246
405. Zhang X, Lin X, Chen S, Chen S (2022) Overexpression of *BpERF*1.1 in *Betula platyphylla* enhanced tolerance to multiple abiotic stresses. Physiol Mol Biol Plants 28:1159–1172
406. Kovtun Y, Chiu WL, Tena G, Sheen J (2000) Functional analysis of oxidative stress activated mitogen-activated protein kinase cascade in plants. Proc Natl Acad Sci USA 97:2940–2945
407. Umezawa T, Yoshida R, Maruyama K, Yamaguchi-Shinozaki K, Shinozaki K (2004) SRK2C, a SNF1-related protein kinase 2, improves drought tolerance by controlling stress-responsive gene expression in *Arabidopsis thaliana*. Proc Natl Acad Sci USA 101:17306–17311
408. Sanghera GS, Wani SH, Hussain W, Singh NB (2011) Engineering cold stress tolerance in crop plants. Curr Genomics 12(1):30
409. Suzuki N, Rizhsky L, Liang H, Shuman J, Shulaev V, Mittler R (2005) Enhanced tolerance to environmental stress in transgenic plants expressing the transcriptional co-activator Multiprotein Bridging Factor 1c. Plant Physiol 139:1313–1322
410. Berrocal-Lobo M, Molina A, Solano R (2002) Constitutive expression of ETHYLENE-RESPONSE-FACTOR1 in *Arabidopsis* confers resistance to several necrotrophic fungi. Plant J 29(1):23–32
411. Cheng MC, Liao PM, Kuo WW, Lin TP (2013) The *Arabidopsis* ETHYLENE RESPONSE FACTOR1 regulates abiotic stress-responsive gene expression by binding to different cis-acting elements in response to different stress signals. Plant Physiol 162(3):1566–1582
412. Xu ZS, Xia LQ, Chen M, Cheng XG, Zhang RY, Li LC, Zhao YX (2007) Isolation and molecular characterization of the *Triticum aestivum* L. ethylene-responsive factor 1 (TaERF1) that increases multiple stress tolerance. Plant Mol Biol 65(6):719–732
413. Zhang GY, Chen M, Li LC, Xu ZS, Chen XP, Guo JM, Ma YZ (2009) Overexpression of the soybean GmERF3 gene, an AP2/ERF type transcription factor for increased tolerances to salt, drought, and diseases in transgenic tobacco. J Exp Bot 60(13):3781–3796
414. Bünemann EK, Bongiorno G, Bai Z, Creamer RE, De Deyn G, de Goede R, Fleskens L, Geissen V, Kuyper TW, Mäder P (2018) Soil quality—a critical review. Soil Biol Biochem 120:105–125
415. Kim K, Samaddar S, Chatterjee P, Krishnamoorthy R, Jeon S, Sa T (2019) Structural and functional responses of microbial community with respect to salinity levels in a coastal reclamation land. Appl Soil Ecol 137:96–105
416. Compant S, Van Der Heijden M, Sessitsch A (2010) Climate change effects on beneficial plant–microorganism interactions. FEMS Microbiol Ecol 73:197–214

417. Kumar A, Patel JS, Meena VS, Srivastava R (2019) Recent advances of PGPR based approaches for stress tolerance in plants for sustainable agriculture. Biocatal Agric Biotechnol 20:101271
418. Jansson JK, Hofmockelm KS (2020) Soil microbiomes and climate change. Nat Rev Microbiol 18:35–46
419. Woo SL, Pepe O (2018) Microbial consortia: promising probiotics as plant biostimulants for sustainable agriculture. Front Plant Sci 9:1801
420. Chanratana M, Joe MM, Choudhury AR, Anandham R, Krishnamoorthy R, Kim K, Jeon S, Choi J, Choi J, Sa T (2019) Physiological response of tomato plant to chitosan-immobilized aggregated *Methylobacterium oryzae* CBMB20 inoculation under salinity stress. Biotech 9:397
421. Li YB, Shi HW, Zhang HW, Chen SF (2019) Amelioration of drought effects in wheat and cucumber by the combined application of super absorbent polymer and potential biofertilizer. Peer J 7:e6073
422. Del Buono D (2020) Can biostimulants be used to mitigate the effect of anthropogenic climate change on agriculture? It is time to respond. Sci Total Environ 751:141763
423. Ahlawat OP, Yadav D, Kashyap PL, Khippal A, Singh G (2021) Wheat endophytes and their potential role in managing abiotic stress under changing climate. J Appl Microbiol 1–20
424. Santos SS, Rask KA, Vestergard M, Johansen JL, Prime A, Frosley TG, Martin Gonzalez AM, He H, Ekelund F (2021) Specialized microbiomes facilitate natural rhizosphere microbiome interactions counteracting high salinity stress in plants. Environ Expt Bot 186:104430

# Chapter 14
# Over View of Symbiosis Mechanisms and Soil Quality Management Practices to Combat Environmental Changes

**Fadime Karabulut, Nowsheen Shameem, Nusrat Shafi, Javid A. Parray, Abeer Hashem, and Elsayed Fathi Abd-Allah**

**Abstract** Higher input requirements for high yields result in environmental issues and the depletion of natural resources in agricultural systems. The wide genetic variation in microbial species reveals that microorganisms with high potential that can adapt to different environmental conditions can be identified. Soil quality is defined as a soil feature that promotes biological activity, protects and maintains environmental quality, and fulfills the function of plant production within the boundaries of an ecosystem. The transformation of phosphorus and nitrogen, which are the building blocks of living cells, from organic form to inorganic and useful form is necessary for plants to be taken up by the microorganisms in the soil. The fixation of elemental nitrogen in the atmosphere takes place by microorganisms living symbiotically as well as non-symbiotically. Plant growth promoting bacteria (PGPR), on the other hand, colonize the rhizosphere and provide the potential to be a biological fertilizer in plant production as well as a biological control agent. Additionally, PGPR exhibits synergistic or antagonistic interactions with the soil and rhizosphere microorganisms

F. Karabulut (✉)
Department of Biology, Firat University, Elazıg 23119, Turkey
e-mail: karabulutfadime9@gmail.com

N. Shameem
Department of Environmental Sciecne, SP College Srinagar Cluster University, Srinagar, JK, India

N. Shafi
Department of Chemistry, GDC Eidgah, Srinagar, JK, India

J. A. Parray
Department of Environmental Sciecne, GDC Eidgah, Srinagar, JK, India

A. Hashem
Botany and Microbiology Department, College of Science, King Saud University, Riyadh 11451, Saudi Arabia

Mycology and Plant Disease Survey Department, Plant Pathology Research Institute, Giza, Egypt

E. F. Abd-Allah
Plant Production Department, College of Food and Agricultural Sciences, King Saud University, Riyadh 11451, Saudi Arabia

Seed Pathology Research Department, Plant Pathology Research Institute, Giza, Egypt

 
J. A. Parray (ed.), *Climate Change and Microbiome Dynamics*, Climate Change Management, https://doi.org/10.1007/978-3-031-21079-2_14

that help or speed up plant growth. Biochar, a carbonaceous substance, is being used more frequently to clean up anthropogenically contaminated soils and restore their ecological functions. In addition, due to its high surface area, porosity, functional groups on its surface, and surface charge, biochar is an effective additive for the removal of inorganic and organic pollutants in water and soil. We think that this chapter will help answer questions about biochar's feasibility, effectiveness, and safety so that it can be used to make the soil more fertile and get rid of pollutants in the soil.

## Introduction

The soil, as one of the most complex habitats on the planet, supports a diverse range of life. It is the primary route by which most nutrients are obtained from the soil and essential nutrients such as trace elements (Cu, Zn, etc.) phosphorus, and nitrogen reach humans via plants [1]. Because of this, soil organisms are incredibly diverse and provide a variety of ecosystem services necessary for both natural and managed ecosystems to function sustainably. Soil biota is a diverse group of macroorganisms (micro-and macro-arthropods, earthworms, and termites) and microorganisms that participate in the global cycle of organic matter, energy, and nutrients (bacteria, fungi, algae, protozoa, and some nematodes). This is a crucial sign that soil biodiversity is reflected in the diversity of living things in the soil [2]. Most soil microorganisms can break down cellulose, hemicellulose, other polysaccharides, hydrocarbons, and lignin derivatives and convert them into usable forms. Furthermore, it converts the nitrogen, sulfur, and some minerals ($Fe^{+3}$, $Fe^{+2}$, and so on) in the soil into usable forms, allowing the plant nutrients required by the plant to become useful [3]. Biodiversity and soil microbiome functioning balance many complex processes, such as soil organisms that help crop and livestock production, harbor antibiotics and pathogens, control nutrient loads in soil and water, and balance the greenhouse gas cycle and climate change are examples [4]. The physical structures and products of these organisms, on the other hand, contribute significantly to the soil structure. Due to this, the sustainability of soil quality and food production is critical for the survival and continuity of microorganisms' activities, which also play a role in the ecosystem cycle. Plants and microorganisms both obtain nutrients from the soil via organic waste accumulation and metabolic activities, and thus alter soil properties. Microorganisms have some direct effect on plants, such as hormone signal stimulation and pathogen resistance. Metabolites released from the roots allow the plant to communicate with microorganisms. Exudates and metabolic activities, in particular, must be thoroughly investigated in order to comprehend the plant-microbe interaction mechanism in the rhizosphere. Plants have a significant impact on the soil as well as the processes and conditions that affect plant life. A plant is a living organism that cannot actively move through the soil throughout its biological cycle. Plants have different tissues, each with a specific function, soil biota, and the plant-soil biota relationship can be positive or negative [5]. In other words, plant roots and microorganisms communicate with one another in the rhizomicrobiome, the most active area of the soil.

Rhizobacteria that colonize plant roots or the rhizosphere and support plant growth by directly immobilizing nutrients while also serving as defense regulators are known as plant growth-promoting rhizobacteria (PGPR). These rhizobacteria encourage plant growth and development in various environments through a number of mechanisms [6]. PGPRs, which interact with plants via these mechanisms and provide them with numerous benefits, attract metabolites that are normally secreted by plant roots and used as nutrients. Plants introduce various inorganic ions (protons and other ions) and organic compounds (phenolics, carbohydrates, amino acids, carboxylic acids, e.g.) identified as potentially important low molecular weight metabolites or exudate into the rhizosphere to alter the biology and chemistry of the root microenvironment [7]. Some of the most complex chemical, physical, and biological interactions that terrestrial plants experience take place at the root and in the surrounding soil environment (also known as the rhizosphere). These interactions take the form of root-microbe, root-insect, and root-root interactions [8]. These interactions, on the other hand, are classified as positive, neutral, or negative, and there is usually an antagonistic and symbiotic relationship between root and PGPR [9]. The term "beneficial plant-microbe interactions" refers to symbiotic relationships in which plants and microorganisms both benefit and suffer negative consequences [10]. Rhizospheric soil microorganisms, for example, contribute to plant nutrient absorption and utilization; it also promotes plant growth and development by decomposing soil nutrients and converting them into usable forms [11]. Furthermore, a compound synthesized by PGPR and delivered to the plant facilitates the absorption of phytohormones or some nutrients from the environment, reduces or prevents the harmful effects of one or more phytopathogenic organisms, atmospheric nitrogen fixation, dissolution of some minerals (phosphorus) in the soil, HCN, antibiotics, siderophores, exopolysaccharides, and volatile compounds that stimulate plant growth, and so on. It is involved in a variety of processes, including the production of various metabolites. As a result, PGPRs tend to advocate for soil sustainability and the health of dormant plants in the soil in an environmentally friendly manner [12]. It is becoming increasingly important as one of the most effective factors for removing biotic (bacteria, viruses, fungi, etc.) and abiotic (heavy metal, drought, salinity, temperature, etc.) stress and ensuring long-term agricultural production. In this context, soil microbial biomass is a crucial metric for assessing soil nutrition and fertility and is a vital part of a sizable ecosystem (rhizosphere microbiome). Biochar promotes plant growth and boosts crop yields. When applied and mixed at specific rates on problematic soils, it improves the physical, biological, and chemical quality criteria of the soil as well as its nutritional values. It aids in the reduction of chemical and organic fertilizer use, as well as the use of compost. It reduces greenhouse gas emissions from crop production, aids in carbon storage in the soil, and ensures that the carbon stored in the soil remains stable for years. By adsorbing agricultural chemicals, it prevents them from entering streams and groundwater, thereby promoting sustainable agricultural production. It prevents nutrient washing in low porosity soils, especially during rainy weather. It aids in water conservation by increasing the soil's moisture retention capacity. It keeps pesticides from being absorbed by plants while they are growing in the soil. It promotes some fungi's growth in the soil, which helps plants grow and enhances the soil's quality. It reduces $CO_2$, $CH_4$, and $N_2O$ gas emissions into the atmosphere from cultivated soils, thereby contributing to global warming. To reap the full benefits of biochar, it should be mixed with the soil's lower layers using appropriate methods. Biochar is typically added to the furrows opened to the tree roots and covered with soil in the biochar application in gardens. Thus, adding biochar to the soil by mixing it with organic fertilizers or different composts improves its efficiency [13].

## Biomass Structure and Sources

All living or non-viable biological resources are considered biomass. Biomass is defined as "non-fossilized and biodegradable organic matter of plant, animal, and microorganism origin" by the United Nations Framework Convention on Climate Change (UNFCCC). Any material that contains organic carbon is considered to be biomass and comes from a variety of sources. Biomass includes plant materials, cellulosic materials, lignin substances, animal products, organic wastes, solid wastes, ocean/sea wastes, agricultural wastes, wastes of animal or human origin, and other natural carbon sources [14]. Green plants produce biomass by converting sunlight into a vegetative substance via photosynthesis. As a result, biomass is defined as organic matter that stores sunlight energy in its chemical bonds. This chemical energy is released when the bonds between C, H, and O are ruptured during combustion, dissolution, or decomposition reactions. In the natural cycle, $CO_2$ released by biomass combustion or thermal decomposition mixes with the atmosphere and then returns to plants via photosynthesis. As a result, the net $CO_2$ emissions of biomass are nearly zero [15]. Ash, protein, extractives (non-soluble substances, nitrogenous compounds, chlorophyll, and waxes), and hemicellulose, cellulose, lignin, and pectin are among the constituents of biomass, each in varying amounts. The most important components that provide information about the type of biomass are the cellulose and lignin ratios. For example, while hard woody structures contain more cellulose, soft woody structures contain more lignin, and vegetative structures such as tree leaves and wheat straw contain more hemicellulose. Biomass resources are classified into four categories: woody plants (from forests and industries), herbaceous plants (from agricultural resources), aquatic plants, and animal manures (from animal origin). Aquatic plants and animal manures have a high moisture content, making them good raw materials for wet processes. Woody plants, on the other hand, are better suited and more cost-effective for processes such as gasification and pyrolysis [15].

## Biochar and Other Additives

Lehmann asserts that biochar, a material rich in carbon with a smooth pore structure, is a byproduct of the thermal decomposition of biomass in an oxygen-free environment. In addition, biochar is a largely stable (recalcitrant) organic carbon compound created by pyrolyzing biomass at temperatures between 300 °C and 1000 °C in an oxygen-depleted environment. According to the researchers, in an oxygen-restricted environment, at temperatures between 300 °C and 600 °C, biomass from plants or animals is converted into biochar. As an environmentally friendly product, biochar is used in four basic areas: soil improvement, carbon sequestration in soil, climate change mitigation, energy production, and pollutant removal from soil and water [16]. Furthermore, biochar is a highly aromatic material with a C content ranging from 400 to 800 g/kg [17]. Biochar—an organic substance with varying electrical charges—captures interest because of its pH, cation exchange capacity, surface adsorption capacity, and nutrient content. The soil's capacity to hold water is increased by increasing its surface area and pore volume. These properties are directly related to the biochar's production temperature [17].

Pyrolysis, hydrothermal carbonization, gasification, and torrefaction are included in biochar production technologies. Char formation is only possible through the thermal conversion of biomass. Thermal conversion yields biochar with an average energy density of 28 kJ/kg [18].

Biochar does not decompose quickly and remains stable in the soil for a long time because of its large surface area (300–2000 $m^2\ g^{-1}$) and lack of nitrogen. In the semi-arid Mediterranean climate zone, carbonaceous compounds with a large surface area and a long soil retention time will be essential for soil structure and fertility due to the rapid decomposition of organic matter caused by heat and the disordered structure of the soil. Current research areas include the mechanisms used by plant root-mycorrhizal fungi to keep carbon bound more tightly in plant tissues and in the soil for a longer period of time for soil fertility (by doing more photosynthesis). Because charred plant material (biochar) contains little nitrogen and will not be decomposed by microorganisms, the carbon material will remain in the soil for a long time, increasing the organic carbon level. Because it retains more water and nutrients, biochar improves plant growth and soil structure. Biochar, or charred plant material with a large surface area, will contribute significantly to the retention of heavy metals in the soil, the retention of nutrients that will mix with drainage and groundwater, and the protection of the food web and human health in the soil. It is significant and unique that the carbon, which is linked by the plant path, is applied to the soil in a charred form without decomposing in the soil or mixing into the atmosphere.

Priority one for science:

1. How precisely can this amount of carbon be measured? How much carbon is actually released from the soil into the atmosphere?
2. Is it possible to reintroduce carbon into the soil that has been released into the atmosphere?

3. How much extra carbon is stored in plant tissues as a result of plant-mycorrhizal cooperation? How is this bound carbon in the soil preserved, and by what mechanism? How effective is biochar at storing carbon?
4. Is it possible to increase carbon sequestration potential by improving soil structure?

Scientists are in desperate need of answers to these questions. In this case, the research is mostly about how to make biochar and how to use it in soil.

A key tactic for increasing soil organic matter and carbon bonding with plants is the use of plant residues as compost or animal manure. However, it is well known that when animal manure and compost decompose quickly in the soil, $CO_2$ is released quickly into the atmosphere, and the carbon source added to the soil in a short period of time is lost in a significant amount. Small amounts of humus and other stable compounds are still present in the soil as a result of the quick decomposition of the organic fertilizer sources that are still applied to the soil today. Organic compounds that must be added to the soil as a result of burning agricultural stubbles and wastes are also burned and released as carbon into the atmosphere. As a result, the soil deteriorates on the one hand while the concentration of greenhouse gases that contribute to global climate change increases on the other. Carbonization (biochar) and re-use of plants that have been used in agriculture for a long time have been frequently discussed approaches in recent years in order to keep the carbon source that should be added to the soil in question in the soil for a long time.

As a result, in recent years, it has been used as charred plant material with a high stable organic carbon content rather than material with a high decomposition property. Biochar is one of these, as it contains a lot of carbon and can persist for a long time in the soil. Biochar has long been used to meet energy demands. However, because of the effect of rising greenhouse gas levels on climate change, biochar is said to be an effective carbon binder. Biochar (terra preta de Indio), or black soil, is thought to have been used in the Amazon region 2500 years ago. According to research, the soils in this area are rich in organic matter, and the plants grow three times as quickly [19]. The ability of biochar's carbon source to persist for a long time in the soil without decomposing is its most crucial characteristic. Biochar has a high nutrient content, a highly stable carbon content, a lime effect on the soil (for acid soils), a reactive surface area, and a redox potential, among other characteristics. It's also a good fit for a high cation exchange capacity, bulk weight, porosity/microbial habitat, and porosity/water holding capacity.

Biochar application reduced ammonium losses by 10% in a study [20]. It is ensured that soil pH is increased, Al toxicity is reduced, the soil tension resistance is reduced, and a habitat for soil organisms is created in acidic soils. According to Australian studies, soil tillage over carbon can be facilitated by the high biochar and water-holding capacity of hard-tillage-resistant soils. The study found that using biochar improved the effectiveness of fertilizer use. Biochar has been found to increase fertilizer efficiency due to its large surface area. The way and method of applying biochar to the soil are determined by the agricultural farm's production plan and structure. It is suggested that it be buried in the soil 10–20 cm deep as a scattering or tape method. Biochar application increased corn and wheat yields by 46–70% in the first year in a study conducted in field conditions in Australia [21].

Composting adds plant biomass to the soil as an organic source, thereby boosting the soil's declining organic matter. Microorganisms decompose compost quickly due to its high nitrogen content. While the plant composting process decomposes the material quickly, biocharification can leave hundreds of compounds in the soil. For soil management, a carbon source that persists in the soil for a longer time is preferable to one that disappears over time. Depending on the state of the biomass, biochar is the preparation and application of a material containing 50–80% carbon. When the significant loss of carbon is compared to the composting of materials that can be produced through thermochemical or hydrothermal (hydrothermal carbonization) processes, biochar is seen as an important sustainable carbon source.

In recent years, scientific organizations have been conducting serious basic research on biochar systems. The application and effects of biochar production have been the subject of research. The subject of biochar necessitates interdisciplinary research because it spans a wide range of disciplines, from chemistry to soil science. The following are some possible impacts of biochar on soil composition and crop yield:

- Increases nutrient efficiency while reducing fertilizer use.
- Increases the capacity and efficiency of water storage.
- Reduces the washing effect and fertilizer gas losses.
- Denitrification is reduced.
- Reduces the toxicity of Al
- Heavy metal's availability is reduced.
- Increases the availability and retention of phosphorus in soil.
- Establishes an environment that is favorable for the development of mycorrhizae and N2 friction.
- Ensures long-term C accumulation in the soil.

The qualities and preparation of the biochar that will be applied have a significant impact on the physical fertility of the soil. Biochar is used up to 5–40 tons per hectare in many studies. Depending on the organic materials used and the technology employed, biochar can have a diverse range of biological, chemical, and physical properties. Biochar is expected to improve soil fertility and contribute to agricultural productivity by ensuring its long-term viability, thanks to its large porosity and surface area [22]. Biochar's physical and chemical properties will enhance soil retention and nutrient uptake [23], reduce soil loss and greenhouse gas (SG) emissions [24], and bind pesticides, toxic substances, and heavy metals. Because of its substantial surface area, carbonized carbon source biochar is expected to have a high capacity to hold nutrients and water. Biochar application had a positive effect on spore germination, according to a study [25]. The capacity of the Biocar is increased by increasing the soil base saturation and water retention [26]. Biochar is frequently used in the purification of toxic substances and water due to its large surface area. At the same time, it facilitates plant root nutrient uptake by keeping nutrients in the soil column [27]. Because Biochar keeps nutrients in the soil column and prevents washing, it's a good idea to use it. As a result, groundwater pollution is kept to a minimum.

## Important Criteria of Quality Soil

### *Soil Physical Quality Criteria*

The ability of aggregates to withstand the relaxing and disintegrating effects of water and mechanical elements, such as active organs, is known as aggregate stability. The structural development of the soil is the result of aggregate formation. The arrangement, stability, size, and continuity of the pores in the soil are all determined by aggregate size. The amount of soil clay, the concentration of electrolytes, and the amount of organic matter all have an impact on how the soil aggregates. Tillage, organic matter and organisms, texture, and rotation all affect aggregate stability. Soil aeration and aggregate stability are both important for keeping water in the soil and making it productive.

Factors such as deterioration in soil aggregation, infiltration, and erosion in the field have an impact on it. Erosion losses are reduced by increasing the percentage of stable aggregates. As a result, aggregate stability is one of the criteria used to gauge soil quality. Due to intensive soil processing for crop production, soil aggregation degrades. Increased tillage intensity, according to research, results in a rapid loss of soil organic matter, little biological activity, and a decline in aggregate stability [28]. Reduced tillage is more effective than conventional tillage at improving soil aggregation [29]. In general, it is asserted that adding plant residues to soil increases based on the volume of organic matter, which enhances the soil's ability to bind together [30]. When used as a soil amendment, biochar can enhance the physicochemical characteristics of degraded or nutrient-poor soils. The porosity and surface functionality of biochar determine its ability to retain soil water [31]. Increased soil porosity from biochar results in more soil surface area and easier water penetration, thanks to its porous internal structure. Previous research has demonstrated that adding biochar to unproductive soils can reduce bulk density, increase total pore volume, and improve water holding capacity [32]. Coal field soils have a lower bulk density than adjacent field soils 9% [33]. Total porosity rose to 50.6% in earth ovens from 45.7% in nearby field soils. Under coal furnaces, there was an 88% relative increase in soil saturated hydraulic conductivity, going from 6.1 to 11.4 cm $h^{-1}$. In the coal furnaces, the hue, value, and chroma all changed by 8, 20, and 20%, respectively, darkening the color of the soil. The dark color of biochar increased the soil surface temperature by an average of 4 °C, while the surface albedo decreased by 37% in charcoal soils. Higher leaching rates have been found in coalfield soils, indicating that surface runoff and erosion at these furnace sites may be reduced. Different research results show that this is the most important thing about Terra Preta soil [34].

The dry soil weight per unit volume is defined as the soil bulk weight. The increase in soil volume weight results in a reduction in the void volume, infiltration rate, and moisture content in the soil, as well as less aeration and a more resistant soil layer to plant roots. Plant root development slows as soil volume weight rises. As a result, bulk weight is thought to be a trustworthy indicator of soil quality. Depending on the type of plant grown, the texture of the soil, and the previous use of the soil, the volume weight of the soil has a detrimental effect on root and plant growth. If the bulk weight of the soil, which is an indicator of compaction, is too high, the nitrogen cycle slows down, runoff increases, soil temperature drops, and plant root growth slows down. While a medium-textured soil's volume weight for plant growth is 1.3 g/cm$^3$, the volume weight that stops plant root and stem development is 2 g/cm$^3$ [35]. However, depending on the soil type, texture, and mineral substance content, the recommended volume weight value in agricultural soils is between 1.1 and 1.4 g/cm$^3$. Other research found that the volume weight for the mulch direct sowing method was 1.25–1.4 g/cm$^3$, 1.11–1.22 g/cm$^3$ for conventional tillage, and 1.20–1.33 g/cm$^3$ for reduced tillage over a 23-year period [36]. It varies between cm3, according to them. The volume weight of the mulch direct sowing method was higher than the other two methods, but it was still below the plant root development limits.

Tillage changes the soil's pore characteristics, which has an impact on water retention and infiltration. The quantity of macro pores in the soil reduces its ability to hold water and infiltration rate, whereas the quantity of micro pores boosts both of these properties. As a result, the increase in tillage density is accompanied by an increase in field traffic and soil compaction, which lowers the infiltration rate. The large number of small aggregates that are produced have a moisture content that is defined as the ideal soil moisture level for tillage. A seed bed made up of small aggregates can store moisture better than a seed bed made up of large aggregates [37]. The right amount of water and air capacity in the soil, as well as the right compaction rate, are all necessary for good root development. However, the amount of water required by the plant to complete its development should be greater than 0.20 m$^3$/m$^3$, or between 0.15 and 0.25 m$^3$/m$^3$ [38].

According to the findings of the study, reduced tillage methods that allow seedbed preparation in one pass or direct sowing to the stubble are more beneficial than traditional tillage methods for increasing the infiltration rate and water holding capacity of the soil.

## *Soil Chemical Quality Criteria*

Soil organic matter is an integral component of soil and is important for the nitrogen cycle, biological activity, cation exchange capacity, and aggregate stability. By increasing nitrogen content, enhancing physical characteristics, and lowering the risk of erosion, soil organic matter enhances soil quality [39]. The quantity and activity of the soil's microbial mass determine how the organic matter cycle behaves. Therefore, the biological and biochemical characteristics of the soil are crucial in the development of the soil's ecology [40]. The nitrogen cycle and soil's ability to hold water are both greatly improved by increased organic matter content [41]. Organic matter depletion lowers cation exchange capacity, aggregate stability, product yield, and, as a result, soil quality. In agricultural soils with tropical and semi-tropical climate characteristics, intensive soil cultivation reduces soil organic matter [28]. Tillage causes a loss of organic matter, which varies depending on soil type, climate, and crop rotation [41].

Organic matter accumulates on the topsoil as a result of topsoil tillage and minimal tillage [42]. The organic matter is dispersed along the base of the plow the due to deep tillage of the soil. When comparing the effects of traditional tillage and the stubble direct sowing method on soil organic matter content, the parcels where direct sowing is applied to the stubble accumulate more than 130% more organic matter on the soil surface than the parcels where conventional tillage is applied [43]. Reduced tillage techniques, according to various studies, raise the soil's organic carbon, nitrogen, and phosphorus contents [42]. It has been discovered that tillage intensity decreases organic matter content in the 0–5 cm soil depth and that plots using the mulch direct sowing method accumulate 33% more organic matter [40]. By mixing the stubble into the soil, more organic matter is added to the soil at the depth of cultivation. Stubble accumulation in the top layer of agricultural soils improves soil quality by increasing organic matter content, particularly in plots where direct sowing is used. Furthermore, using minimum tillage and mulching techniques improves soil quality and crop productivity [44]. Soil tillage systems have a direct or indirect impact on soil pH, cation exchange capacity, and electrical conductivity. The pH of the soil decreases as soil organic matter content rises. Similar to this, the electrical conductivity of the top layer of soil rises as the rate of infiltration of the soil does. Washing the clay-sized soil particles from top to bottom lowers the soil's ability to exchange cations. However, as organic matter breaks down as a result of soil cultivation, the rate of oxidation affects the soil's ability to exchange cations.

The pH range in which the majority of plants can grow and produce their highest yields is their preferred range. Depending on the source of the fertilizer and the differential uptake and distribution of positively and negatively charged ions, crop harvesting, fertilizer application, and plant growth can all acidify the soil [45].

After other requirements, like the availability of water and nutrients, have been satisfied, acidic soils are typically amended by adding agricultural lime to raise the pH. Other conditions, such as the availability of water and nutrients, can promote the growth of plants. Previous research has demonstrated that high pH biochar lowers the toxicity of aluminum in red ferralitic soils, raises the pH of the soil by about one-third of a lime, and raises calcium levels [46]. Different biochar types had different effects on soil pH, as was seen when they were applied. The pH of sandy soil rose from 7.1 to 8.1 when biochar made from 39 t ha-1 herbaceous feedstock was added, according to study. Depending on the pyrolysis temperature and the type of raw material utilized, the pH of the biochars used in this study ranges from 6.0 to 9.6. For biochars made from woody raw materials, the pH rise was less pronounced. When the biochars used in the study were added to silt loam soils at rates up to 39 t $ha^{-1}$, there was a less dramatic overall pH increase. The higher buffering capacity of silt-loam soils is thought to be the reason for the smaller pH increases. The salinity of soil was recently studied, scientists found that adding co-composed biochar, poultry manure, and pyrolignous solution to saline soil significantly reduced its salinity and raised the pH by 3.6 $g.kg^{-1}$. [47]. Spontaneous surface oxidation reactions take place when fresh biochar is exposed to oxygen and water in the soil, increasing the net negative charge and subsequently the CEC. It has been discovered that high-negative-charge biochar particles enhance soil aggregation and plant nutrient availability [48]. The high reactivity of biochar particle surfaces can be attributed in part to pH [49]. The actual CEC of the biochar varies with the feedstock and pyrolysis temperature because these functional groups act as the main sites for pH-dependent charges. In soils, biochar aging results in the development of quinine functional groups while increasing hydroxyl and carboxyl groups [50, 51]. Functional groups containing oxygen build up on the surface of biochar as it ages. When defining these properties, it is thought that both the aromaticity brought on by the H:C ratio and the oxidation state brought on by the O:C ratio are essential. Biochars that were fresh or artificially aged had a much higher negative charge than biochars that were naturally aged [52]. In the pH range of 7.0–11.0, fresh biochar had a very low surface negative charge and only a small positive charge [53], and after artificial oxidation, the surface negative charge increased up to pH 3.5 [54]. In contrast, fresh or artificially aged biochar had a lower negative surface charge than naturally aged biochar [52]. Rapid H+ consumption and mineral dissolution reaction set it apart in terms of element release kinetics. Biochar composition and structure are influenced by pyrolysis conditions and biomass type [55]. As a result, there are significant differences in biochar properties linked to changes in nutrient content and retention [56]. Also, because biochars have different physicochemical properties, the availability of nutrients to plants varies from one biochar to the next. Biochars made from fertilizer and animal product feedstocks have a higher nutrient content than those made from plant materials, particularly wood [57]. As opposed to being a primary source of nutrients, biochars may be more useful as a soil conditioner and a catalyst for nutrient conversion [58].

## *Biological Quality Criteria of Soil*

Soil organisms with biological activity in the soil contribute to soil quality improvement by regulating the breakdown of waste materials with plant and animal origins, the biochemical cycle, and soil structure formation. Organisms in the soil; It is split into two categories: macro and micro. In the nitrogen cycle, soil aggregation, plant pathology, and plant development, microorganisms play a critical role [59]. Crop rotation, fertilization, and tillage are examples of soil management techniques that have an impact on microbial activity and macroorganism diversity. The physical and chemical properties of the soil have an immediate effect on the microbial activity that forms the basis for soil quality criteria. Changes in soil management are more quickly reflected in microbiological properties such as soil enzyme activity. In many studies, intensive soil processing is claimed to reduce the number of macro organism nests in the soil, allowing for an increase in infiltration rate and better aeration, and thus lowering soil quality [60].

Recent research has discovered that soil microbial mass and activity are influenced by soil cultivation, mulching, product type, rotation, fertilization, pesticide applications, and drainage applications. According to their research, direct sowing produced 60% more microbial carbon mass in February, 140% more in May, and 75% more in October than conventional tillage [43].

It has been discovered that plots where direct sowing is applied to the stubble at a soil depth of 0–10 cm have a higher rate of microbial mass than plots where conventional tillage is used [61]. They also discovered that soils with stubble contained 61–96% more microbial carbon and nitrogen than soils without stubble. The total organic carbon and nitrogen fractions in stubble fields are higher than in fields without stubble application, as can be seen [62]. According to studies, the addition of green manure to the soil boosts its N and P content while retaining organic matter and the physical, chemical, and biological properties [36]. Enzymes, which have a crucial function in soil microbial activities, allow chemical reactions to start and progress quickly in the soil. Because of their close relationship with soil microflora, enzyme activities are effective in changing soil properties and are thus considered a good soil quality criterion. When compared to other tillage systems, there is a higher amount of water-soluble carbon and enzyme activity at 5 cm of soil below the surface in soils where the stubble is directly sown [63]. Similarly, in stubble field conditions, it has been determined that the stubble direct sowing method has a higher amount of microbial mass and enzyme activity than other tillage systems. It has been discovered that soils where organic agriculture is used have higher enzyme activity than soils where conventional tillage systems are used. The activity and longevity of the soil microbial mass are impacted by the carbon ingress into the soil. In continuous production areas, the direct sowing method applied to stubble and the presence of stubble on the soil surface improves microbial mass and activity. Reduced tillage, he claims, boosts the number of macro-organisms in the soil. They unearthed that the amount of microbial activity, total nitrogen and phosphorus in the soil, microbial mass, and enzyme activity were all significantly higher in soils from organic agriculture [64]. In some stubble field studies, it was discovered that areas, where the direct sowing method was used, had higher microbial activity and microbial mass than areas where conventional tillage was used [65].

As a soil amendment, biochar must enhance soil health because, once added, it cannot be removed from the soil [66]. In light of the soil's properties, the climate, management practices, and especially the incorporation of organic matter, soils are complex communities of organisms that constantly change [67]. Applying biochar to soil is probably going to have a different impact on the soil biota than adding fresh organic matter; these effects might affect the variety, activity, and abundance of biotic communities in the soil [68]. The variations are brought on by biochar's relative stability and the fact that it has a lower bioavailable carbon content than recently formed organic matter. Biochar's ability to be porous has been demonstrated to alter biological functionality [69]. It also changes the availability of substrate and the activity of enzymes on or near the biochar particles. Instead of acting as the primary food source for microbes, biochar is thought to enhance the physical and chemical conditions in soils [68]. Using slow-pyrolyzed wood biochar and phosphorus-dissolving microbes (PSM) in various soil conditions in three different countries, it is more likely to be determined the impact of soil characteristics and crop type on the particular crop output of biochar (India, Thailand, and the United Kingdom). These results provided an explanation for why biochar significantly increased crop yields in P-deficient soils but was ineffective in boosting PSM activity for P mobilization in phosphate-rich soils [70].

Microbes can change the biomass and composition of microorganisms, and microbes can change the characteristics of biochar [68]. Its large surface area, porous structure, and capacity to adsorb soluble organic matter and inorganic nutrients, biochar is the ideal environment for microbes to thrive [71]. According to their physical and chemical characteristics, bacteria, actinomycetes, and arbuscular mycorrhizal fungi can all preferentially colonize biochar. They claimed that the addition of biochar increased microbial abundance [72]. This applies to bacteria, actinomycetes, and arbuscular mycorrhizal fungi, all of which have a preference for colonizing biochar depending on their physicochemical properties when peanut shell biochar is used. Similar to the above, applying 23.2 and 116.1 t C ha-1 of mango wood biochar increased P availability in soils by 163 and 208%, respectively, but reduced AMF abundance by 43 and 77% [73]. When compared to mycorrhizal fungi and high N applications, biochar, mycorrhizal fungi, and high N decreased above-ground plant biomass by 42% while encouraging mycorrhizal root colonization. This is proof that biochar and nitrogen are causing mycorrhizal fungus parasitism. In mycorrhizal-rich but nitrogen-deficient soils, biochar increased surface oxidation [74].

Studies [75] show that using fertilizer and biochar increases microbial biomass in comparison to mineral fertilizer. Microbial immobilization is a key mechanism for retaining nitrogen in soils that have been affected by leaching [76]. Microbial activity is stimulated by increased C availability, which leads to increased N demand, promoting $NO_3^-$ immobilization and recycling. Although there was no evidence of higher soil respiration rate, after adding glucose to biochar-modified soils, microbial growth rates increased, indicating low levels of biodegradable SOM but adequate soil nutrients to support microbial population growth [75]. The production of biochar increased crop yield, soil microbial biomass, plant tissue K concentration, total soil C and N, soil P and K, and nodulation with beans, red clover, soybeans, broad bean, and BNF [77]. Numerous researchers have looked into how biochar affects the structure of the soil and the activity of the microorganisms (Table 14.1).

## Increasing the Organic Matter Content in Quality Soil

The production of biogas from decomposing animal wastes in an airless environment, obtaining energy from biogas, and using the remaining materials as organomineral fertilizers are all considered good waste management practices. Vegetable waste,

**Table 14.1** Biochar's effects on soil structure and microbial activity

| Biochar's forerunner | Temperature of pyrolysis (°C) | Distribution of particles and soil type | The impact on microbial communities and soil aggregation | References |
|---|---|---|---|---|
| Conocarpus wood | 400 | Light loamy soil characteristic of a Torriorthents | The use of 1.5 and 2% biochar increased the share of 1–0.5 mm and 0.5–0.25 mm water-resistant aggregates, while 0.5% biochar increased the share of 1–0.5 mm and 0.5–0.25 mm water-resistant aggregates | [78] |
| Corn stems | ND | Clay loamy soil, or vertisol | On aggregation and microbiological activity, between the biochar variant and the control, there were no differences | [79] |
| Corn stems | 350–500 | Fluvisol clay | Microaggregate stability has increased as a result of the interaction of biochar with clay minerals, a rise in microbial biomass, and a decline in carbon mineralization. An increase in the types of bacteria that help to keep microaggregates stable (Actinobacteria, Acidobacteria.) | [80] |

(continued)

**Table 14.1** (continued)

| Biochar's forerunner | Temperature of pyrolysis (°C) | Distribution of particles and soil type | The impact on microbial communities and soil aggregation | References |
|---|---|---|---|---|
| Manure | 500 | the soils of farms, clay loam | The proportion of macro-aggregates significantly increased when 2% biochar was added to both types of soil. The reaction was amplified when biochar was added to light loam | [81] |
| Pine (*Pinus sylvestris*) wood and fir (*Picea abies* wood | 550 | Endogleyic Stagnosol clay loamy Soil Cutanic Vertic Luvisol clay | Depending on the incubation mode, an increase in the proportion of water-resistant aggregates was observed; loam aggregated better when it was wet while clay soil aggregated better when it was dried | [82] |
| Agricultural grain waste (cereal and sunflower husks) | 600 | Silty clay loam | It seems contradictory that soil microorganisms' respiration activity did not increase as soil content and microbial biomass increased | [83] |
| Quercus Phillyraeoides, Oak wood and bamboo *Phyllostachy edulis* | 600 | Clay loamy soil, plinthustults | The application of 0.5 and 1% biochar resulted in a significant increase in the aggregates' water resistance. Increased beta-glucosidase and dehydrogenase activity and microbial biomass | [84] |

(continued)

**Table 14.1** (continued)

| Biochar's forerunner | Temperature of pyrolysis (°C) | Distribution of particles and soil type | The impact on microbial communities and soil aggregation | References |
|---|---|---|---|---|
| *Leucaena leucocephala* (Lam De Wit) wood | 700 | Clay loamy soil and typical Paleudults | The aggregates' stability is significantly improved after 63 days of 5% biochar incubation. At the same time, the carbon content of microbial biomass increases | [85] |

food waste, and organic waste from cities; In parks and gardens, tree leaves, pruning wastes, mowed grass wastes, greenhouse production wastes, plant stems and residues from which fruit or seeds have been removed, spoiled feed, straw, and silage wastes can all be composted and applied to the soil. Approximately 12.8 million tons of organic waste are released after plant production, with the majority of it being wasted. With the organic character of half of this discarded material, high-quality compost can be made.

Some sources that can be used to increase the organic matter content of our soils include stoops, animal manures, vegetable and food industry wastes, wastewater treatment sludge, green manures, leonardite, and biochar. Stubble is one of the most significant sources of organic matter in the soil. The most prevalent sources of organic matter are animal wastes from animals like cattle, sheep, and poultry. They are an important source of nutrients in addition to being organic matter. After a largely anaerobic and long-term decomposition process, animal wastes are typically turned into uncontrolled heaps and used as organic fertilizers in agricultural areas. Uncontrolled anaerobic heaps release methane, one of the greenhouse gases blamed for global warming, plant nutrients are lost due to long-term decomposition and washing, and microbiological disinfection is impossible. As a result, it loses its effectiveness as a fertilizer and pollutes the environment. A significant amount of nutrients are lost due to washing and evaporation as a result of these wastes being stored in unsuitable conditions or applied to the land at random, and the expected benefit in terms of agricultural production and soil fertility cannot be fully realized. Furthermore, it has the potential to pollute both surface and underground water resources.

A significant amount of economic gain will be achieved by applying compost made from plant and city wastes to soils, increasing the soil organic matter content, besides protecting or improving soil health. This is because the soil's increased ability to hold water will result in water savings, and the nutrients it contains will result in less need for chemical fertilizers. Plant nutrients such as sulfur (S), potassium (K), phosphorus (P), nitrogen (N), zinc (Zn), humic-fulvic acid, and compost-derived organic matter are found together and used as a base fertilizer in organomineral fertilizers. Organomineral fertilizers, which are made as "organic matter+mineral fertilizer" by combining the beneficial effects of organic materials on soil fertility, reduce nutrient loss through washing while also increasing the efficiency of the minerals used by improving the soil's fertility elements.

Waste water treatment sludges; As public awareness of environmental issues grows, the amount of treatment sludge left over from waste water treatment in treatment plants, which are now required to be built and operated, is gradually increasing. Today, it is critical to dispose of treatment sludge in an environmentally responsible manner in order to protect the natural environment and ensure its long-term viability. When organic resources are in short supply and these resources are scarce, wastewater treatment sludge appears to be a viable alternative. Due to the organic matter it contains, sewage sludge initially increases the soils' low levels of organic matter. Furthermore, it is considered to be a suitable material for enhancing plant growth and soil fertility due to the presence of some nutrients, particularly N and P. In addition to plant nutrients, sewage sludge contains toxic elements, pathogenic microorganisms, and parasitic organism eggs, the contents of which vary depending on the characteristics of the wastewater and the processes used to obtain it. As a result, the chemical properties of the treatment sludge must be determined before it is applied to the soil. The total and useful N and P content of the sludge must be taken into account, especially when determining which soils to apply it to and at what rate. Depending on the sludge properties, there are some drawbacks and limitations to applying sewage sludge to soils. In order to avoid negative effects on soil ecosystems, the environment, and human health, irregular and uncontrolled use of sewage sludge should be avoided, and multi-year trials on the use of sewage sludge in soils should be conducted [86].

Green Fertilization: Plants can get the nitrogen they need by adding mineral fertilizers to the soil or by bacteria binding atmospheric nitrogen to the soil. $N_2$ gas, which accounts for 78% of the atmosphere, provides no direct benefit to plants or microorganisms. Some microorganisms, on the other hand, bind free nitrogen gas in the atmosphere and convert it to ammonia, which plants can use. Biological nitrogen fixation is the term for this phenomenon. Natural nitrogen-fixing microorganisms, particularly *Rhizobium spp.* The value of biological nitrogen fixation, which is realized through bacterial symbiosis with leguminous plants, is growing by the day. Microorganisms that play a role in nitrogen fixation reduce mineral nitrogen input, providing nitrogen to the soil at a lower cost while also reducing the problems that mineral nitrogen can cause. Legumes fix 70–300 kg of nitrogen per hectare each year in the soil. Forage crops such as clover, vetch, and clover attract attention as prominent plants in green manure. In 3–5-year crop rotation systems with clover species in arid areas with low organic matter, where the grain-fallow system is used, the ratio of nitrogen and organic matter in the soil increases from year to year [86].

Leonardite is a stratified clayey organic sedimentary rock that formed in prehistoric times as a result of the decomposition, humification, oxidation, and metamorphosis of plant and animal remains in aquatic environments such as lakes and swamps, as well as the influence of volcanic movements under pressure, temperature, and anaerobic conditions. It contains a lot of humic acid, and every lignite deposit could be a source of leonardite. One liter of liquid humic acid is said to be equivalent to 8 tons of animal manure, while one kilogram of solid humic acid is said to be equivalent to 30 tons of animal manure. Because of this, leonardite is often used in organic farming to improve the soil and add nutrients.

Organic materials formed by barn manure, compost, and mulching to increase soil organic matter are mineralized over time, depending on soil cultivation, climate characteristics such as temperature and precipitation, and microorganism activity. As a result, it has been suggested in recent years that instead of composting, plant material be charred (biochar) and used as an organic carbon source in agriculture. Biochar has a high stable carbon content, a high cation exchange capacity, increases the water holding capacity of the soil due to its structure, and provides a good habitat for microbial organisms. There are studies that show how to increase corn, wheat, and alfalfa yields. When combined with compost, biochar is said to increase microbial activity and productivity, decrease nitrogen oxide ($N_2O$) and methane ($CH_4$)

emissions from the soil, and increase the soil's ability to store carbon. However, because the use of biochar in agricultural applications is still relatively new, plant materials are carbonized using a variety of methods, ranging from traditional charcoal production to a stove system and advanced automatic control systems, and the material produced varies accordingly. So, before biochar can be used in agricultural soils, long-term gravel trials need to be done to look at the types of materials made, the soil, and the crop yield [86].

## Critical Symbiosis Mechanisms in Soil

PAHs, which are widely distributed in various types of soils, contribute to biochar's contribution to stable organic pollutant degradation. The problem with PAHs generated during the production of biochar and their impact on microbial communities [87]. There are two ways that biochar can be used to destroy organic pollutants. The activation of the natural microbial community is the first, and the second is the application of biochar to stimulate the soil self-cleaning process. Biochar application promotes the destruction of PAHs in soils contaminated by PAHs and HM due to HM adsorption [88]. Biochar made from sawdust and wheat straw has been shown to significantly reduce PAH levels in soils contaminated with petroleum products in studies. In comparison to biochar produced at a lower temperature (300 °C), biochar produced at a high pyrolysis temperature (500 °C) is more efficient. It also doesn't appear to reduce the PAH content of the initial raw material. In general, for soil remediation, biochar produced at higher temperatures has a higher sorption efficiency for organic pollutants. The high surface area and microporosity of biochars are most likely to blame for this [89]. The increased surface area of biochar is likely caused in large part by the pore size distribution, according to the positive relationship between micropore volume and surface area [90]. Additionally, the composition of soil microbial communities has significantly changed as a result of the rising proportion of PAH-degrading taxa [91]. The biochar from bamboo (*Bambusa vulgaris*) showed decreased PAH bioavailability due to adsorption when heated to 700 °C, whereas the biochar from corn stalks (*Zea mays*) induced PAH degradation activation using microbial communities, the study cautions. The percentage of bacteria from the *Arthrobacter* and *Flavobacterium* genera has increased concurrently, as have the genes linked to PAH degradation [92]. The same group obtained similar results in rice field soils [93]. In the subsequent interaction with microorganism exterminators, the type of biochar (feedstock and pyrolysis conditions) can now take center stage. Biochar with a low pyrolysis temperature will interact closely with PAHs, making them inaccessible to both plants and microorganisms. It affects things. The best biochars for absorbing non-polar pollutants are said to be those made at high pyrolysis temperatures because of their high structural aromaticity [90]. The second strategy is to use biochar as a delivery system for microbially destructive strains. Biochar is used to immobilize Pseudomonas putida, which has different starting raw materials and pyrolysis temperatures. As a result, the degradation of PAHs is significantly increased. PAHs with 4–5 rings have a stronger effect than those with 3 and 6 rings [94]. *Mycobacterium gilvum* immobilized on rice straw biochar led to a significantly higher rate of phenanthrene, floranthene, and pyrene degradation in soil that had previously been contaminated with PAHs when compared to biochar application alone or bacterial biomass alone [95]. One proposed mechanism is PAH adsorption on biochar followed by immobilized bacteria biodegradation. In cases of alginate adsorbed on biochar, the use of retained microbial consortia effectively reduces soil toxicity in combined pollution with Cr (VI) salt and pyrene [96]. Unknown microorganism-biochar interactions may contribute to the breakdown of persistent organic pollutants. In addition to sorption, it also entails the chemical interaction of pollutants with biochar, which can make the biochar more available to bacterial destroyers. Most likely, mechanisms involving free radicals are involved in this interaction [97]. PFR on the biochar surface, in particular, has been shown to interact with hydrogen peroxide, resulting in the formation of hydroxyl radicals that decompose 2-chlorobiphenyl [97]. Chlorobenzene degradation in rice husk obtained at 550 °C has been demonstrated to be caused by the oxygen-active form formation mechanism [98]. Additionally, information on the function of biochar PFR in the p-degradation nitrophenol's was discovered in tests carried out without the aid of microorganisms [99]. However, it is thought to be a strong possibility that these mechanisms also exist in nature. During the incubation of the PAH-reducing strain Achromobacter xylosoxidans with various PAHs, information is also available regarding the accumulation of hydrogen peroxide in the medium [100].

Further interaction between hydrogen peroxide and biochar PFR can lead to the formation of hydroxyl radicals, which can destroy PAH aromatic structures and increase the availability of the material for bacterial transformation. Both the characteristics of the carbon sorbent and the characteristics of the soil in which it is incorporated have an impact on the complicated system that governs the interaction of microorganisms with biochar. This system becomes even more complicated when heavy metals and polyaromatic hydrocarbons are present in the soil. The system's components all interact with one another simultaneously through a variety of physicochemical and biological processes. Figure 14.1. depicts the interaction of microorganisms and soil with biochar.

## Conclusions and Future Perspectives

Biochar has a variety of effects on soil properties (chemical, biological and physical). Increased soil pH and buffering capacity from biochar improve acidic soils. In order to do so, you'll need to know the pH and salinity of the biochar you're using in acidic soils. In fine-grained soils, biochar can improve infiltration and hydraulic conductivity. In addition, it appears that biochar has a stronger impact on hydraulic conductivity in coarse-textured soils than in fine-textured soils. By adding biochar to the surface, you can enhance particle transport by both water and wind (dust).

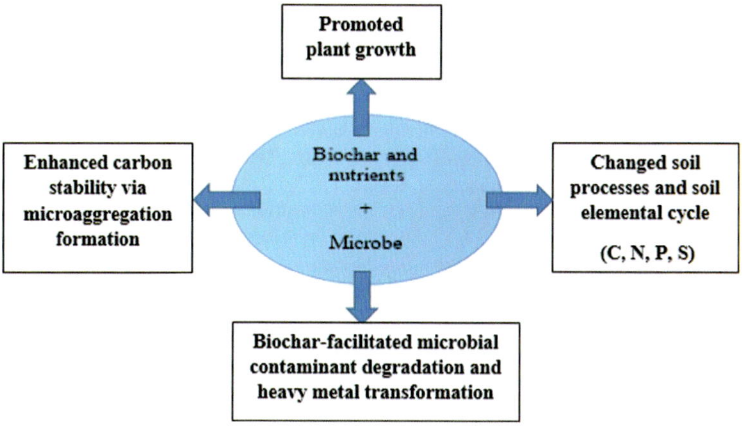


**Fig. 14.1** The relationship between biochar and soil-based microorganisms [101]

Factors such as the characteristics of the soil and biochar, the type of crop, and any potential costs all play a role in determining the rate at which biochar will be mixed into the soil. Use of biochar as an environmentally friendly sorbent for soil immobilization and agricultural soil improvement is critical. Pyrolysis conditions, biochar precursors, and soil properties all have an impact on its utility. The pH changes caused by biochar application, as well as the potential toxicity of biochar due to volatile pyrolysis product emissions, have a big impact on the way soil microbial communities work and what they eat. The complexities of biochar's effect on soil, as well as differences in soil properties, can lead to conflicting data, making it difficult to compare experiment results. Biochar's action on soil and microorganisms requires more research. This necessitates the development of a single biochar test model with a set of parameters to investigate.

## References


1. Steffan JJ, Brevik EC, Burgess LC, Cerda A (2018) The effect of soil on human health: an overview. Eur J Soil Sci 69(1):159–171. https://doi.org/10.1111/ejss.12451
2. Orgiazzi A, Bardgett RD, Barrios E, Behan-Pelletier V, Briones MJI, Chotte JL, De Deyn GB, Eggleton P, Fierer N, Fraser T, Hedlund K, Jeffrey S, Johnson NC, Jones A, Kandeler E, Ortas I (2018) Organomineral fertilizer workshop, papers, 1st edn, May, Istanbul. ISBN: 978-975-7169-89-5
3. Ipek M, Eşitken A (2017) The actions of PGPR on micronutrient availability in soil and plant under calcareous soil conditions: an evaluation over Fe nutrition. Plant-microbe interactions in agro-ecological perspectives, pp 81–100. https://doi.org/10.1007/978-981-10-6593-4_4
4. Jacoby R, Peukert M, Succurro A, Koprivova A, Kopriva S (2017) The role of soil microorganisms in plant mineral nutrition—current knowledge and future directions. Front Plant Sci 8:1617. https://doi.org/10.3389/fpls.2017.01617

5. Bever JD (2003) Soil community feedback and the coexistence of competitors: conceptual frameworks and empirical tests. New Phytol 157(3):465–473. https://doi.org/10.1046/j.1469-8137.2003.00714.x
6. Gupta G, Parihar SS, Ahirwar NK, Snehi SK, Singh V (2015) Plant growth promoting rhizobacteria (PGPR): current and future prospects for development of sustainable agriculture. J Microb Biochem Technol 7:096–102. https://doi.org/10.4172/1948-5948.1000188
7. Prasad M, Chaudhary M, Choudhary M, Kumar TK, Kumar Jat L (2017) Rhizosphere microorganisms towards soil sustainability and nutrient acquisition. agriculturally important microbes for sustainable agriculture, pp 31–49. https://doi.org/10.1007/978-981-10-5589-8_2
8. Bais HP, Weir TL, Perry LG, Gilroy S, Vivanco JM (2006) The role of root exudates in rhizosphere interactions with plants and other organisms. Annu Rev Plant Biol 57(1):233–266. https://doi.org/10.1146/annurev.arplant.57.032905.105159
9. Vacheron J, Desbrosses G, Bouffaud ML, Touraine B, Moënne-Loccoz Y, Legendre L, Wisniewski-Dyé F, Prigent-Combaret C (2013) Plant growth-promoting rhizobacteria and root system functioning. Front Plant Sci 4:356. https://doi.org/10.3389/fpls.2013.00356
10. Bulgarelli D, Schlaeppi K, Spaepen S, van Themaat EVL, Schulze-Lefert P (2013) Structure and functions of the bacterial microbiota of plants. Annu Rev Plant Biol 64(1):807–838. https://doi.org/10.1146/annurevarplant-050312-120106
11. Buée M, De Boer W, Martin F, Van OL, Jurkevitch E (2009) The rhizosphere zoo: an overview of plant-associated communities of microorganisms, including phages, bacteria, archaea, and fungi, and of some of their structuring factors. Plant Soil 321(1–2):189–212. https://doi.org/10.1007/s11104-009-9991-3
12. Akhtar N, Qureshi MA, Iqbal A, Ahmad MJ, Khan KH (2012) Influence of Azotobacter and IAA on symbiotic performance of Rhizobium and yield parameters of lentil. J Agric Res 50(3):361–372
13. Alma MH, Altıkat A (2021) Biochar and soil physical properties. J Inst Sci Technol 11(4):2599–2612
14. Shearer D, Gaunt J, Peacocke GVC (2013) US patent no 8,361,186. US Patent and Trademark Office, Washington, DC
15. McKendry P (2002) Energy production from biomass (part 1), Overview of biomass. Biores Technol 83(1):37–46
16. Antal MJ, Grønli M (2003) The art, science, and technology of charcoal production. Ind Eng Chem Res 42(8):1619–1640
17. Enders A, Hanley K, Whitman T, Joseph S, Lehmann J (2012) Characterization of biochars to evaluate recalcitrance and agronomic performance. Biores Technol 114:644–653
18. Mohan D, Pittman CU, Steele PH (2006) Pyrolysis of wood/biomass for bio-oil, a critical review. Energy Fuels 20(3):848–889
19. Duku MH, Gu S, Ben Hagan E (2011) Biochar production potential in Ghana—a review. Renew Sustain Energy Rev 15:3539–3551
20. Lehmann J, Da Silva JP, Steiner C, Nehls T, Zech W, Glaser B (2003) Nutrient availability and leaching in an archaeological Anthrosol and a Ferralsol of the Central Amazon basin: fertilizer, manure and charcoal amendments. Plant Soil 249:343–357
21. Karer J, Wımmer B, Zehetner F, Kloss S, Soja G (2013) Biochar application to temperate soils: effects on nutrient uptake and crop yield under field conditions. Agr Food Sci 22:390–403
22. Laird DA (2008) The charcoal vision: a win-win-win scenario for simultaneously producing bioenergy, permanently sequestering carbon, while improving soil and water quality. Agron J 100:178–181
23. Case SDC, Mcnamara NP, Reay DS, Whitaker J (2012) The effect of biochar addition on $N_2O$ and $CO_2$ emissions from a sandy loam soil—the role of soil aeration. Soil Biol Biochem 51:125–134
24. Singh BP, Hatton BJ, Singh B, Cowie AL, Kathuria A (2010) Influence of biochars on nitrous oxide emission and nitrogen leaching from two contrasting soils. J Environ Qual 39:1224–1235

25. Rillig MC, Wagner M, Salem M, Antunes PM, George C, Ramke HG, Tıtırıcı MM, Antonietti M (2010) Material derived from hydrothermal carbonization: effects on plant growth and arbuscular mycorrhiza. Appl Soil Ecol 45:238–242
26. Glaser B, Lehmann J, Zech W (2002) Ameliorating physical and chemical properties of highly weathered soils in the tropics with charcoal—a review. Biol Fertil Soils 35:219–230
27. Chan KY, Van Zwieten L, Meszaros I, Downie A, Joseph S (2007) Agronomic values of greenwaste biochar as a soil amendment. Aust J Soil Res 45:629–634
28. Bayer C, Martin-Neto L, Mielniczuk J, Pillon CN, Sangoi L (2001) Changes in soil organic matter fractions under subtropical no-till cropping systems. Soil Sci Soc Am J 65:1473–1478
29. Lal R, Mahboubi AA, Fausey NR (1994) Long-term tillage and rotation effects on properties of a central Ohio soil. Soil Sci Soc Am J 58:517–522
30. Unger PW (1997) Aggregate and organic carbon concentration interrelationships of a Torrertic Paleustoll. Soil Tillage Res 42:95–113
31. Suliman W, Harsh JB, Abu-Lail NI, Fortuna AM, Dallmeyer I, Garcia-Pérez M (2017) The role of biochar porosity and surface functionality in augmenting hydrologic properties of a sandy soil. Sci Total Environ 574:139–147
32. Abel S, Peters A, Trinks S, Schonsky H, Facklam M, Wessolek G (2013) Impact of biochar and hydrochar addition on water retention and water repellency of sandy soil. Geoderma 202–203:183–191
33. Oguntunde PG, Abiodun BJ, Ajayi AE, van de Giesen N (2008) Effects of charcoal production on soil physical properties in Ghana. J Plant Nutrit Soil Sci 171:591–596
34. Sombroek W, Ruivo MDL, Fearnside PM, Glaser B, Lehmann J (2003) Amazonian Dark Earths as carbon stores and sinks, Amazonian Dark Earths. Springer, pp 125–139
35. Singht KK, Colvin TS, Erbach DC, Mughal AQ (1992) Tilth index: an approach to quantifying soil tilth. Trans ASAE 35(6):1777–1785
36. McVay KA, Budde JA, Fabrizzi K, Mikha MM, Rice CW, Schlegel AJ, Peterson DE, Sweeney DW, Thompson C (2006) Management effects on soil physical properties in long-term tillage studies in Kansas. Soil Sci Soc Am J 70:434–438
37. Dexter AR (2004) Soil physical quality: Part II. Friability, tillage, tilth and hard-setting. Geoderma 120:215–225
38. Craul PJ (1999) Urban soils: applications and practices. Wiley, Toronto
39. Janzen HH, Campbell CA, Ellert BH, Bremer E (1997) Soil organic matter dynamics and their relationship to soil quality. Dev Soil Sci 25:277–292
40. Rolda´n A, Caravaca F, Herna´ndez MT, Garcı´a C, Sa´nchez-Brito C, Vela´squez M, Tiscaren˜o M (2003) No-tillage, crop residue additions, and legume cover cropping effects on soil quality characteristics under maize in Patzcuaro watershed (Mexico). Soil Tillage Res 72:65–73
41. Lal R, Kimble J, Follett RF (1998) Need for research and need for action. CRC Press, Boca Raton, FL, pp 447–454
42. Campbell CA, Mc Conkey BG, Bierderbeck VO, Zenner RP, Curtin D, Peru MR (1998) Long-term effects of tillage and fallow-frequency on soil quality attributes in a clay soil in semiarid southwestern Saskatchewan. Soil Tillage Res 46:135–144
43. Feng Y, Motta AC, Reeves DW, Burmester CH, Van S, Osborne JA (2003) Soil microbial communities under conventional-till and no-till continuous cotton systems. Soil Biol Biochem 35:1693–1703
44. Ghuman BS, Sur HS (2001) Tillage and residue management effects on soil properties and yields of rainfed maize and wheat in a subhumid subtropical climate. Soil Tillage Res 58:1–10
45. Fageria N, Baligar V (2008) Ameliorating soil acidity of tropical Oxisols by liming for sustainable crop production. Adv Agron 99:345–399
46. Steiner C, Teixeira WG, Lehmann J, Nehls T, de Macêdo JLV, Blum WE, Zech W (2007) Long term effects of manure, charcoal and mineral fertilization on crop production and fertility on a highly weathered Central Amazonian upland soil. Plant Soil 291:275–290

47. Lashari MS, Liu Y, Li L, Pan W, Fu J, Pan G, Zheng J, Zheng J, Zhang X, Yu X (2013) Effects of amendment of biochar-manure compost in conjunction with pyroligneous solution on soil quality and wheat yield of a salt-stressed cropland from Central China Great Plain. Field Crops Res 144:113–118
48. Joseph S, Peacocke C, Lehmann J, Munroe P (2009) Developing a biochar classification and test methods. Biochar Environ Manage Sci Technol 107–126
49. Cheng C-H, Lehmann J (2009) Ageing of black carbon along a temperature gradient. Chemosphere 75:1021–1027
50. Lehmann J, Joseph S (2015) Biochar for environmental management: science, technology and implementation, 2nd edn. Routledge, London and New York
51. Mukome FN, Kilcoyne AL, Parikh SJ (2014) Alteration of biochar carbon chemistry during soil incubations: SR-FTIR and NEXAFS investigation. Soil Sci Soc Am J 78:1632–1640
52. Cheng CH, Lehmann J, Engelhard MH (2008) Natural oxidation of black carbon in soils: changes in molecular forms and surface charge along a climosequence. Geochemica et Cosmochimica Acta 72:1598–1610
53. Li M, Liu Q, Lou Z, Wang Y, Zhang Y, Qian G (2014) Method to characterize acidbase behaviour of biochar: site modelling and theoretical simulation. Sustain Chem Eng 2:2501–2509
54. Silber A, Levkovitch I, Graber ER (2010) pH-dependent mineral release and surface properties of cornstraw biochar: agronomic implications. Environ Sci Technol 44:19318–19323
55. Subedi R, Taupe N, Pelissetti S, Petruzzelli L, Bertora C, Leahy JJ, Grignani C (2016) Greenhouse gas emissions and soil properties following amendment with manure-derived biochars: influence of pyrolysis temperature and feedstock type. J Environ Manage 166:73–83
56. Agegnehu G, Bird M, Nelson P, Bass A (2015) The ameliorating effects of biochar and compost on soil quality and plant growth on a Ferralsol. Soil Res 53:1–12
57. Alburquerque JA, Calero JM, Barrón V, Torrent J, del Campillo MC, Gallardo A, Villar R (2014) Effects of biochars produced from different feedstocks on soil properties and sunflower growth. J Plant Nutrit Soil Sci 177:16–25
58. DeLuca TH, MacKenzie MD, Gundale MJ (2009) Biochar effects on soil nutrient transformations. Biochar For Environmental Management. Earthscan publishing, London, pp 251–270
59. Buckley DH, Schmidt TM (2001) The structure of microbial communities in soil and the lasting impact of cultivation. Microb Ecol 42:11–21
60. Kladivko EJ, Akhouri NM, Weesies G (1997) Earthworm populations and species distributions under notill and conventional tillage in Indiana and Illinois. Soil Biol Biochem 29:613–615
61. Aslam T, Choudhary M, Sagar S (1999) Tillage impacts on soil microbial biomass C, N and P, earthworms and agronomy after two years of cropping following permanent pasture in New Zealand. Soil Tillage Res 51:103–111
62. Malhi SS, Lemke R, Wang ZH, Chhabra BS (2006) Tillage, nitrogen and crop residue effects on crop yield, nutrient uptake, soil quality, and greenhouse gas emissions. Soil Tillage Res 90:171–183
63. Monokrousos N, Papatheodorou E, Diamantopoulos Stamou G (2006) Soil quality variables in organically and conventionally cultivated field sites. Soil Biol Biochem 38:1282–1289
64. Marinari S, Mancinelli R, Campiglia E, Grego S (2006) Chemical and biological indicators of soil quality in organic and conventional farming systems in Central Italy. Ecol Indicat 701–711
65. Salinas-Garcia JR, Hons FM, Matocha JE, Zuberer DA (1997) Soil carbon and nitrogen dynamics as affected by long-term tillage and nitrogen fertilization. Biol Fert Soils 25:182–188
66. Lone AH, Najar GR, Ganie MA, Sofi JA, Ali T (2015) Biochar for sustainable soil health: a review of prospects and concerns. Pedosphere 25:639–653
67. Chen WF, Meng J, Han GM, Zhang WM (2013) Effect of biochar on microorganisms quantity and soil physicochemical property in rhizosphere of spinach (*Spinacia oleracea* L.). Appl Mech Mater 295:210–219

68. Lehmann J, Rillig MC, Thies JAC, Masiello CA, Hockaday WC, Crowley D (2011) Biochar effects on soil biota—a review. Soil Biol Biochem 43:1812–1836
69. Gomez J, Denef K, Stewart C, Zheng J, Cotrufo M (2014) Biochar addition rate influences soil microbial abundance and activity in temperate soils. Eur J Soil Sci 65:28–39
70. Deb D, Kloft M, Lässig J, Walsh S (2016) Variable effects of biochar and P solubilizing microbes on crop productivity in different soil conditions. Agroecol Sustain Food Syst 40:145–168
71. Thies JE, Rillig MC (2009) Characteristics of biochar: biological properties. In: Lehmann J, Joseph S (eds) Biochar for environmental management: science and technology. Earthscan, London, UK, pp 85–105
72. Abujabhah IS, Bound SA, Doyle R, Bowman JP (2016) Effects of biochar and compost amendments on soil physico-chemical properties and the total community within a temperate agricultural soil. Appl Soil Ecol 98:243–253
73. Warnock DD, Mummey DL, McBride B, Major J, Lehmann J, Rillig MC (2010) Influences of non-herbaceous biochar on arbuscular mycorrhizal fungal abundances in roots and soils: results from growth-chamber and field experiments. Appl Soil Biol 46:450–456
74. LeCroy C, Masiello CA, Rudgers JA, Hockaday WC, Silberg JJ (2013) Nitrogen, biochar, and mycorrhizae: alteration of the symbiosis and oxidation of the char surface. Soil Biol Biochem 58:248–254
75. Birk J, Steiner C, Teixiera W, Zech W, Glaser B (2009) Microbial response to charcoal amendments and fertilization of a highly weathered tropical soil Amazonian dark earths. Wim Sombroek's Vision, Springer, pp 309–324
76. Burger M, Jackson LE (2003) Microbial immobilization of ammonium and nitrate in relation to ammonification and nitrification rates in organic and conventional cropping systems. Soil Biol Biochem 35:29–36
77. Biederman LA, Harpole SW (2013) Biochar and its effects on plant productivity and nutrient cycling: a meta-analysis. GCB Bioenergy 5:202–214
78. Ibrahim HM, Al-Wabel MI, Usman AR, Al-Omran A (2013) Effect of Conocarpus biochar application on the hydraulic properties of a sandy loam soil. Soil Sci 178(4):165–173
79. Rahman MT, Zhu QH, Zhang ZB, Zhou H, Peng X (2017) The roles of organic amendments and microbial community in the improvement of soil structure of a Vertisol. Appl Soil Ecol 111:84–93
80. Zheng H, Wang X, Luo X, Wang Z, Xing B (2018) Biochar-induced negative carbon mineralization priming effects in a coastal wetland soil: Roles of soil aggregation and microbial modulation. Sci Total Environ 610:951–960
81. Ouyang L, Wang F, Tang J, Yu L, Zhang R (2013) Effects of biochar amendment on soil aggregates and hydraulic properties. J Soil Sci Plant Nutr 13(4):991–1002
82. Soinne H, Hovi J, Tammeorg P, Turtola E (2014) Effect of biochar on phosphorus sorption and clay soil aggregate stability. Geoderma 219:162–167
83. Brtnicky M, Hammerschmiedt T, Elbl J, Kintl A, Skulcova L, Radziemska M, Latal O, Baltazar T, Kobzova E, Holatko J (2021) The potential of biochar made from agricultural residues to increase soil fertility and microbial activity: impacts on soils with varying sand content. Agronomy 11(6):1174
84. Demisie W, Liu Z, Zhang M (2014) Effect of biochar on carbon fractions and enzyme activity of red soil. CATENA 121:214–221
85. Jien SH, Wang CS (2013) Effects of biochar on soil properties and erosion potential in a highly weathered soil. CATENA 110:225–233
86. Ortas I (2018) Organomineral fertilizer workshop, papers, 1st edn, May, Istanbul. ISBN: 978-975-7169-89-5
87. Kołtowski M, Oleszczuk P (2015) Toxicity of biochars after polycyclic aromatic hydrocarbons removal by thermal treatment. Ecol Eng 75:79–85
88. Wawra A, Friesl-Hanl W, Puschenreiter M, Soja G, Reichenauer T, Roithner C et al (2018) Degradation of polycyclic aromatic hydrocarbons in a mixed contaminated soil supported by phytostabilisation, organic and inorganic soil additives. Sci Total Environ 628:1287–1295

89. Uchimiya M, Wartelle LH, Lima IM, Klasson KT (2010) Sorption of deisopropylatrazine on broiler litter biochars. J Agric Food Chem 58:12350–12356
90. Ahmad M, Rajapaksha AU, Lim JE, Zhang M, Bolan N, Mohan D et al (2014) Biochar as a sorbent for contaminant management in soil and water: a review. Chemosphere 99:19–33
91. Kong L, Gao Y, Zhou Q, Zhao X, Sun Z (2018) Biochar accelerates PAHs biodegradation in petroleum-polluted soil by biostimulation strategy. J Hazard Mater 343:276–284
92. Ni N, Song Y, Shi R, Liu Z, Bian Y, Wang F et al (2017) Biochar reduces the bioaccumulation of PAHs from soil to carrot (Daucus carota L.) in the rhizosphere: a mechanism study. Sci Total Environ 601:1015–1023
93. Ni N, Wang F, Song Y, Bian Y, Shi R, Yang X et al (2018) Mechanisms of biochar reducing the bioaccumulation of PAHs in rice from soil: degradation stimulation vs immobilization. Chemosphere 196:288–296
94. Chen B, Yuan M, Qian L (2012) Enhanced bioremediation of PAH-contaminated soil by immobilized bacteria with plant residue and biochar as carriers. J Soils Sediments 12(9):1350–1359
95. Xiong B, Zhang Y, Hou Y, Arp HPH, Reid BJ, Cai C (2017) Enhanced biodegradation of PAHs in historically contaminated soil by M. gilvum inoculated biochar. Chemosphere 182:316–324
96. Wang C, Chen S, Wu L, Zhang F, Cui J (2018) Wheat straw-derived biochar enhanced nitrification in a calcareous clay soil. Polish J Environ Stud. https://doi.org/10.15244/pjoes/76504
97. Fang G, Gao J, Liu C, Dionysiou DD, Wang Y, Zhou D (2014) Key role of persistent free radicals in hydrogen peroxide activation by biochar: implications to organic contaminant degradation. Environ Sci Technol 48(3):1902–1910
98. Zhang Y, Xu X, Cao L, Ok YS, Cao X (2018) Characterization and quantification of electron donating capacity and its structure dependence in biochar derived from three waste biomasses. Chemosphere 211:1073–1081
99. Yang J, Pan B, Li H, Liao S, Zhang D, Wu M et al (2015) Degradation of p-nitrophenol on biochars: role of persistent free radicals. Environ Sci Technol 50(2):694–700
100. Sazykin IS, Sazykina MA, Khmelevtsova LE, Seliverstova EY, Karchava KS, Zhuravleva MV (2018) Antioxidant enzymes and reactive oxygen species level of the Achromobacter xylosoxidans bacteria during hydrocarbons biotransformation. Arch Microbiol 200:1057–1065
101. Gorovtsov AV, Minkina TM, Mandzhieva SS, Perelomov LV, Soja G, Zamulina IV, Rajput VD, Sushkova SN, Mohan D, Yao J (2020) The mechanisms of biochar interactions with microorganisms in soil. Environ Geochem Health 42(8):2495–2518. https://doi.org/10.1007/s10653-019-00412-5

# Chapter 15
# Symbiosis Mechanisms and Usage of Other Additives Like Biochar in Soil Quality Management

**Soheila Aghaei Dargiri and Ali Movahedi**

**Abstract** Major improvements in farm management are required to establish further stable industry systems and strengthen poor regional economies. In global agriculture, soil deterioration, including decreased fecundity and enhanced deterioration, is a serious worry. The impact of biochar on soil microbial populations is closely tied to agricultural food production. The complex interactions between plant roots and microorganisms take place in the plant rhizosphere. Biochar has the potential to be a new and valuable fertilizer, either directly or indirectly. This is because of their low fertility and the environmental and economic benefits they provide. In addition, previous studies/meta-analyses synthesized only microbial community responses to biochar based mainly on traditional techniques (such as PLFA and DGGE). With the rapid development of analytical methods (e.g., high throughput sequencing), in this study, we can examine the diversity and abundance of microorganisms with higher classification accuracy (such as bacteria and fungi) in biochar-modified soils. Conditions or has the potential for targeted soil management. Although there is growing interest in utilizing biochar for soil management, some studies have found detrimental effects. There are still several research gaps and ambiguities to be addressed in this chapter. In future research, further relevant investigations, particularly long-term tests, will be required to close these information gaps.

## Knowledge Objectives

1. The accurate service life of biochar is yet sometimes understood. We must fee rather a consideration to the decomposition rate of biochars in soil. Thus, we can choose biochar correctly and administer resources suitably.

S. A. Dargiri (✉)
Department of Horticulture, Faculty of Agriculture and Natural Resources, University of Hormozgan, Bandar Abbas, Iran
e-mail: S.aghaei.phD@hormozgan.ac.ir

A. Movahedi
Co-Innovation Center for Sustainable Forestry in Southern China, Ministry of Education, College of Biology and the Environment, Nanjing Forestry University, Nanjing 210037, China
e-mail: ali_movahedi@njfu.edu.cn

 
J. A. Parray (ed.), *Climate Change and Microbiome Dynamics*, Climate Change Management, https://doi.org/10.1007/978-3-031-21079-2_15

2. Comprehensive the interaction systems among biochar and soil microbes to disclose the systems of heterogeneous impacts of biochar on soil improvement.
3. The effect of biochar on the functional ecology of microorganisms and its effects on soil were investigated.

## Introduction

The requirements to expand rather supportable agriculture mechanisms and cure faint village economies necessitate the main alternation in agriculture management. Soil degradation, which contains reduced fertility and enhanced erosion, is relevant in global agriculture [1]. The world population is expected to reach 8 billion people by 2024 [2], so food security and the distribution of human carbon dioxide ($CO_2$) will be significant issues in sustainable human progress [3]. Biochar generation from agricultural remains has the possibility of reducing both problems at the long time. Pyrolysis in the shortage of oxygen in organic substances [4] creates a yield with a high value of turbulent carbon, which has a long lifetime in soil [5]. Biochar modification of soils is as well as probably a strategy for enhancing plant efficiency [6–10], which maybe represent other requirements for the achievement and extension of the technology. Biochar has many permeable physical structures, which enhance the maintenance of soil humidity and nutrients [6, 9]. In addition, its main section of C, biochar as well as includes hydrogen (H), oxygen (O), magnesium (Mg), and macronutrients such as N, phosphorus (P), and potassium (K) that can enhance crop manufacturing for most crops around the world [11–16]. Biochar has added vital interest over the last two decades because of its possibilities as a C analysis, bioremediation, soil fertility, wastewater, and general environmental administration mechanism in agriculture [17]. Biochar addition in the soil has shown useful results in increasing nutrient persistence, giving refuge to microorganisms, enhancing soil structure, and increasing the attraction of nutrients by plants, which eventually resulted in improvements in plant development and product [18, 19].

## What is Symbiosis?

Symbiosis is a phenomenon in which two or more organisms with distinct genealogical histories live in close association with each other [20]. In the last decades, symbiosis, ‘the living together of unlike organisms’ [20], has moved from the outskirts of biology to a central location. The phenomenon is now regarded as a ubiquitous ecological power and main driver of progress among the tree of life [21, 22]. Possibly the maximum joint symbioses are those among multicellular eukaryotes and microorganisms, containing bacteria, fungi, protozoa, and even viruses. Insects are the maximum varied and plentiful animals in earthly ecosystems and, owing to their numerical advantage, forsooth busy in the maximum microbial symbioses. While all insects encode endogenous systems expanding resource inception (e.g., digestion, nutrition, and detoxification), and position their own systems for replication, their inhabitant microbiota have been mostly co-opted to support these functions and to, sometimes, confer fully new property [23].

Plenty of specialized microbes construct their living through changes to host insect fitness. Universal symbionts have been shown to administer insect breeding and alter sex ratios—effects not ever to the hosts' benefit [24, 25].

## Background and Biochar Definition

Biochar is known as "black gold" [26–28]. Biochar is a recalcitrant C that reduces slowly in the soil and can take thousands of years to damage [29, 30]. Biochar is a dark carbon-rich solid made by thermal analysis of biomass under oxygen-bounded surroundings at temperatures usually between 300 and 700 °C [31–33]. In this chapter, we critically considered the impact of biochar on soil attributes, featuring soil physicochemical and biological attributes. Furthermore, the biochar systems in enhancing soil fertility were also chaptered. The instruction to further comprehend the interactions among biochar and soil, four appendix issues are subjected in which chapter (Fig. 15.1): (i) biochar as an origin of nutrients; (ii) attraction and diffusion of nutrients on biochar; (iii) the impression of biochar on attributes of soils; and (iv) the influence of biochar on biota in soil. Many studies have shown that biochar has great external areas [34], large charge densities [35], down bulk compression [36, 37], stable porous structures, and numerous organic carbon contents [38–40], which may reduce soil bulk density (SBD) and gain large tissue soil water holding capacity (SWHC) due to its large surface area [41]. Biochar is as well as known as a much important implement of environmental management [34].

Biochar is a carbon-rich crop pyrolysis organized under oxygen-confined environments and used purposely in soil used as an alternative to amend agronomic and environmental interests [4, 5, 42–47]. Similar to charcoal in key specifications containing the combination of permanent, rebellious forms of organic carbon [48], biochar is outstanding among the same substances of its predesignate application as a soil modification [49] and a long-term C storage strategy [50]. Feedstocks for biochar manufacture contain a large confine of substances such as agricultural crop and forestry residues, municipal wastes, and animal manures, among others [51, 52]. Biochars key attributes, that is up pH, porosity, particular level region, and CEC, are mainly associated with feedstock and manufacturing methods [53]. These attributes affect how the material's interacts with soil's physical, chemical, and biological elements as well as how the substance will behave in an ecosystem. [54, 55].

Biochar as a soil modification may increase soil productivity [56, 57] and maintain yield fertility [58, 59] by improving nutrient accessibility and decreasing leaching waste. This may reduce fertilizer needs [60–62] and even enhance plant nutrient provision [63]. Biochar as well as stimulates microbial activity and variety [31, 64–66]. In addition, biochar may increase oil water property valence [67–69] and

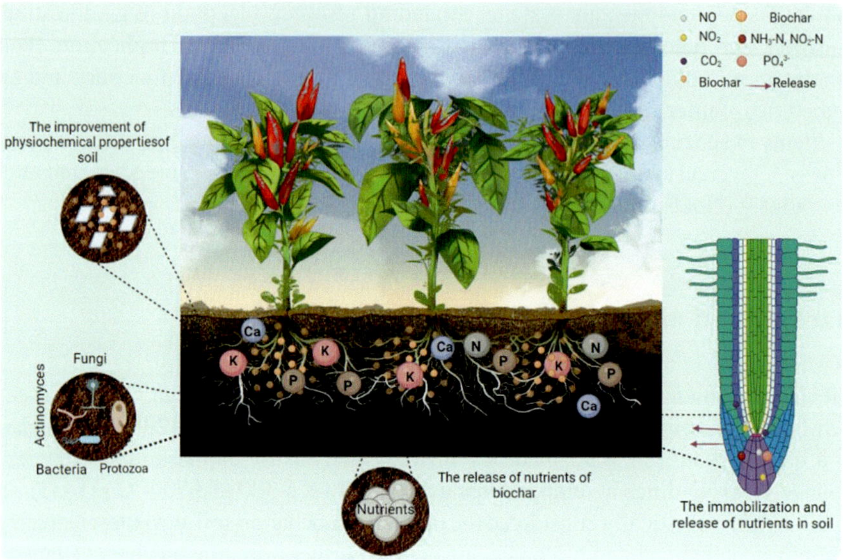


**Fig. 15.1** The probable mechanisms for progress soil fertility

decrease emissions of greenhouse gases [40, 70, 71], also control the stimulus, bioaccessibility, and toxicity of contaminants [34, 72–74]. Biochar usage as well as may enhance soil carbon analysis possible for universal warming mitigation [49, 75, 76] by carbon dioxide removal from the atmosphere. However, biochars long-period compatibility for detain C are combined with permanent and rebellious forms of organic C after plant organic material has undergone pyrolysis. Likewise, crop answers to biochar use can differences by soil kind, which can change by charcoal origin. In several instants, no useful or even harmful impacts on soil nutrient condition and Plant performance is highlighted [77].

## *Biochar Impacts on Soil Attribute*

Biochar may increase plant development by physical improvement of soil specification (bulk density, level region, water property valence, permeation [58, 68], and soil chemical specification (considerable salt, nutrient maintenance, accessibility, CEC, and pH) [78]. Besides, biochar amends soil biological attributes by enhancing variety and providing an appropriate environment for soil microbial communities [31, 79, 80]. Biochar's rebellion against chemical and biological activities supports its long-time agronomic and environmental interests' environment with a habitation period spanning hundreds to thousands of years [48, 81, 82].

## *Biochar Impacts on Plant Development and Yield Fertility*

Metanalyses show biochar use can enhance upper land plant fertility by ~10 or even ~25% [42, 83]. As described in prior parts, the improvements in plant development and crop yields with biochar use result from the amendment of physical, chemical, and biological attributes of soils. Nevertheless, the impacts of biochar use are not included useful. Jeffery [5] introduced 28–39% various in crop fertility (crop production and aboveground biomass) Below biochar modification to soils. Important crop benefits from biochar use to soils have been presented for different crop varieties in several surroundings [45]. However, as might be expected, higher yield and fertility have been observed in humid areas. Minus impacts of biochar modification on crop fertility have been introduced in peat soils.

Considerable produce crop reduction in biochar-improved soils has significantly enhanced soil C/N proportions that result in nitrogen immovability [84, 85]. The effectiveness of biochar in enhancing plant fertility is changing [83] and is impacted by climate, soil attributes, yield type, and experimental conditions [86]. Answer diversity as well as may be described by biochar feedstock and pyrolysis activities, along with the interactions that occur with soil use between biochar and the soil's Biological and non-biological components [52]. Positive yield fertility has risen mostly in a vase than in ground experiments, in acidic than neutral soils, and sandy than in loam and silt soils [5, 87].

A significant frame of investigation has examined and discovered useful impacts of biochar use on salt-impacts soils [92], which are joint in the arid area. Hammer [93] recommended that the interaction of biochar and symbiotic microorganisms would be a foundation for common handling in agricultural mechanism (p 114). While these materials' proposal promises, they point to suitable feedstock original and manufacturing, as numerous prices of several char may enhance soil salinity and sodicity [88].

## *Biochar Relationship of Microorganisms in Fertility*

Biochar has been displayed not alone to modify soil physicochemical attributes but to convert soil biological features [31, 55, 89–92]. These adjustments could enhance soil mechanism, including rising organic/mineral collection (aggregates) and bore region [93]. Increase nutrient cycles, which contain the gain of nutrient maintenance and immobilization, and the rise of nutrient reduction [66], thus promoting plant development [94]. Furthermore, microorganisms, similar rhizosphere bacteria and fungi, may comfort plant development immediately [95, 96]. Brief, conversion in microbial community combination or activity obliged by biochar can enhance nutrient terms and plant development additionally the cycling of soil organic matter [55, 97, 98].

## *Effect of Biochar on Microorganisms' Community*

There are expanding specialties in using biochar as an alternative to administering in soil biota, and low adjustments of soil biota stimulated by biochar usage are equally powerful. Several systems can illustrate how biochar could influence microorganisms in soils: (1) adjustments in nutrient accessibility; (2) additions in other microbial communities; (3) modifications in plant-microbe signaling; and (4) environment establishment and defecation from hyphal grazers. Microbial attributes are major affected by the soil food web. In addition, the trophic mechanisms of the soil food web many depend on the amount, modality, and diffusion of organic matter. Although the slow rates of manufacturing soil organic matter compared with other carbon cycle flows, its comparative resistance to microbial analysis promotes the accumulation of organic materials in soil [99, 100].

## *Effect of Biochar on Microbial Plenty*

Moreover, nutrient and carbon accessibility may impact microbial plenty. This impact varied significantly from the similar figures of biochar and the specific microorganisms group. It can be apparent that symbiotic connections with biota through altering nutrient provisions were divided from the similar demands of the plant. The effect of increasing C accumulation by important properties or root function in the rhizosphere and C as energy material for heterotrophic microorganisms has been reported. [31].

Therefore, the effect on microbial plenty was comparable with the several spheres of biochar changes containing rhizosphere and mass soil. On the other hand, under nutrient-limiting surroundings, microbial plenty can be enhanced due to the larger nutrient accessibility after biochar implementation [101]. The possible causes were biochar-driven changes in nutrient persistence or the distribution of nutrients by the biochar [31]. Several previous types of research appear to show that the appendix features may overcome the effect of nutrient and C accesses on microbial biomass, (i) the available nutrient and C accessibility in soil; (ii) the increasable extent of nutrient and C; and (iii) the attributes of microorganisms.

Microbial plenty could be enhanced after microorganism's sorb to biochar regions, which simulate them less sensitive to leaching in soil. Hydrophobic appeal, electrostatic elements, and induced expansion are included in the principal diffusion activities of biochar [102]. Furthermore, biochar, including a well-created hold structure, can supply a Microorganisms' dwelling environment. Even bacteria and fungi are considered major preserved versus predators or competitors by climbing hold habitats in biochar [103–105]. Biochar could be used to reduce toxins and chemical signals that might prevent microbial development. Pollock (1947) designated that biochar could release the development-limiting compounds.

Additionally, high-temperature biochars have been discovered to have a tougher absorption on elements that are toxic to microorganisms [106, 107]. Furthermore, moisture can affect major microbial plenty. Microorganisms would be painful in the soil of intermittent cleaning, which can enhance the torpid or even cause death [108]. Biochar has a large water supporting capacity for the large level region that could advertise the development of microorganisms. Nevertheless, major argument cannot be acquired only from the initial resources and property of biochar. There is a conjecture that bacterial cells or development-controlling elements can play a significant key in absorption.

## *Effect of Biochar on Microbial Composition and Structure*

The total of biochar can reason several modifications in microbial community structure, so trophic interactions are probably altered. Fortunately, few researchers have concentrated on the biological importance of the change in pH increased by biochar. Fortunately, some researchers have concentrated on the biological significance of the conversion in pH influenced by biochar. Sometimes, the diversity of microorganisms could be reduced or reduced after adding biochar to soil. For example, bacterial diversity was influenced by as many as 25% in biochar-rich *Terra preta* soils compared to unmodified soils in both culture-independent [90] and culture-dependent [91] studies.

Nonetheless, when compared to unaltered soils, Terra preta and a biochar-amended temperate soil had less diversity of archaea [113] and fungi [114]. This information suggests that numerous microbial populations respond in various ways following biochar application into the soil. The mechanism of the soil microbial community in biochar-improved soils has been explored using down, medium-to-high-resolution techniques such as PLFA, qPCR, DGGE, TGGE, and DNA and RNA studies. (Fig. 15.2).

## *Effect of Biochar on Microbial Activity*

In agroecosystems, decomposer microorganisms could raise nutrient distribution from soil organic substances to the rhizosphere of the crop, which is necessary for the entry of nutrients and the trouble in crop production [109]. Several indexes, such as enzymes and metabolism prices, may be utilized as an alternative to distinguishing the soil biological activity. With the influences of biological activities and community changes, the persistence of N and P was enhanced [31, 89, 105]; then, these activities can gain plant nutrient accessibility in nutrient-confined agroecosystems [110].

Domene [111] featured no important adjustments in microbial activity when divided as basal movement and feeding prices, noting that net microbial machining of organic C did not change with biochar application but with similarities in soil

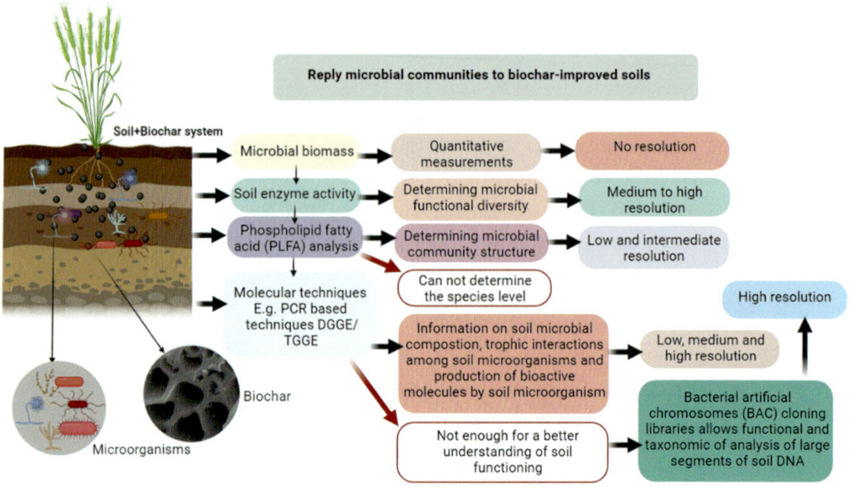


**Fig. 15.2** The process for detecting microbial community combinations in biochar-improved soils and their comparative impactiveness during separability is difficult

texture. This conclusion followed other long-time studies below area surroundings with no change or fewer break prices [112]. Thus, the enhanced microbial activity is feasible based on the mineralizable organic extent of fresh biochars.

## *Effect of Biochar on Functional Ecology of Microorganisms*

Adjustments of biochar can either gain or reeducation plenty of soil activities, thus C mineralization [55, 98], denitrification and methane oxidation [113, 114], and nutrient alternations [115]. Many causes can be accountable for these factors, thus, modified C sources or nutrient accessibility and absorption of inorganic and organic competition. Furthermore, many enzyme activities, water retention, and infiltration properties or changes in hold architecture can impact functional microbial ecology. In other words, modifications of soil activities could be appearing as a result of the modifications of microbial community structure, plenty, actually, and metabolism. The mineralization or oxidation of biochar itself will be impressed by the modifications of microbial attributes.

Nevertheless, these soil activities apparition on several features, containing the quantities of available present C sources, the absorption of organic C of simple deterioration, the current of stable biochar, or the impact of pH and phenolic materials on the microbial community. Furthermore, biochar can enable the microbially induced alternations of nutrients in the soil. Moreover, microorganisms could create ethylene in fresh biochar, related to reducing $N_2O$ and $CO_2$ emissions [71]. So, after biochar treatment, the improvements of microbial functional operations could decrease dictation of gaseous nutrient emissions, preserve nutrients, and facilitate nutrient cycling.

## *The Impact of Biochar on Beneficial Soil Organisms*

Biochar has been the carbon-rich byproduct produced when biomass is heated in a sealed container with little or no accessible air with the goal of modifying soil and resources to intercept carbon (C) and hold or improve soil functions [59]. Biochar addition to soil has a major effect on crop yield and root colonization by microorganisms (e.g., mycorrhizal fungi) and nematodes [116]. Interactions among biochar, soil, microbes, and plant roots were known to arise within a bit after usage in the soil [59]. Apparently, to [59], Dissolution, hydrolysis, carbonation, decarbonization, hydration, and redox reactions are the main methods affecting soil biochar weathering and interactions by soil microbiota. The prices at which these responses arise are related to the nature of the comments, kind of biochar, and climatic circumstances. Biochar can impression physical and chemical attributes and also useful soil microorganisms similar to bacteria, fungi, and invertebrates in field and laboratory surroundings [116]. Biochar has too been shown to raise nutrient accessibility at a more prolonged period rate by improving nitrogen (N) mineralization or nitrification [117, 118] as a result of enhancing microbial development and activity [31] and by decreasing soil nutrient losses due to its great ion interchange inclusion [119]. Several prior research have demonstrated that biochar has a good impact on soil fertility and can boost plant development [42, 120, 121], thereby having a devious positive impact on net ecosystem C perception.

As a soil repair, biochar can increase microbial biomass [128], increase soil microbial activity [35], and change the microbial community in soil [94]. Biochars utilized in soil may have an impact on soil microbial community structure due to their high attraction valence [35], changing soil pH [129], and microbial environment adjustment. According to Lehmann [35], biochars include polycyclic aromatic hydrocarbons and other hazardous carbonyl chemicals that may have bactericidal or fungicidal properties.

## Biochar Impact on Rhizosphere Microorganisms

The effect of biochar on the issue and biomass of microorganisms and their productiveness in colonizing plant roots were maximum. It may be related to the kind of soil which has been established. Biochar may enhance the biomass of microorganisms and their activity in soils. Kolb [122] noticed that enhancing doses of charcoal gain the populations of soil microbes as measured by their break activity.

## *Biochar—Microorganism Interaction*

Biochar impacts the soil microbial actuality and biomass, alters the bacteria in the soil to fungal relationship and soil enzyme activity, and transforms the microbial community [123]. Biochar application may significantly alter the microbial community structure even when it does not change the overall microbial activity and biomass. To understand the microbial responses to biochar, use in soils, gene version numbers serve as a more sensitive metric than microbial biomass [131]. Biochar exposes synergistic interactions to microorganisms by performing as an original of nutrients, enabling microbial colonization, giving microbial region, and removing/reducing contaminant toxicity from the nearby environment [124]. During the same period, several antagonistic impacts of biochar, such as distribution of remaining adverse elements/chemicals and immobilization of chosen nutrients, are also introduced. The efficiency of biochar to increase microbial remediation of organic contaminants would thus belong on the pure impact of the upper synergistic and antagonistic impacts and change from condition to condition (Fig. 15.3).

Several techniques were used to experiment with microbial activity and community structure, including fluorescence in situ hybridization (FISH), phospholipids fatty acid quantitation (PLFA), and the molecular fingerprinting of 16S rRNA

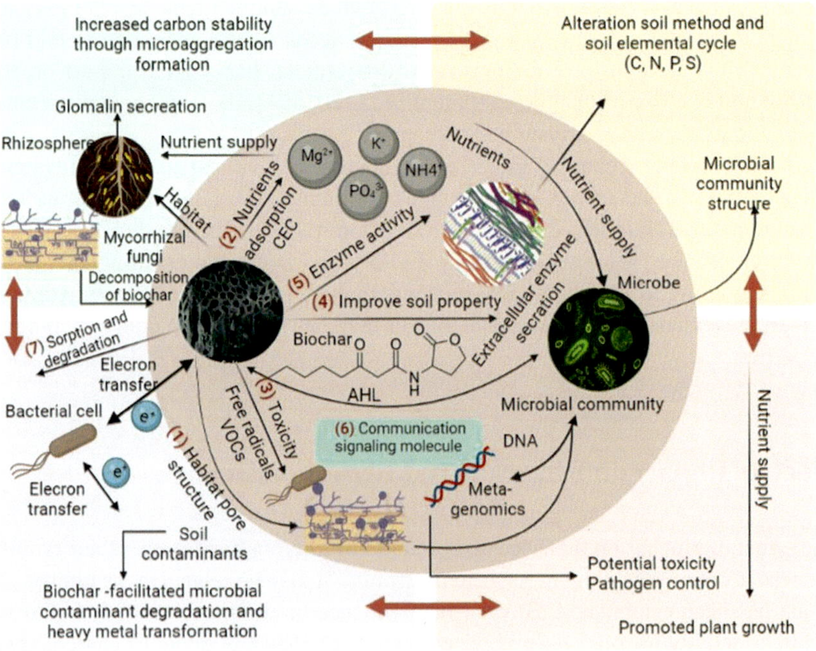


**Fig. 15.3** Suggest mechanisms of biochar-microbe interactions and the environmental effects of biochar

gene fragments. Alternation in the comparative plenty of Acidobacteria, Actinobacteria, Gemmatimonadetes, and Verrucomicrobia was frequently discovered using numerous-by sequencing, under treatment with biochar [125, 126].

The connection between biochar and microbes is shown in the middle round region, while the wall four boxes illustrate the effects of their interaction on carbon analysis, soil activities (elemental cycling), pollutant degradation, and plant development. Interactions among the biochar and the microbes and its impacts contain the following: (1) biochar may function as a microbial refuge with its pore mechanism; (2) via absorption of nutrient cations through functional groups, biochar may amend soil cation exchange valence and hold nutrients for microbial development; (3) Biochar's free radicals and volatile organic chemicals may be poisonous to numerous soil bacteria, preventing soil-borne diseases, and paying attention to plant development; (4) Biochar has the potential to affect soil properties (such as pH, water value, and aeration conditions) as well as the growth template of soil bacteria; (5) Biochar has the potential to adsorb enzyme molecules and boost soil enzyme operations and elemental durations; (6) Biochar may adsorb and increase the hydrolysis of signaling molecules, disrupting microbial relationships and altering microbial community mechanisms; (7) biochar may raise the absorption (via biochar level functional groups) and degradation of soil contaminants (facilitated via electron conduction among biochar, microbes, and contaminants), which may decrease the toxicity of contaminants to soil microbes. The interactions among biochar and soil microbes may change the microbial community and their metabolic pathways (which may be revealed by metagenomics resolution of microbial DNA sequencing), resulting in variable soil activities. There are interactions between various environmental impacts as good.

## *The Microorganism Pattern in Soil Health Progress*

Soil microorganisms are active soil engineers, positioning the soil for plant development by making nutrients available and key development regulators efficient. They also help with organic matter transformation and xenobiotic breakdown in the soil [127]. Inherent microbial communities provide various functional roles in adhering and absorbing mineral nutrients to physical levels, as well as decomposing organic wastes, to produce a section of soil [128–130]. The full roles of plants and microbes are property to the combability of soil for agriculture and farming [131]. It is outstanding that even little human interventions, such as the excess of sewage mud provided to gain the soil inhabitant microbial crowd of *Proteobacteria* and *Bacteroidetes* in bauxite productive access regions and increased the producer of soil organization [132]. Another from the soil establishment, the process of nutrient cycling, a necessary section to retain soil fertility, is steered by microbes in several biogeochemical cycles [133].

The application of rhizosphere bacteria to amend soil fertility instead of chemical fertilizers has been encouraged to achieve supportable plant development [134].

The amelioration of plant efficiency is an assembled procedure, including interaction with particular microbes or consortiums. Novel approaches import symbiotic engineering relationships to the construction of nonlegumes and other main crops to make nitrogen [135, 136], thereby converting them into soil fertility-contributing plants. This will importantly amend the global food provisions and assistance to meet sustainability goals.

The achievement of a chosen microbial inoculum relies on its might to prosper and function along with the autochthonous microbes and the abiotic ingredients of that habitat [128]. The duration and strangeness of the microbe in the soil hinge on how it interacts with other biotic ingredients in the ecosystem, and frequently, plant interactions with microbial consortia are rather impressive than signal microbes [137, 138]. So, soil fertility is undoubtedly associated with microbial diversity and its development-promoting qualities [139].

## Microorganism Bioengineering for Soil Health Improvement Through Remediation

Genetic engineered ones could be engaged for further efficiency due to the damage to native microbes in acclimatizing to the novel environment and performing depression of pollutants efficiently [140]. These engineered microorganisms may efficiently remediate most contaminants, which natural native microbes cannot degrade. A confine of molecular tools is accessible for making GMOs like biolistic change, electroporation, conjugation, horizontal conduction of bacterial DNA, molecular cloning, and shift in protoplast. Transfer and expression of new genes with great degradation valency minimize the remediation period. Engineered microbes may remediate a variety of substances similar to toluene, octane, and amplitude of microorganisms in charcoal enhanced soil naphthalene, salicylate, and xylene by expressing genes encoded in the bacterial plasmid [141].

### *Interactions of Biochar and Microorganisms in Soil*

Biochar affects soil microbial activity and biomass, converts soil bacteria to fungus, increases soil enzyme activity, and changes the microbial community [134, 150, 151]. Even when microbial activity and biomass are not alternated, the use of biochar can modify the microbial community mechanism. To more effectively translate microbial responses to biochar use in soils, gene version concerns may serve as a more sensitive metric than microbial biomass [142].

### *Biochar Attribute as a Possible Effective Microbial Transport*

Biofertilizers (rhizospheric beneficial microorganisms) have emerged as a feasible supplement to fertilizers in improved soil productivity in supportable agricultural systems. Plant development-promoting microorganisms can be incorporated into agricultural soils with the help of a suitable carrier matter capable of deploying enough viable populations of the microorganisms to carry out strategic patterns like phosphate solubilization, nitrogen fixation, phytohormone synthesis, humification, and plant conversion. Characteristics of a good carrier (simple processing and sterilization (autoclave, irradiation); non-toxicity for microbial and/or plant inoculum; moisture absorption; availability in sufficient quantity; high organic matter and nitrogen value; low cost; pH buffering capacity granular particles, porosity, surface characteristics, carrier-microbe mixture consistency) [143].

## Microorganisms as Biofertilizers

Due to the upper-mentioned subjects relevant to chemical fertilizers and pesticides, there has been a significant growth in tolerable agriculture using rather ecological and obvious ways, such as biopesticides and biofertilizers. Under optimal conditions, biofertilizers can also be inoculated on grains in the roots of various production plants, and they can also be applied to the soil immediately [144]. Biofertilizer is a material that includes habitats microorganisms that, when practical to seed, plant levels, or soil, mobilize the accessibility of nutrients, particularly by their biological activity, and advance plant development [145]. Biofertilizers improve nutrients by naturally fixing atmospheric nitrogen, solubilizing phosphorus, and stimulating plant growth through the incorporation of growth-promoting substances [146, 147]. They may be grouped in several routes, supported by their nature and subordinate.

In this sense, the microorganisms, when practical to the soil or the plant, that aid enhance the accessibility of nutrients to production plants are known as biofertilizers, which are eco-friendly and inexpensively means to chemical fertilizers [148]. Several microorganisms utilize different strategies such as stabilization /mobilizing/recycling nutrients in the agricultural ecosystem to be useful for the crops, improving plant development and fertility [149].

The plant rhizosphere, the capillary area of soil comprehensive the root mechanism of growing plants, is colonized by a large confine of microbial taxa, out of which bacteria and fungi contain the most many groups [150]. Free-living soil bacteria that prosper in the rhizosphere colonize plant roots and comfort plant development are designated as plant-development-promoting rhizobacteria that produce and hide different regulatory chemicals in the plant roots' presence assist in plant development promotion [151, 152].

Bacteria and fungi that inhabit the rhizosphere may subordinate as bio fertilizers that cultivate plants' development and growth by comforting biotic and abiotic stress tolerance and suffering host plants' nutrition. They may subordinate biopesticides because many microorganisms kill insects and other pests that threaten crops. Moreover, microorganisms have the capability to reduce and resolve adverse organic also mineral composed that stack in the soil as contaminating matters, which are the result of plenty of processes containing agriculture practices. They use the bioremediation function, gaining soil and plant safety [153].

Bacterial biofertilizers are a type of bacteria that aid in the stabilization of various nutrients required for plant development in soil [154]. They may repair nitrogen, solubilize phosphorus, potassium, or other micronutrients, and conceal organic substances that suppress plant diseases or promote plant development. Examples of the most favorite bacterial biofertilizers that have been practical are *Azotobacter*, *Azospirillum*, *Rhizobium*, and *Bacillus*, among others, as shown in Fig. 15.4 [155, 156].

*Rhizobium* is utilized in legume crops, *while Azotobacter* and *Azospirillum* are employed in non-legume crops. *Acetobacter* has a strong preference for sugar [157]. Using these bacteria as biofertilizers to promote plant development and crop efficiency, improve soil productivity, and control phytopathogens promotes supportable agriculture by showing eco-friendly means to synthetic agrochemicals, such as chemical products and pesticides.

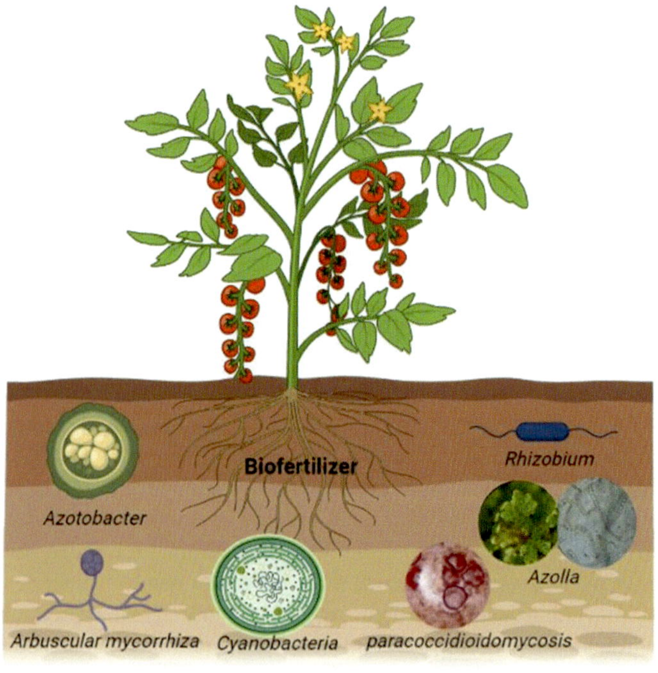


**Fig. 15.4** Different types of organic fertilizers

The fungal biofertilizers form a symbiotic communication within the plant roots. Such communication is called mycorrhiza, which allows the distribution and attraction of nutrients, mainly phosphorus. Certain nutrients cannot spread easily into the soil, and the roots empty these nutrients from the comprehensive area. Arbuscular mycorrhiza is useful soil fungi that form a symbiotic communication with plants and plenty of crops through the roots of vascular plants [158]. The hyphae of these fungi develop in the evacuation area, enhancing the attraction level of plants and improving the availability of nutrients [159]. The symbiosis of arbuscular mycorrhiza fungi improves the plant rhizosphere microenvironment, gain the attraction of mineral elements by the plant, enhances stress and disease opposition, and cultivates plant development [160].

The usage of microbial biofertilizers has various benefits, as mentioned above, such as their simple application and down cost and their use impacts on soil and plants. However, several competitors have prevented their wide and prosperous application. Firstly, a primary good laboratory screening is necessary to search for a good and particular biofertilizer strain. In addition, making and quality control of biofertilizers import artificial technology and eligible and trained human resources, together with loss of sufferance financial resources to spread and the unacceptability of suitable transportation services along with storage facilities, construction it an involved method from the starting to the end. It must be highlighted between the basic matters that may be found, containing the needy kind of crops, the application of unproper strains, the little shelf life, the loss of qualified technical staff, the loss of awareness between farmers, and environmental restrictions, etc. [161]. Microbial strains shall be good to survive in soil, become with the production on which they are inoculated, and interact with native microflora in soil and abiotic effects to be effective and prosperous bio inoculants.

## Biochar Amendment with Microorganism

The biological amendment of biochar may be achieved by pre-treating the feedstock with anaerobic digestion and making a film on the inner and outside levels of biochar [162]. Digestion of damaged matter by aerobic and anaerobic bacteria gains the economy by generating bio-fertilizers and biofuel. Biochar generated from bacterial digestion action a key pattern in improving hydrophobicity, CEC, and level region and is frequently employed to delete heavy metals, pharmaceuticals, and contaminants from polluted soils by expanding biofilms [163, 164]. Biochar-changed bio asphalt improves biomass usage and increases environmental conversion [165].

## Biochar Quality Variations as a Soil Modification

Biochar crops from various sources change largely in characteristics and functions valence as a soil modification. Biochar is created from biomass matters using the thermochemical technique pyrolysis, through which organic residues are heated in $O_2$-graties or many finites, ambient pressure environments for some time to be carbonized into charcoal, with the efficiency of pyrolysis bio-oil and syngas as by crops [166]. Forest waste, production debris, food processing losses, and manures containing sewage muck and biosolids are all used as joint biochar feedstock. These biomass matters are important variations in organic and ash compositions, attributing to the notable modality conversions of the resulting biochar crops. Carbonization (pyrolysis) causes significant penetration of biochar quality attributes. Three parameters are generally applied to administer the carbonization situations: pyrolysis (peak) temperature, solid habitation period, and heating rate, stretching to a large confine of values [167]. A high temperature speeds the carbonization process, allowing the pyrolytic transformation of biomass to achieve a deeper surface and be perfect in a short amount of time [168, 169]. Biochar crops result from incomplete pyrolysis and contain considerable amounts of uncarbonized carbon (i.e., with the crystalline identity of the pioneer matters) [170, 171]. Biochar is the principal crop of slow pyrolysis and is still the crop of fast pyrolysis (pyrolysis bio-oil) and gasification (pyrolysis with mild oxidation—syngas). Carbonization conditions (temperature, considerable occupancy period, and heating rate) can be rectified using any of the three thermochemical strategies to enhance main crop output. Even with several feedstocks, gasification and rapid pyrolysis biochars have less OC and a higher cinder value than products from slow pyrolysis.

## Plant Development and Soil Microflora Stimulation

Many reports show that biochar can stimulate the soil microflora, resulting in greater carbon accumulation in the soil. Besides adsorbing organic materials, nutrients, and gases, biochars may suggest a region for bacteria, actinomycetes, and fungi [105]. It has been claimed that rapid heating of biomass (fast pyrolysis) will result in biochar with fewer microorganisms, smaller pores, and relatively liquid and gas components [172]. Water containment growth after biochar application in soil has been successfully established [182], which can affect soil microbial communities. Biochar creates an ideal environment for important and diverse groups of soil microbes. The interaction of biochar with soil microbes, on the other hand, is an ongoing phenomenon.

Applying biochar enhanced mycorrhizal production in clover bioassay plants by providing the appropriate situations for colonization of plant roots [173]. Warnock [119] summarized four systems through which biochar may influence the functioning of mycorrhizal fungi: (i) variation in the physical and chemical properties of soil,

(ii) devious effects on mycorrhizae via offer to other soil microbes, (iii) plant fungus signaling interposition and detoxification of toxic chemicals on biochar, and (iv) providing shelter from mushroom browsers.

## *Biochar-Microbe Interaction Mechanisms in Soil*

Biochar has an effect on soil microbial activity and biomass, changes the soil bacteria-fungi connection and soil enzyme activity, and changes the microbial association mechanism [130, 133, 134, 150, 184]. The use of biochar can change the mechanism of the microbial community even if it does not affect the microbial activity or biomass. Concerns about gene version may be a more sensitive metric than microbial biomass in interpreting the microbial response to biochar application in soils [142]. Several techniques, including ergosterol production, quantitative actual-period polymerase chain reaction (q-PCR), fluorescence in situ hybridization (FISH), phospholipid fatty acid quantitation (PLFA), molecular fingerprinting of 16S rRNA gene fragments using denaturing gradient gel electrophoresis (DGGE) and terminal restriction fragment length polymorphism (TRFLP), and high-throughput sequencing, are used to investigate microbial activity and community mechanisms [126, 142, 174–176]. Changes in the relative abundance of Acidobacteria, Actinobacteria, Gemmatimonadetes, and Verrucomicrobia with biochar treatment are largely detected utilizing high-throughput sequencing [125, 126]. By using these techniques, the effects of biochar in soil improvement can be investigated [126, 142, 174–176].

## *Biochar Provides a Haven for Microorganisms*

The advantages of biochar for microorganisms is that biochar may act as a shelter for microbes due to their mechanism [65]. The benefits of biochar for microorganisms include the ability of biochar to act as a sanctuary for germs due to its mechanism [177]. However, the colonization of bacterial cells and fungal hyphae is spatially heterogeneous among the biochar's outside and inner pores [65, 178]. Three possible mechanisms have explained several patterns of microbial colonization in biochar surfaces and pores: (1) biochar pores have better nutrient availability than natural soil pores, (2) biochar pores may interact with soil organic matter (such as humic acids) be closed (3) Hazardous substances such as PAHs can be found in biochar (especially in fresh biochar) [65, 107, 179]. Microbial colonization on the surfaces and pores of biochar is also related to the aging process of biochar, which can be considered as temporal heterogeneity [65].

### *Biochar Provides Nutrients for Soil Microorganisms*

Biochar contains nutrients (such as potassium, magnesium, sodium, nitrogen, and phosphorus) [191, 192] and enhances soil nutrients due to its large surface area, large pores, and negative charge [193Cation exchange valency (CEC) is an important indicator of a soil's ability to retain cationic ions and accumulate nutrients to support microbiological activity. The modified soil CEC that occurs from biochar application reflects a superior nutrient maintenance ability and a decreased nutrient loss via leaching, which is beneficial for soil microbial activity [126], particularly for microorganisms living in soils with a low organic matter content [64, 194–196].

Biochar provides nutrients to soil bacteria by absorbing nutritional cations and inorganic anions through its area functional groups, notably oxygen-containing groups such as the carboxylate group [180–185].

## Studies Have Noted the Positive Effect of Biochar

CEC at low and medium pyrolysis temperatures Several studies have shown that CEC of biochar increases with pyrolysis temperature [120, 184, 186]. Species and pyrolysis design parameters, including temperature, heating rate, and holding length, primarily distinguish biochar functional groups and, consequently, biochar's potential to increase soil CEC [186–189]. In one study, biochar CEC was shown to be pH-dependent, increasing from low to neutral pH values [126], indicating possible interactions between pH and CEC transformation in biochar-treated soils. Furthermore, interaction between biochars and soil minerals may be responsible for the high-period retention of minerals during biochar aging [190]. Biochars are often lower in available carbon for microbial use because they have a better C/N ratio than their feedstocks and are difficult to reduce with microbes due to the loss in N accumulation. Bacteria and fungi are distinguished by their carbon origins and different tolerances to environmental factors such as pH and water position [176, 191]. Some biochar compounds are known as microbial repressors, and they include benzene (the dominant product of pyrolysis prior to glowing combustion of char), methoxyphenols and phenols (the crop of pyrolysis of hemicelluloses and lignin), carboxylic acids, ketones, furans (which are commonly presented as sorbet VOCs on biochar), and PAHs [192–194].

### *Biochar Modifies Microbial Habitats*

Biochar may improve microbial habitats by increasing the physical properties of the soil. Biochar porosity may reduce soil bulk compaction, increase soil aeration [82], and control the transport of soil microorganisms in biochar-amended soil [177].

Biochar may enhance the accessible water amount that penetration nutrient availability to microbial cells [78]. In addition, the biochar may enhance water value at the constant wilting part, which displays the ability of biochar, Due to its high porosity, it is difficult for plants to retain water. Water conservation in this strategy is especially valuable in sandy and damaged soils [78]. In addition, biochar is an alternative to water holding capacity, which has a stronger ability in soil to retain water compared to dry and wet cycles in the natural environment, which may encourage the maintenance of a constant microbial activity [195]. Pyrolysis parameters (especially temperature, heating rate, and time) and raw material compositions (eg, lignin and lipid concentrations) used to create biochar govern porosity, carbon stability, and nutrient uptake [177, 187, 196]. The role of biochar in improving soil properties and microbial habitats can be linked to the feedstock types and pyrolysis procedures employed in biochar production.

## *Biochar Changes Soil Enzyme Activity*

Enzymes catalyze the majority of the elemental efficiency in soil, which describes nutrient bioaccessibility and contains a yield of C, N, P, and S. Soil enzymatic activities respond faster to soil management than other soil changes, and soil problem is a sign of biological changes and soil quality [197]. In the organic material analysis, decreased microbial abundance and soil enzyme activity may enhance C breakdown [198]. Possible systems involved in biochar influence on enzyme activity (1) Biochar adsorbs extracellular enzyme molecules and/or layers on the level or limits enzyme responses [215], thereby reducing their external dependence on layers [199]; (2) biochar penetrations enzyme activity with alters in soil physiochemical attributes (especially pH) [200]; and (3) Biochar produces a number of small compounds that are thought to serve as allosteric regulators or inhibitors of specific enzymes (for example, putative up-regulation of -N-acetylglucosaminidase activity with ethylene) [201]. The absorption (binding) of enzymes on biochar and soil organic matter can change the kinetic properties of enzyme activity [218, 219], and this is the most important system regulating soil enzyme activity [200].

The sorption efficiency of the enzyme and layers operations on the biochar mechanism: sorption of enzyme molecules on biochar levels is considered to be driven by non-coulombic forces among the primrose areas of the protein and the primrose areas of the biochar levels, and the sorption of little molecular polar layer (e.g., a disaccharide) on charred fractions (mainly activated carbon) is stabilized through hydrogen bonding to polar level groups (e.g., COOH, SO4H, PO4H) on the sorbents [202]. Alternations in level functional groups in aged biochar change the sorption valence of enzyme and layer, thus impacting enzyme activity [203]. Biochar may reduce the activation energy (Ea, which is related to an enzyme's temperature sensitivity) of an enzyme-catalyzed response and adjust the enzymatic sensitivity to temperature changes (in terms of Q10), resulting in higher b-glucosidase and arylsulfatase activity [199].

Soil enzymes, on the other hand, respond quickly to soil management (e.g., organic material modification) [213], therefore changes in soil characteristics caused by biochar use should be considered For the third mechanism, biochar inhibitors may participate in enzyme-catalyzed responses as well: for example, following pyrolysis, plant biochars may liberate an issue of benzofurans, polycyclic fragrant hydrocarbons, and heterocyclic compounds, which are inhibitory compounds to soil enzymes [202].

## *Biochar Reduces the Toxicity of Pollutants for Soil Microorganisms*

As a soil conditioner, biochar may reduce the toxicity of soil pollutants to soil microbes [221]. Immobilization of soil pollutants (containing hard elements such as Al, Cd, Co, Cr, Mn and Ni as well as biological pollutants and PAHs) on biochar, and thus reducing their bioavailability, may be the main reason for reducing the toxicity of pollutants. soil to microbes and increase microbial biomass [204–206].

## *Biochar for Sustainable Soil Management*

Soil depression is a critical menace to the global environment and the United Nations Sustainable Development Goals [207, 208]. Sustainable soil management is called for by many stakeholders [209–211]. Biochar is constructed from the pyrolysis of biomass under an oxygen-confined environment. The sense was brought about a decade forward, but its factual application may date behind pre-Columbian Amazonians [212].

### Biochar for Soil Remediation

Soil contamination by different heavy metals and metalloids is largely divided [213, 214], offending the public and creating disproportionate safety matters for disadvantaged groups [215, 216]. Biochar is impressive in immobilizing heavy metals containing Cd, Pb, etc. [217, 218]. Different amendment strategies have been prospected to strengthen the immobilization ability of biochar manufactured from a diverse feedstock [219, 220]. Besides the remediation of heavy metal polluted soil, biochar has as well as been a prospect to address different kinds of degraded ground. Biochar was used to comfort the rehabilitation of coal mine spoils [221]. Therefore, supportable soil management will need biochar matter to be high-tough and sustainable.

### Biochar for Nutrient Management

Biochar is created from biomass containing many nutrient elements, such as nitrogen, phosphorus, Sulphur, and potassium. Pending pyrolysis and/or weathering operations, these elements may be transformed into mineral forms instead of bioaccessibility. Much research has focused on applying biochar as a nutrient enhancer or another nutrient preparatory. Moreover, biochar may maintain some nutrients, thus decreasing nutrient damage through leaching or gaseous transpiration. The last meta-analysis showed that biochar only does not gain production in crops. However, when combined with mineral fertilizers, biochar could achieve a production yield of 15% compared to inorganic fertilizers [222]. Biochar may as well as change nutrient interaction, explaining the feasibility of nutrient optimization [223]. Biochar maintenances many promises for this matter may be constructed by a decentralized plain complex-up in one's backyard or farm field [224], similar to what ancient people have accomplished. Research advance on this forefront may profit millions of smallholder farmers [225, 226].

### Biochar for Soil Health

Healthy soil and supportable agricultural action advance biodiversity [227, 228], which major increases necessary ecological services [229]. Biochar may change the physicochemical attributes of soil in many manners, thus improving soil health. For instance, biochar may improve soil addition release, water supply capacity, and soil compression. It is essential to comprehend further the effects impacting the period of biochar's impact, and plan optimized use strategies accordingly.

### Biochar for Climate Alteration Reduce

Soil shows the more incredible earthly carbon pool [230]. Soil carbon storage is impacted by farm management strategies [141, 231, 232], and soil microbial activities may as well as affect the transpiration of $N_2O$ [219, 233, 234], a greenhouse gas with 298 periods of atmospheric heat-trapping capability of $CO_2$ [235]. Biochar use enhancement soil organic value in soil, resulting in carbon analysis [236]. Biochar surplus could reduce $N_2O$ transpiration induced by chaff reflux [237];. However, the biochar dosage needs to be optimized for great biochar dosage was found to decrease nitrogen maintenance and nitrogen application by productions [238].

## Response of Microbial Populations to Soils Amended with Biochar

Biochar exhibits a range of physicochemical properties due to feedstock plantings, pyrolysis circumstances, and amendment processes (such as activation, magnetic amendment, and acid/basis treatments) [35, 239, 240]. Despite extensive research into the chemical and physical properties of biochar, the effects of biochar on soil biological functions remain unknown. Comprehensive effects of biochar on soil biological activities would necessitate long-term monitoring and investigation of changes in natural science properties in biochar-improved soils. The use of biochar will have cumulative effects on the natural science properties of the soil, including interactions between living and non-living factors and increasing their activities in the soil [124]. Over the preceding two years, studies have revealed that biochar-soil use could alter soil biological properties by enhancing soil microbial functional activities [241], (Fig. 15.5). Furthermore, the effects of biochar on soil biological characteristics as influenced by other soil organisms and crops were investigated [199, 242]. Because soil microorganisms play an important role in soil ecosystem functions and services (e.g., driving biogeochemical cycles, suppressing pathogens, and maintaining soil growth and health), the next phase of biochar research should focus on long-term effects. The use of biochar should focus on soil biota and soil health. It is critical to investigate the potential of biochar to improve soil quality in the face of future environmental changes [243]. Bacteria, fungi, nematodes, algae, archaea, actinomycetes, bacteriophages, and protozoa are all found in soil. These bacteria are involved in a variety of beneficial soil processes, including nutrient recycling, organic material recombination, soil-mechanism organization, discharge of plant development advancements, organic pollutant degradation, and disease suppression [244]. Soil microbial functional processes and community mechanisms may be useful in differentiating the impacts of biochar on soil biological characteristics.

## Future Research Directions

Considering the physical, chemical and biological effects of biochar on soil discussed in this chapter, we suggest the following areas for further research:

1. The majority of studies have focused on the possible quantities of biochar employed in modifying soil fertility in relation to changing soil physicochemical characteristics. It is also important to test the value of which C-rich matter in modifying soil health via its effects on microbial variety and operation.
2. By revealing the type of biochar as well as the soil species and composition of microorganisms, microbial interactions with soil and plants can be dramatically altered. Consequently, investigating the interactions of microorganisms with different biochar processes, different prices of biochar use, and different

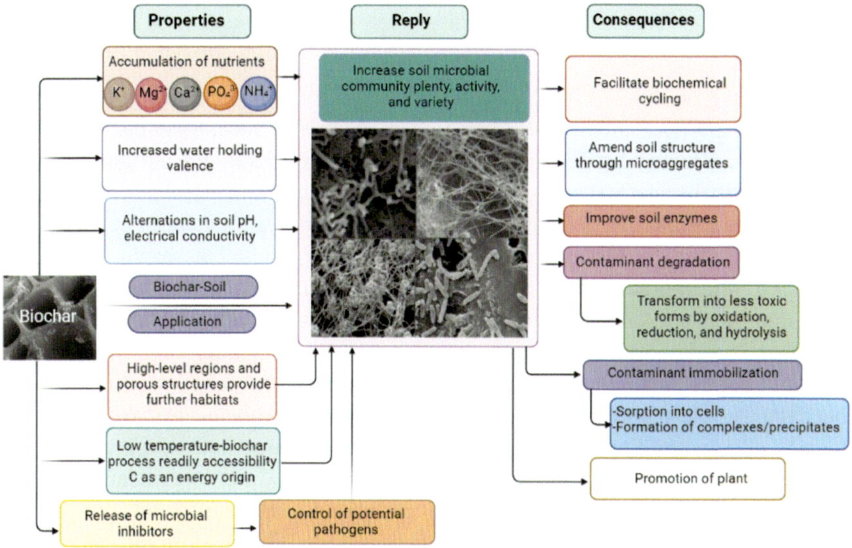


**Fig. 15.5** The effect of biochar on soil microorganisms and the microbial response to biochar use is shown schematically

types of plants in the above period is critical to recognize the value of biochar effects on soil microorganisms over time and under different conditions.

3. So far, research on biochar and microbial activities and interactions in soil has relied on small-scale laboratory incubations and greenhouse pot observations. It is recommended that a large-scale field experiment be conducted to study high-periodic soil-plant interactions with microbes as affected by biochar application, with temporal variations in such high-periodic research.
4. Based on this chapter and other articles, a major study using biochar as a growth promoter of specific soil microorganisms to achieve a desired goal (such as promoting soil nutrient cycling) should use customized biochar (actively Select biochar raw materials and production status).
5. It is difficult to isolate the impacts of biochar on a specific soil biological exclusivity or a specific soil microorganism in a microbial relationship. Artificial and sectioning-border analytical procedures like fluorescence in situ hybridization (FISH) and nanoscale secondary ion mass spectrometry (NanoSIMs) may be adopted to help improve theoretical science in this regard.
6. The adoption of high-resolution molecular-based techniques such as PLFA, PCR, DGGE, TGGE, and DNA and RNA analyzes are needed to identify families, genera, or even surface types, which will be useful for developing comprehensive microbial mechanisms with biochar in improving soils.

## Conclusions

Biochar may have direct effects on microbial and biomass development and help reduce pollutant risks in water and soil to a level suitable for human health and the environment. Biochar usage for the recovery of agricultural soil attributes and as an ecologically secure sorbent for the polluted soil immobilization has considerable possible. The effectiveness of its application to a significant extent depends on the pyrolysis situation, the biochar precursors, and soil attributes. The surplus of biochar may impact the soil attribute to a great extent. For a further comprehensive biochar effect system on the soil and microorganisms, it is essential to expand only the pattern of biochar experiments containing the list of parameters that much be studied. Before the beginning of current biochar application in agricultural function, it is essential to expand the international standards on possibly toxic pyrolysis yield value also the manners of removing possibly negative impacts by the alternative of pre-acting of biochar. The major research on biochar interactions with microorganisms and their composed extension in the soil will permit the use of many useful and ecologically safe instruments for soil remediation if acknowledge biochar of great modality is used.

## References

1. Jianping Z (1999) Soil erosion in Guizhou province of China: a case study in Bijie prefecture. Soil Use Manage 15(1):68–70
2. Sachs JD (2012) From millennium development goals to sustainable development goals. 379(9832):2206–2211
3. Biermann F, Kanie N, Kim RE (2017) Global governance by goal-setting: the novel approach of the UN sustainable development goals. Curr Opinion Environ Sustain 26:26–31
4. Sohi SP, Krull E, Lopez-Capel E, Bol R (2010) A review of biochar and its use and function in soil. Adv Agron 105:47–82
5. Jeffery S, Verheijen FG, van der Velde M, Bastos AC (2011) A quantitative review of the effects of biochar application to soils on crop productivity using meta-analysis. Agric Ecosyst Environ 144(1):175–187
6. Ajayi AE, Rainer H (2017) Biochar-induced changes in soil resilience: effects of soil texture and biochar dosage. J Plant Nutrit 27(2):236–247
7. Clough TJ, Condron LM, Kammann C, Müller C (2013) A review of biochar and soil nitrogen dynamics. Argonomy 3(2):275–293
8. Hammer EC, Balogh-Brunstad Z, Jakobsen I, Olsson PA, Stipp SL, Rillig MC (2014) A mycorrhizal fungus grows on biochar and captures phosphorus from its surfaces. Soil Bio Biochem 77:252–260.
9. Laird D, Fleming P, Wang B, Horton R, Karlen D (2010) Biochar impact on nutrient leaching from a Midwestern agricultural soil. Geoderma 158(3–4):436–442
10. Vaccari F, Maienza A, Miglietta F, Baronti S, Di Lonardo S, Giagnoni L, Lagomarsino A, Pozzi A, Pusceddu E, Ranieri R (2015) Biochar stimulates plant growth but not fruit yield of processing tomato in a fertile soil. Agric Environ Ecosyst 207:163–170
11. Diatta AA, Thomason WE, Abaye O, Thompson TL (2020) Assessment of nitrogen fixation by mungbean genotypes in different soil textures using 15N natural abundance method. J Soil Sci Plant Nutrit 20(4):2230–2240

12. Adnan M, Fahad S, Zamin M, Shah S, Mian IA, Danish S, Zafar-ul-Hye M, Battaglia ML, Naz RMM, Saeed B (2020) Coupling phosphate-solubilizing bacteria with phosphorus supplements improve maize phosphorus acquisition and growth under lime induced salinity stress. Plants 9(7):900
13. Seleiman MF, Kheir AMS (2018) Maize productivity, heavy metals uptake and their availability in contaminated clay and sandy alkaline soils as affected by inorganic and organic amendments. Chemosphere 204:514–522
14. Seleiman MF, Alotaibi MA, Alhammad BA, Alharbi BM, Refay Y, Badawy SA (2020) Effects of ZnO nanoparticles and biochar of rice straw and cow manure on characteristics of contaminated soil and sunflower productivity, oil quality, and heavy metals uptake. Argonomy 10(6):790
15. Seleiman MF, Almutairi KF, Alotaibi M, Shami A, Alhammad BA, Battaglia ML (2020) Nano-fertilization as an emerging fertilization technique: why can modern agriculture benefit from its use? Plants 10(1):2
16. Adeyemi O, Keshavarz-Afshar R, Jahanzad E, Battaglia ML, Luo Y, Sadeghpour A (2020) Effect of wheat cover crop and split nitrogen application on corn yield and nitrogen use efficiency. Argonomy 10(8):1081
17. Diatta AA, Fike JH, Battaglia ML, Galbraith JM, Baig MB (2020) Effects of biochar on soil fertility and crop productivity in arid regions: a review. Arab J Geosci 13(14):1–17
18. Bonanomi G, Ippolito F, Cesarano G, Nanni B, Lombardi N, Rita A, Saracino A, Scala F (2017) Biochar as plant growth promoter: better off alone or mixed with organic amendments? Frontiers 8:1570
19. Abrol V, Sharma P (2019) Biochar: an imperative amendment for soil and the environment. IntechOpen
20. De Bary A (1878) Cassel, Ueber symbiose—Tageblatt 51 Versamml 1878:121–126
21. McFall-Ngai M, Hadfield MG, Bosch TC, Carey HV, Domazet-Lošo T, Douglas AE, Dubilier N, Eberl G, Fukami T, Gilbert SF (2013) Animals in a bacterial world, a new imperative for the life sciences. Proc Natl Acad Sci USA 110(9):3229–3236
22. Parfrey LW, Moreau CS, Russell JA (2018) Introduction: the host-associated microbiome: pattern, process and function. Mol Ecol 27(8):1749–1765
23. Oliver KM, Martinez AJ (2014) How resident microbes modulate ecologically-important traits of insects. 4:1–7
24. Charlat S, Hurst GD, Merçot HJ (2003) Evolutionary consequences of Wolbachia infections. Trends Genet 19(4):217–223
25. Werren JH, Baldo L, Clark ME (2008) Wolbachia: master manipulators of invertebrate biology. Nat Rev Microbiol 6(10):741–751
26. Marris E (2006) Putting the carbon back: black is the new green. Nature 442(7103):624–626
27. Nguyen V-T, Nguyen T-B, Chen C-W, Hung C-M, Chang J-H, Dong C-D (2019) Influence of pyrolysis temperature on polycyclic aromatic hydrocarbons production and tetracycline adsorption behavior of biochar derived from spent coffee ground. Bioresour Technol 284:197–203
28. Parmar A, Nema PK, Agarwal T (2014) Biochar production from agro-food industry residues: a sustainable approach for soil and environmental management. 107(10):1673–1682
29. Weber K, Quicker P (2018) Properties of biochar. Fuel 217:240–261
30. Pariyar P, Kumari K, Jain MK, Jadhao PS (2020) Evaluation of change in biochar properties derived from different feedstock and pyrolysis temperature for environmental and agricultural application. Sci Total Environ 713:136433
31. Lehmann J, Rillig MC, Thies J, Masiello CA, Hockaday WC, Crowley DJ (2011) Biochar effects on soil biota—a review. Sol Biol Biochem 43(9):1812–1836
32. Shin J, Lee S-I, Park W-K, Choi Y-S, Hong S-G, Park S-WJ (2014) Carbon sequestration in soil cooperated with organic composts and bio-char during corn (Zea mays) cultivation. J Agric Chem Environ 3(04):151
33. Peng X, Deng Y, Peng Y, Yue K (2018) Effects of biochar addition on toxic element concentrations in plants: a meta-analysis. Sci Total Environ 616:970–977

34. Ahmad M, Rajapaksha AU, Lim JE, Zhang M, Bolan N, Mohan D, Vithanage M, Lee SS, Ok YS (2014) Biochar as a sorbent for contaminant management in soil and water: a review. Chemosphere 99:19–33
35. Rajapaksha AU, Chen SS, Tsang DC, Zhang M, Vithanage M, Mandal S, Gao B, Bolan NS, Ok YS (2016) Engineered/designer biochar for contaminant removal/immobilization from soil and water: potential and implication of biochar modification. Chemosphere 148:276–291
36. Jain S, Singh A, Khare P, Chanda D, Mishra D, Shanker K, Karak T (2017) Toxicity assessment of Bacopa monnieri L. grown in biochar amended extremely acidic coal mine spoils. Ecol Eng 108:211–219
37. Liu Y, Zhu J, Ye C, Zhu P, Ba Q, Pang J, Shu L (2018) Effects of biochar application on the abundance and community composition of denitrifying bacteria in a reclaimed soil from coal mining subsidence area. Sci Total Environ 625:1218–1224
38. Herath H, Camps-Arbestain M, Hedley M (2013) Effect of biochar on soil physical properties in two contrasting soils: an Alfisol and an Andisol. Geoderma 209:188–197
39. Jones BE, Haynes R, Phillips IR (2010) Effect of amendment of bauxite processing sand with organic materials on its chemical, physical and microbial properties. J Environ Manage 91(11):2281–2288
40. Singh B, Singh BP, Cowie AL (2010) Characterisation and evaluation of biochars for their application as a soil amendment. Aust J Soil Res 48(7):516–525
41. Villagra-Mendoza K, Horn R (2018) Effect of biochar addition on hydraulic functions of two textural soils. Geoderma 326:88–95
42. Biederman LA, Harpole WS (2013) Biochar and its effects on plant productivity and nutrient cycling: a meta-analysis. GCB Bioenergy 5(2):202–214
43. Enders A, Hanley K, Whitman T, Joseph S, Lehmann J (2012) Characterization of biochars to evaluate recalcitrance and agronomic performance. Bioresour Technol 114:644–653
44. López-Valdez F, Fernández-Luqueño F (2014) Fertilizers: components, uses in agriculture and environmental impacts. Nova Science Publishers, Inc
45. Lehmann J, Joseph S (2015) Biochar for environmental management: science, technology and implementation. Routledge
46. Thomas SC, Gale N (2015) Biochar and forest restoration: a review and meta-analysis of tree growth responses. New Forests 46(5):931–946
47. Wang J, Xiong Z, Kuzyakov Y (2016) Biochar stability in soil: meta-analysis of decomposition and priming effects. GCB Bioenergy 8(3):512–523
48. Zimmerman AR (2010) Abiotic and microbial oxidation of laboratory-produced black carbon (biochar). Technology 44(4):1295–1301
49. Lehmann J, Gaunt J, Rondon M (2006) Bio-char sequestration in terrestrial ecosystems—a review. Mitig Adap Strateg Glob Change 11(2):403–427
50. Mašek O, Brownsort P, Cross A, Sohi S (2013) Influence of production conditions on the yield and environmental stability of biochar. Fuel 103:151–155
51. Duku MH, Gu S, Hagan EB (2011) Biochar production potential in Ghana—a review. Renew Sustain Energy Rev 15(8):3539–3551
52. Sohi S, Lopez-Capel E, Krull E, Bol R (2009) Biochar, climate change and soil: a review to guide future research. 5(09):17–31
53. Joseph S, Taylor P (2014) The production and application of biochar in soils. Advances in biorefineries. Elsevier, pp 525–555
54. Joseph S, Graber E, Chia C, Munroe P, Donne S, Thomas T, Nielsen S, Marjo C, Rutlidge H, Pan G-X (2013) Shifting paradigms: development of high-efficiency biochar fertilizers based on nano-structures and soluble components. Carbon Manage 4(3):323–343
55. Liang B, Lehmann J, Sohi SP, Thies JE, O'Neill B, Trujillo L, Gaunt J, Solomon D, Grossman J, Neves EJ (2010) Black carbon affects the cycling of non-black carbon in soil. Org Geochem 41(2):206–213
56. Kloss S, Zehetner F, Wimmer B, Buecker J, Rempt F, Soja G (2014) Biochar application to temperate soils: effects on soil fertility and crop growth under greenhouse conditions. J Plant Nutr Soil Sci 177(1):3–15

57. Lehmann J, Pereira da Silva J, Steiner C, Nehls T (2003) Nutrient availability and leaching in an archaeological Anthrosol and a Ferralsol of the Central Amazon basin: fertilizer, manure and charcoal amendments. Plant Soil 249(2):343–357
58. Asai H, Samson BK, Stephan HM, Songyikhangsuthor K, Homma K, Kiyono Y, Inoue Y, Shiraiwa T, Horie T (2009) Biochar amendment techniques for upland rice production in Northern Laos: 1. Soil physical properties, leaf SPAD and grain yield. J Field Crops Res 111(1–2):81–84
59. Lehmann J, Joseph S (2009) Biochar for environmental management, L. technology. Earthscan, Biochar for environmental management: an introduction, pp 1–12. In: Lehmann J, Joseph S (ed) Biochar for environmental management, science and technology. Earthscan, London
60. Laird DA, Fleming P, Davis DD, Horton R, Wang B, Karlen DL (2010) Impact of biochar amendments on the quality of a typical Midwestern agricultural soil. Geoderma 158(3–4):443–449
61. Woolf D, Amonette JE, Street-Perrott FA, Lehmann J, Joseph S (2010) Sustainable biochar to mitigate global climate change. J Nat Commun 1(1):1–9
62. Yao Y, Gao B, Zhang M, Inyang M, Zimmerman AR (2012) Effect of biochar amendment on sorption and leaching of nitrate, ammonium, and phosphate in a sandy soil. J Chemosph 89(11):1467–1471
63. Glaser B, Lehmann J, Zech W (2002) Ameliorating physical and chemical properties of highly weathered soils in the tropics with charcoal—a review. Biol Fertil Soils 35(4):219–230
64. Gomez J, Denef K, Stewart C, Zheng J, Cotrufo MF (2014) Biochar addition rate influences soil microbial abundance and activity in temperate soils. Eur J Soil Sci 65(1):28–39
65. Quilliam RS, Glanville HC, Wade SC, Jones DL (2013) Life in the 'charosphere'–Does biochar in agricultural soil provide a significant habitat for microorganisms? Soil Biol Biochem 65:287–293
66. Steiner C, Glaser B, Geraldes Teixeira W, Lehmann J, Blum WE, Zech W (2008) Nitrogen retention and plant uptake on a highly weathered central Amazonian Ferralsol amended with compost and charcoal. J Plant Nutr Soil Sci 171(6):893–899
67. Karhu K, Mattila T, Bergström I, Regina K (2011) Biochar addition to agricultural soil increased CH4 uptake and water holding capacity—results from a short-term pilot field study. Agric Ecosyst Environ 140(1–2):309–313
68. Sun F, Lu S (2014) Biochars improve aggregate stability, water retention, and pore-space properties of clayey soil. J Plant Nutr Soil Sci 177(1):26–33
69. Wang D, Zhang W, Hao X, Zhou D (2013) Transport of biochar particles in saturated granular media: effects of pyrolysis temperature and particle size. Environ Sci Technol 47(2):821–828
70. Spokas K, Koskinen W, Baker J, Reicosky D (2009) Impacts of woodchip biochar additions on greenhouse gas production and sorption/degradation of two herbicides in a Minnesota soil. Chemosphere 77(4):574–581
71. Spokas KA, Baker JM, Reicosky DC (2010) Ethylene: potential key for biochar amendment impacts. J Plant Soil 333(1):443–452
72. Beesley L, Moreno-Jiménez E, Gomez-Eyles JL (2010) Effects of biochar and greenwaste compost amendments on mobility, bioavailability and toxicity of inorganic and organic contaminants in a multi-element polluted soil. Environ Pollut 158(6):2282–2287
73. Hale S, Hanley K, Lehmann J, Zimmerman A, Cornelissen G (2011) Effects of chemical, biological, and physical aging as well as soil addition on the sorption of pyrene to activated carbon and biochar. Technology 45(24):10445–10453
74. Uchimiya M, Wartelle LH, Klasson KT, Fortier CA, Lima IM (2011) Influence of pyrolysis temperature on biochar property and function as a heavy metal sorbent in soil. J Agric Food Chem 59(6):2501–2510
75. Barrow CJ (2012) Biochar: potential for countering land degradation and for improving agriculture. Appl Geogr 34:21–28
76. Sohi SP (2012) Carbon storage with benefits. Science 338(6110):1034–1035
77. Blackwell P, Joseph S, Munroe P, Anawar HM, Storer P, Gilkes RJ, Solaiman ZM (2015) Influences of biochar and biochar-mineral complex on mycorrhizal colonisation and nutrition of wheat and sorghum. Plant 25(5):686–695

78. Abel S, Peters A, Trinks S, Schonsky H, Facklam M, Wessolek G (2013) Impact of biochar and hydrochar addition on water retention and water repellency of sandy soil. Geoderma 202:183–191
79. Abujabhah IS, Bound SA, Doyle R, Bowman JP (2016) Effects of biochar and compost amendments on soil physico-chemical properties and the total community within a temperate agricultural soil. Appl Soil Ecol 98:243–253
80. Tong H, Hu M, Li F, Liu C, Chen M (2014) Biochar enhances the microbial and chemical transformation of pentachlorophenol in paddy soil. Soil Biol Biochem 70:142–150
81. Fang Y, Singh B, Singh B, Krull E (2014) Biochar carbon stability in four contrasting soils. Eur J Soil Sci 65(1):60–71
82. Whitman T, Lehmann J (2009) Biochar—one way forward for soil carbon in offset mechanisms in Africa? Environ Sci Policy 12(7):1024–1027
83. Liu X, Zhang A, Ji C, Joseph S, Bian R, Li L, Pan G, Paz-Ferreiro J (2013) Biochar's effect on crop productivity and the dependence on experimental conditions—a meta-analysis of literature data. Plant Soil 373(1):583–594
84. Bridle T, Pritchard D (2004) Energy and nutrient recovery from sewage sludge via pyrolysis. Water Sci Technol 50(9):169–175
85. Chan KY, Van Zwieten L, Meszaros I, Downie A, Joseph S (2007) Agronomic values of greenwaste biochar as a soil amendment. Aust J Soil Res 45(8):629–634
86. Wang J, Pan X, Liu Y, Zhang X, Xiong Z (2012) Effects of biochar amendment in two soils on greenhouse gas emissions and crop production. Plant Soil 360(1):287–298
87. Crane-Droesch A, Abiven S, Jeffery S, Torn MS (2013) Heterogeneous global crop yield response to biochar: a meta-regression analysis. Environ Res Lett 8(4):044049
88. Dahlawi S, Naeem A, Iqbal M, Farooq MA, Bibi S, Rengel Z (2018) Opportunities and challenges in the use of mineral nutrition for minimizing arsenic toxicity and accumulation in rice: a critical review. Chemosphere 194:171–188
89. Pietikäinen J, Kiikkilä O, Fritze H (2000) Charcoal as a habitat for microbes and its effect on the microbial community of the underlying humus. 89(2):231–242
90. Kim J-S, Sparovek G, Longo RM, De Melo WJ, Crowley D (2007) Bacterial diversity of terra preta and pristine forest soil from the Western Amazon. Soil Biol Biochem 39(2):684–690
91. O'neill B, Grossman J, Tsai M, Gomes JE, Lehmann J, Peterson J, Neves E, Thies JE (2009) Bacterial community composition in Brazilian Anthrosols and adjacent soils characterized using culturing and molecular identification. Microb Ecol 58(1):23–35
92. Grossman JM, O'Neill BE, Tsai SM, Liang B, Neves E, Lehmann J, Thies JE (2010) Amazonian anthrosols support similar microbial communities that differ distinctly from those extant in adjacent, unmodified soils of the same mineralogy. Microb Ecol 60(1):192–205
93. Rillig MC, Mummey DLJNP (2006) Mycorrhizas and soil structure. 171(1):41–53
94. Warnock DD, Lehmann J, Kuyper TW, Rillig MC (2007) Mycorrhizal responses to biochar in soil–concepts and mechanisms. Soil 300(1):9–20
95. Schwartz MW, Hoeksema JD, Gehring CA, Johnson NC, Klironomos JN, Abbott LK, Pringle A (2006) The promise and the potential consequences of the global transport of mycorrhizal fungal inoculum. Ecol Lett 9(5):501–515
96. Compant S, Clément C, Sessitsch A (2010) Plant growth-promoting bacteria in the rhizo- and endosphere of plants: their role, colonization, mechanisms involved and prospects for utilization. Soil Biol Biochem 42(5):669–678
97. Wardle DA, Nilsson M-C, Zackrisson O (2008) Fire-derived charcoal causes loss of forest humus. Science 320(5876):629–629
98. Kuzyakov Y, Subbotina I, Chen H, Bogomolova I, Xu X (2009) Black carbon decomposition and incorporation into soil microbial biomass estimated by 14C labeling. Soil Biol Biochem 41(2):210–219
99. Ding Y, Liu Y, Liu S, Li Z, Tan X, Huang X, Zeng G, Zhou L, Zheng B (2016) Biochar to improve soil fertility. A review. Agron Sustain Devolep 36(2):1–18
100. Dhuldhaj UP, Malik N (2022) Global perspective of phosphate soliloquizing microbes and phosphatase for improvement of soil, food and human health. Cell Mol Biomed Rep 2(3):173–186. https://doi.org/10.55705/cmbr.2022.347523.1048

101. Taylor C (1951) The nutritional requirements of the predominant bacterial flora of the soil. In: Proceedings of the society for applied bacteriology. Wiley Online Library, pp 101–111
102. George N, Davies JT (1988) Parameters affecting adsorption of microorganisms on activated charcoal cloth. J Chem Technol Biotechnol 43(3):173–186
103. Ezawa T, Yamamoto K, Yoshida S (2002) Enhancement of the effectiveness of indigenous arbuscular mycorrhizal fungi by inorganic soil amendments. Soil Sci Plant Nutrit 48(6):897–900
104. Saito M, Marumoto T (2002) Inoculation with arbuscular mycorrhizal fungi: the status quo in Japan and the future prospects. Plant Soil 273–279
105. Thies JE, Rillig MC (2012) Characteristics of biochar: biological properties, Biochar for environmental management. Routledge, pp 117–138
106. Chen H, Yao J, Wang F, Choi MM, Bramanti E, Zaray G (2009) Study on the toxic effects of diphenol compounds on soil microbial activity by a combination of methods. J Hazard Mater 167(1–3):846–851
107. Kasozi GN, Zimmerman AR, Nkedi-Kizza P, Gao B (2010) Catechol and humic acid sorption onto a range of laboratory-produced black carbons (biochars). J Environ Technol 44(16):6189–6195
108. Schimel J, Balser TC, Wallenstein M (2007) Microbial stress-response physiology and its implications for ecosystem function. Ecology 88(6):1386–1394
109. Bardgett R (2005) The biology of soil: a community and ecosystem approach. Oxford university press
110. Major J, Rondon M, Molina D, Riha SJ, Lehmann J (2010) Maize yield and nutrition during 4 years after biochar application to a Colombian savanna oxisol. Plant Soil 333(1):117–128
111. Domene X, Mattana S, Hanley K, Enders A, Lehmann J (2014) Medium-term effects of corn biochar addition on soil biota activities and functions in a temperate soil cropped to corn. Soil Biol Biochem 72:152–162
112. Woolf D, Lehmann J (2012) Modelling the long-term response to positive and negative priming of soil organic carbon by black carbon. Biogeochemistry 111(1):83–95
113. Yanai Y, Toyota K, Okazaki M (2007) Effects of charcoal addition on N2O emissions from soil resulting from rewetting air-dried soil in short-term laboratory experiments. Soil Sci Plant Nutr 53(2):181–188
114. Van Zwieten L, Singh B, Joseph S, Kimber S, Cowie A, Chan KY (2012) Biochar and emissions of non-$CO_2$ greenhouse gases from soil, Biochar for environmental management. Routledge, pp 259–282
115. DeLuca TH, Gundale MJ, MacKenzie MD, Jones DL (2015) Biochar effects on soil nutrient transformations, Biochar for environmental management. Routledge, pp 453–486
116. Ajema L (2018) Effects of biochar application on beneficial soil organism. Int J Res Stud Sci Eng Technol 5(5):9–18
117. Abdu H, Robinson D, Jones SB (2007) Comparing bulk soil electrical conductivity determination using the DUALEM-1S and EM38-DD electromagnetic induction instruments. Soil Water Manage Conserv 71(1):189–196
118. Ameloot N, Graber ER, Verheijen FG, De Neve S (S) Interactions between biochar stability and soil organisms: review and research needs. Eur J Siol Sci 64(4):379–390
119. Atkinson CJ, Fitzgerald JD, Hipps NA (2010) Potential mechanisms for achieving agricultural benefits from biochar application to temperate soils: a review. Plant Soil 337(1):1–18
120. Lehmann J (2007) Bio-energy in the black. Front Ecol Environ 5(7):381–387
121. Jeffery S, Bezemer TM, Cornelissen G, Kuyper TW, Lehmann J, Mommer L, Sohi SP, van de Voorde TF, Wardle DA, van Groenigen JW (2015) The way forward in biochar research: targeting trade-offs between the potential wins. GCB Bioenergy 7(1):1–13
122. Kolb SE, Fermanich KJ, Dornbush ME (2009) Effect of charcoal quantity on microbial biomass and activity in temperate soils. Soil Sci Soc Am J 73(4):1173–1181
123. Ahmad M, Ok YS, Kim B-Y, Ahn J-H, Lee YH, Zhang M, Moon DH, Al-Wabel MI, Lee SS (2016) Impact of soybean stover-and pine needle-derived biochars on Pb and As mobility, microbial community, and carbon stability in a contaminated agricultural soil. J Environ Manage 166:131–139

124. Zhu X, Chen B, Zhu L, Xing B (2017) Effects and mechanisms of biochar-microbe interactions in soil improvement and pollution remediation: a review. Environ Pollut 227:98–115
125. Nielsen S, Minchin T, Kimber S, van Zwieten L, Gilbert J, Munroe P, Joseph S, Thomas T (2014) Comparative analysis of the microbial communities in agricultural soil amended with enhanced biochars or traditional fertilisers. Agric Ecosyst Environ 191:73–82
126. Mackie K, Marhan S, Ditterich F, Schmidt H, Kandeler E (2015) The effects of biochar and compost amendments on copper immobilization and soil microorganisms in a temperate vineyard. Agric Ecosyst Environ 201:58–69
127. Tejada M, Benítez C, Parrado J (2011) Application of biostimulants in benzo (a) pyrene polluted soils: short-time effects on soil biochemical properties. Appl Soil Ecol 50:21–26
128. Finkel OM, Castrillo G, Paredes SH, González IS, Dangl JL (2017) Understanding and exploiting plant beneficial microbes. Curr Opinion Plant Biol 38:155–163
129. Lakshmanan V, Selvaraj G, Bais HP (2014) Functional soil microbiome: belowground solutions to an aboveground problem. Plant Physiol 166(2):689–700
130. Kumar A, Dubey A (2020) Rhizosphere microbiome: Engineering bacterial competitiveness for enhancing crop production. J Adv Res 24:337–352
131. Lyu D, Msimbira LA, Nazari M, Antar M, Pagé A, Shah A, Monjezi N, Zajonc J, Tanney CA, Backer R (2021) The coevolution of plants and microbes underpins sustainable agriculture. Microorganisms 9(5):1036
132. Ke W, Zhang X, Zhu F, Wu H, Zhang Y, Shi Y, Hartley W, Xue S (2021) Appropriate human intervention stimulates the development of microbial communities and soil formation at a long-term weathered bauxite residue disposal area. J Hazard Mater 405:124689
133. Basu S, Kumar G, Chhabra S, Prasad R (2021) Role of soil microbes in biogeochemical cycle for enhancing soil fertility, New and future developments in microbial biotechnology and bioengineering. Elsevier, pp 149–157
134. Nehal N, Rathore US, Sharma N (2021) Microbes and soil health for sustainable crop production, Microbial metatranscriptomics belowground. Springer, pp 581–613
135. Rogers C, Oldroyd GED (2014) Synthetic biology approaches to engineering the nitrogen symbiosis in cereals. J Exp Bot 65(8):1939–1946
136. Ryu M-H, Zhang J, Toth T, Khokhani D, Geddes BA, Mus F, Garcia-Costas A, Peters JW, Poole PS, Ané J-M, Voigt CA (2020) Control of nitrogen fixation in bacteria that associate with cereals. Nat Microbiol 5(2):314–330
137. Verbruggen E, Van Der Heijden MG, Weedon JT, Kowalchuk GA, Röling WFM (2012) Community assembly, species richness and nestedness of arbuscular mycorrhizal fungi in agricultural soils. Mol Ecol 21(10):2341–2353
138. Nemergut DR, Schmidt SK, Fukami T, O'Neill SP, Bilinski TM, Stanish LF, Knelman JE, Darcy JL, Lynch RC, Wickey P (2013) Patterns and processes of microbial community assembly. Microbio Biol Mol Rev 77(3):342–356
139. Cordero OX, Polz MF (2014) Explaining microbial genomic diversity in light of evolutionary ecology. Nat Rev Microbiol 12(4):263–273
140. Yuanfan H, Jin Z, Qing H, Qian W, Jiandong J, Shunpeng L (2010) Characterization of a fenpropathrin-degrading strain and construction of a genetically engineered microorganism for simultaneous degradation of methyl parathion and fenpropathrin. J Environ Manage 91(11):2295–2300
141. Kumar NM, Muthukumaran C, Sharmila G, Gurunathan B (2018) Genetically modified organisms and its impact on the enhancement of bioremediation. Bioremediation: applications for environmental protection and management. Springer, pp 53–76
142. Chen J, Liu X, Zheng J, Zhang B, Lu H, Chi Z, Pan G, Li L, Zheng J, Zhang X (2013) Biochar soil amendment increased bacterial but decreased fungal gene abundance with shifts in community structure in a slightly acid rice paddy from Southwest China. Appl Soil Ecol 71:33–44
143. Wong JW, Ogbonnaya UO (2021) Biochar porosity: a nature-based dependent parameter to deliver microorganisms to soils for land restoration. Environ Sci Pollut Res Int 28(34):46894–46909

144. Abbasniayzare SK, Sedaghathoor S, Dahkaei MNP (2012) Effect of biofertilizer application on growth parameters of Spathiphyllum illusion. 12(5):669–673
145. Ortiz A, Sansinenea E (2021) Recent advancements for microorganisms and their natural compounds useful in agriculture. Appl Microbiol Biotechnol 105(3):891–897
146. Panpatte DG, Jhala YK, Vyas RV, Shelat HN (2017) Microorganisms for green revolution. Springer
147. Wong CKF, Teh CY (2021) Impact of biofertilizers on horticultural crops. 39–103
148. Husson O, Sarthou J-P, Bousset L, Ratnadass A, Schmidt H-P, Kempf J, Husson B, Tingry S, Aubertot J-N, Deguine J-P (2021) Soil and plant health in relation to dynamic sustainment of Eh and pH homeostasis: a review. Plant Soil 466(1):391–447
149. Meena RS, Das A, Yadav GS, Lal R (2018) Legumes for soil health and sustainable management. Springer
150. de la Fuente Cantó C, Simonin M, King E, Moulin L, Bennett MJ, Castrillo G, Laplaze L (2020) An extended root phenotype: the rhizosphere, its formation and impacts on plant fitness. Plant J 103(3):951–964
151. Basu A, Prasad P, Das SN, Kalam S, Sayyed R, Reddy M, El Enshasy H (2021) Plant growth promoting rhizobacteria (PGPR) as green bioinoculants: recent developments, constraints, and prospects. Sustainability 13(3):1140
152. Khoshru B, Mitra D, Khoshmanzar E, Myo EM, Uniyal N, Mahakur B, Mohapatra PKD, Panneerselvam P, Boutaj H, Alizadeh M (2020) Current scenario and future prospects of plant growth-promoting rhizobacteria: an economic valuable resource for the agriculture revival under stressful conditions. J Plant Nutrit 43(20):3062–3092
153. Medfu Tarekegn M, Zewdu Salilih F, Ishetu AI (2020) Agriculture, Microbes used as a tool for bioremediation of heavy metal from the environment. 6(1):1783174
154. Fasusi OA, Cruz C, Babalola O (2021) Agricultural sustainability: microbial biofertilizers in rhizosphere management. Agriculture 11(2):163
155. Bhardwaj D, Ansari MW, Sahoo RK, Tuteja N (2014) Biofertilizers function as key player in sustainable agriculture by improving soil fertility, plant tolerance and crop productivity. Microb Cell Fact 13(1):1–10
156. Gupta G, Parihar SS, Ahirwar NK, Snehi SK, Singh V (2015) Plant growth promoting rhizobacteria (PGPR): current and future prospects for development of sustainable agriculture. J Microb Biochem Technol 7(2):096–102
157. Zambrano-Mendoza JL, Sangoquiza-Caiza CA, Campaña-Cruz DF, Yánez-Guzmán CFY (2021) Use of biofertilizers in agricultural production, vol 193
158. Liu R-C, Xiao Z-Y, Hashem A, Abd_Allah EF, Wu Q-S (2021) Mycorrhizal fungal diversity and its relationship with soil properties in Camellia oleifera. Agriculture 11(6):470
159. Diagne N, Ngom M, Djighaly PI, Fall D, Hocher V, Svistoonoff S (2020) Roles of arbuscular mycorrhizal fungi on plant growth and performance: importance in biotic and abiotic stressed regulation. Diversity 12(10):370
160. Mitra D, Djebaili R, Pellegrini M, Mahakur B, Sarker A, Chaudhary P, Khoshru B, Gallo MD, Kitouni M, Barik DP (2021) Arbuscular mycorrhizal symbiosis: plant growth improvement and induction of resistance under stressful conditions. 44(13):1993–2028
161. Itelima J, Bang W, Onyimba I, Sila M, Egbere O (2018) Bio-fertilizers as key player in enhancing soil fertility and crop productivity: a review
162. Islam T, Li Y, Cheng H (2021) Biochars and engineered biochars for water and soil remediation: a review. Sustainability 13(17):9932
163. Hou X, Cui J, Liu W, Jiang N, Zhou X, Qi H, Meng J, Luan Y (2020) LncRNA39026 enhances tomato resistance to Phytophthora infestans by decoying miR168a and inducing PR gene expression. Phytopathology 110(4):873–880
164. Wang B, Jiang Y-S, Li F-Y, Yang D-Y (2017) Preparation of biochar by simultaneous carbonization, magnetization and activation for norfloxacin removal in water. Boresour Technol 233:159–165
165. Zhou X, Moghaddam TB, Chen M, Wu S, Adhikari S, Xu S, Yang C (2020) Life cycle assessment of biochar modified bioasphalt derived from biomass. ACS Sustain Chem Eng 8(38):14568–14575

166. Guo M, He Z, Uchimiya SM (2016) Introduction to biochar as an agricultural and environmental amendment. 63:1–14
167. Guo M (2020) The 3R principles for applying biochar to improve soil health. Soil Syst 4(1):9
168. Song W, Guo M (2012) Quality variations of poultry litter biochar generated at different pyrolysis temperatures. J Anal Appl Porolys 94:138–145
169. Chen Y, Zhang X, Chen W, Yang H, Chen H (2017) The structure evolution of biochar from biomass pyrolysis and its correlation with gas pollutant adsorption performance. Bioresour Technol 246:101–109
170. Chun Y, Sheng G, Chiou CT, Xing B (2004) Compositions and sorptive properties of crop residue-derived chars. Environ Sci Technol 38(17):4649–4655
171. Marco Keiluweit M, Nico P, Johnson M, Kleber MG (2010) Dynamic molecular structure of plant biomass-derived black carbon (biochar). Environ Sci 44:1247–1253
172. Nartey OD, Zhao B (2014) Biochar preparation, characterization, and adsorptive capacity and its effect on bioavailability of contaminants: an overview. Adv Mater Sci Eng 2014
173. Solaiman ZM, Blackwell P, Abbott LK, Storer P (2010) Direct and residual effect of biochar application on mycorrhizal root colonisation, growth and nutrition of wheat. Aust J Soil Sci 48(7):546–554
174. Hale L, Luth M, Kenney R, Crowley D (2014) Evaluation of pinewood biochar as a carrier of bacterial strain Enterobacter cloacae UW5 for soil inoculation. Appl Soil Ecol 84:192–199
175. Kolton M, Meller Harel Y, Pasternak Z, Graber ER, Elad Y, Cytryn E (2011) Impact of biochar application to soil on the root-associated bacterial community structure of fully developed greenhouse pepper plants. Appl Environ Microbiol 77(14):4924–4930
176. Rousk J, Brookes PC, Bååth E (2009) Contrasting soil pH effects on fungal and bacterial growth suggest functional redundancy in carbon mineralization. Appl Environ Microbiol 75(6):1589–1596
177. Abit SM, Bolster CH, Cai P, Walker SL (2012) Influence of feedstock and pyrolysis temperature of biochar amendments on transport of Escherichia coli in saturated and unsaturated soil. Environ Sci Technol 46(15):8097–8105
178. kheyrodin H, Jami R, Rehman FU (2022) Cellular structure and molecular functions of plants, animals, bacteria, and viruses. Cell Mol Biomed Rep 2(1):33–41. https://doi.org/10.55705/cmbr.2022.330941.1021
179. Quilliam RS, Rangecroft S, Emmett BA, Deluca TH, Jones DL (2013) Is biochar a source or sink for polycyclic aromatic hydrocarbon (PAH) compounds in agricultural soils? GCB Bioenergy 5(2):96–103
180. Chen Z, Xiao X, Chen B, Zhu L (2015) Quantification of chemical states, dissociation constants and contents of oxygen-containing groups on the surface of biochars produced at different temperatures. Environ Sci Technol 49(1):309–317
181. El-Naggar AH, Usman AR, Al-Omran A, Ok YS, Ahmad M, Al-Wabel MIJ (2015) Carbon mineralization and nutrient availability in calcareous sandy soils amended with woody waste biochar. Chemosphere 138:67–73
182. Jien S-H, Wang C-S (2013) Effects of biochar on soil properties and erosion potential in a highly weathered soil. Chemosphere 110:225–233
183. Liang B, Lehmann J, Solomon D, Kinyangi J, Grossman J, O'Neill B, Skjemstad JO, Thies J, Luizão FJ, Petersen J (2006) Black carbon increases cation exchange capacity in soils. Soil Sci Soc Am J 70(5):1719–1730
184. Mukherjee A, Zimmerman A, Harris W (2011) Surface chemistry variations among a series of laboratory-produced biochars. Geoderma 163(3–4):247–255
185. Yuan H, Lu T, Wang Y, Chen Y, Lei T (2016) Sewage sludge biochar: Nutrient composition and its effect on the leaching of soil nutrients. Geoderma 267:17–23
186. Yuan J-H, Xu R-K, Zhang H (2011) The forms of alkalis in the biochar produced from crop residues at different temperatures. Boresour Technol 102(3):3488–3497
187. Chen B, Zhou D, Zhu L (2008) Transitional adsorption and partition of nonpolar and polar aromatic contaminants by biochars of pine needles with different pyrolytic temperatures. Environ Sci Technol 42(14):5137–5143

188. Jeong CY, Dodla SK, Wang JJ (2016) Fundamental and molecular composition characteristics of biochars produced from sugarcane and rice crop residues and by-products. Chemosphere 142:4–13
189. Mukherjee A, Zimmerman AR (2013) Organic carbon and nutrient release from a range of laboratory-produced biochars and biochar–soil mixtures. Geoderma 193:122–130
190. Novak JM, Busscher WJ, Laird DL, Ahmedna M, Watts DW, Niandou MAS (2009) Impact of biochar amendment on fertility of a southeastern coastal plain soil. Soil Sci 174(2):105–112
191. Zhang Q, Zhou W, Liang G, Sun J, Wang X, He P (2015) Distribution of soil nutrients, extracellular enzyme activities and microbial communities across particle-size fractions in a long-term fertilizer experiment. Appl Soil Ecol 94:59–71
192. Ghidotti M, Fabbri D, Hornung A (2017) Engineering, Profiles of volatile organic compounds in biochar: insights into process conditions and quality assessment. ACS Sustain Chem Eng 5(1):510–517
193. Lyu H, He Y, Tang J, Hecker M, Liu Q, Jones PD, Codling G, Giesy JP (2016) Effect of pyrolysis temperature on potential toxicity of biochar if applied to the environment. Environ Pollut 218:1–7
194. Spokas KA, Novak JM, Stewart CE, Cantrell KB, Uchimiya M, DuSaire MG, Ro KS (2011) Qualitative analysis of volatile organic compounds on biochar. Chemosphere 85(5):869–882
195. Liang C, Zhu X, Fu S, Méndez A, Gascó G, Paz-Ferreiro J (2014) Biochar alters the resistance and resilience to drought in a tropical soil. Environ Res Lett 9(6):064013
196. Cantrell KB, Hunt PG, Uchimiya M, Novak JM, Ro KS (2012) Impact of pyrolysis temperature and manure source on physicochemical characteristics of biochar. Bioresour Technol 107:419–428
197. Bandick AK, Dick RP (1999) Field management effects on soil enzyme activities. Soil Biol Biochem 31(11):1471–1479
198. Luo L, Gu J-D (2016) Alteration of extracellular enzyme activity and microbial abundance by biochar addition: implication for carbon sequestration in subtropical mangrove sediment. J Environ Manage 182:29–36
199. Paz-Ferreiro J, Fu S, Méndez A, Gascó G (2015) Biochar modifies the thermodynamic parameters of soil enzyme activity in a tropical soil. J Soils Sediments 15(3):578–583
200. Zimmerman AR, Ahn M-Y (2010) Organo-mineral–enzyme interaction and soil enzyme activity, Soil enzymology. Springer, pp 271–292
201. Bailey VL, Fansler SJ, Smith JL, Bolton Jr H (2011) Reconciling apparent variability in effects of biochar amendment on soil enzyme activities by assay optimization. Soil Biol Biochem 43(2):296–301
202. Lammirato C, Miltner A, Kaestner M 2011) Effects of wood char and activated carbon on the hydrolysis of cellobiose by β-glucosidase from Aspergillus niger. Soil Biol Biochem 43(9):1936–1942
203. Gibson C, Berry TD, Wang R, Spencer JA, Johnston CT, Jiang Y, Bird JA, Filley TR (2016) Weathering of pyrogenic organic matter induces fungal oxidative enzyme response in single culture inoculation experiments. Org Geochem 92:32–41
204. Qian L, Chen B (2013) Dual role of biochars as adsorbents for aluminum: the effects of oxygen-containing organic components and the scattering of silicate particles. Environ Sci Technol 47(15):8759–8768
205. Qian L, Chen M, Chen B (2015) Competitive adsorption of cadmium and aluminum onto fresh and oxidized biochars during aging processes. J Soils and Sediments 15(5):1130–1138
206. Zielińska A, Oleszczuk P (2016) Bioavailability and bioaccessibility of polycyclic aromatic hydrocarbons (PAHs) in historically contaminated soils after lab incubation with sewage sludge-derived biochars. Chemosphere 163:480–489
207. Amundson R, Berhe A, Hopmans J, Olson C, Sztein A, Sparks D (2015) Soil and human security in the twenty-first century. Science 348:1261071
208. Dubois O (2011) The state of the world's land and water resources for food and agriculture: managing systems at risk. Earthscan

209. Bryan BA, Gao L, Ye Y, Sun X, Connor JD, Crossman ND, Stafford-Smith M, Wu J, He C, Yu D (2018) China's response to a national land-system sustainability emergency. Nature 559(7713):193–204
210. Pennock D, McKenzie N, Montanarella L (2015) Rome, Italy, Status of the world's soil resources
211. Hou D, O'Connor D (2020) Green and sustainable remediation: concepts, principles, and pertaining research, Sustainable remediation of contaminated soil and groundwater. Elsevier, pp 1–17
212. Wang L, Ok YS, Tsang DC, Alessi DS, Rinklebe J, Wang H, Mašek O, Hou R, O'Connor D, Hou D (2020) New trends in biochar pyrolysis and modification strategies: feedstock, pyrolysis conditions, sustainability concerns and implications for soil amendment. Soil Use Manage 36(3):358–386
213. Rodríguez-Eugenio N, McLaughlin M, Pennock D (2018) Soil pollution: a hidden reality, FAO
214. Hou D, O'Connor D, Igalavithana AD, Alessi DS, Luo J, Tsang DC, Sparks DL, Yamauchi Y, Rinklebe J, Ok YS (2020) Metal contamination and bioremediation of agricultural soils for food safety and sustainability. Nat Rev Earth Environ 1(7):366–381
215. O'Connor D, Hou D, Ok YS, Lanphear BP (2020) The effects of iniquitous lead exposure on health. Nat Sustain 3(2):77–79
216. Zhang J, Hou D, Shen Z, Jin F, O'Connor D, Pan S, Ok YS, Tsang DC, Bolan NS, Alessi DS (2020) Effects of excessive impregnation, magnesium content, and pyrolysis temperature on MgO-coated watermelon rind biochar and its lead removal capacity. Environ Res 183:109152
217. Jing F, Chen X, Wen X, Liu W, Hu S, Yang Z, Guo B, Luo Y, Yu Q, Xu Y (2020) Biochar effects on soil chemical properties and mobilization of cadmium (Cd) and lead (Pb) in paddy soil. Soil Use Manage 36(2):320–327
218. O'Connor D, Peng T, Zhang J, Tsang DC, Alessi DS, Shen Z, Bolan NS, Hou D (2018) Biochar application for the remediation of heavy metal polluted land: a review of in situ field trials. Sci Environ Manage 619:815–826
219. Wang L, Bolan NS, Tsang DC, Hou D (2020) Green immobilization of toxic metals using alkaline enhanced rice husk biochar: effects of pyrolysis temperature and KOH concentration. Sci Total Environ 720:137584
220. Hou R, Wang L, O'Connor D, Tsang DC, Rinklebe J, Hou D (2020) Effect of immobilizing reagents on soil Cd and Pb lability under freeze-thaw cycles: implications for sustainable agricultural management in seasonally frozen land. Environ Int 144:106040
221. Ghosh D, Masto RE, Maiti SK (2020) Ameliorative effect of Lantana camara biochar on coal mine spoil and growth of maize (Zea mays). Soil Use Manage 36(4):726–739
222. Ye L, Camps-Arbestain M, Shen Q, Lehmann J, Singh B, Sabir M (2020) Biochar effects on crop yields with and without fertilizer: a meta-analysis of field studies using separate controls. Soil Use Manage 36(1):2–18
223. Fonseca AA, Santos DA, Passos RR, Andrade FV, Rangel OJP (2020) Phosphorus availability and grass growth in biochar-modified acid soil: a study excluding the effects of soil pH. Soil Use Manage 36(4):714–725
224. Maroušek J, Strunecký O, Stehel V (2019) Biochar farming: defining economically perspective applications. Clean Technol Environ Policy 21(7):1389–1395
225. Cui Z, Zhang H, Chen X, Zhang C, Ma W, Huang C, Zhang W, Mi G, Miao Y, Li X (2018) Pursuing sustainable productivity with millions of smallholder farmers. Nature 555(7696):363–366
226. Hazell PB, Poulton C, Wiggins S, Dorward A (2007) The future of small farms for poverty reduction and growth. Int Food Policy Res Inst
227. Kremen C, Merenlender AM (2018) Landscapes that work for biodiversity and people. Science 362(6412):eaau6020
228. Mäder PJA, Dubois D, Gunst L, Fried P, Niggli U (2002) Flie? Bach 1694–1697
229. Isbell F, Calcagno V, Hector A, Connolly J, Harpole WS, Reich PB, Scherer-Lorenzen M, Schmid B, Tilman D, Van Ruijven J (2011) High plant diversity is needed to maintain ecosystem services. Nature 477(7363):199–202

230. Quéré C, Andrew RM, Friedlingstein P, Sitch S, Hauck J, Pongratz J, Pickers PA, Ivar Korsbakken J, Peters GP, Canadell JG (2018) Global carbon budget
231. Chowaniak M, Głąb T, Klima K, Niemiec M, Zaleski T, Zuzek D (2020) Effect of tillage and crop management on runoff, soil erosion and organic carbon loss. Soil Use Manage 36(4):581–593
232. Singh K, Whelan B (2020) Soil carbon change across ten New South Wales farms under different farm management regimes in Australia. Soil Use Manage 36(4):616–632
233. Bhatia A, Sasmal S, Jain N, Pathak H, Kumar R, Singh A (2010) Mitigating nitrous oxide emission from soil under conventional and no-tillage in wheat using nitrification inhibitors. Agric Ecosyst Environ 136(3–4):247–253
234. Hu J, Inglett KS, Wright AL, Clark MW, Reddy KR (2020) Nitrous oxide dynamics during denitrification along a hydrological gradient of subtropical grasslands. Soil Use Manage 36(4):682–692
235. Livingston JE, Lövbrand E, Olsson JA (2018) From climates multiple to climate singular: maintaining policy-relevance in the IPCC synthesis report. Environ Sci Policy 90:83–90
236. Zhao B, O'Connor D, Shen Z, Tsang DC, Rinklebe J, Hou D (2020) Sulfur-modified biochar as a soil amendment to stabilize mercury pollution: an accelerated simulation of long-term aging effects. Environ Pollit 264:114687
237. Li N, Ma X, Xu H, Feng Y, Ren G, Yang G, Han X, Wang X, Ren C (2020) Biochar addition mitigates nitrogen loss induced by straw incorporation and nitrogen fertilizer application. Soil Use Manage 36(4):751–765
238. Ma R, Guan S, Dou S, Wu D, Xie S, Ndzelu BS (2020) Different rates of biochar application change 15N retention in soil and 15N utilization by maize. Soil Use Manage 36(4):773–782
239. Lou K, Rajapaksha AU, Ok YS, Chang SX(2016) Sorption of copper (II) from synthetic oil sands process-affected water (OSPW) by pine sawdust biochars: effects of pyrolysis temperature and steam activation. J Soils Sediments 16(8):2081–2089
240. Mohan D, Sarswat A, Ok YS, Pittman Jr CU (2014) Organic and inorganic contaminants removal from water with biochar, a renewable, low cost and sustainable adsorbent—a critical review. Bioresour Technol 160:191–202
241. Li Y, Li Y, Chang SX, Yang Y, Fu S, Jiang P, Luo Y, Yang M, Chen Z, Hu S (2018) Biochar reduces soil heterotrophic respiration in a subtropical plantation through increasing soil organic carbon recalcitrancy and decreasing carbon-degrading microbial activity. Soil Biol Biochem 122:173–185
242. Palansooriya KN, Ok YS, Awad YM, Lee SS, Sung J-K, Koutsospyros A, Moon DH (2019) Impacts of biochar application on upland agriculture: a review. J Environ Manage 234:52–64
243. Griffiths BS, Philippot L (2013) Insights into the resistance and resilience of the soil microbial community. FEMS Microbiol Rev 37(2):112–129
244. Kirchman DL (2018) Processes in microbial ecology. Oxford University Press

# Chapter 16
# Methanogenesis and Its Role in Climate-Change Alleviation

**Shahnaz Anjum, Asma Nazir, Bisma Farooq, Madeha Farooq, and Shayista Yousuf**

**Abstract** Methanogenesis is the biological generation of methane ($CH_4$) by anaerobic microbes belonging to the Archaea domain, also known as methanogens. Understanding how microbial methanogenesis reacts to temperature is crucial for anticipating how this powerful greenhouse gas will interact with climate change. Microorganisms in the environment play a significant role in both global and terrestrial methane emissions and sinks. Climate change mitigation efforts strive to reduce and prevent the emission of harmful greenhouse gases. Researchers have expanded on the importance of methylotrophic communities in global carbon cycle and reducing the influence of greenhouse gases such as methane, carbon dioxide, water vapours, and indirectly carbon derivatives in the environment because of their function in climate change mitigation. The positive response of the methylotrophic community is therefore changing the warm ground surface to cooler temperatures, resulting in a more adaptable habitat for species to survive. The reaction of respiratory carbon (C) emission to temperature change can be reduced over time by a compensatory thermal response in microbial activity. The mass-specific $CH_4$ respiration rates of the methanogens drop with warming and rise with cooling, implying that microbial methanogenesis has temperature-dependent compensatory responses. However, a complete mechanistic understanding of the reaction of methane cycle to global

S. Anjum (✉)
Department of Botany, Lovely Professional University, Phagwara, Punjab 144411, India
e-mail: shinaz.anjum@gmail.com

A. Nazir
Department of Clinical Biochemistry, University of Kashmir, Hazratbal, Srinagar, J&K 190006, India

B. Farooq
Department of Biomedical Engineering, Satyabhama Institute of Science and Technology, Chennai 600119, India

M. Farooq
Departement of Microbiology, Asian International University, Imphal, Manipur 795113, India

S. Yousuf
Department of Zoology, University of Kashmir, Hazratbal, Srinagar, J&K 190006, India

 
J. A. Parray (ed.), *Climate Change and Microbiome Dynamics*, Climate Change Management, https://doi.org/10.1007/978-3-031-21079-2_16

warming is still deficient. This chapter discusses the role of the methylotrophic community in reducing greenhouse gas emissions that cause climate change.

## Introduction

For more than 12,000 years, the global climate has been steady, and this stability is essential to human survival [1]. However, throughout the past century, the average global temperature surged up by 1.5 °F, and within the next 100 years, it is predicted to rise by an additional 0.5–8.6 °F. This is a critical problem since even little changes in the average global temperature can lead to significant changes in the climate and weather [2]. According to the IPCC's most recent Fifth Assessment Report, it is very likely that human activity is to blame for the phenomena of climate change that have been seen over the past few decades. Without a doubt, since the 1950s, the atmosphere and the seas have warmed, the amount of snow and ice has decreased, the sea level has risen, and greenhouse gas concentrations have pitched in a way that hasn't happened in centuries or millennia [3]. Emitted greenhouse gases are the primary determinants of anthropogenic radiative forcing. Together, $CO_2$, $CH_4$, and $N_2O$ account for more than 80% of the total radiative forcing (the cause of the greenhouse effect), and their present concentrations and rates of growth are higher than those seen in the previous 800,000 and 20,000 years, respectively [4]. While $CH_4$ (1.804 ppm) and $N_2O$ (0.324 ppm) have far higher warming potential than $CO_2$, which is by far the most prevalent greenhouse gas (GHG) in the atmosphere (390 ppm; without accounting for $H_2O$), this has moved research focus and potential mitigation techniques towards these non-$CO_2$ GHGs [5]. At the moment, one of the most complicated challenges in the world is climate change, which has implications for the scientific, economic, social, political, moral, and ethical realms [49]. It is primarily brought about by the impacts of four greenhouse gases—carbon dioxide, methane, nitrous oxide, and chlorofluorocarbons—having greater atmospheric concentrations [6]. The first three gases that are released as a result of microbial activity have a 1, 12, and 298 year atmospheric lifespan and a 100, 25, and 114 year global warming potential, respectively (Center for Climate and Energy Solutions, USA Web site). Natural ecosystems are seen to be carbon sinks, like the ocean and forest, and protecting them through silviculture and green technology is seen as another strategy to mitigate the problem. Through its efforts to mitigate climate change, United Nations Environment Protection (UNEP) plays a significant role in maintaining a low-carbon society on a global scale. To reduce greenhouse gas emissions, a variety of innovative technologies are used, including solar power, tidal power, hydrogen fuel cells, wind power, and geothermal power [7]. Processes like the flow of greenhouse gases are impacted by climate change, particularly changes in temperature and moisture content, in one of two ways: by altering the physiology of already existing microbial populations, or by altering the makeup of the microbial community. It is commonly acknowledged that microbes influence the concentration of GHGs such as $CO_2$, $CH_4$, and $N_2O$ [8]. The microbial world is extremely significant in this context because it plays a crucial role in the carbon and nitrogen cycles and is engaged in the emission and removal of gases that contribute to climate change, such as carbon dioxide and methane [9–11]. In 2005, the average global $CO_2$ concentration was roughly 380 ppm, which was nearly 80 ppm higher than the previous record high over the previous 650,000 years [12]. Numerous changes in the global environment brought about by microorganisms have also impacted them [13–15]. In reality, a number of microbes may be impacted by climate change, which might have an adverse effect on the environment, the economy, and society [16, 17]. While heterotrophic microorganisms break down organic substances to release greenhouse gases, photosynthetic microbes consume atmospheric carbon dioxide. The net carbon flow is primarily determined by the balance between the two processes, and it varies across different ecosystems based on climatic factors like temperature. As a result, microbial reactions play a critical role in the earth's carbon cycle since they not only lock up large amounts of carbon but also release it [18–20]. It is important to emphasise that most greenhouse gases, including $CO_2$, $CH_4$, and $N_2O$, are produced by bacteria [21]. Methane ($CH_4$) is a GHG that is released into the environment by some microbial communities, including those found in termite guts, rumens, marshes, and seas. As a carbon source, methane may be used by microorganisms like methanotrophs, which helps to lower the amount of GHGs in the environment. There is a knowledge deficit about the major reactions of soil bacterial and fungal populations to climate change, despite their active participation in terrestrial ecosystem function. Microbes that use reduced carbon substrates without a carbon–carbon bond are known as methylotrophs. Methanotrophic bacteria include both methylotrophs, which do not consume methane, and methanotrophs (which consume reduced carbon substrates other than methane). Apart from methane, this functional group may use substances like methanol, methylamine, dimethylamine, formate, and formaldehyde as its only sources of carbon and energy, and it frequently participates in the global carbon cycle [22, 23]. Only 5% of the world's atmospheric $CH_4$ sink is accounted for by methanotrophs' biological oxidation of $CH_4$ [20]. Prior to being released into the atmosphere, up to 90% of the $CH_4$ generated in the soil is additionally oxidised by methanotrophs [24]. Since there is less microbial variety and less evaluation of bacterial and fungal communities, there is a vacuum in our understanding of dryland environments in particular. By discussing and describing the impact of aridity change (a sort of climate change) to soil bacterial and fungal diversity, this gap is partly narrowed [25]. They examined the composition and abundance of distinct dryland ecosystems across all continents, with the exception of Antarctica, and came to the conclusion that as aridity increased, bacterial and fungal populations shrank. The composition and number of Chloroflexi and Proteobacteria increased as a result of this sort of climatic change, while Verrucomicrobia and Acidobacteria dropped. A potentially effective method of reducing the effects of global climate change is the management of the microbial ecosystem. The ecology and function of beneficial microbial communities must be understood in order to be managed. Due to the simplified $CH_4$ pathway and the involvement of specialist bacteria, the $CH_4$ biocycle is easier to understand than other GHG cycles.

## Methane

Methane ($CH_4$) is one of the three primary greenhouse gases, along with carbon dioxide ($CO_2$) and nitrous oxide ($N_2O$), and it has a 25-fold greater potential to cause global warming than $CO_2$. The ozone layer's deterioration is also impacted by $CH_4$ [26, 27]. About two thirds of the worldwide $CH_4$ emissions, or total anthropogenic methane, are caused by men [25]. According to a study, agriculture is responsible for 47–56% of all anthropogenic $CH_4$ emissions, of which 12–37% may be of enteric origin [6, 28–30]. After stabilising for a while, methane concentrations have been rising again since 2007, which is now ascribed to changes in climate-induced methane releases from natural wetlands. Methane contributes 17% of radiative forcing [9]. The primary sources of human-related methane emissions include domestic ruminants, rice fields, carbon mines, landfills, and the use of fossil fuels [25]. On the other hand, methane is also released naturally from sources including termites, wetlands, and seas [31]. Ruminants are the main producers of $CH_4$ among animals. Their huge fore stomach, or rumen, features an ongoing fermentation mechanism. More than 70% of the stomach's capacity is taken up by the rumen, which has a volume of 15 L in sheep and 100–150 L in cattle [32]. The primary source of methane synthesis is microbial fermentation of hydrolyzed carbohydrates, which is seen as an energy loss for the animal [33–35]. Ruminant $CH_4$ generation is influenced by a variety of variables, including ruminant intake, feed quality and type, energy intake, animal size, growth rate, output level, genetics, and ambient temperature [36]. Ruminant methane emission lowers the effectiveness of nutrient uptake. Therefore, one of the most significant objectives for animal nutritionists is to manipulate the rumen microbial environment to reduce methane emission by ruminants and to increase their performance. Reducing ruminant methane emissions improves production, increases nutrient use efficiency, and lessens the impact of methane on global warming [8].

## Carbon Cycling and Climate Change

The global carbon cycle of different ecosystems on earth provides the best explanation for the fluxes of carbon in the environment. As a component of life and one of the most plentiful substances on earth, carbon is also a key factor in determining the world's climate, its unpredictability, and the availability of energy for humanity. In the end, $CO_2$ is used by plants during the process of photosynthesis after being removed from the atmosphere by the bacterial and fungal breakdown of dead tissues and organic components. A crucial class of bacteria known as methylotrophs uses greenhouse gases like $CO_2$ and $CH_4$ to reduce the effects of global warming [37]. Along with the many other autotrophs, including plants and bacteria that can make photosynthetic material, methanogens are among the organisms that use $CO_2$ as a source of energy. Heterotrophs use organic substances for growth as well, converting them to $CO_2$. Through a variety of chemical processes, including methanogenesis, methanotrophism, carbon dioxide fixation, anaerobic respiration, and fermentation, the equilibrium in carbon cycling is maintained. Methylotrophic bacteria oxidise methane, the second most prevalent and strong greenhouse gas, together with its derivatives (methanol, formaldehyde, methylamine, dimethylamine, trimethylamine, and formic acids) [24, 36, 37]. Methane is the second most important gas after $CO_2$ in terms of its contribution to global warming and the destruction of the ozone layer. Methanogenesis in animals and the decomposition of organic matter are significant contributors to global warming since it is a powerful greenhouse gas with a global warming potential 25 times greater than carbon dioxide [38]. Although it may not be a net contributor since it uses ambient carbon dioxide to form organic material, its ultimate impact is to turn that carbon dioxide into methane, a considerably more powerful greenhouse gas. Degradation and decomposition are processes that methylotrophic bacteria use to keep the environment's carbon cycle in check. Organic molecules undergo biodegradation, which releases $CO_2$ into the atmosphere [1]. Prokaryotes, such as Actinomycetes, Arthrobacters, Pseudomonads, and Fermicutes, in addition to methylotrophic bacteria, play a critical role in the biodegradation of hazardous carbon and carbon derivatives. These microbial communities react to environmental change sensitively by looking at the various microbial populations of soil, which are markers of climate change. Numerous anthropogenic activities and interferences, such as, deforestation, construction of industries, combustion of fossil fuels by vehicles, air and water pollution have an impact on climate change or unanticipated environmental variation [39]. Changes in the cycle of carbon and nitrogen across the globe have been impacted by these interferences. Climate change is caused by both the rise in greenhouse gases and the sum of all these atmospheric changes. Microbes have long had an impact on humanity, and we play a part in changing the energy balance and atmospheric composition. Methane, carbon dioxide, and nitrous oxide have been brought into the atmosphere as a result of human meddling and activity, and this induction predominates over greenhouse gas fluxes brought about by microorganisms [40]. Researchers also looked at the idea that bacterial and fungal communities expand more quickly in response to global warming. As they expand quickly, their respiration increases the amount of $CO_2$ in the atmosphere, which warms the climate [41, 42]. In this way, microbial organisms contribute to and have an impact on climate change. Additionally, complicated metabolic processes in the carbon and nitrogen cycles are impacted by inorganic nutrients [43]. In the past, several methylotrophic strains have been described as actively contributing to climate change and lowering greenhouse gas emissions [2, 29–31]. On an individual level, action is required to combat global climate change across all nations. By using other fuels and adopting low-carbon lifestyles, GHG emissions may be minimised. Mitigation studies show that the amount of GHGs in the atmosphere is decreasing which slows down climate change. This reduction in GHGs is made possible by using less energy. Numerous bacteria are also contributing to the lowering and decrease of these hazardous gases.

## Methanogenesis

Methanogenesis, also known as biomethanation, is a multi-step process involving several microorganisms, including those that are hydrolytic, fermentative, acetogenic, and most importantly, methanogenic. The term "methanogens" refers to anaerobic bacteria from the domain Archaea that are involved in the biological synthesis of methane. The sole metabolic process carried out by methanogens is methanogenesis. Methanogens are only able to employ a few number of substrates that are derived from the anaerobic basement of the organic matter by hydrolytic and fermentative bacteria for this metabolism [18]. That suggests that methanogens accept a terminal place in the trophic chains of microbes. These methanogens vary from bacteria and eukarya because they lack the peptidoglycan that bacteria and eukarya have in their cell walls [39]. Based on the substrate used for methane generation, there are three main routes for producing the gas: hydrogenotropic, acetoclastic, and methylotropic. The most common route among them is hydrogenotropic and acetoclastic. The majority of rumen methanogenesis is produced by hydrogenotrophic methanogens, which turn $CO_2$ into $CH_4$ [16]. Methanogenesis, or the process of producing methane, depends on alkyl radical-containing substances such formate, acetate, methanol, methyl sulphides, and methylamines. Substrate-specific methyltransferases convert the alkyl radical in these substances into $CH_4$. Other bacteria and fungi found in the local microbial communities largely create these substrates by decomposing organic materials. Aerobic methanotrophic bacteria can utilise methane that escapes from anaerobic environments as a source of carbon and energy, or it can escape into the atmosphere, where it participates heavily in atmospheric chemical processes and is a significant greenhouse gas [44]. Methane generation is a significant and common kind of microbial metabolism. It is the last stage of the breakdown of biomass in anoxic settings. The majority of natural gas accumulations are due to thermogenesis, with methanogenesis accounting for a sizeable portion of them [10, 32, 33]. The methyl-oxidation route, similar to the first, is used to further oxidise an alkyl radical into $CO_2$, which causes the hydrogenotrophic pathway to operate in the opposite direction. This results in the abbreviation equivalents for this methanogenesis. Without oxygen and other electron acceptors like nitrate, sulphate, and iron, methanogenesis takes place. The release of ATP for numerous cellular functions results from the synthesis of methane. The methyl-coenzyme M reductase (Mcr) complex, which catalyses the last step of reducing methyl-coenzyme M to methane, is the essential enzyme in methanogenesis. As an alternative to the reducing equivalents produced by the methyl-oxidation route, this mechanism makes absolute use of the $H_2$ that is already available in the environment and is associated with an electron donor. It appears that the methanogens limited to this other pathway start to bond with the surroundings found in the gut. Acetate is a smart substrate for methanogenesis used by a few archaea that are connected to the Methanosarcinales [45]. Methanogens produce methane from $H_2 + CO_2$ (hydrogenotrophic), acetate (acetotrophic), or methanol and methylamines to provide energy (methylotrophic). These substrates are a byproduct of the decomposition of organic matter in anoxic habitats (such as wetlands, sediments, permafrost, and landfills), which is facilitated by a network of bacteria hydrolyzing polymers into monomers that may then be fermented. Temperature, quantity, and type of organic matter are all regulated by physical variables (such as water table/flooding in wetlands) or other microorganisms or plants, which in turn govern concentrations of oxygen and alternative electron acceptors (e.g., $NO_3$, $NO_2$, $Fe_3+$, $SO_4$) [7, 44]. In general, nitrogen is thought to hinder the production of methane, either directly or indirectly, through hazardous denitrifying intermediates ($NO_2$, $N_2O$, and NO) or as an oxidant for denitrifiers ($NO_3$, $NO_2$ that can compete with methanogens for substrate [3, 4]. Methanogens also require nitrogen as a nutrient, which they can obtain either by fixing $N_2$ or by absorbing $NH_4+$ or $NO_3$. For the latter two, they must contend with plants and other bacteria (such denitrifiers), a relationship that has received little research.

## Methylotrophs Mitigating Methane

Methane is the second most significant greenhouse gas after carbon dioxide in terms of its impact on short-term climate change. Future climatic harmony may be threatened by the ongoing release of methane from many sources, whether from immediate anthropogenic sources or perhaps quickly from the Arctic. As a result, there is a considerable worry about using different ways to reduce methane emission. Numerous anthropogenic and Arctic-related causes have given rise to the development of a wide range of mitigating methods, but they still need to be improved upon before being used more widely. However, there are still a lot of unknowns regarding the precise processes, scope, and techniques of the Arctic's fast methane emission. Being a significant GHG, methane has a variety of paths and mechanisms for release into the environment, including wetlands, lakes, and oceans. It may also be distributed equally across wide regions or concentrated in tiny patches [46]. However, one of the most important processes for methane emission into the atmosphere is bubbles that are produced from the sediments of Arctic sources. A few sources in the Arctic, where methane is concentrated in pockets, may be used with the methane release mitigation technologies, even though most of them are based on restricted gas streams of 0.1% methane or greater. In addition to other methods, a few mitigating techniques designed specifically for rice fields and agricultural soils have also demonstrated promise for Arctic wetlands and thawing permafrost. However, a number of additional Arctic-specific mitigation techniques have been proposed; they need more research. In order to address current methane sources and prospective Arctic sources, experts have so far identified four relevant research and development areas: (1) Methane emission detection and measurement; (2) Small and distant methane stream mitigation; (3) Dilute (1000 ppm) methane stream mitigation; and (4) Methanotroph and methanogen ecology understanding. Additionally, the use of methylotrophs and a thorough explanation of soil methanotrophy might be a useful tool to address methane emissions naturally released from closed landfills and a significant drop in waste-related GHG emissions after methanotrophic reactions [22]. Methanotrophs have developed and gained the ability to use $CH_4$ as their only source of carbon and energy to grow aerobically. These bacteria are crucial in converting $CH_4$ into organic compounds and releasing $CO_2$ for use by autotrophs [40]. Additionally, the major component breakdown that results from a number of photochemical processes is the oxidation of methane in the atmosphere in the presence of hydroxyl (OH) radicals. The primary reactive species in the trophosphere is the hydroxyl radical, which is created photochemically in the atmosphere and interacts with many types of organic molecules [20]. A study on the biodegradation of methane and the buildup of polyhydroxybutyrate (PHB) utilising an isolated strain and a methanotrophic consortia has produced encouraging findings for the reduction of methane. It went on to explain that the isolate and the consortium had specific methane consumption rates of 100 and 17 mg $CH_4$ g $h^{-1}$, respectively. Additionally, the two-phase partitioning bioreactor (TPPB) was tested for its ability to remove methane from an air stream while containing 10% volume-to-volume silicon oil. The TPPB encouraged PHB production at rates of 34 and 38% w/w and advocated a 33–45% rise in methane removal. Under these circumstances, the consortium's particular methane degradation rate reduced to that of the isolated strain while remaining unchanged for the collaboration. According to the study, strain CZ2 of the bacterium *Methylobacterium organophilum* is able to use methane and accumulate up to 57% (w/w) of PHB when nitrogen is scarce. Additionally, it was shown that *Methylobacterium organophilum* CZ2 and *Methylosinus trichosporium* OB3b had similar specific $CH_4$ (methane) consumption rates and capacities for accumulating PHB. So, methylotrophs contribute to reducing GHS emission into the environment and have enormous potential for producing PHB industrially from waste gases [47]. Since it is known that methylotrophic bacteria may use C1 chemicals, such as methane, there is a persistent effort to identify and describe new species of methane-degrading bacteria. Therefore, by effectively using methane, such new methylotrophic bacteria may contribute to lessening the effects of global warming. Additionally, identifying and assessing specific plant growth-promoting (PGPR) strains for their capacity to decompose methane would undoubtedly open new doors for many uses of such cultures, including the promotion of plant growth, the tolerance of abiotic stress, and methane mitigation [30, 48]. The simplest spectrophotometric assay for methane screening using microbial strains was recently studied and compared to other methods available, including the traditional gas liquid chromatographic technique, assay of specific enzymes, and molecular analysis of the genes encoding methane monooxygenase and methanol dehydrogenase (mmo and mxaF) respectively. Jhala and associates were able to effectively restore bacterial cultures that degrade methane by enriching soil with water and using methane as the only carbon source [29]. Additionally, colorimetric plates assay identified the existence of soluble methane monooxygenase (sMMO) enzyme and measured their survival in evacuated tubes containing methane. By finding the genes encoding the enzymes (methane monooxygenase and ethanol dehydrogenase) and qualitatively estimating the enzyme activity in the isolates, it was possible to further confirm the ability of the isolates to degrade methane. Research on the slurry material taken from the Herman Pit, a former mercury mine, showed the importance of methanotropic bacteria in the aerobic removal of $CH_4$ from sediments. Furthermore, the existence of acidophilic or acid-tolerant methanotrophs was shown by the methanogenic activity that was carried out under artificially acidic circumstances. Thus, maximal activity at pH 4.5 with incubated slurries was used to validate acid-tolerant methanotrophs. Such methanotrophs also had their sterol and hopanoid lipids extracted, which is a feature of methanotrophs, and their abundance was augmented by a rise in sediment methane consumption. Additionally, the genomic DNA isolated from methane-oxidizing enrichment cultures revealed an amplified sequence for the pmo A gene that matched methanotrophic Gammaproteobacteria. An enrichment culture was created under acidic conditions (pH 4.5) using methane oxidation [2]. Another important worry of the scientific community is the environment's rising $CO_2$ concentration, and much focus is currently being placed on determining how methylotrophs contribute to $CO_2$ mitigation. Since it is anticipated that waste-related biomass will be harvested sustainably and there would not be any net $CO_2$ emissions because it is believed that $CO_2$ produced by food waste decomposition can be absorbed by the following year's crop, most biomass or biomass-based waste degradation is typically not included in domestic or international greenhouse gas inventory totals. GHG inventories, however, also include methane emissions from waste caused by anaerobic decomposition [22]. Formaldehyde (HCOH) and $CO_2$ are typically two C1 oxidation products involved in methanotrophic activities. Additionally, there are two mechanisms for assimilating carbon during methanotrophic metabolism: the serine pathway and the RuMP system. During methanotrophic metabolism, the serine route uses two moles of HCOH and one mole of $CO_2$ to create a three-carbon intermediate. In the RuMP route, three moles of HCOH are used up, resulting in the generation of three major metabolic carbon intermediates. The RuMP route is therefore more effective than the serine pathway. Additionally, the RuMP route is superior than the serine pathway for both ATP consumption and molar yield values (g of cell dry weight/mol of substrate consumed), where bacteria utilise C1 compounds [23]. Because all methanogens are capable of removing $CO_2$ from the air, they do so by converting it to cell material and $CH_4$. Methanotrophs have little effect on the carbon cycle, but they do have an impact on the amount of plentiful greenhouse gases in the atmosphere due to their metabolism.

## Methylotrophs Mitigating Methane in Paddy Fields

One effect of the methane imbalance throughout the atmosphere is the global shift in the physiochemical characteristics of the climate. The finest illustration of significant methane sources is a rice field [12, 49, 50]. Since methane is produced in large quantities in rice fields, methanotrophic bacteria play a significant role in reducing methane through biodegradation. In the paddy field, there is a cycle of microbial activity wherein flooding circumstances encourage the methanogens, which produce methane gas. The methanotrophic bacteria there then trap the methane gas, converting it to methanol and biomass in the process. Methane monooxygenase (mmo) enzyme is a necessary component for methanotroph activity, and oxygen is needed to make it reactive. This methane oxidation enzyme system is stimulated by aerobic methanotrophs. The green algae that cover the surface of the flooded rice field typically cause this aerobic situation [51]. Methylotrophic isolates with functioning enzyme systems were collected from Gujarati wetland paddy fields, and upon biochemical and molecular analysis, they were identified as several species of *Bacillus* and *Penibacillus*. The existence of the particulate methane monooxygenase (pmoA) genes that encode the subunits in gene cluster is demonstrated by the working enzyme system. While the mmoX gene encodes (part of the hydroxylase component) in *Methylobacterium extrorquens*, the presence of the pmoA gene implies methane use by bacteria like *P. illinoisensis*, *B. aerius*, *B. subtilis*, and *Rhizobium* sp. In a research, communities that are effective at using methane, such as *P. illinoisensis* and *Rhizobium* sp., were shown to have the mxaF gene, which codes for the subunit of the methanol dehydrogenase enzyme. A recognised bacterial group that promotes plant development was found to have methane breakdown enzymes and genes in the methane reducing communities isolated from wetland rice fields [30]. These particular methylotrophic communities are systematically arranged over the soil surface in paddy fields, with the capacity and power to digest the greenhouse gas methane, resulting in aerobic soil surface conditions. This well-organized film is related to the algal populations that are mostly seen in rice fields. By driving the activity of methane oxidation, the algal communities play a significant part in reducing greenhouse gas emissions in the environment. A thin coating of algae reduced methane emission in a microcosm experiment without rice plants. In addition, the presence of algae on the surface of submerged rice fields encouraged methanotrophs and constrained the number of methanogens. According to a study, in the presence of rice, $CH_4$ emission occurs mostly through aerechyma [52, 53]. Studies confirm the involvement of methylotrophs in the reduction of greenhouse gas emissions in the environment.

## Enzymes Involved in Methane Production

The complexity and uniqueness of methanogenesis as a type of anaerobic respiration lies in the need for six exceptional coenzymes, including methanofuran, ferredoxin, methanopterin, coenzyme M, coenzyme B and coenzyme F420: a pathway and several specific membrane-bound enzyme complexes coupled to the creation of a proton gradient driving ATP synthesis [15]. $CO_2$, acetate, and substances containing methyl groups, such as methanol, methylated amines, and methylated sulphides, are the three main substrates for the production of methane. Due to this, there are three separate routes for the formation of $CH_4$: hydrogenotrophic, acetoclastic, and methylotrophic [11, 14]. Although the three routes have different intermediates and enzyme processes, they nonetheless have common characteristics in the ultimate stages of $CH_4$ synthesis. The yield of a carrier-bound methyl intermediate is influenced by both the hydrogenotrophic and acetoclastic processes. Methanopterin, a product of the hydrogenotrophic route, and sarcinapterin, a product of the acetoclastic pathway, are the carrier proteins. All three processes include the addition of the methyl group to coenzyme M via a particular, membrane-bound methyltransferase and the consequent decrease of methyl coenzyme M to $CH_4$ via the crucial enzyme methyl coenzyme M reductase [54]. The three methanogenic processes are further explained in the supporting information in small print. Methyl coenzyme M reductase is made up of a dimer of the three subunits (McrA), (McrB), and (McrG), and it has a special active site termed coenzyme F430 that includes porphinoid nickel [19]. About 300 kDa is the apparent molecular mass of the enzyme. Methyl coenzyme M reductase has two specific isoenzymes that have been found [66]. The second enzyme has a different substrate affinity and is known as methyltransferase for methyl reductase two [5]. The mcrBDCGA operon codes for methyl coenzyme M reductase activity, whereas the MRT is encoded by the mrtBDGA operon [55, 56]. The mrt operon lacks the identical counterpart of gene mcrC [55]. The byproducts of the genes mcrC (McrC), mcrD (McrD), and mrtD (MrtD) are under 20 kDa are the. Their purpose is yet unknown and it is still unclear how primary sensors and signal transduction cascades work [57]. However, evidence for regulation was found in the availability of trace elements [58]. This is because many methanogenesis-related enzymes have trace metals (such as molybdenum, tungsten, selenium, and nickel) in their active sites. It was discovered that the abundance of the substrate $H_2$ regulates the synthesis of various important methanogenesis-related enzymes together with MRC. The two isoenzymes of methyl coenzyme M reductase are differently expressed in Methanothermobacter species with the help of $H_2$ availability, with isoenzyme I (methyl coenzyme M reductase) being predominately expressed in $H_2$ limiting environments [47–56]. Control of gene expression of the methanogens is still poorly understood, necessitating more research.

## Current Status and Future Perspective

The use of DNA extraction, PCR, sequencing, and probe biases, and a lack of bioinformatics support for next-generation sequencing and metaproteomics, continue to limit innovative technologies. The development of bioinformatics tools, however, has led to a noteworthy advancement in this sector in recent years. The current dispute will create quantitative information for bacteria involved in the $CH_4$ cycle and to parameterize this data for substantial use in climate and ecological models. Because their metabolic capacities are not well known, many methanogens and methanotrophs are not cultivable. This is a crucial need for the accurate integration of microbiological data in the prediction forms. Stable isotope probing and methods like DNA and RNA analysis can help determine the physiological capacities of different animals. Due to information gaps about DNA and RNA, stable isotope probing methods with a relatively high substrate concentration are required to label DNA sufficiently [57, 58]. PLFA-SIP, which combines stable isotope probing with PLFA, may detect active bacteria at ecologically relevant concentrations. This method, however, is unable to precisely identify microorganisms at the species level due to a lack of phylogenetic precision. Environmentally substantial amounts of substrate may be used for metagenomic and metaproteomic investigations thanks to technological advancements in SIP and associated apparatus [59–65]. Additionally, it is necessary to classify the habitats used by populations of methanogens and methanotrophs. Therefore, a demonstration of niche adaptation in methanogens and methanotrophs was provided before [38, 66–72]. However, in the next three millimetres of water-saturated soils, Reim and colleagues discovered vertical niche divergence in gamma proteobacterial methanotrophs [73]. Given the local commerce that may be identified on a small scale, this is very significant and indicates the necessity for specific niche identification.

## Conclusion

Methanogenesis is the anaerobic production of methane by methanogenic Archaea. Methanogenesis can come from a variety of anthropogenic and natural sources (human sources). Methylotrophic bacteria use and break down reduced carbon molecules like methane, contributing significantly and significantly to climate change. This particular bacterial group is unusual in that it helps to maintain the climate by lowering greenhouse gas emissions. The rice field is the most prevalent environment for methanotrophs, where enzymatic activities are aided by other species including methanogens and algae. Although methane ($CH_4$) emissions are projected to vary due to climate change, the dynamics of methanogens and methanotrophs under this transition have not yet been thoroughly studied. Agriculture, particularly the rearing of cattle, is the largest anthropogenic source of methanogenesis. Methanogenesis from the production of animals and organic matter decomposition contributes significantly to global warming. The inclusion of microbial knowledge into the development of prediction models will be greatly aided if we can identify the niche separation for certain microbial groups with specified physiological capabilities and their control. Furthermore, such information may be used to investigate extensive data on the generation of methane and the use of particular unidentified genes as a molecular pathway.

## References

1. Allison SD, Wallenstein MD, Bradford MA (2010) Soil-carbon response to warming dependent on microbial physiology. Nat Geosci 3:336–340
2. Baesman SM, Miller LG, Wei JH, Cho Y, Matys ED, Summons RE, Welander PV, Oremland RS (2015) Methane oxidation and molecular characterization of methanotrophs from a former mercury mine impoundment. Microorganisms 3:290–309
3. Bodelier PLE (2011) Interactions between nitrogenous fertilizers and methane cycling in wetland and upland soils. Curr Opin Environ Sustain 3:379–388

4. Bodelier PLE, Steenbrergh AK (2014) Interactions between methane and nitrogen cycling; current metagenomic studies and future trends. In: Marco D, Caister (eds) Metagenomics of the microbial nitrogen cycle: theory methods and applications. Academic Press, pp 33–85
5. Bonacker LG, Baudner S, Mörschel E, Böcher R, Thauer RK (1993) Properties of the two isoenzymes of methyl-coenzyme M reductase in Methanobacterium thermoautotrophicum. Eur J Biochem 217(2):587–595
6. Brask M, Lund P, Weisbjerg MR, Hellwing ALF, Poulsen M, Larsen MK, Hvelplund T (2013) Methane production and digestion of different physical forms of rapeseed as fat supplements in dairy cows. J Dairy Sci 96(4):2356–2365
7. Bridgham SD, Cadillo-Quiroz H, Keller JK, Zhuang QL (2013) Methane emissions from wetlands: biogeochemical, microbial, and modeling perspectives from local to global scales. Glob Change Biol 19:1325–1346
8. Bunglavan SJ (2014) Methanogenesis and recent techniques for mitigation of methanogenesis in ruminants. J Livest Sci 5:35–48
9. Ciais P, Sabine C, Bala G, Bopp L, Brovkin V, Canadell J, Chhabra A, DeFries R, Galloway J, Heimann M, Jones C (2014) Carbon and other biogeochemical cycles. In: Climate change 2013: the physical science basis. Contribution of working group I to the fifth assessment report of the intergovernmental panel on climate change. Cambridge University Press, pp 465–570
10. Cramer B, Franke D (2005) Indications for an active petroleum system in the Laptev Sea, NE Siberia. J Pet Geol 28(4):369–384
11. Deppenmeier U (2002) The unique biochemistry of methanogenesis. Prog Nucleic Acid Res Mol Biol 71:223–283
12. Dubey SK (2005) Microbial ecology of methane emission in rice agroecosystem: a review. Appl Ecol Environ Res 3(2):1–27
13. Edenhofer O, Pichs-Madruga R, Sokona Y (2012) Renewable energy sources and climate change mitigation: special report of the intergovernmental panel on climate change. Intergovernmental Panel on Climate Change. ISBN 978-92-9169-131-9
14. Ferry JG (1999) Enzymology of one-carbon metabolism in methanogenic pathways. FEMS Microbiol Rev 23(1):13–38
15. Ferry JG (2010) Biochemistry of acetotrophic methanogenesis. In: Handbook of hydrocarbon and lipid microbiology, pp 357–367
16. Fonty G, Morvan B (1996) Ruminal methanogenesis and its alternatives. Ann Zootech 313–318
17. French S, Levy-Booth D, Samarajeewa A, Shannon KE, Smith J, Trevors JT (2009) Elevated temperatures and carbon dioxide concentrations: effects on selected microbial activities in temperate agricultural soils. World J Microbiol Biotechnol 25:1887–1900
18. Garcia JL, Patel BK, Ollivier B (2000) Taxonomic, phylogenetic, and ecological diversity of methanogenic Archaea. Anaerobe 6(4):205–226
19. Gunsalus RP, Wolfe RS (1980) Methyl coenzyme M reductase from Methanobacterium thermoautotrophicum. Resolution and properties of the components. J Biol Chem 255(5):1891–1895
20. Hanson R, Hanson T (1996) Methanotrohic bacteria. Microbiol Rev 60(2):439–471
21. Hedderich R, Whitman W (2006) Physiology and biochemistry of the methane-producing Archaea. In: Dworkin M, Falkow S, Rosenberg E, Schleifer KH, Stackebrandt E (eds) The prokaryotes. Springer, New York, USA, pp 1050–1079
22. Hettiaratchi J, del Castillo Sternenfels U (2013) Mitigation/reduction of GHG emissions in solid/hazardous waste management. In: Climate change modeling, mitigation, and adaptation, pp 600–620
23. Hilger HA, Humer M (2003) Biotic landfill cover treatments for mitigating methane emissions. Environ Monit Assess 84(1–2):71–84
24. Holland MA, Polacco JC (1994) PPFMs and other contaminants: is there more to plant physiology than just plant? Annu Rev Plant Physiol Plant Mol Biol 45:197–209
25. Huarte A, Cifuentes V, Gratton R, Clausse A (2010) Correlation of methane emissions with cattle population in Argentine Pampas. Atmos Environ 44(23):2780–2786

26. Iguchi H, Yurimoto H, Sakai Y (2015) Interactions of methylotrophs with plants and other heterotrophic bacteria. Microorganisms 3(2):137–151
27. IPCC (2007) Climate change 2007: the physical science basis. Cambridge University Press, Cambridge
28. IPCC (2013) In: Stocker TF, Qin D, Plattner GK, Tignor M, Allen SK, Boschung J, Nauels A, Xia Y, Bex V, Midgley PM (eds) Climate change: the physical science basis. Contribution of working group I to the fifth assessment report of the intergovernmental panel on climate change. Cambridge University Press, p 1535
29. Jhala YK, Rajababu VV, Panpatte Deepak G, Shelat Harsha N (2015) Rapid methods for isolation and screening of methane degrading bacteria. J Biorem Biodegrad 7:322
30. Jhala YK, Rajababu VV, Shelat Harsha N, Patel HK, Patel HK, Patel KT (2014) Isolation and characterization of methane utilizing bacteria from wetland paddy ecosystem. World J Microbiol Biotechnol 30(6):1845–1860
31. Kappler U, Nouwens AS (2013) Metabolic adaptation and trophic strategies of soil bacteria C1-metabolism and sulfur chemolithotrophy in Starkeya novella. Front Microbiol 4:1–12
32. Katz BJ (2011) Microbial processes and natural gas accumulations. Open Geol J 5(1)
33. Kietäväinen R, Purkamo L (2015) The origin, source, and cycling of methane in deep crystalline rock biosphere. Front Microbiol 6:725
34. Kolb S, Stacheter A (2013) Prerequisites for amplicon pyrosequencing of microbial methanol utilizers in the environment. Front Microbiol 4:268
35. Kristensen T, Mogensen L, Knudsen MT, Hermansen JE (2011) Effect of production system and farming strategy on greenhouse gas emissions from commercial dairy farms in a life cycle approach. Livest Sci 140(1–3):136–148
36. Kumar M, Srivastava AK, Pandey AK (2015) Biocontrol activity of some potent methylotrophs isolated from Bhitarkanika mangrove sediment. Int J Curr Res Biosci Plant Biol 2(6):101–106
37. Kumar M, Tomar RS, Paul D, Lade H (2016) Methylotrophic bacteria in sustainable agriculture. World J Microbiol Biotechnol 32:120
38. Kumaresan D, Héry M, Bodrossy L, Singer AC, Stralis-Pavese N, Thompson IP, Murrell JC (2011) Earthworm activity in a simulated landfill cover soil shifts the community composition of active methanotrophs. Res Microbiol 162(10):1027–1032
39. Lakhani N, Lakhani P, Sheikh AA, Bhagat R, Dar RR, Dogra P (2017) Methanogenesis: are ruminants only responsible: a review. J Pharmacogn Phytochem 6(6):2347–2352
40. Large PJ (1983) Methylotrophy and methanogenesis. Aspects of microbiology, vol 8. American Society for Microbiology, Van Nostrand Reinhold, Wokingham
41. Maestre FT, Delgado-Baquerizo M, Jeffries TC, Eldridge DJ, Ochoa V, Gozalo B, Quero JL, García-Gómez M, Gallardo A, Ulrich W, Bowker MA (2015) Increasing aridity reduces soil microbial diversity and abundance in global drylands. Proc Natl Acad Sci 112(51):15684–15689
42. Manzoni S, Taylor P, Richter A, Porporato A, Ågren GI (2012) Environmental and stoichiometric controls on microbial carbon-use efficiency in soils. New Phytol 196:79–91
43. Meena KK, Kumar M, Kalyuzhnaya MG, Yandigeri MS, Singh DP, Saxena AK, Arora DK (2012) Epiphytic pink-pigmented methylotrophic bacteria enhance germination and seedling growth of wheat (*Triticum aestivum*) by producing phytohormone. Antonie Leeuwenhoek 101(4):777–786
44. Megonigal JP, Hines ME, Visscher PT (2004) Anaerobic metabolism: linkages to trace gases and aerobic processes. In: Schlesinger WH (ed) Biogeochemistry. Elsevier–Pergamon, pp 317–424
45. Microbiology Online (2015) Microbes and climate change. http://www.microbiologyonline.org.uk/aboutmicrobiology/microbesandclimatechange. Accessed 15 Dec 2015
46. Montzka SA, Dlugokencky EJ, Butler JH (2011) Non-$CO_2$ greenhouse gases and climate change. Nature 476:43–50
47. Morgan RM, Pihl TD, Nölling J, Reeve JN (1997) Hydrogen regulation of growth, growth yields, and methane gene transcription in Methanobacterium thermoautotrophicum deltaH. J Bacteriol 179(3):889–898

48. Murrell JC, Whiteley AS (2011) Stable isotope probing and related technologies. ASM Press, Washington, DC, USA
49. NASA (2015) http://climate.nasa.gov/solutions/adaptation-mitigation/. Accessed 15 Dec 2015
50. Nazaries L, Tate KR, Ross DJ, Singh J, Dando J, Saggar S, Baggs EM, Millard P, Murrell JC, Singh BK (2011) Response of methanotrophic communities to afforestation and reforestation in New Zealand. ISME J 5(11):1832–1836
51. Neufeld JD, Chen Y, Dumont MG, Murrell JC (2008) Marine methylotrophs revealed by stable-isotope probing, multiple displacement amplification and metagenomics. Environ Microbiol 10(6):1526–1535
52. Neufeld JD, Dumont MG, Vohra J, Murrell JC (2007) Methodological considerations for the use of stable isotope probing in microbial ecology. Microb Ecol 53(3):435–442
53. Oremland RS, Culbertson CW (1992) Importance of methane-oxidizing bacteria in the methane budget as revealed by the use of a specific inhibitor. Nature 356:421–423
54. Oshkin IY, Beck DAC, Lamb AE, Tchesnokova V, Benuska G, Benuska G, McTaggart TL, Kalyuzhnaya MG, Dedysh SN, Lidstrom ME, Chistoserdova L (2014) Methane-fed microbial microcosms show differential community dynamics and pinpoint taxa involved in communal response. ISME J 9(5):1119–1129
55. Palut MPJ, Canziani OF (2007) Contribution of working group II to the fourth assessment report of the intergovernmental panel on climate change. Cambridge University Press
56. Peng S, Ingram KT, Neue H-U, Ziska LH (1995) Climate change and rice. IRRI, Springer, New York, pp 81–91
57. Pihl TD, Sharma S, Reeve JN (1994) Growth phase-dependent transcription of the genes that encode the two-methyl coenzyme M reductase isoenzymes and N5-methyltetrahydromethanopterin: coenzyme M methyltransferase in Methanobacterium thermoautotrophicum delta H. J Bacteriol 176(20):6384–6391
58. Reeve JN, Nölling J, Morgan RM, Smith DR (1997) Methanogenesis: genes, genomes, and who's on first? J Bacteriol 179(19):5975–5986
59. Reim A, Lüke C, Krause S, Pratscher J, Frenzel P (2012) One millimetre makes the difference: high-resolution analysis of methane-oxidizing bacteria and their specific activity at the oxic–anoxic interface in a flooded paddy soil. ISME J 6(11):2128–2139
60. Ross PM, Adam P (2013) Climate change and intertidal wetlands. Biology (Basel) 2(1):445–480
61. Shibata M, Terada T (2010) Factors affecting methane production and mitigation in ruminants. Anim Sci J 81(1):2–10
62. Singh A, Dubey SK (2012) Temporal variations in methanogenic community structure and methane production potential of tropical rice ecosystems. Soil Biol Biochem 48:162–166
63. Singh BK, Bardgett RD, Smith P, Reay DS (2010) Microorganisms and climate change: terrestrial feedbacks and mitigation options. Nat Rev Microbiol 8:779–790
64. Six J, Frey SD, Thiet RK, Batten KM (2006) Bacterial and fungal contributions to carbon sequestration in agroecosystems. Soil Sci Soc Am J 70:555–569
65. Stackhouse KR, Pan Y, Zhao Y, Mitloehner FM (2011) Greenhouse gas and alcohol emissions from feedlot steers and calves. J Environ Qual 40(3):899–906
66. Steigerwald VJ, Stroup D, Hennigan AN, Palmer JR, Pihl TD, Daniels CJ, Reeve JN (1993) Methyl coenzyme-M reductase II genes and their close linkage to the methyl viologen-reducing hydrogenase-polyferredoxin operon in the genomes of Methanobacterium thermoautotrophicum and Methanothermus fervidus. In: Baltz RH, Hegeman GD, Skatrud PL (eds) Industrial microorganisms: basic and applied molecular genetics. American Society for Microbiology Press, Washington, DC, USA, pp 109–115
67. Steinfeld H, Wassenaar T (2007) The role of livestock production in carbon and nitrogen cycles. Annu Rev Environ Resour 32:271–294
68. Thauer RK (1998) Biochemistry of methanogenesis: a tribute to Marjory Stephenson: 1998 Marjory Stephenson prize lecture. Microbiology 144(9):2377–2406
69. UCAR (2011) Biogeochemical cycles. https://spark.ucar.edu/longcontent/biogeochemical-cycle

70. Udakis L (2013) Microbes and climate change. Society for General Microbiology, Reading
71. US EPA (2015) Climate change: basic information. http://www3.epa.gov/climatechange/basics/. Accessed 15 Dec 2015
72. US EPA (2016) Climate change: greenhouse gas emissions: greenhouse gases overview. https://www3.epa.gov/climatechange/ghgemissions/gases.html. Accessed 20 Mar 2016
73. Verma P, Saxena R, Tomar RS (2016) Rhizobacteria: a promising tool for drought tolerance in crop plants. Int J Pharma Biosci 116–125

# Chapter 17
# Potency of Three Cruciferous Plants Extracts as Agro-Phyto-Remidiator Against Root Knot Nematode *Meloidogyne spp.* in *Daucus carota* (Carrot) Under Climate Stress Conditions

**Baby Tabassum, Mohammad Hashim, and Jagriti Madan Dhingra**

**Abstract** Carrot, *Daucas carota* is another important crop that is most cultivated throughout India and consumed by human beings and animals. The root knot nematode (RKN) *Meloidogyne incognita* infestation significantly reduces the yield of carrot at initial inoculums of 230–2300 $J_2$/g soil. One strategy to address these concerns is to develop an effective agro-phyto-remediator to these tiny enemies that have zero toxicity to non-target organisms and can be applied at very low cost. Biochemical studies reveals that in certain cruciferous plants like *Brassica rapa*, *Brassica botrytis* and *Raphanus sativous* having nematicidal principle as α tetraethylene and 5-1-3-butenyl 2,2 bithienyl, polyacetylene compounds like trans 3,11-trideca-1-3,11-triene 5,7,9 trizene etc. targeted the percent mortality of *Meloidogyne incognita* juveniles increased almost equally from higher 100% upto 6.25% dilution after 24, 48, and 72 h exposure period of *Raphanus sativus* leaf extract, while *Brassica botrytis* caused significant percent mortality of *Meloidogyne* juveniles i.e. 100% was observed within 24 h exposure with leaf extract in its 100 and 50% concentrations whereas leaves extract of turnip was most effective and showed 100% $J_2$ killed followed by 85.67–96.75% mortality with 50–6.25% dilation after 72 h exposure. Histo-pathological and molecular studies show infection of *Meloidogyne* incognita increased transpiration, photosynthesis or water content and decreased the level of sugars, ascorbic acid and fruit quality. In present study, observed high metabolic activities with intense cytoplasm and nuclei in giant cells produced by nematodes in the carrot.

B. Tabassum (✉) · M. Hashim · J. M. Dhingra
Department of Zoology, Toxicology Lab, Govt. Raza P.G. College, Rampur 244901, U.P., India
e-mail: dr.btabassum@gmail.com

Mahatma Jyotiba Phule Rohilkhand University, Bareilly 243005, U.P., India

M. Hashim
Department of Biochemistry, Mohammad Ali Jauhar University, Rampur 244901, U.P., India

 
J. A. Parray (ed.), *Climate Change and Microbiome Dynamics*, Climate Change Management, https://doi.org/10.1007/978-3-031-21079-2_17

## Introduction

Nematodes show tremendous structural diversities and occur in almost all kinds of biotypes in enormous numbers. An acre of cultivable land contains 3,000,000,000 nematodes while marine beach sand may contain approximately 11–18,000 or sometimes even 90,000 specimens of *Anguina tritici* Jairajpuri [1]. Upto 1930 approximately 4,500 species of nematodes had been described which rose upto 9,000 by 1950. The latter day numbers of investigated species are almost 15,000 but the estimates of subsist species are around more or less 500,000 or more. It reminds the remark of the late eminent nematologist Dr. N. A. Cob of the US department of Agriculture, "If all the matter in the universe except the nematodes were swept away, our world would still be recognizable, we would find its mountains, hills, valleys, rivers, lakes and ocean represented by a film of nematodes".

Evaluated comprehensive average easy yield loss is 12.3% by plant parasitic nematodes in total prime crops. Annual 14% yield loss evaluated in 20 crops and average deprivations for 42 crops in advanced countries are reported almost 8.9% when contrast to 15.7% of developing countries. Uttar Pradesh is one of the most fertile states of India, where almost all types of crops and vegetables are extensively grown. The state leads in total production of a variety of crops though in many cases yield per acre is rather low but the farmers who are mostly ignorant of these microscopic nematode pests inhabiting the soil and attacking their crops, fail to understand the reason for crop failure. 102 known species belonging to 33 genera of *Tylenchida* and 59 species belonging to *Dorylimids* had been noticed in Utter Pradesh Sehgal et al. [2]. Moens and Wesemael [3] also reported that carrots (*Daucus carota L.*) great loss occurred by the RK nematode *Meloidogyne chitwoodi.*

Carrot, *Daucus carota* is another important crop which is most cultivated throughout India and consumed by human beings as well as by animals. It is rich in carotene and is used in various ways of coloring butter and other food articles. Out of the major groups of carrot, Asiatic and Temperate groups are rich in carotenoids which contain appreciable quantity riboflavin and thiamine while the Asiatic types have more anthocyanin pigments and less of carotenoid pigments Gill and Kataria [4]; Rebecca et al. [5]; Raees-ul and Prasad [6].

In temperate regions carrot is seriously affected by *Heterodera carotae.* Greco and Brandonisio [7]; Moens and Wesemael [3] estimated 100% crop loss by the nematode. Other important nematode pests found on carrots are carrot cyst nematode and RKN, *M. javanica* and *M. incognita* respectively. *M. javanica* on carrot exhibits constriction, digitation and cracking in the tap root system. The RKN infection significantly reduces the yield loss in carrot at initial inoculum of 230–2300 J/g soil. Ribonuclease activity also decreases in carrot plants, tolerant to *Meloidogyne hapla*, whereas, increase in the secondary phloem and xylem tissues of susceptible plants have been noticed by Krypl and Janas [8]; Phillips [9] resulting in reduced functional metabolism.

Nowadays, crop scientists are searching for simple, eco-healthy, economically low tactics which integrate into an overall nematode management system. In the current study also an endeavor has been made to calculate the impact of attract various parts of the cruciferous plants like *Raphanus sativus*, *Brassica rapa* and *Brassica botrytis* against *Meloidogyne* second stage juvenilies in in vitro and in vivo on *D. carota* plants.

## *Histo-Pathological and Molecular Studies*

Histopathological and molecular studies reveal that there is an increase of total DNA and RNA in *M. incognita* infected regions of the host plant Masood and Saxena [10]; Phani et al. [11]; Bayani et al. [12]. Infection of *M. incognita* increased transpiration, photosynthesis or water content and decreased the level of sugars, ascorbic acid and fruit quality Tyagi and Rehman [13]; Ahmed et al. [14] recorded increased protein and ascorbic acid devoid of lignin in giant cells because of the *M. incognita* infection. Chlorophyll content also become low because of *M. incognita* Ganguly and Dasgupta [15]; Lu et al. [16] estimated low protein and Indole acetic acid activity and high auxin peroxidase activity in RKN infected gall than healthy roots.

Bruno et al. [17]; Meidani et al. [18] observed high metabolic activity with intense cytoplasm and nuclei of giant cells produced due to nematodes in carrots. The cultivars susceptibility of *M. javanica* on carrot in terms of penetration development was observed by Debia et al. [19]. Also noticed that the symptoms produced by *M. javanica* on carrot include constriction, digitation and cracking in the tap root system. The RKN infection significantly yields loss of carrot at initial inoculum of 230–2300 $J_2$/g soil Huang and Charchar [20]. Such symptoms on root; as well quantitative estimation of yield loss was studied by Hay et al. [21] at different: inoculum levels and indicated 50 yield loss at 10 $J_2$/g soil and no yield at 30–50 $J_2$/g soil.

Singh et al. [22] observed the physio-biochemical changes in carrot root caused by *M. hapla* and also reported reduction in protein synthesis and Protein Amino Lipid (PAL) levels in susceptible plants and increased level of RNA. The phenol level also increased in infected plants but was more resistant than susceptible plants. However, studies on tolerance and resistance of carrot to RKN by Meidani et al. [18] indicated a link, with high number of foliage and low soluble polysaccharide. The acidic fraction of pectins obtained from carrot contains 74% galacturonic acid. The oligogalacturonides containing fraction with the lower molecular weight turned out to be the most active in blocking the adherence of bacteria and epithelial cells in a biological test system by preparing oligo-galacturonic acids Therefore, two oligo-galacturonoides, produced by partial hydrolysis of carrot pectin in stomach are responsible for the anti-diarrheal activity of carrot soup by blocking the adherence of bacteria to epithelial cell Follrich et al. [23]. Agarwal and Ghosh [24] reported, carrot juice contains an alkaloid, pyrolidine, and daucine and is a refrigerant, a tonic and useful in the kitchen in many other ways.

Krishnamurthy and Murthy [25] and Dhaliwal and Arora [26] reported economic losses due to pets between 6,000 and 29,000 crores, while Van Burkum and Sheshadri [27] probably for the first time accounted annual losses of *Anguina tritici* caused about

10 million in wheat, 3 million in coffee by *pratylenchus coffeae* and *Heterodera* caused disease of Molya 8 million annually in Rajasthan, India. It is estimated that in South Asia 89,000 tonnes of chickpea are lost due to nematode infestation Cunha et al. [28].

Nematode damage is so insidious that it is highly devastating to crop. More than 2000 plant parasitic species of nematodes are recorded and they tenanted every possible métier the plant offers. Thus, all the parts of the plants over and beneath the ground seem to be attacked by nematodes, which may be specifically ectoparasitic or endoparasitic.

## *Chemical Control*

Chemical control of nematodes in soil dates back to 1881 when Kuhn applied carbon di-sulphide ($CS_2$) to control sugar-beet cyst nematodes in Germany. After that, Bessey [29] also observed its efficacy against RKN. Then Mathews [30] found nematicidal qualities of chloropicrin (tear gas) and surplus chloropicrin of World War I was used in greenhouses, seed beds and special crops. With the commencement of World War II its use was deflected into war efforts and commercial soil fumigation terminated until Taylor and McBeth [31] manifested nematode control by methyl-bromide (MBr), a broad spectrum biocide. The introduction of 1,2-Dichloropropane 1,3-Dichloropropane and Ethylene Di-Bromide Haydock et al. [32] led to the acceptance and verify of the significance of phyto pathogenic nematodes in yield losses which increased nematode management options. This catalyzed the development of the phyto-nematology and fumigation industry as well. The problem of phyto-toxicity of DD and EDB was overcome by the development of 1,2-Di Bromo,3 Chloro-propane D'errico et al. [33]. In (2020) Talavera-Rubia et al. [34] reported nematicidal efficacy of milbemectin sodium.

## *Nematode Control by Fumigants*

The rapidity and extent of the use of fumigant were the most interesting and surprising responses in the history of pesticides. Widespread use of fumigants started somewhere in 1950 as crop insurance and after having dominated an era of two decades, the fumigant nematicides gave way to nonvolatile non-fumigants organophosphates and carbamates in 1970s in due course of programs designed for insecticides. The non-fumigants were advantageous over fumigants being less phytotoxic Van Burkum and Hoestra [35] VC-13 (dichlofenthion), the first organophosphate nematicide was used to protect ornamental and turfs Perry et al. [36]; Gad [37]. Thionazin was the next important: nematicide used by Jenkin and Guengerich [38].

Other environmental impacts include phytotoxic effects to non-target organisms and residues in soil and crops. Some of them are carcinogenic and also produce suppressing effects on nitrifying bacteria Castro and Beiser [39] and Mckerny [40]. There is also risk to livestock in consumption of produce from pesticide treated soil Young et al. [41]. However, despite of their well-founded concerns about their impacts on unwanted elements, usable water, air quality, and food safety measurement of the crop protection chemicals are very likely to important contrivance in agriculture well into 21st because of the pivotal role in modern global food production Beyer [42], One strategy to inscription these concerns to develop practically effectual agro-phyto-chemicals remediator that have minimum mephitic to non-target organisms and can be applied at economic rate.

## *Agro-Phytoremediator*

Pest control agents from natural sources had evolved eco-healthy, economic as well as suppressed pest populations reported by many workers like Waterfield and Zilberman [43] and Zaki and Bhatti [44]. As plant products being naturally evolved ingredients, they preserved the natural equilibrium in the ecosystem.

There are also several reports that cellulose when integrated in the soil reduced the percentage of plant parasitic nematodes (PPN). The population of *P. penetrans* and *Heterodera tobacum* was considerably inhibited by the application of chopped paper and white pine saw dust as reported by Miller and Edgington [45] and Miller and Weihrmenn [46] respectively. Mankau and Das [47] observed that addition of pure chitin to the soil inhibited the percentage of *M. incognita* and also the development of knot in root. Soil amended with the hydrated extract of sawdust reduced the salvation of eggs in *M. javanica* Sitaramiah [48]. In the various parts of the world there is a common use of oil cake as fertilizers. Lear [49] had reported reduction in *M. javanica* and *Heterodera schactii* by amending with Castor pomace. Hundred percent reductions of *T. semi-penetrans* were reported by Szczygłowska et al. [50]. In India exhaustive work had been done by Singh and Sitaramiah [51] who found oil cakes of *Azadirachta indica, Ricinus communicus, Brassica, peanuts linseed, Madhuca indica* etc. capable of reducing *Meloidogyne* population in field/plots. Tarla et al. [52] found oil cakes and its extracts harmful to the nematodes. Many other unusual amendments had been shown to reduce nematode percentage, however the related function is poorly defined till date but some of them may offer an effective means of nematode control only in small plots.

Many weeds like *Catharanthus, Chenopodium, Argemone, Datura, Ricinus* and many more having phyto-therapeutic value had been reported by Abid and Maqbod [53]; Vats and Nandal [54] reported that the percentage loss of carrots *(Daucus carota L.)* damaged by the RKN *Meloidogyme chitwoodi*. Various effect of chemicals and their mode of action had also been studied in detail by many workers Douda et al. [55]; Pinheiro et al. [56]; Ahmad et al. [14]; Cunha et al. [28] from plants and had been proved toxic to nematodes.

Green synthesis of silver nanoparticles by *Cnidoscolus aconitifolius* extract was experimented by Fabiyi [57] in plants of carrots infected by *M. incognita* juveniles in soil as reducing elements, whereas silver nitrate is the metal precursor. AgNPs treated carrot plants showed higher yield and inhibition of *M. incognita* as well.

**Management of nematodes by modern techniques**

Traditionally, physical, chemical, biological, cultural and regulatory methods are adopted for the management of these tiny nematodes. But modern biology is influenced by ultra-modern techniques like gene cloning, genetic engineering; gene splicing and recombinant DNA used as resistance factors against RKN in egg plants. Another unusual approach to the plant genetic transformation is introducing foreign DNA by micro-projectile bombardment. Enormous amount of work is done in the identification of gene loci in nematodes pests in numbers of crops. Pireda et al. [58] derived head towards the location of chromosomal resistance to *G. rostochiensis* in the potato crops with RKN, *Maloidogyne spp.* (Klein) Anna et al. [59]; Rybczyński et al. [60]. The single dimensional Polyacrylamide Gel Electrophoresis (PAGE) on a fixed pH or pH gradient gel is more commonly used for characterization of nematodes. The pH gradient gels obtained with the incorporation of suitable ampholytes are used for isoelectric focusing of proteins Michael et al. [61].

The techniques of hybridization using specific primers, DNA polymerase enzymes, thermal cycling leading to Polymerase Chain Reaction (PCR) for DNA synthesis and use of Restriction Fragment Length Polymorphism (RFLP) among different population/species of a nematode taxon is gaining popularity in nematology these days. RFLP of mtDNA has been performed to convert several *Meloidogyne* species and genetic divergence in mt DNA was observed by Powers and Harris [62]. Castagnone-Sereno et al. [63]; Bairwa et al. [64] studied the phylogenetic relationships between the amphimictic and pathogenetic species of Meloidogyne using DNA analysis.

# Material and Methods

## *Cobb's Technique Modified by Barker [65]*

The extractions of nematodes from the soil or roots were held by "Cobb's sieving and decanting method" water was mixed in the soil by passing supernatant through 100, 200 and 400 mesh sieves. Nematode suspensions thus collected were used to study the population dynamics and rate of infection.

Calculations for population dynamics and rate of infection have been done by using following formulae:

**Norton's Formulae** [66]**:**

1. Relative Abundance (RA)

$$RA = \frac{Number\ of\ samples\ containing\ different\ species}{Number\ of\ samples\ collected} \times 100$$

2. Relative Density (RD)

$$RD = \frac{Number\ of\ individual\ of\ a\ species\ in\ a\ sample}{Total\ number\ of\ all\ individuals\ in\ a\ sample} \times 100$$

3. Relative Frequency (RF)

$$RF = \frac{Frequency\ of\ one\ species}{Total\ number\ of\ all\ individuals\ in\ a\ sample} \times 100$$

4. Dominance Value Index (DVI)

$$DVI = \frac{RA + RF + RD}{3}$$

**Johnson** [67]**:**

a. **For Histopathology**

Selected root parts of host plant 1–2 cm long washed properly then bleached in NaOCl (sodium hypochlorite) for one-two minute. After proper washing root parts were transferred into acid fuchsin stain then heated upto boiling point and cooled down at room temperature. Finally root parts were mounted in glycerin and microphotographs were taken for histo-pathological studies.

b. **For Histology**

Galled roots were preserved in 4% formalin for histology of *M. incognita* female by following procedure—Took the infected root parts of the carrot. Passed through ethanol series:

50%—3 changes (30 min each)
60%—for 30 min
70%—overnight
80%—for 30 min
90%—for 30 min
90%—15 min (2 changes)

- Cleared the material in methyl-benzoate (50–60) min and transferred to 20% celloidin solution in methyl-benzoate for at least 3 days.
- Passed through three changes of benzene, 10 min each.
- Passed through two changes, paraffin warmed at 70 °C, 10 min each.
- Embedded in clean paraffin.
- The ribbons were made with the help of a microtome and kept for all night at 35–40 °C in the incubator.
- Mounted in DPX.

- Sections were studied under microscope and suitable microphotographs had been taken.

## *Details of Experimental Plants (In-Vitro)*

Fresh parts: of ten experimental plants categorized into three parts were taken.

**Cruciferous Plants:**

In India substantial study perform to inhibit the nematodes (Plant Parasitic) by the use of several cruciferous plants. Stahmann et al. [68] found the presence of antinemic phenyl isothiocyanate in crucifers.

i. ***Brassica rapa*** **(Turnip)**: It is a well-known vegetable which belongs to Family Cruciferae. It is largely cultivated for the sake of its leaves as well as the thickened roots Mathur [69].
ii. ***Brassica botrytis*** **(Cauliflower)**: The cauliflower belongs to Family Cruciferae and is eaten for its inflorescence. The leaves are applied in gout and rheumatism Mathur [69].
iii. ***Raphanus sativus*** **(Radish)**: Another member of Family Cruciferae is annual or biennial plants mostly cultivated during winter months for the fleshy tuberous roots. The juice of the fresh leaves is diuretic and laxative. The seeds are carminative and also yield an essential oil. The roots are used as drugs for urinary complaints, piles and gastrodynia pains Mathur [69].

A. **Reddy et al.** [70]: It is used for the Mean Gall Index value **(MGI) Scale**
   1 = 1–25 galls without egg masses
   2 = 26–50 galls without egg masses
   3 = with numerous egg masses

$$\mathbf{MGI} = \frac{Number\ of\ total\ galls\ counted\ in\ each\ replicate}{3}$$

B. **Atwal and Balraj** [71]: It is used for in vivo yield loss of *D. carota.*
C. **Statistical Calculation**: Statistical calculation like minimum value, maximum value, average median value, standard deviation, correlation coefficient and root squared value were taken with the help of a computer package.

# Results and Discussion

## *Histology of* M. incognita *Female*

Cross section of female body of *Meloidogyne incognita* through the anterior side showed esophageal gland lobe and intestine Whereas, in the posterior end sections showed various organs like ovary, oviduct, oviduct with oocytes, spermatheca, and uterus without eggs while, in some cases uterus with eggs as well as rectal gland had also been noticed (Figs. 17.1, 17.2, and 17.3).

Almost in all cases ovaries, uterus and rectal glands had been noticed from the posterior end. Eisenback [72] observed that in the sections of female *Meloidogyne*, a large portion of the body cavity is filled by a pair of tubular, highly convoluted gonads. Approximately 60% of the gonad was occupied by ovaries. Spermatheca was located

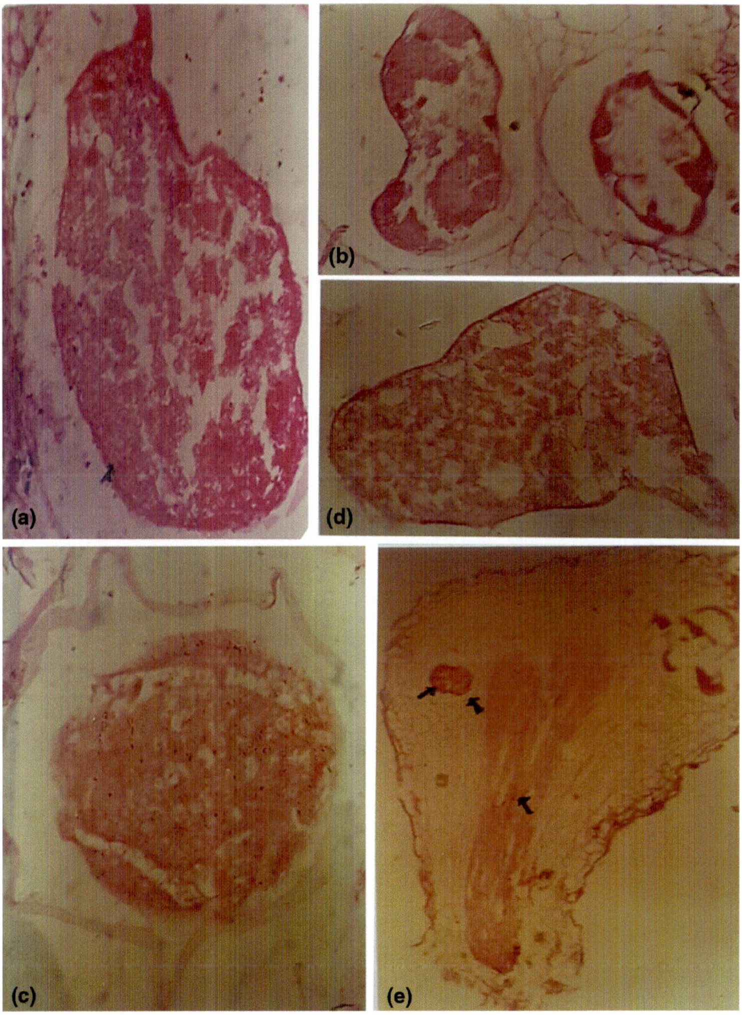


**Fig. 17.1** **a** Anterior posterior region of female *M. incognita* (L.S.), **b** female *M. incognita* through neck region (C.S.), **c** anterior lateral region of egg laying female *M. incognita* (C.S.), **d** posterior lateral region of female *M. incognita (L.S.)*. **e** *D. carota* showing giant cell adjacent to vascular bundles and abnormal growth of tissues after 26 days inoculation of *M. incognita* $J_2$ (L.S.)

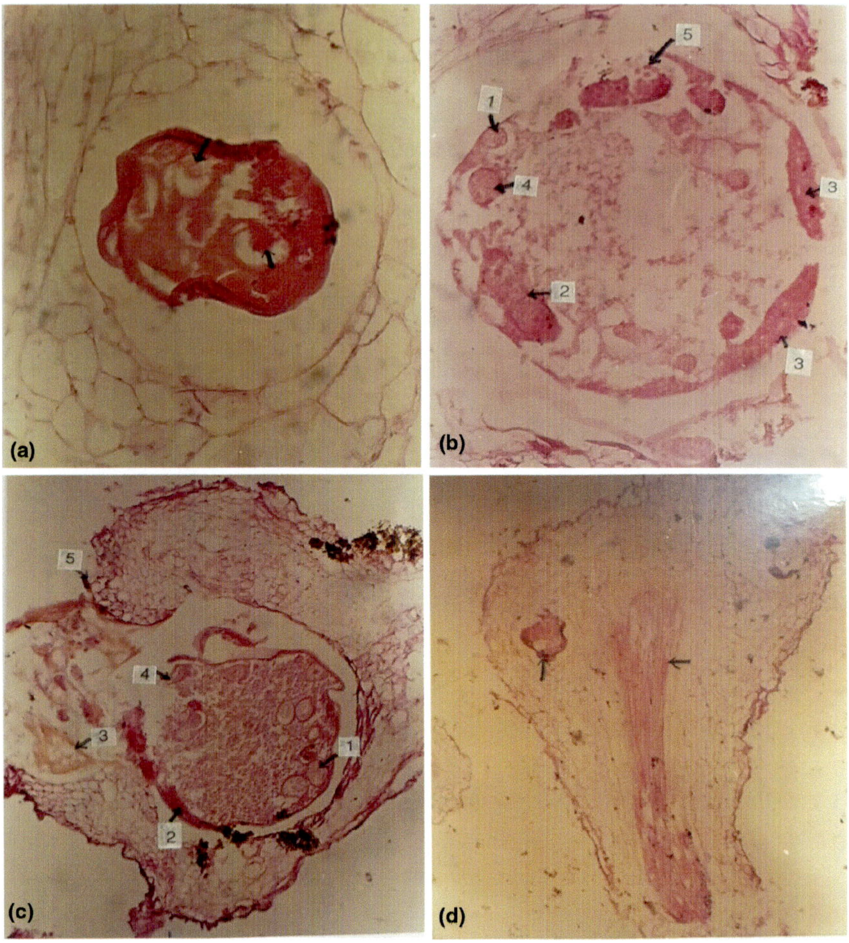


**Fig. 17.2** **a** Anterior part of female *M. incognit*a showing esophageal lobe and intestine (C.S.). **b** Posterior region of egg laying female of *M. incognit*a showing (C.S.): (1) ovary, (2) oviduct with oocyte, (4) uterus, (3) rectal gland, (5) spermatheca. **c** Posterior region of female *M. incognita* showing (C.S.): (1) uterus, (2) rectal gland, (3) hyaline portion of the gelatinous sheath, (4) ruptured cell wall and cortical cells of *D. carot*a, (5) abnormal vascular bundle of infected *D. carota.* **d** Hyperplasia and hypertrophy in infected *D. carota* after 15 days of the inoculation of $J_2$ (L.S.)

posterior to the oviduct. Posterior to the spermatheca, the uterus was differentiated. The two uteri of the female reproductive tract fuse to form one common duct posterior to which laid a large rectal gland. The present observations were in confirmation of Viglierchio [73]; Nguyen and Duong [74] findings.

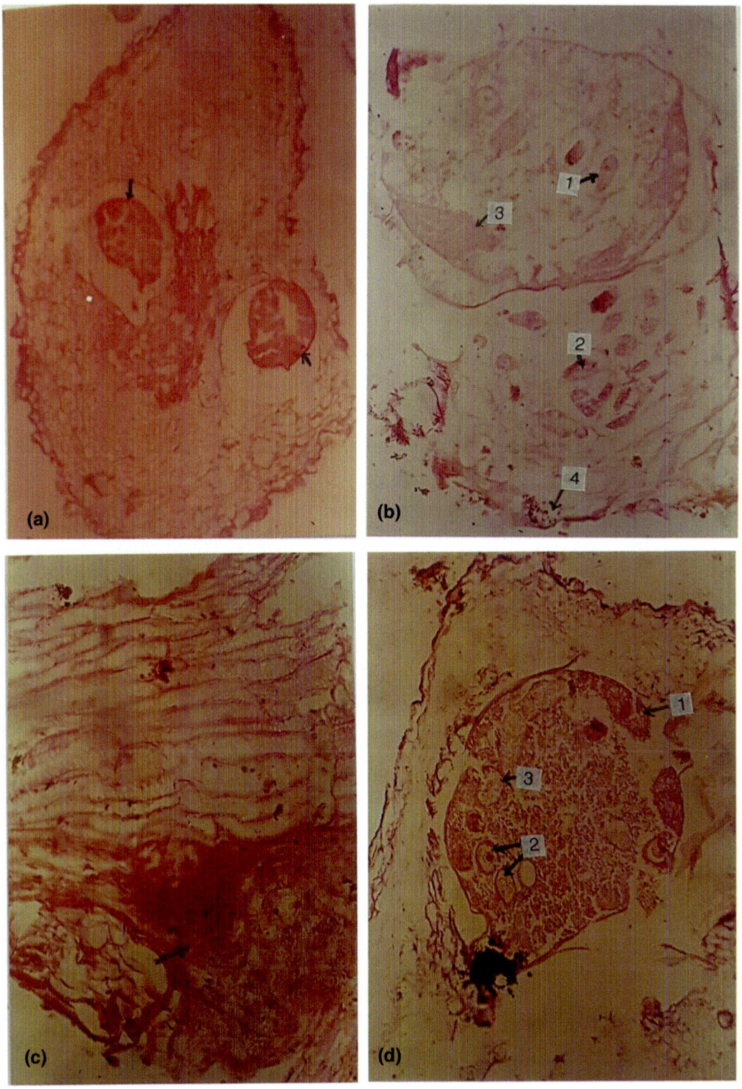


**Fig. 17.3** **a** Giant cells in the infected root of *D. carota* (20 days) after inoculation with *M. incognita* $J_2$ sections through the posterior part of body. **b** Posterior end of female *M. incognita* showing (C.S.): (1) uterus with few eggs, (2) spermatheca and ovary, (3) rectal gland, (4) hyaline sheath with few eggs. **c** Infected root of *D. carota* showing feeding side in cortical cells and thick dense granular protoplasm in phloem and nurse cells (L.S.). **d** Posterior end of egg laying female of *M. incognita* showing (C.S.): (1) rectal gland, (3) uterus, (2) ovary and oviduct

## *Histopathology of Carrot*

No visual symptoms were generally observed above the ground parts of the carrot. However, the nematode infestation resulted in the formation of cracks on tubers, forking of tap root accompanied by beads like galls on secondary roots and extensively reduced plant size.

After gaining entry through the root cap epidermis, the second stage juveniles penetrate secondary roots inter-cellular as well as intra-cellular through the cortex, endodermis and pericycle and reach the phloem. Soon-after penetration, $J_2$ began to feed, increased in size and became oriented perpendicularly towards the longitudinal perpendicular axis of the root having the posterior portion outside from the root (Fig. 17.4a, b). A slightly wider passage than the nematode bodies with thick walls was formed by the destruction of cortical cells. Soon after the infection, the juveniles were found to enlarge in size perhaps due to the pressure exerted on the cortical cells. The nematodes feed in the cortex as well as in the phloem. In the cortex, the cells at the feeding site stain pink red with lactophenol. The giant cells formed by *M. incognita* in carrot differed from those in roots of other susceptible crops, like tomato, by their thin walls and smaller size. Characteristic wound healing responses i.e. formation of callus like tissues or wound periderm and their precipitated constituents had also been observed. The proliferation of phloem cells at the feeding site was not so marked, though these cells had thick and dense granular protoplasm. In several sections hyperplasia of cortical and hypertrophy of pericycle cells and nurse cells had also been observed (Figs. 17.5 and 17.6).

Highly infected roots reveal histo-pathological changes which conduct to the element conjation as reported by Sudha and Prabhoo [75]. Whereas, in the histopathological studies Charles and Venkitesan [76] reported rupturing of cortical cells and formation of syncytial cells with thick end walls in the stellar region. Khan and Khan [77] observed reduced plant growth due to low and small size of stomata and trachoma. Procinai and Ambroguini [78] observed high metabolic activity with intense cytoplasm and nuclei in giant cells, produced by nematodes in carrots. Abnormal xylem and parenchyma with thickened cell walls were observed in all root knot nematode infected tissues except in rhizome meristems Routaray et al. [79]. Lanjewar and Shukla [80] found *M. incognita* was entering the cortex and stellar regions converting into giant cells. These giant cells showed karyotin nuclear divisions and had thickened cell walls. Sasser and Carter [81] presumed that giant cells produced by parasitic activity were chiefly nurse cells in the vascular tissues, which had cell wall impressions to soak nutrients from nearby cells. These were produced by mitosis without cytokinesis Dropkin [82]. Haseeb et al. [83] observed greater oxidase and peroxidase activity in vascular bundles which might be responsible for delaying lignification. Corky wounds were found at infection sites in differentiated rhizomes and fresh roots Shah and Raju [84]. Whereby, characteristic wound healing responses like formation of callus like tissues or wound periderm at the wound site observed in present study had also been reported by Stobbe [85] in yam tubers who presumed that it might be due to the production of resin, gum, latex or callose and

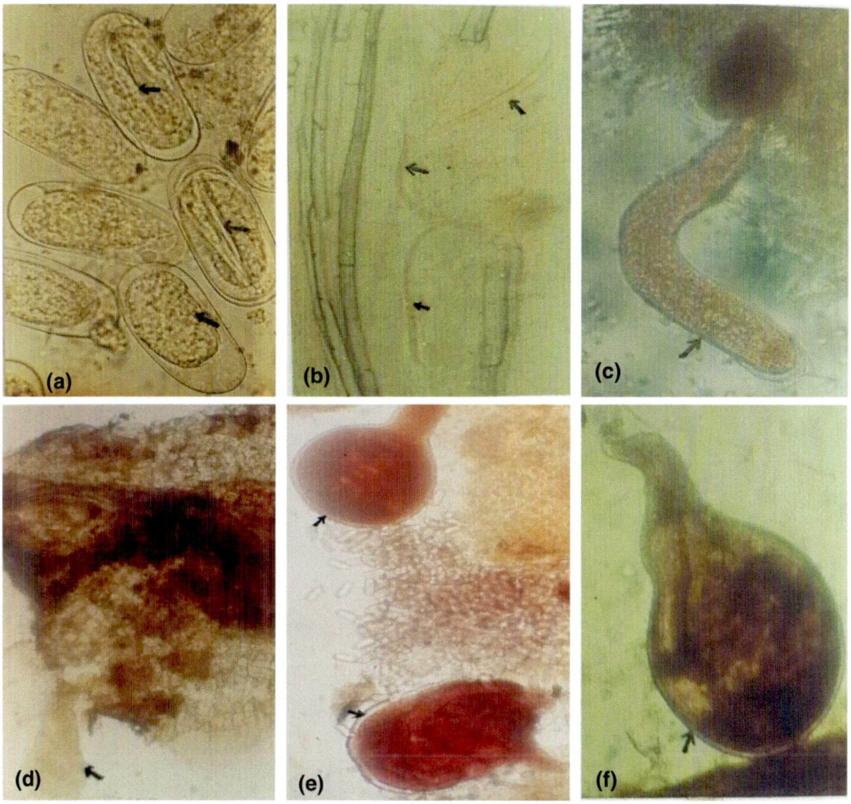


**Fig. 17.4** **a** Embryogenesis in the *M. incognita* eggs and juvenile of the previous stage just before hatching. **b** Second stage juveniles ($J_2$) of *M. incognita*. **c** Spike tailed or sausage shaped J, larvae of *M. incognita*. **d** Fourth stage of female *M. incognita*. **e** Developing female of *M. incognita*. **f** Mature female of *M. incognita*

intense suberization in the wound area. Vilsoni et al. [86]; Valette et al. [87]; Alamgir [88] reported that the burrowing nematodes migrate intra-cellular which leads into the giant galleries in rhizomes by infestation of nematodes.

Hence, the infestation of nematodes somehow, disturbed the metabolic activities of infected plants and in infected plants stellar regions of roots were occupied by developing females. Cortical cells in areas where females occurred showed rupture. The epidermis disintegrated, thus, allowing the body of the female to protrude out of the roots. The damage caused to the root tissues may suppress the flow of food materials to various parts of the plants. Moreover, the nutritive value of the tubers was lowered to a considerable extent. Additionally, dwarf tap root, constriction and formation of crack on tap root affected the yield reducing the market value of carrot.

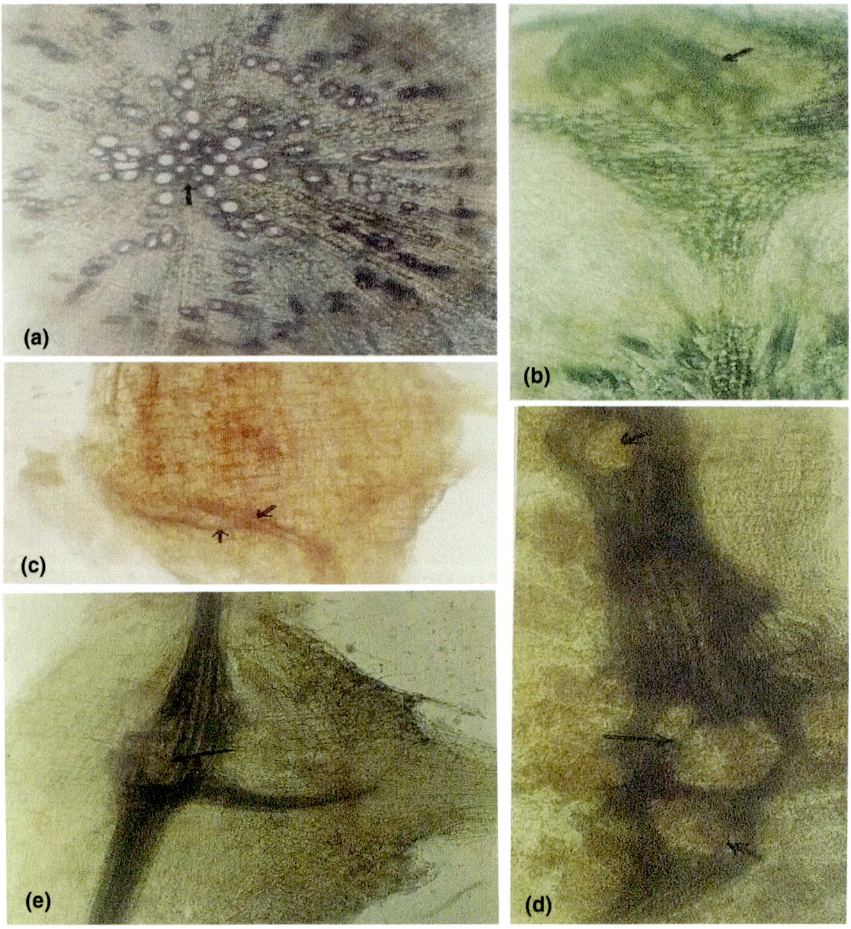


**Fig. 17.5** **a** Normal root of *D. carota* showing exarch type of xylem (T.S.). **b** Infected root of *D. carota* showing characteristic "Callus" like tissue formation (T.S.). **c** Gall with $J_2$ in secondary root of infected *D. carota* (W.M.). **d** Abnormal xylem and phloem cells in infected root of *D. carota* (W.M.). **e** Abnormal growth of vascular bundle in infected root of *D. carota* (W.M.)

## Brassica rapa *(Turnip)*

The nematostatic effect of *Brassica rapa* leaves, petiole and roots extracts on juvenile mortality showed in Tables 17.1 and 17.2. Leaves extract of turnip was most effective and showed 100% $J_2$ killed followed by 85.67–96.75% mortality with 50–6.25% dilutions after 72 h exposure. In all the cases at lower concentrations the nematicidal activity started diminishing as less percent of juveniles' mortality in root extract had been noticed. However, the efficacy of the stock solution of petiole and root extract was noticed to be 51.10–89.17%. The percent mortality increased from 31.10 to

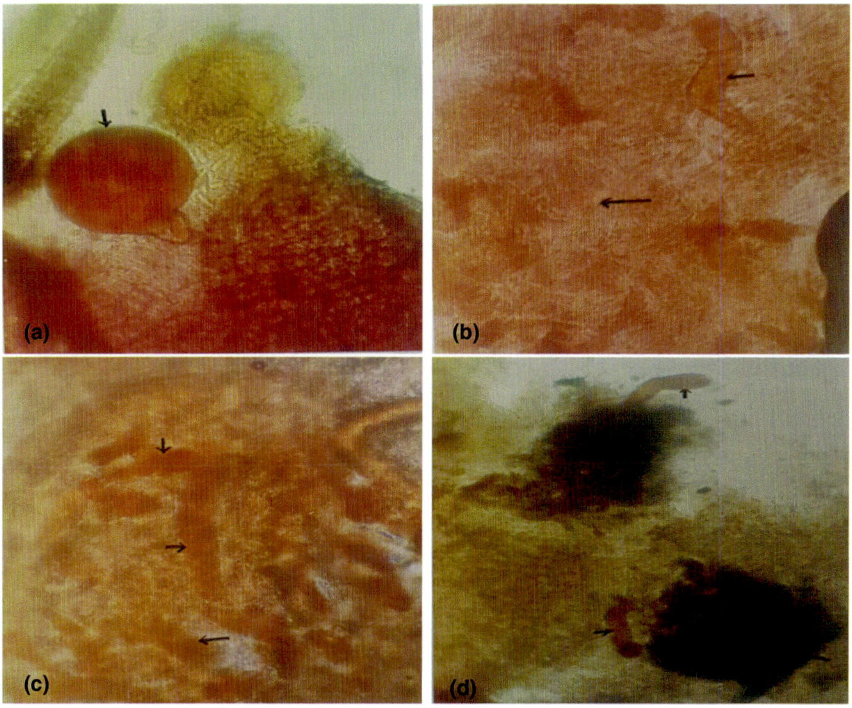


**Fig. 17.6** **a** Developing female of M. incognita in the root of *D. carota* plant tissues (W.M.). **b** Infected secondary root with third stage larvae of *M. incognita* and abrupted stelar region of infected *D. carota* (W.M.). **c** Infected root of *D. carota* showing dense granular protoplasm and abundance of larvae of *M. incognita* in cortical region (T.S.). **d** Sausage shaped larvae of *M. incognita* along with damaged plant cells and precipitated constituents of infected *D. carota* (W.M.)

40.58% after 72 h exposure with 12.5% and 25% dose respectively but remained remarkably lower than the leaf extract in which the mortality also showed an upward trend with exposure timing.

Rao et al. [89] recorded 42.80% $J_2$ mortality of *M. incognita* with root exudates and 55.07% with leaf extract of *Brassica campestris* after 48 h respectively which supported present findings. On the other hand *Brassica rapa* showed least susceptibility to *M. incognita* among ten different vegetables examined by Kihika-Opanda et al. [90]. In India, Ahuja and Mukhopadhyay [91] also reported least 10–25% susceptibility of *Brassica rapa* against *M. incognita* in field and micro-plots experiments among twenty three vegetables examined. Hence, may presume *Brassica rapa* to possess some antagonistic properties against *M. incognita*. Further proved by seedlings of tomato roots were dipped in the water extracts of oil seed cakes of *Brassica rapa* by Vijayalakshmi and Goswami [92] and then disclosed to *M. incognita* (1000 $J_2$/pot). After 45 days, the most significant enhancement in plant growth and marked inhibition in nematode infestation had been recorded with *Brassica rapa* aqueous extract. Feyisa et al. [93] also tested *Brassica campestris* aqueous leaf extract

**Table 17.1** In-vitro experiments of plant extracts

| Plant extract | Exposure time (h) | Doses % | | | | | | |
|---|---|---|---|---|---|---|---|---|
| *Barassica rapa* | | 100 | 50 | 25 | 12.50 | 6.50 | 3.13 | 1 |
| Leaf | 24 | 68.27 | 61.9 | 78.2 | 56.79 | 48.78 | 10.98 | |
| | 48 | 92.39 | 76.75 | 65.22 | 68.83 | 63.43 | 24.39 | 4.45 |
| | 72 | 100 | 93.33 | 96.75 | 93.18 | 85.67 | 42.86 | 5.7 |
| Petiole | 24 | 58.89 | 37.66 | 28.8 | 27.23 | 12.98 | 0 | 0 |
| | 48 | 83.37 | 49.38 | 34.44 | 29.52 | 12.55 | 2.18 | 0 |
| | 72 | 89.17 | 71.74 | 64.18 | 40.58 | 23.77 | 2.9 | 0 |
| Root | 24 | 51.1 | 22.25 | 13.23 | 0 | 0 | 0 | 0 |
| | 48 | 75.83 | 48.16 | 17.16 | 6.16 | 0 | 0 | 0 |
| | 72 | 88.12 | 84.5 | 50.2 | 31.1 | 2.2 | 0 | 0 |
| *Barassica botrytis* | | | | | | | | |
| Leaf | 24 | 100 | 100 | 68.51 | 58.8 | 35.57 | 26.69 | 2.2 |
| | 48 | 100 | 100 | 100 | 93.35 | 86.5 | 28.53 | 6.65 |
| | 72 | 100 | 100 | 100 | 100 | 93.42 | 31.5 | 8.91 |
| Petiole | 24 | 100 | 77.71 | 35.55 | 22.21 | 13.33 | 0 | 0 |
| | 48 | 100 | 78.5 | 42.56 | 26.68 | 17.77 | 2.22 | 0 |
| | 72 | 100 | 100 | 57.8 | 37.41 | 26.6 | 4.45 | 0 |
| Shoot | 24 | 86.67 | 80 | 64.42 | 44.45 | 28.85 | 0 | 0 |
| | 48 | 100 | 82.14 | 73.3 | 46.2 | 37.59 | 0 | 0 |
| | 72 | 100 | 100 | 75.17 | 55.5 | 48.8 | 0 | 0 |
| Root | 24 | 51.1 | 22.25 | 13.23 | 0 | 0 | 0 | 0 |
| | 48 | 75.83 | 48.16 | 17.16 | 6.16 | 0 | 0 | 0 |
| | 72 | 88.12 | 84.5 | 50.2 | 31.1 | 2.2 | 0 | 0 |
| *Raphanus sativus* | | | | | | | | |
| Leaf | 24 | 100 | 78.69 | 52.94 | 47.71 | 40.02 | 34 | 20.85 |
| | 48 | 100 | 87.89 | 65.22 | 65.1 | 56.54 | 44.54 | 24.78 |
| | 72 | 100 | 93.35 | 91.84 | 86.36 | 73.91 | 59.74 | 29.44 |
| Petiole | 24 | 100 | 78.62 | 41.44 | 32.75 | 30.3 | 8.09 | 2.1 |
| | 48 | 100 | 85.89 | 53.21 | 50.54 | 48.83 | 11.22 | 3.5 |
| | 72 | 100 | 100 | 75.4 | 64.59 | 62 | 14.73 | 3.5 |
| Root | 24 | 91.11 | 77.75 | 66.06 | 55.5 | 37.73 | 24.4 | 0 |
| | 48 | 100 | 88.51 | 71.01 | 64.44 | 44.2 | 35.55 | 6.1 |
| | 72 | 100 | 100 | 84.2 | 73.57 | 55.55 | 40 | 17.7 |

**Table 17.2** Statistical analysis of in-vitro experiments of *Cruciferous* plant extract with exposure hrs and doses against $J_2$ of *Meloidogyne*

| Statistical analysis of in-vitro experiment | | | | | | | | |
|---|---|---|---|---|---|---|---|---|
| Plant extracts | Exposure time (h) | Minimum value | Maximum value | Average | Median value | Standard deviation | Correlation coefficient | R-squared |
| *Brassica rapa* | | | | | | | | |
| Leaf | 24 | 2 | 68.27 | 48.84 | 56.79 | 25.0583 | −0.9133 | 0.8341 |
| | 48 | 4.45 | 92.39 | 58.34 | 68.83 | 29.4846 | −0.9321 | 0.8688 |
| | 72 | 5.07 | 100 | 73.92 | 93.15 | 33.1787 | −0.8501 | 0.7228 |
| Petiole | 24 | 0 | 58.89 | 23.65 | 27.23 | 19.6758 | −0.9122 | 0.9452 |
| | 48 | 0 | 84.17 | 41.76 | 40.58 | 32.134 | −0.9904 | 0.981 |
| | 72 | 0 | 87.17 | 41.76 | 40.58 | 32.134 | −0.9904 | 0.981 |
| Root | 24 | 0 | 51.1 | 12.36 | 0 | 17.7731 | −0.8425 | 0.7195 |
| | 48 | 0 | 75.83 | 21.04 | 6.16 | 27.5212 | −0.8849 | 0.7831 |
| | 72 | 0 | 88.12 | 36.58 | 31.1 | 35.9094 | −0.9574 | 0.9167 |
| *Brassica botrytis* | | | | | | | | |
| Leaf | 24 | 2.2 | 100 | 55.96 | 58.8 | 34.2412 | −0.9866 | 0.9734 |
| | 48 | 6.65 | 100 | 73.57 | 93.35 | 36.176 | −0.8618 | 0.7427 |
| | 72 | 8.91 | 100 | 76.26 | 100 | 36.0323 | −0.8263 | 0.6828 |
| Petiole | 24 | 0 | 100 | 35.54 | 22.21 | 36.1159 | −0.9446 | 0.8923 |
| | 48 | 0 | 100 | 37.81 | 26.68 | 34.9129 | −0.9643 | 0.9299 |
| | 72 | 0 | 100 | 46.49 | 37.41 | 38.2544 | −0.9737 | 0.9481 |
| Shoot | 24 | 0 | 86.67 | 43.48 | 44.45 | 33.0228 | −0.9854 | 0.971 |
| | 48 | 0 | 100 | 48.46 | 46.2 | 36.3077 | −0.9836 | 0.9675 |
| | 72 | 0 | 100 | 54.21 | 55.5 | 38.8056 | −0.9688 | 0.9387 |
| Root | 24 | 0 | 51.1 | 12.36 | 0 | 17.77 | −0.8482 | 0.7195 |
| | 48 | 0 | 75.83 | 21.04 | 6.16 | 27.5212 | −0.8849 | 0.7831 |
| | 72 | 0 | 88.12 | 36.58 | 31.1 | 35.9094 | −0.9574 | 0.9167 |
| *Raphanus sativus* | | | | | | | | |
| Leaf | 24 | 20.85 | 100 | 53.45 | 47.71 | 25.2377 | −0.9615 | 0.9246 |
| | 48 | 24.78 | 100 | 63.43 | 65.1 | 23.4209 | −0.9791 | 0.9586 |
| | 72 | 29.44 | 100 | 76.37 | 86.36 | 22.8999 | −0.9258 | 0.8572 |
| Petiole | 24 | 2.1 | 100 | 41.5 | 32.75 | 33.1007 | −0.9622 | 0.9258 |
| | 48 | 3.33 | 100 | 50.43 | 50.54 | 32.6906 | −0.9695 | 0.94 |
| | 72 | 3.5 | 100 | 60.03 | 64.59 | 35.2619 | −0.959 | 0.9197 |
| Root | 24 | 0 | 91.11 | 50.36 | 55.5 | 29.4066 | −0.9919 | 0.9838 |
| | 48 | 6.1 | 100 | 58.54 | 64.44 | 29.9799 | −0.9873 | 0.9749 |
| | 72 | 17.7 | 100 | 67.28 | 73.57 | 28.8023 | −0.9809 | 0.9699 |

and found marked reduction in the hatching of egg in *M. incognita.* Aqueous extracts of *Brassica rapa* inhibition of hatching from mass eggs and penetration of juveniles of *M. incognita* into *V. radiata* as reported by Majumdar and Mishra [94].

## Brassica botrytis *(Cauliflower)*

The data present in Tables 17.1 and 17.2 showed that leaf extract of *Brassica botrytis* caused significant percent mortality of *M. incognita* juveniles than the other tested parts of this plant. Highest mortality i.e. 100% was observed within 24 h exposure with leaf extract in its 100 and 50% concentrations while with petiole extract in stock solution the same observation had been noticed. With 100 and 50% shoot extract 100% mortality occurred after 48 h exposure. Higher concentration of root extract of *B. botrytis* suppressed mortality in comparison to other part's extract, whereas, lowest 1% dilution of petiole, shoot and root extract was totally a failure to cause mortality. It was also proved statistically (Tables 17.1 and 17.2). Percent mortality in the leaf extract was less at 3.125 and 1% doses and in the petiole extract at 6.25 and 3.125% dilution while in root extract at 12.5 and 6.25% doses when compared to higher concentrations.

These findings confirmed the observations by Abid et al. [95] who noticed ethanol extract of *Brassica botrytis* causing 4, 10, and 23% juveniles' mortality of *M. javanica* after 24, 48, and 72 h of exposure respectively in lower concentrations. Alam [96] reported that minced leaves of cauliflower inhibited the percentage of plant parasitic nematodes. Aisha et al. [97] reported seeds and oil cakes of *Brassica* species to be extremely nematicidal against *Heterodera schachtii.* The chopped floral parts of *Brassica botrytis* against *Tylenchids* had been reported very effective by Haseeb and Alam [98] while Chandravadana et al. [99] and Abid et al. [95] had also reported *Brassica botrytis* possessing nematicidal potential against *M. incognita.* Ahuja and Mukhopadhyay [91] reported *Brassica botrytis* to be resistant against *M. incognita* infestation. Whereby, Thies [100] studies that marked inhibition in root gall index of *M. incognita* with the treatment of oil cakes of Brassica species infecting different vegetables in the field trials.

## Raphanus sativus *(Radish)*

The percent mortality of *M. incognita* juveniles increased almost equally from highest 100% upto 6.25% dilutions after the exposure 24, 48, and 72 h. Root extract was interestingly more effective than petiole extract except the initial low mortality after 24 h exposures was 91.11% instead of 100% mortality like petiole extract 100% dose. It was discernible that in the juveniles, exposed to 3.125 and 1% concentrations, mortality occurred from 8.09 to 59.74% and 2.10 to 29.44% respectively for 24, 48, and 72 h exposure in all the leaves, petiole and root extracts. In all cases a marked increment in percent mortality with the increment of treatment time (Tables 17.1 and 17.2). No $J_2$ mortality had been recorded in root extract at 24 h.

The above findings support the work of Nandal and Bhatti [101] also confirming the result of root extract of seven alkaloids bearing plants including *Raphanus, Brassica botrytis, B. compestris and Mentha* etc. were more inhibitory than shoot extracts for hatching of juveniles of *M. incognita* observed by Haseeb et al. [102]. According to Kerakalamatti et al. [103] experimented aqueous extracts of oil cakes R. sativus against *Hoplolaimus indicus* reported nematicidal efficacy. Gardner and Caswell-Chen [104] tested cultivars of *R. sativus* against *M. javanica* and *M. incognita* finding all vascular plants to be exposed. On the contrary Belair [105] did not find *R. sativus* as the host for *M. hapla.* Muller [106] also reported that *R. sativus* was implicated in inhibiting the percentage of *M. incognita* studied under greenhouse and micro-plot conditions.

Some of the workers detected certain toxic principles like ricinine ($C_8H_8M_2O_2$), sinigrin (Glycoside), quercetin ($C_{15}H_{10}O_7$), arachin and conarchin, nimbidin and thiniomone which had been isolated from castor, mustard, mahua, groundnut and neem oil cakes respectively. On the other hand menthol and menthone were extracted from *M. arvensis* and certain alkaloids like ajmalicine, serpentine and reserpine from *C. roseus.* Agarwal and Ghosh [24] observed that all these compounds had inhibited nematode percentage. Decline in nematode percentage population probably appears due to production of fatty acids as suggested by Johnson [107]; Klemens and Gerard [108]. Whereas, Khan [109] proposed that probably the position of the "OH" group present in hydroquinone, arbutin, phloroglucinol, orcinol and resorcinol and in some pyrogallol, catechol and gallic acid, some precursor and compounds evinced in plants determine the toxicity against nematodes.

## Conclusions

As far as the mechanism is concerned this is understood that the efficacy of plant extract is governed by composition of the compound present in plant parts and the degree of decomposition as influenced by the biological and physical environment of the soil. By and large, the following explanations have been given by different workers: The products from decomposition of plant matter are directly toxic to nematodes Ntalli et al. [110]. Organic matter present in plants on decomposition brings changes in the abiotic and biotic factor of plants surrounding it which results in the host-parasitic equation Vander Laan [111]. Organic amendments facilitate the soil array for higher root length, resulting in more absorption of the nutrients of the soil, and minimizing the distraction of nematodes. Widmer et al. [112]; Ansari et al. [113] suggested that biological management, uses of botanical and topsoil modification techniques record high among others practices of nematode control in environmental safety point of view. As almost 2400 plant species around the World known as pesticidal or nematicidal tendency, but now a days caution should be taken because many report shows phytochemicals contain many toxicants which may cause eco-toxicity, hepatotoxicity, cytotoxicity and even cause carcinogenicity.

## References


1. Jairajpuri MS (1986) Morphology of nematodes. In: Swarup G, Dasgupta (eds) Plant parasitic nematodes of India - problems and progress. IARI, New Delhi, pp 10–57
2. Sehgal M, Srivastava DS, Malik M, Singh A (2021) Root-knot nematode (*Meloidogyne incognita*) an emerging problem in pointed gourd in Sitapur, Uttar Pradesh, India: a serious threat. Int J Agric Appl Sci 2(1):123–125. https://doi.org/10.52804/ijaas2021.2113
3. Moens M, Wesemael W (2008) Quality damage on carrots (*Daucus carota L.*) caused by the root-knot nematode *Meloidogyne chitwoodi*. Nematology 10(2):261–270. https://doi.org/10.1163/156854108783476368
4. Gill HS, Kataria AS (1974) Some biochemical studies in European and Asiatic varieties of carrots (*Daucus carota*). Curr Sci 43(6):184–185
5. Rebecca LJ, Sharmila S, Paul MD, Seshiah C (2014) Extraction and purification of carotenoids from vegetables. J Chem Pharm Res 6(4):594–598
6. Raees-ul H, Prasad K (2018) Role of various coloring pigments within different varieties of carrot (*Daucus carota*). Adv Plants Agric Res 8(5):386–387. https://doi.org/10.15406/apar.2018.08.00341
7. Greco N, Brandonisio A (1980) Relationship between *Heterodera carotae* and carrot yield. Nematologica 26:497–510
8. Krypl JS, Janas KM (1988) Synthesis of RNA and protein with ribonuclease activity in carrot infested with *Meloidogyne hapla*. Physiol Plant Pathol 7:213–220
9. Phillips B (2019) When it comes to carrots: want not, root-knot. Michigan State University Extension - October 3
10. Masood A, Saxena SK (1980) Nucleic acid change in three cultivars of tomato infected with *Meloidogyne incognita*. Indian J Nematol 10:102–104
11. Phani V, Bishnoi S, Sharma A, Davies KG, Rao U (2018) Characterization of *Meloidogyne indica* (Nematoda: Meloidogynidae) parasitizing neem in India, with a molecular phylogeny of the species. J Nematol 50(3):387–398. https://doi.org/10.21307/jofnem-2018-015
12. Bayani U, Singh A, Zamboni P, Mahajan R (2009) Oxidative stress and neurodegenerative diseases: a review of upstream and downstream antioxidant therapeutic options. Curr Neuropharmacol 7(1):65–74. https://doi.org/10.2174/157015909787602823
13. Tyagi A, Rehman BV (1981) Histopathological studies of to mato root galls caused by *Meloidogyne incognita*. Proc Indian Acad Sci 84:109–115
14. Ahmad L, Siddiqui ZA, Abd_Allah EF (2019) Effects of interaction of *Meloidogyne incognita*, *Alternaria dauci* and *Rhizoctonia solani* on the growth, chlorophyll, carotenoid and proline contents of carrot in three types of soil. Acta Agric Scand, Sect B—Soil Plant Sci 1–8. https://doi.org/10.1080/09064710.2019.1568541
15. Ganguly AK, Dasgupta DR (1987) Comparison of proteins and enzymes from galls and non-gall part of some root system of tomato cultivar pusa ruby infected with root knot nematode. Indian J Nematol 17:343–345
16. Lu P, Davis RF, Kemerait RC, van IMW, Scherm H (2014) Physiological effects of *Meloidogyne incognita* infection on cotton genotypes with differing levels of resistance in the greenhouse. J Nematol 46(4):352–359
17. Bruno F, Quentin M, Jaubert-Possamai S, Abad P (2016) Gall-forming root-knot nematodes hijack key plant cellular functions to induce multinucleate and hypertrophied feeding cells. J Insect Physiol 84:60–69. https://doi.org/10.1016/j.jinsphys.2015.07.013
18. Meidani C, Ntalli NG, Giannoutsou E, Adamakis IS (2019) Cell wall modifications in giant cells induced by the plant parasitic nematode *Meloidogyne incognita* in wild-type (Col-0) and

the *fra2 Arabidopsis thaliana* katanin mutant. Int J Mol Sci 20(21):54–65. https://doi.org/10.3390/ijms20215465
19. Debia PJG, Barros BCB, Puerari HH, Dias-Arieira CR (2020) *Meloidogyne javanica* parasitism on the vegetative growth and nutritional quality of carrots. Cienc Rural. FapUNIFESP (SciELO) 50(9). https://doi.org/10.1590/0103-8478cr20190585
20. Huang CS, Charchar JM (1982) Preplanning inoculum densities of root knot nematode related to carrot yield in green house. Plant Dis 66:1064–1066
21. Hay FS, Ophel-Keller K, Hartley DM, Pethybridge SJ (2016) Prediction of potato tuber damage by root-knot nematodes using quantitative DNA assay of soil. Plant Dis 100(3):592–600. https://doi.org/10.1094/PDIS-05-15-0537-RE
22. Singh G, Kanwar RS, Sharma L, Neeraj LKC, Kaushik P (2020) Biochemical changes induced by *Meloidogyne graminicola* in resistant and susceptible pearl millet (*Pennisetum glaucum L.*) hybrids. Plant Pathol J 19:132–139
23. Follrich B, De Bettignies Dutz A, Meissner P, Shellmoser S, Jurenistsch J, Guggenbichler JP (1998) About oligosaccharides from carrots with antibacterial activity. Phytomedic 3(1):302
24. Agarwal VS, Ghosh B (1985) Drug plants of India (root drugs). CSIR Publications, New Delhi
25. Krishnamurthy RBH, Murthy KSRK (1983) Proceedings of national seminar on crop losses due to insect pest. Indian J Ent Special issue 1–2, Hyderabad, pp 1–12
26. Dhaliwal GS, Arora R (1996) Principal of insect pest management. National Agricultural Technology Information Centre, Ludhiya, p 374
27. Van-Burkum JA, Seshadri AR (1970) Some important nematode problems in India X Inter. In: Nematology symposium. E.S.N., Pescara, pp 17–19
28. Cunha TG, Visôtto LE, Pinheiro LM, God PIVG, Rosa JMO, Oliveira CMG, Lopes EA (2021) Distribution of *Meloidogyne* species in carrot in Brazil. Ciênc Rural, St Maria 51:5. https://doi.org/10.1590/0103-8478cr20200552
29. Bessey EA (1911) Root-knot and its control. US Dep Agric Bull Cart 217–289
30. Mathews JD (1919) Reports on the W.B. Randall research assistant. Nursery and market garden experiment research station, Chechunt, Herts. Annu Rep 5:18–21
31. McBeth CV (1954) Tests on the susceptibility and resistance of several southern grasses to the root knot nematode *Heterodera marioni*. Proc Helminthol Soc Wash 12:41–44
32. Haydock PPJ, Woods SR, Grove IG, Hare MC (2006) Chemical control of nematodes. In: Plant nematology, vol 3, chap 16, pp 92–102. https://doi.org/10.1079/9781845930561.0392
33. D'errico G, Marra R, Vinale F, Landi S, Roversi PF, Woo SL (2017) Nematicidal efficacy of new abamectin-based products used alone and in combination with indolebutyric acid against the root-knot nematode *Meloidogyne incognita*. REDIA 100:95–101. https://doi.org/10.19263/REDIA-100.17.12
34. Talavera-Rubia M, Vela-Delgado MD, Verdejo-Lucas S (2020) Nematicidal efficacy of milbemectin against root-knot nematodes. Plants 9(7):839. https://doi.org/10.3390/plants9070839
35. Van-Burkum JA, Hoestra H (1979) Practical aspects of the chemical control of nematodes in soil. In: Muller D (eds) Soil disinfestation. Elsevier, Amsterdam, pp 53–59
36. Perry AS, Yamamoto I, Ishaaya I, Perry R (1998) Fumigants and nematicides. Insectic Agric Environ 19:130–136. https://doi.org/10.1007/978-3-662-03656-3_19
37. Gad SC (2014) Encyclopedia of toxicology. Nematicides 3:473–474. https://doi.org/10.1016/b978-0-12-386454-3.00888-5
38. Jenkin L, Guengerich HW (1959) Chemical dips for control of nematodes on bare root nursery stock. Plant Dis Report 43.1095–1097
39. Castro CE, Belser MO (1968) Reductive dehalogenation of the biotic ethylene dibromide, 1-2-dibromo-3-chloropropane and 2, 3 dibromobutane in soil. Environ Sci Technol 2:779–783
40. Mckenry MV (1980) Nature, mode of action and biological activity of nematicides. In: CRC handbook of pest management in agriculture. CRC Press, Cleveland, pp 59–73
41. Young RW, Miller LI, Harrison WA, Engel RW (1959) Bromide level of cow's milk as influenced by feeding peanut vines produced on soil fumigated with ethylene dibromide. Toxicol Appl Pharmacol 1:384–390

42. Beyer EM (1991) Crop protection: meeting the challenge. Bringhton Crop Prot Conf Weeds 1:3–22
43. Waterfield G, Zilberman D (2012) Pest management in food systems: an economic perspective. Annu Rev Environ Resour 37(1):223–245. https://doi.org/10.1146/annurev-environ-040911-105628
44. Zaki FA, Bhatti DS (1997) Phytotherapeutic effect of some plant leaves on *Meloidogyne javanica* infecting tomato. Nematol Mediterr 17:71–72
45. Miller PM, Edlington LV (1962) Swarup G, Dasgupta DS, Gill JS (eds) Soil amendments in nematode management. An appraisal of ecofriendly approaches. I.C.A.R., New Delhi, pp 106–114
46. Miller PM, Wihrheim S (1966) Effect of paper and sawdust soil amendments on meodow nematode and subsequent verticillium wilt of tomatoes. Plant Dis Report 66:745–747
47. Mankau R, Das S (1969) The influence of chitin amendments on *Meloidogyne incognita*. J Nematol 1:15
48. Sitaramiah K (1974) Effect of extracts and distillates of amended soil on the activity of *Meloidogyne javanica*. Indian J Mycol Plant Pathol 4:138–144
49. Lear B (1959) Application of castor pomace and cropping of castor bean to soil to reduce nematode populations. Plant Dis Report 43:459–460
50. Szczygłowska M, Piekarska A, Konieczka P, Namieśnik J (2011) Use of brassica plants in the phytoremediation and biofumigation processes. Int J Mol Sci 12(11):7760–7771. https://doi.org/10.3390/ijms12117760
51. Singh RS, Sitaramiah K (1973) Control of plant parasitic nematodes with organic amendments of soil. Final Technical Report PL 480 scheme, Project no A7 CR 223
52. Tarla DN, Erickson LE, Hettiarachchi GM, Amadi SI, Galkaduwa M, Davis LC, Nurzhanova A, Pidlisnyuk V (2020) Phytoremediation and bioremediation of pesticide-contaminated soil. Appl Sci 10:1217. https://doi.org/10.3390/app10041217
53. Abid M, Maqbool MA (1991) Effect of bare root dip treatment in amoilcakes and neem leaf extract on the root-knot development and growth of tomato and eggplant. Pak J Nematol 9:13–16
54. Vats R, Nandal SN (1993) Effect of different concentrations of leaves extracts of neem and Eucalyptus used as bare root dip treatment of tomato seedlings against *Meloidogyne javanica*. Curr Nematol 4:15–18
55. Douda O, Zouhar M, Nováková E, Mazáková J (2012) Alternative methods of carrot (*Daucus carota*) protection against the northern root knot nematode (*Meloidogyne hapla*). Acta Agric Scand, Sect B - Soil Plant Sci 62(1):91–93. https://doi.org/10.1080/09064710.2011.570373
56. Pinheiro JB, Carvalho ADF, Rodrigues CS, Cruz EM, Pereira RB, Vieira JV (2019) Establishment of carrot populations (*Daucus carota L.*) in areas naturally infested by root-knot nematodes. Int Soc Hortic Sci 1:1249. https://doi.org/10.17660/ActaHortic.2019.1249.23
57. Fabiyi OS (2021) Sustainable management of *Meloidogyne incognita* infecting carrot (*Daucus carota*): green synthesis of silver nanoparticles with *Cnidoscolus aconitifolius*. Vegetos 34:277–285. https://doi.org/10.1007/s42535-021-00216-y
58. Pireda O, Bonierbale MW, Plaisted RL, Brodie BB, Tanksley SD (1998) Identification of RFLP markers linked to the H. gene conferring resistance to the potato cyst nematode. Genome 36:152–156
59. Anna C, Ramon M, Adela G, Montserrat P (1999) Plant responses to drought, from ABA signal transduction events to the action of the induced proteins. Plant Physiol Biochem 37(5):327–340. https://doi.org/10.1016/s0981-9428(99)80039-4
60. Rybczyński JJ, Davey MR, Tomiczak K, Niedziela A, Mikuła A (2015) Systems of plant regeneration in gentian in vitro cultures. In: Rybczyński JJ, Davey MR, Mikuła A (eds) The Gentianaceae - volume 2: biotechnology and applications. Springer-Verlag Berlin Heidelberg, pp 1–38. https://doi.org/10.1007/978-3-642-54102-5
61. Michael RD, Paul A, Power BJ, Kenneth CL (2005) Plant protoplasts: status and biotechnological perspectives. 23(2):131–171. https://doi.org/10.1016/j.biotechadv.2004.09.008

62. Powers TO, Harris TS (1996) A polymerase chain reaction method for identification of five major Meloidogyne species. J Nematol 25:1–6
63. Castagnone-Sereno P, Danchin GJE, Deleury E, Guillemaud T, Malausa T, Abad P (2010) Genome-wide survey and analysis of microsatellites in nematodes, with a focus on the plant-parasitic species *Meloidogyne incognita*. 11(1):598. https://doi.org/10.1186/1471-2164-11-598
64. Bairwa A, Venkatasalam EP, Sudha R, Umamaheswari R, Singh BP (2017) Techniques for characterization and eradication of potato cyst nematode: a review. J Parasit Dis 41:607–620. https://doi.org/10.1007/s12639-016-0873-3
65. Barker KR (1985) Nematode extraction and bioassay. In: Barker KR, Carter CC, Sasser JN (eds) An advanced treatise on Meloidogyne. Vol. II. Methodology. Dept. of Pathol. and the U.S. Agency for Intll. Develop., pp 19–38
66. Norton DC (1978) Ecology of plant parasitic nematodes. Wiley, N.Y., USA, p 268
67. Johnson AW (1996) Managing nematode population in crop production. In: Rigghs RD (ed) Nematology in the southern region of the United States, pp 193–203. South Coop Ser Bull 50:276–206
68. Stahmann MA, Link KP, Walker JC (1943) Presence of antinemic phenyiso-thiocyanate in roots of cruciferous plants. J Agric Res 67:49
69. Mathur RC (1972) Systematic botany (angiosperm). CBS, Delhi, pp 244, 276, 267, 328, 356, 397, 224, 276, 313
70. Reddy BMR, Krishnappa K, Karuna K (1997) Seeding bare root dip with chemicals for the management of root knot nematode in brinjal. Indian J Nematol 27(1):55–59
71. Atwal AS, Balraj S (1990) Pest population and assessment of crop losses. Publication and Information Division, Indian Council of Agriculture Research, IARI, Pusa
72. Eisenback JD (1985) Morphological comparison of head shape and stylet morphology of second stage juveniles of *Meloidogyne* species. J Nematol 14:339–343
73. Viglirchio DR (1979) Selected aspects of root knot nematode physiology in root knot nematodes (*Meloidogyne*) systemic biology and control. Academic Press, New York, pp 115–154
74. Nguyen JD, Duong H (2022) Anatomy, abdomen and pelvis, female external genitalia. Stat Pearls Publishing, Treasure Island (FL)
75. Sudha SS, Prabhoo NR (1983) *Meloidogyne* (Nematoda: Meloidogynidae) induced root galls of the banana plantMusa paradisiaca—a study of histopathology. Proc: Anim Sci 92:467–473
76. Charles JSK, Venkitesan TS (1994) Control of cyst nematode, *Heterodera oryzicola* in banana cv. Nendran with nematicides. Banana Research Station, Kerala Agricultural University, Kannara, Thrissur, India. J Ann Plant Prot Sci 2(2):49–51
77. Khan AA, Khan MW (1996) Distribution of root knot nematodes species and races infecting vegetable crops in eastern U.P. Indian J Nematol 26(2):238–244
78. Procinai MG, Ambroguini L (1979) The ultrastructure of giant cells induced in carrot by *Heterodera carotae*. J Nematol 30:470–474
79. Routaray BN, Sahoo H, Das SN (1987) Resistance of blackgram varieties against *Meloidogyne incognica*. Indian J Nematol 17(2):3–33
80. Lanjewar RD, Shukla VN (1988) Parasitism and interaction between *Pythium moriotylum* and *Meloidogyne incognita* in soft rot complex of ginger. Indian J Nematol 15(2):170–173
81. Sasser JN, Carter CC (1984) An advanced treatise on Meloidogyne. Vol. I. Biology and control. U.S. Agency for International Dev. North Carolina, pp 21–27
82. Dropkin VH (1969) Cellular response of plants to nematode infection. Ann Rev Phytopathol 7:101–122
83. Haseeb A, Pandey R, Husain A (1986) Proceeding of the national conference on plant parasitic nematodes of India: problems and progress, New Delhi, pp 21–22
84. Shah JJ, Raju EC (1997) Histopathology of ginger (*Zingiber officinale*) infected by soil nematode *Meloidogyne spp.*. Phyton 16:79–84
85. Stobbe H (2002) Developmental stages and fine structure of surface callus formed after debarking of living lime trees (*Tilia sp.*). Ann Bot 89(6):773–782. https://doi.org/10.1093/aob/mcf137

86. Vilsoni F, Mac Clure MA, Butler LD (1976) Occurrence, host range and histopathology of *Radopholus similis* in ginger *Zingiber officinale*. Plant Dis Report 60:417–420
87. Valette C, Andary C, Geiger JP, Sarah JL, Nicole M (1998) Histochemical and cytochemical investigations of phenols in roots of banana infected by the burrowing nematode *Radopholus similis*. Phytopathology 88(11):1141–1148. https://doi.org/10.1094/PHYTO.1998.88.11.1141
88. Alamgir ANM (2017) Therapeutic use of medicinal plants and their extracts: volume 1. Prog Drug Res 1–17. https://doi.org/10.1007/978-3-319-63862-1
89. Rao MS, Reddy PP, Mittal A, Chandravadana MV, Nagesh M (1996) Effect of some secondary plant metabolites as seed treatment agent against *Meloidogyne incognita* on tomato. Nematol Mediterr 24(1):49–51
90. Kihika-Opanda R, Tchouassi PD, Ng'ang'a MM, Beck JJ, Torto B (2022) Chemo-ecological insights into the use of the non-host plant vegetable black-jack to protect two susceptible solanaceous crops from root-knot nematode parasitism. J Agric Food Chem 70(22):6658–6669. https://doi.org/10.1021/acs.jafc.2c01748
91. Ahuja S, Mukhopadhyay MC (1985) Effect of nematode population of *Meloidogyne incognita* on the reproduction and growth of radish and carrot. Bull Ent 26:214–217
92. Vijaylakshmi K, Goswami BK (1987) Effect of root tip dip treat ment of tomato seedlings of aqueous extracts of some oil seed cakes on root knot nematode infestation. Ann Agric Res 8(1):168–171
93. Feyisa B, Lencho A, Selvaraj T, Getaneh G (2015) Evaluation of some botanicals and *Trichoderma harzianum* for the management of tomato root-knot nematode (*Meloidogyne incognita* (Kofoid and White) Chit Wood). Adv Crop Sci Technol 04(01):1–10. https://doi.org/10.4172/2329-8863.1000201
94. Majumdar V, Mishra SP (1991) Nematicidal efficacy of some plant products and management of *Meloidogyne incognita* in pulse crops by soaking seeds in their aqueous extracts. Curr Nematol 2(1):27–32
95. Abid M, Choudhary MI, Maqbool MA, Atta-Ur-Rehman (1997) Preliminary screening of some plants for their nematicidal activity against *Meloidogyne javanica*. Nematol Mediterr 25:155–157
96. Alam MM (1976) Organic amendments in relation to nematodes. PhD thesis, A.M.U., Aligarh, pp 52–56
97. Aisha S, Rose R, Irshad M, Rizwan AA (2015) Oil cake amendments: useful tools for the management of phytonematodes. Asian J Plant Pathol 9:91–111. https://doi.org/10.17311/ajppaj.2015.91.111
98. Haseeb A, Alam MM (1984) Control of plant parasitic nematodes with chopped plant leaves. Indian J Plant Pathol 2:180–181
99. Chandravadana MV, Nidiry ESJ, Khan RM, Rao MS (1994) Nematicidal activity of serpentine against *M. incognita*. Fundam Appl Nematol 177:185–186
100. Thies JA (2011) Virulence of *Meloidogyne incognita* to expression of N gene in pepper. J Nematol 43(2):90
101. Nandal SN, Bhatti DS (1983) Preliminary screening of some weeds/shrubs for their nematicidal activity against *Meloidogyne javanica*. Indian J Nematol 13:123–127
102. Haseeb A, Khan AM, Saxena SK (1981) Effect of certain alkaloid bearing plants on the mortality and larval hatching of *Meloidogyne incognita*. Geobios 7:3–5
103. Kerakalamatti M, Mesta RK, Kiran KC, Rudresh DL (2020) Evaluation of nematicides and oil cakes against root knot nematode caused by *Meloidogyne incognita* in pomegranate. Int J Curr Microbiol App Sci 9(7):1763–1775
104. Gardner J, Caswell-Chen EP (1994) *Raphanus sativus*, *Sinapis alba*, and *Fagopyrum esculentum* as hosts to *Meloidogyne incognita*, *Meloidogyne javanica*, and *Plasmodiophora brassicae*. Suppl J Nematol 26(4S):756–760
105. Belair G (1992) Effects of cropping sequences on population densities of *M. hapla* and carrot yield in organic soil. J Nematol 24(3):450–456

106. Muller R (1996) Organic amendments in nematode control an examination of the literature. Nematropica 12:319–326
107. Johnson LF (1959) Effect of the addition of organic amendments to soil on root knot of tomatoes II. Relation of soil temperature, moisture and pH. Phytopathology 52:410–413
108. Klemens E, Gerard WK (2006) Nematodes as sentinels of heavy metals and organic toxicants in the soil. J Nematol 38(1):13–19
109. Khan AM (1969) Studies on plant parasitic nematodes associated with vegetables crops in U.P. Final Technical Report PL-480 scheme; 1-11, 1964 to 31-10
110. Ntalli N, Adamski Z, Doula M, Monokrousos N (2020) Nematicidal amendments and soil remediation. Plants 9(4):429. https://doi.org/10.3390/plants9040429
111. Vander Laan PA (1956) The influence of organic maturing on de elopement of potato eelworm *Heterodera rostochiensis*. Nematologica 1:112–115
112. Widmer T, Mitkowski NA, Abawi G (2003) Soil organic matter and management of plant-parasitic nematodes. J Nematol 34:289–295
113. Ansari RA, Rizvi R, Mahmood I (2020) Management of phytonematodes: recent advances and future challenges. In: Plant parasitic nematodes management through natural products: current progress and challenges, chap 13, pp 297–315. https://doi.org/10.1007/978-981-15-4087-5_13

# Chapter 18
# Heavy Metals Pollution and Role of Soil PGPR: A Mitigation Approach

**Smita Patil, Abullais Ansari, Ashwini Sarje, and Ashok Bankar**

**Abstract** Heavy metal pollution is a serious threat to human health and the environment. It is severely augmented by several industrial activities. The main causes of metal pollution include several industrial processes such as metal forging, smelting, mining, fossil fuel burning, and the use of sewage sludge on agricultural sites. Toxic heavy metals discharged from these sources adversely affect the population of soil microorganisms and the physicochemical properties of the soil, reducing soil fertility and crop productivity. These heavy metals are not biodegradable and remain in the environment. Several conventional methods are used for removal or detoxification of heavy metals that have several drawbacks such as high cost, difficult to operate and toxic in nature. Therefore, bioremediation techniques have emerged as an alternative technique for remediation of heavy metals that have polluted soils. In metal-contaminated soil, the natural role of metal-tolerant plant growth-promoting rhizobacteria (PGPR) in maintaining soil fertility is fading with increasing use of pesticides. In addition to its role in detoxifying or removing toxic metals, rhizobacteria also promote plant growth via other mechanisms such as the production of growth promoting substances and siderophores. Phytoremediation is another new, low-cost in situ technology used to remove toxic pollutants from contaminated soil. The efficiency of phytoremediation can be enhanced by heavy-metal tolerant PGPR. In this book chapter, the significance of the PGPR for direct application to metal contaminated soil under a wide range of agro-ecological conditions has been discussed. The chapter also gives insight on re-establishment of metal contaminated soils and consequently, promotes crop productivity and their significance in phytoremediation. Thus, in the future bioremediation can be an effective technology for treatment of metal polluted environments.

S. Patil · A. Ansari · A. Sarje
Department of Microbiology, Camp Education Society's Arvind B. Telang Senior College of Arts, Commerce and Science, Savitribai Phule Pune University, Pune, India

A. Bankar (✉)
Department of Microbiology, MES Abasaheb Garware College Affiliated to Savitribai Phule Pune University, Pune, Maharashtra, India
e-mail: ashokbankar@gmail.com



J. A. Parray (ed.), *Climate Change and Microbiome Dynamics*, Climate Change Management, https://doi.org/10.1007/978-3-031-21079-2_18

## Introduction

Some heavy metals are essential for living organisms at low concentrations but can be harmful at high concentrations [1, 2]. Toxic heavy metals are those which are not essential to life and are often toxic at lower concentrations [3]. Heavy metals have several physicochemical properties such as ubiquity, toxicity, accumulation, non-biodegradability and persistence. Due to rapid urbanisation and several industrial activities a variety of toxic heavy metals are discharged into the soil environment [4, 5]. Heavy metals are constantly released into the environment through several human activities like mining, smelting, long-term use of mineral fertilizers, sewage sludge, pesticides, fuel and energy use, and wastewater [6, 7]. Most importantly, Cr, As, Cd, Ni, Cu, Pb, Co and Zn are commonly found in soil environment [8]. Heavy metal pollution has received special attention worldwide due to their negative impact on public health and the environment [6]. Heavy metals are accumulated in the human body through the food chain [2, 5, 9]. They have detrimental effects on various human body organs such as the digestive tract, kidneys, nervous system, skin, vascular muscles, and immune system. They can even cause congenital deficiencies and cancer [10]. The combined effects of several metals on humans can lead to complex stress regimes. Serious complications such as abdominal colitis, bloody diarrhoea, and renal failure due to high doses of heavy metals have been observed, but low dose exposure may be diagnosed as fatigue, anxiety, and neuropsychiatric disorders [11, 12]. Heavy metal soil pollution can reduce soil quality, soil fertility, microbial biodiversity, and plant productivity [13]. Accumulation of heavy metals in soil is a concern for the agricultural production sector, as increased uptake by plants can compromise food quality and quantity [14]. Management of heavy metal pollution is an important issue, as agricultural exports are sold internationally on the basis of environmental safety and sustainability [15].

Several methods have been used to remediate heavy metal-polluted soil and restore soil properties [6]. The suitable remediation techniques are selected based on the site characteristics, the nature of contaminants, the level of contamination, and the final use of the polluted soil. In general, physicochemical methods are widely used to remove heavy metals from polluted soil [6]. Traditional methods of heavy metal soil clean-up include extraction and immobilization of heavy metals, leading to excavation of land [16]. The conventional physicochemical techniques used to remove heavy metals are simple, quick, and effective. However, these techniques are costly, consume large amounts of energy, produce toxic by-products, and are not eco-friendly [17, 18]. In addition, these methods affect the physicochemical properties of the soil, affect the microbial biodiversity and can make the soil unsuitable for agriculture.

Therefore, to effectively manage heavy metal soil pollution, scientists have developed alternative biological approaches by using microorganisms [6, 17]. These microorganisms have some morphological, physiological, metabolic, and molecular characteristics to combat heavy metal toxicity. These properties can be used to remove heavy metals from polluted soil [17, 18]. Microbial remediation involves several microorganisms such as bacteria, microalgae, yeast and fungi to remove, transform, and detoxify heavy metals that remain in the environment [19–21]. Endogenous and exogenous microorganisms have several mechanisms to combat heavy metal toxicity. Microbial mechanisms such as extracellular or intracellular sequestration, metal chelating agent production, precipitation, enzymatic detoxification, and volatilization play important roles in bioremediation of heavy metal-polluted soils [20–24]. These biological approaches are chosen over physicochemical methods because they are simple, easy to implement, widely applicable, reliable, inexpensive, non-destructive, and eco-friendly [25]. Biological-based approaches are dependent on the type of microorganisms, the ability to resist metals, the degree of pollution, and the physicochemical properties of the soil. However, these limitations can be overcome by developing new microbial species that express specific genes of interest [6, 17, 26].

## Significance of Heavy Metal Tolerance Mechanisms in PGPR

PGPR are soil bacteria that grow in the rhizosphere of plants and promote plant growth through several mechanisms. Plant roots interact with a number of different microorganisms, which affect the plant growth as well as soil conditions. Rhizosphere bacterial colonization is known to be beneficial to bacteria, but their presence may also be useful to plants. PGPR are found beneficial for several agricultural systems to enhance crop yield and quality [27, 28]. Heavy metal stress has been reduced by PGPR because they have various mechanisms to tolerate and allow the uptake of heavy metal ions inside cells. Such mechanisms include (1) metal transport through the plasma membrane (2) intracellular metal ion accumulation and sequestration (3) heavy metal precipitation (4) detoxification of heavy metals and (5) adsorption or desorption of metals as shown in Fig. 18.1 and metal tolerating PGPRs are listed in Table 18.1 [29–31].

The minimum inhibitory concentrations (MIC) of Cu, Cr, Ni, and Cd were 186.9 $\pm$ 29.60, 88.0 $\pm$ 12.36, 153.81 $\pm$ 34.38, and 130.54 $\pm$ 28.21 μg/mL for *P. aeruginosa*, respectively [32]. It was reported that 32 bacterial isolates were obtained from metal-contaminated soil samples. Among these bacterial isolates, *C. oceanosedimentum* showed high resistance to cadmium (18 mM) [34]. Similarly, *Stenotrophomonas rhizophila* was highly resistant to Cr (VI). This bacterial isolate completely reduced 50 mg/L Cr (VI) within 48 h [33]. It was found that 27 rhizobacterial isolates were tested against Cr (VI). NT 15, NT19, NT20, and NT27 isolates were found to exhibit high Cr (VI) resistance in the presence of Cr (VI) at concentrations of 100–200 mg/L without loss of PGPR trait [36]. Six strains of rhizobacteria were isolated from heavy metal-contaminated soil in abandoned mines. These strains used were multi-tolerant to heavy metals and had some plant growth-promoting properties [46]. The PGPR have been used as seed inoculants to intentionally metal-treated or modified soils or already contaminated soils. The obtained results have shown a significant reduction in metal toxicity [47]. The PGPR are known to protect plants from metal toxicity, as well as to improve soil fertility and promote plant productivity by providing essential nutrients and growth regulators [48–50].

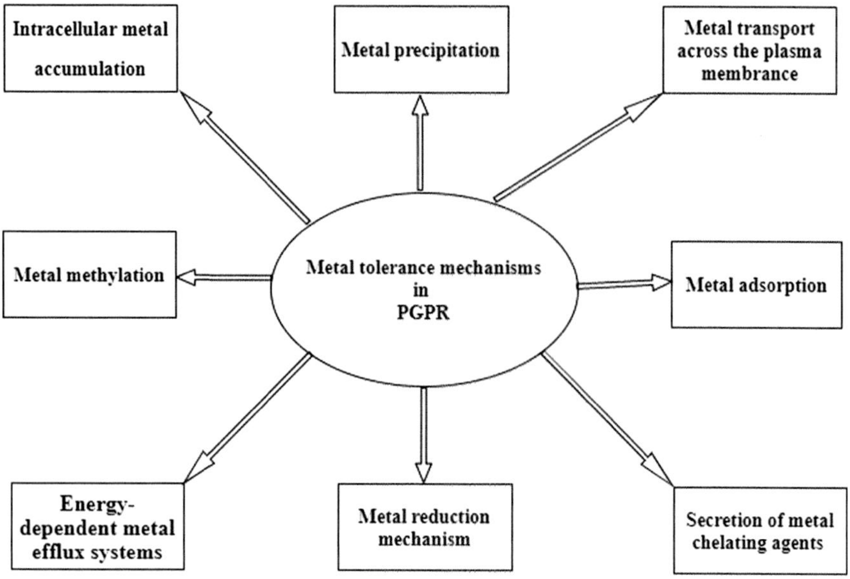


**Fig. 18.1** Possible metal tolerance mechanisms in PGPR [29–31]

**Table 18.1** List of heavy metal tolerating PGPR

| PGPR | Metal tolerated | Reference |
|---|---|---|
| *P. aeruginosa* | Cu, Cr, Ni, and Cd | [32] |
| *Stenotrophomonas rhizophila* | Cr (VI) | [33] |
| *C. oceanosedimentum* | Cd | [34] |
| *P. aeruginosa* and *B. gladioli* | Cd | [35] |
| *Pseudomonas* sp | Cr (VI) | [36] |
| *Bacillus* spp | Cr | [37] |
| *B. subtilis* SJ101 | Ni | [38] |
| *B. licheniformis*, *M. luteus*, and *P. fluorescens* | As | [39] |
| *Pseudomonas Sp*, *Bacillus Sp*, *Cupriavidus Sp*, and *Acinetobacter Sp* | Pb, Cd, and Cu | [40, 41] |
| *P. fluorescens* | Cd and Pb | [42] |
| *Rhizobium sp. RP5* | Zn and Ni | [43] |
| *Rhizobacterium sp. D14* | As | [44] |
| *Sinorhizobium sp. Pb002* | Pb | [45] |

Heavy metals adhere to extracellular polymeric substances (EPSs) that are naturally secreted by several bacterial cells, such as proteins, nucleic acids, fatty acids, polysaccharides, and humic substances. These EPSs have a very high binding affinity for heavy metals such as lead, cadmium and copper. Bacteria such as *Staphylococcus aureus*, *Micrococcus luteus*, and *Azotobacter* spp. have been reported for production of exopolymer that show high metal binding affinity [51]. Plant growth is promoted by reducing the stress induced by the ethylene-mediated effects on plants by producing 1-aminocyclopropane-1-carboxylate (ACC) deaminase enzyme [52–54]. Some microbes have the ability to produce low molecular weight siderophores as iron-chelating agents for immobilization of iron. Siderophores also have a binding affinity for other toxic heavy metals. Therefore, siderophores have the ability to minimize the bioavailability of heavy metals and reduce their metal toxicity. Bacterial metabolites are capable of crystallizing or precipitating heavy metals to reduce cellular uptake of heavy metals [55, 56].

The advantages of such microorganisms, with their multiple properties of metal resistance or reduction and the ability to promote plant growth through various mechanisms in metal-contaminated soil, are the most suitable options for bioremediation studies. PGPR can impose various indirect impacts on plants such as plant pathogen inhibition activity by competing for nutrients and space [57, 58]. In addition to the direct and indirect positive effects on biomass production, plant-associated bacteria can also contribute to increased metal availability and uptake, and reduced phytotoxicity of metals [59]. In recent years, PGPR has been shown to be effective in enhancing phytoremediation of petroleum and other pollutants [60, 61]. PGPR interacts with toxic heavy metals in soil, reducing their bioavailability. Energy-dependent metal efflux systems such as ATPases and chemiosmotic ion or proton pumps have been reported for the uptake of Cr and Cd metallothionein by bacterial cells [55]. The mechanism of cytosolic metal sequestration has been previously reported. In this mechanism, metallothionein, a low-molecular weight, bacterial cells to detoxify heavy metals such as Cd, Cu, Hg, and Ag secrete cysteine rich metal binding protein. Methylation of heavy metals by bacterial cells has been reported as an alternative mechanism of bacteria [56, 62]. The metal reduction mechanism has been studied in several bacteria. For example, detoxification of chromium involves the reduction of Cr (VI) to Cr (III) reported previously [63].

PGPR has the ability to produce various metal chelating agents, such as siderophores and organic acids, in the soil environment. They can acidify the microenvironment and induce the changes in redox potential [64, 65]. Due to these inherent mechanisms, the rhizosphere bacterium, which promotes plant growth, is a potential candidate for soil metal remediation. PGPR can also contribute to the reduction of phytotoxicity of metals via biosorption and bioaccumulation mechanisms. Bacterial cells have a very high surface-area-to-volume ratio and may adsorb more heavy metals than inorganic soil components either by a metabolism-independent passive or by a metabolism-dependent active process [66, 67]. Many authors suggest that the bacterial biosorption or bioaccumulation mechanism, along with other plant growth-promoting properties, including ACC deaminase and plant hormone production, is involved in promoting plant growth in metal-contaminated soils [38, 68]. The genes encoding heavy metal resistance of microorganisms need to be identified. Several molecular techniques have been used to identify metal resistance genes in microorganisms [69]. DNA microarray technique has been adopted as a powerful tool for identifying gene regulation under stress heavy metals [70]. The mass spectrometry-based proteomic techniques have been used to investigate the patterns of proteins expression due to intracellular metal accumulation [71]. Whole-genome sequencing method has been shown to help identify genes that play an important role in enhancing metal accumulation process [72]. Similarly, transcriptomics analysis techniques have been used to identify genes responsible for effective metal accumulation processes [73]. In addition, bioinformatics and mathematical modelling have been used to analyse the microbial metal resistance capability [74]. Therefore, advanced techniques have the potential to improve the metal bioaccumulation processes in the future.

## Rhizoremediation of Heavy Metal-Polluted Soil

Rhizoremediation is the remediation of polluted soil by rhizobacteria observed in the rhizosphere of plants. The symbiosis of microorganisms and plants in the plant rhizosphere found to be useful as an effective restoration technique. This is a relatively novel approach and may provide a practical remedy [75, 76]. PGPR, which promote plant growth, are soil bacteria that grow in the rhizosphere of plants and promote plant growth through various mechanisms. Plant roots interact with a number of different microorganisms, which affect plant growth as well as soil conditions. Rhizosphere bacterial colonization is known to be beneficial to bacteria, but their presence may also be beneficial to plants [27, 28, 77]. Some PGPR strains have been applied to plants that grow in poor soils that are heavily contaminated with heavy metals. Under these conditions, uninoculated plants and plants inoculated with the LMR250 strain did not grow, while the other five bacterial inoculants restored plant growth. The best performing strain, *Pseudarthrobacter oxydans* LMR291, has been reported as an excellent biofertilizer or biostimulant that promotes plant growth in contaminated soil [46].

In addition, a pot assay was performed to determine if the *Curtobacterium oceanosedimentum* strain could promote Chili growth under cadmium stress. Bacterial colonization significantly increased root and shoot lengths by up to 58% and 60%, respectively, compared to controls. After inoculation with the cadmium-resistant strain, the plants gained both fresh and dry weight. In both the control and inoculated plants, cadmium accumulates more in the roots than in shoots, indicating that Chili stabilizes Cd levels. In addition to improving plant properties, Cd-resistant strains have also been shown to increase the amount of total plant chlorophyll, total phenol, proline, and ascorbic acid. The PGPR inoculants protect the plants from adverse effects of cadmium [34]. Inoculations of *P. aeruginosa* and *B. gladioli* showed improvements in root length, shoot length, and photosynthetic pigments.

Levels of protein-bound and non-protein bound thiols were also increased in Cd-treated seedlings. Therefore, microorganisms have growth promoting properties that allow them to reduce the metal toxicity in plants [35].

The PGPR NT27 isolate was a strain of the genus *Pseudomonas*. In the presence of Cr (VI), the shoot and root dry weights of *M. sativa* was increased by 97.6 and 95.4%, respectively, compared to uninoculated control plants. Chlorophyll content has also increased significantly, and the stress markers, hydrogen peroxide, malondialdehyde, and proline have decreased. Thus, chromium-tolerant *Pseudomonas* sp had a positive effect on shoots and roots of *M. sativa* plants by reducing chromium toxicity [36]. Six Cr-tolerant PGPR strains were isolated and identified as *Bacillus* spp. The consortium of Cr-tolerant strains was used for the inoculation in combination with Biochar. The highest increase in shoot and root length was (22–23.4%) and the highest increase in chlorophyll and SOD was (28–40%). Similarly, proline and sugar levels improved to 20.5% and 9.6%, respectively. A significant reduction in Cr uptake was recorded in the dry biomass of wheat plants, with Cr concentrations of $0.28 \pm 1.01$ mg/kg compared to controls. Therefore, according to the results, PGPR and biochar are an important tools for protecting plants from chromium toxicity and can be used as inoculum for better crop production [37]. Nearly 180 Cr (VI) resistant PGPRs were isolated, and after screening, 10 efficient bacteria that could function under Cr (VI) stress conditions were selected. Wheat seeds (*Triticum aestivum* L.) were inoculated with selected bacterial isolates and sown in Cr (VI) contaminated (20 mg/kg) pots. The results showed that Cr (VI) contaminated soil significantly suppressed plant growth and development. However, inoculation significantly improved plant growth parameters compared to uninoculated plants. In inoculated pots, soil Cr (VI) levels were reduced by up to 62%. Cr (VI) levels were up to 36% lower in roots and up to 60% lower in shoots than uninoculated plants grown in contaminated pots [78].

The effects of PGPR, which stimulates plant growth under stress, are considered an effective strategy. It has been studied that plant grown in heavy metals polluted areas in the presence of PGPR were able to accumulate significant amounts of heavy metals in some plant parts than plants grown in soils without microbial flora [79]. The IAA-producing strain *B. subtilis* SJ101 promoted the growth of *Brassica juncea* in Ni-contaminated soil [38]. Similarly, Zn, Cu, Ni, and Co tolerant IAA producing strains were found to promote rapid root growth of *B. juncea* in soil contaminated with Cd [53]. Pinter et al. [39] found that siderophore production, phosphate solubilization, and nitrogen fixation activity of As-resistant *B. licheniformis*, *M. luteus*, and *P. fluorescens* increase the biomass of grapevine in the presence of high As concentrations. Environmental adaptability of Cd, Pb, and Cu resistant bacterial strains obtained from rhizospheric soil of *Boehmeria nivea* growing around mine refineries [80]. Scientists revealed rhizosphere bacteria of the genera *Pseudomonas*, *Bacillus*, *Cupriavidus*, and *Acinetobacter* are resistant to Pb, Cd, and Cu. A wide range of plant growth promoting properties of rhizobia including nitrogen fixation, solubilization of insoluble minerals such as phosphate, phytohormones and siderophores production, ACC deaminase synthesis, and volatile compounds such as acetoin and 2, 3-butanediol. Thus, rhizobia are found to be good candidates for detoxification of heavy metals [40, 41].

Of the 58 PGPR isolates, 8 bacterial strains were screened for multiple heavy metal tolerance, salt tolerance, indole-3-acetic acid, phosphate solubilization, and siderophore production, and finally the WW-40 strain was selected as a potent PGPR. Applying this strain under greenhouse conditions, the highest 52% of seed germination, 1078% of vigour index, 68.57% of shoot length, 71% root length, 44.44% of shoot fresh weight, 50% of root fresh weight, 52.38% of shoot biomass, and 66.66% of root biomass increased significantly compared to heavy metal treatment maize seedlings. Chlorophyll content increased by 68.75% in the consortium with Zn compared to the Zn inoculated pot. Similarly, the carotenoid content of Zn consortium pot increased by 57.89% and the xanthophylls content of the Zn consortium pot increased by 65.62% compared to other metal treatment pots. Therefore, the heavy metal resistant isolates that stand out in this study may be PGPR strains for both bioremediation and crop growth promotion [81]. The use of PGPR supports plant growth in contaminated soil, and urea-degrading bacteria can immobilize heavy metals by carbonate precipitation process. Therefore, dual treatment with such bacteria may be useful for plant growth and bioremediation in polluted soil. Pot experiments were carried out to grow radish plants in soil contaminated with Cd and Pb treated with PGPR *P. fluorescens*, and the results were compared with dual inoculation of *P. fluorescens* in combination with ureolytic *S. epidermidis* HJ2. The removal rate of Cd and Pb from the soil was 17% with PGPR alone, and more than 83% was reported with combined treatment [42]. Table 18.2 shows the importance of PGPR in phytoremediation of heavy metal contaminated soil.

**Table 18.2** PGPR-assisted phytoremediation of heavy metal contaminated soil

| PGPR | Plant/s | Heavy metal/s | Impact on plant | Reference |
|---|---|---|---|---|
| *B. licheniformis*, *M. luteus*, and *P. fluorescens* | Grapevine | Pb, Cd, and Cu | Increased the biomass of grapevine | [80] |
| *B. subtilis* SJ101 | *B. juncea* | Ni | Promoted the growth of plant | [38] |
| *Pseudomonas* Sp | *M. sativa* | Cr (VI) | Increased shoot and root length, chlorophyll content enhanced | [36] |
| *Bacillus* Sp with biochar | Wheat plant | Cr | Increased shoot and root length, chlorophyll content enhanced | [37] |
| *C. oceanosedimentum* | Chili | Cd | Significantly increased root and shoot lengths | [34] |

(continued)

**Table 18.2** (continued)

| PGPR | Plant/s | Heavy metal/s | Impact on plant | Reference |
|---|---|---|---|---|
| *B. licheniformis, M. luteus, P. fluorescens* | *Vitis vinifera* | As | *M. luteus* increased plant biomass, protein content, and POX activity<br>*B. licheniformis* increased plant biomass and APX<br>*P. fluorescens* augmented POX activity | [39] |
| *Bacillus megaterium* | *B. campestris* and *B. rapa* | Cd | Inoculation increased biomass, soluble proteins, and vitamin C content | [82] |
| *B. safensis* and *P. fluorescens* | *Helianthus annuus* | Zn and Pb | Inoculation reduced Zn and Pb uptake by plant tissues | [83] |
| *Klebsiella oxytoca* | *H. annuus* | Co, Pb, and Zn | Inoculation enhanced plant growth | [84] |
| *Klebsiella* sp. | *Vigna radiata* | Cd, Cu, and Pb | Inoculation promoted plant growth under HM stress | [85] |
| *Kocuria flava* and *B. vietnamensis* | *Oryza sativa* | As | Inoculation promoted plant growth (shoot and root length and weight) | [86] |

## Possible Rhizobacterial Strategies for Heavy Metals Bioremediation

### *Rhizobacterial Biosorption of Heavy Metals*

Biosorption is a new biological technique that has been employed for the last 20 years. It is an inexpensive approach to remove heavy metals from polluted environments [87]. Biosorption is based on the ionic interactions between the extracellular surface of living cells or dead biomass with metal ions. Therefore, most of the pollutants adhere on the cell surfaces instead of being oxidised by aerobic or anaerobic metabolism. Biosorption is considered as an effective technique for removal of various heavy metals from aqueous solutions [88, 89]. Researchers have shown that charged functional groups act as nucleation sites for the biosorption of various metal-containing precipitates. There are three mechanisms reported by which heavy metals can be adsorbed from contaminated environment: (1) Adsorption on the bacterial cell surfaces (2) Additional surface complexation and precipitation of actinides and (3) Precipitation of actinides with bacterial cell lysates [90]. In microorganisms, heavy metals are accumulated through adsorption or absorption processes reported previously [91–93]. Adsorption is the main mechanism of heavy metal accumulation observed in several microorganisms. Adsorption is an energy-independent process that occurs in both living and non-living bacterial cells. However, absorption is an energy-dependent process that occurs in living bacterial cells [94]. Bacterial cell walls have some specific functional groups such as carboxyl, amine, phosphonate, and hydroxyl groups [95]. These functional groups are involved in metal binding on the cell surfaces [96]. Anionic carboxyl and phosphate groups contribute to overall negative charge on microbial cell walls. Almost all heavy metals are positively charged and easily interact with cell walls. Therefore, metal ions bind or accumulate inside the cell via cell membrane [97]. Thus, the success of the metal adsorption process depends on the diverse structure of the bacterial cell wall. Gram-positive bacterial cell wall consists of a thick layer of peptidoglycan, which has high adsorption capacity [98, 99]. Gram-positive bacteria have the ability to remove heavy metal cations due to their electronegative charges due to the presence of teichoic and teichuronic acids in the cell wall. Thus, metal binding mechanism depends on the chemical nature of cell biomass and ionic strength of metal ions [100, 101] (Fig. 18.2).

Uptake of Cd (II) by biomass of *Sphingomonas paucimobilis* has been reported earlier. The ability of living cells to remove Cd (II) was found to be significantly higher than that of dead cells [104]. Another study also reported that live cells of *Enterobacter cloacae* TU cells were superior in removing Cd (II) compared to dead cells [105]. Huang et al. [106] studied those dead cells have been shown to have higher Cd (II) biosorption capacity than live cells [106]. It has also been shown that live and dead biomass of *P. plecoglossicida* have approximately the same Cd (II) biosorption capacity [107].

However, being biosorbent, little research has been carried out on live and dead cells of PGPR. The use of live or dead biomass to remove heavy metals continues

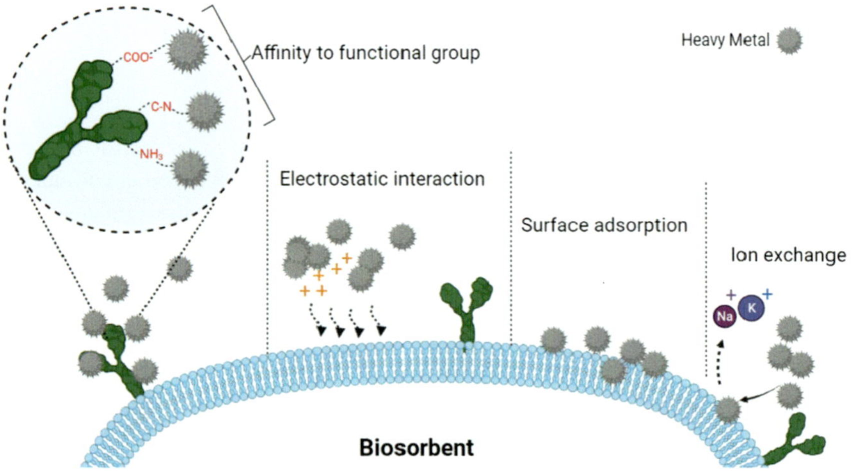


**Fig. 18.2** Biosorption of heavy metals on bacterial cell surface [90, 102, 103]

to be debated. Therefore, living and non-living biomass of *C. necator* GX_5, *Sphingomonas* sp. GX_15, and *Curtobacterium* sp. GX_31 have been used as biosorbents to compare their Cd (II) adsorption capacities [108]. Dead cells showed higher adsorption capacity than the live cells of *Curtobacterium* sp. GX_31. However, in the case of *C. necator* GX_5 and *Sphingomonas* sp. GX_15, the loading capacity of non-living biomass was stronger when compared with living biomass at 20 mg/L of Cd (II). After autoclaving, slight changes in the spectrum were observed, and FTIR analysis showed that more functional groups of the dead biosorbents were involved in Cd (II) binding. FTIR study also revealed that functional groups such as hydroxyl, amino, amide, and carboxyl groups played a vital role in complexation with Cd (II). Thus, it was concluded that dead cells are more effective biosorbents for Cd (II) remediation [108]. In another study, 10 different PGPRs were isolated, and identified as *Arthrobacter globiformis*, *B. megaterium*, *B. cereus*, *B. pumilus*, *S. lentus*, *E. asburiae*, *S. paucimobilis*, *Pantoea* spp., *Rhizobium rhizogenes*, and *R. radiobacter*. These isolates were tested for their arsenic biosorption capability. It was observed that all rhizobacteria showed arsenic biosorption capability. However, *S. paucimobilis* showed the highest biosorption capacity for arsenic ($146.4 \pm 23.4$ mg/g dry cell weight) [109].

Therefore, PGPR not only promotes plant growth, but are also promising biosorbents for removing heavy metals from the environment. However, there is still some debate about the biosorption and bioaccumulation processes, and their role in cadmium adsorption. Therefore, cadmium biosorption and bioaccumulation study was carried out by using three different Cd (II)-resistant PGPR such as *C. necator* GX_5, *Sphingomonas* sp. GX_15, and *Curtobacterium* sp. GX_31. The study found that the highest Cd (II) removal efficiency values for GX_5, GX_15, and GX_31 were 25.05%, 53.88%, and 86.06%, respectively at 20 mg/L of Cd (II) [110]. Recently, several microorganisms are genetically modified to improve the metal sorption capacity [111, 112]. Bacteria such as *S. xylosus* and *S. carnosus* are transgenic strains that express two different polyhistidyl peptides (His3-Glu-His3 and His6) reported earlier [113]. Similarly, *E. coli* and *P. putida* strains have been employed for phosphate biosorption through phosphate-binding protein [114]. *E. coli* was genetically modified to express the Ni21 transport system and at the same time overexpress pea MT as a carboxyl-terminal fusion with glutathione S-transferase (GSTMT). This change improved the Ni21-accumulating capacity of *E. coli* [115].

## Bioaccumulation of Heavy Metals by Rhizobacteria

Uptake of heavy metals by microorganisms occurs in two main stages: (i) metabolism-independent; and (ii) metabolism-dependent [90]. In the first stage, metal binding takes place on the cell surface via various mechanisms such as adsorption, precipitation, complexation, ion-exchange, and crystallization [116]. In the second stage, the metal uptake in microorganisms occurs through bioaccumulation process. Heavy metal ions are adsorbed on the cell surface and slowly enter the cytoplasm of the cell. Therefore, the metal species remain immobilized within the cell cytoplasm of the cell. This process is also known as metal sequestration [69]. This process is slow and dependent on several factors such as metabolic energy, temperature and metabolic inhibitors [90].

Bioaccumulation process in which microorganisms use importer complexes to take up heavy metals into the intracellular space via translocation pathways through the lipid bilayer. Once heavy metals enter cells, they can be sequestrated by several proteins and peptide ligands [69]. Bacteria synthesize metal-binding proteins such as metallothionein (MT) after exposure to high concentrations of metals to enhance their metal-binding capacity [117]. Therefore, MTs have metal-binding capacity and are encoded by genes expressed in a diverse group of rhizobacteria to facilitate the accumulation of heavy metals [118]. Recombinant expression of inner membrane importers from three major transporter classes: (i) channels, (ii) secondary carriers, and (iii) primary active transporters are studied well to enhance heavy metal bioaccumulation by increasing cytoplasmic uptake from the periplasmic membrane [119] as shown in Fig. 18.3.

Microorganisms employed for metal bioaccumulation must be metal tolerant to one or more metal contaminants at high concentrations. They also should have the metal biotransformational potential to convert toxic heavy metals into non-toxic

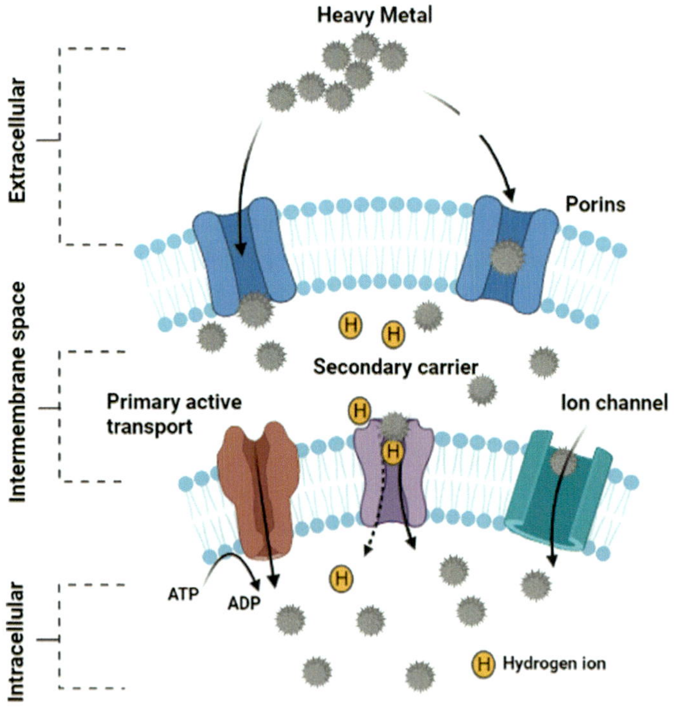


**Fig. 18.3** Bioaccumulation of heavy metals by bacterial cell [90, 119]

forms [120, 121]. Thus, PGPRs not only promote plant growth but also found to be promising agents for heavy metal remediation. Li et al. [110] isolated three cadmium-resistant PGPR namely *Cupriavidus necator* GX_5, *Sphingomonas* sp. GX_15, and *Curtobacterium* sp. GX_31 and used for bioaccumulation study under different Cd (II) concentrations. The study revealed that bioaccumulation was dominant in *C. necator* GX_5 and metal uptake was about 50.66–60.38%. The bioaccumulation study was also evidenced by different techniques such as SEM–EDX, TEM and FTIR spectroscopy. Further bioaccumulation study showed that heavy metals (cadmium and zinc) were mostly adhered on the cell wall instead of accumulating inside the cells [122]. In case of rhizobacteria, heavy metals in soluble and complex form are accumulated by live bacterial cells [123]. Studies on bioaccumulation of heavy metals by PGPR are very less reported and thus there is scope to carry out research in future.

## *Rhizobacterial Exopolysaccharides (EPS) for Heavy Metal Remediation*

EPS is a complex mixture of high molecular weight biopolymer metabolites produced by several microorganisms that protects against harsh environmental conditions. Rhizobacterial EPS has high metal binding capability which composed of polysaccharides, proteins, uronic acid, humic substances, lipids nucleic acid, and glycoproteins. Alginate (EPS) obtained from *Azotobacter* shows a strong metal binding capability. This property of EPS helps in remediation of toxic heavy metals by creating a microenvironment of essential metal ions to maintain the health of soil ecosystem and promotes plant growth [124–127]. EPS can assist in biofilm formation that protect cells in adverse conditions and helping plants by absorbing more water and nutrients [128]. Biofilms have been employed in bioremediation processes because of their inherent ability to thrive in harsh environments. Bacterial biofilms are highly dense biomass embedded in EPS used for metal remediation via biosorption and bioaccumulation processes [129]. EPS of bacterial biofilm have high metal binding affinity. EPS form organometal complexes via electrostatic forces of attraction [129]. Thus, heavy metals are immobilised by bacterial biofilms via EPS and cell membrane components due to their high affinity towards heavy metals [130]. The ionic charges on the EPS of biofilm are due to several functional groups such as carboxyl, amino, phenol, phosphate, and sulfhydryl groups. These functional groups are responsible net negative charges on the EPS surface that assist the formation of organometallic complexes with heavy metals [129, 130]. Three-dimensional excitation-emission matrix (EEM) fluorescence spectroscopy was used to study the interaction of EPS of biofilm and Hg (II). In this study, EPS of biofilm is a class of organic ligands that are important for complexing with Hg (II) and have profound effects on chemical forms, mobility, bioavailability, and ecotoxicity of heavy metals in the aquatic environment [130]. Thus, EPS could be an effective biosorbent for heavy metals. EPS obtained from rhizobacteria exhibited strong heavy metal binding capacity, removing precipitated metal sulfides and oxides, leading to formation of EPS-metal complexes and thus, promoting remediation of heavy metals [131]. Carboxyl and phosphate groups of EPS produced by *P. putida* have been reported for adsorption of $Cd^{2+}$ [132]. EPS of *A. chroococcum* strain XU1 exhibited biosorption capacity about 33.5 and 38.9 mg/g for lead and mercury, respectively [126].

It has been also reported that biofilm-grown cells have showed high resistance to heavy metals. Further study revealed that *Pseudomonas* biofilms was developed in presence of lead and zinc. However, there was no direct evidence provided by authors to prove the metal resistance potential of biofilms [133]. The nitrogen-fixing species *Sinorhizobium meliloti* has the ability to synthesize two different symbiosis-promoting EPSs: (1) succinoglycan and (2) galactoglucan. These EPSs have been studied to play important roles in plant development and protection from environmental stress. Researchers evaluated the role of EPS in bacterial resistance to heavy metals and metalloids, which are known to affect various biological processes. A recent study showed that EPS is essential for protecting bacteria from the toxicity of Hg (II) and As (III) stress. Biofilm formation has also been observed in the presence of heavy metals. Therefore, it was finally concluded that bacterial strain, which produces EPS have higher metal resistance ability compared to non-EPS bacterial strain [134]. PGPR such as *Pseudomonas* sp. H13 and *Brevundomonas* sp. H16 were reported for their ability to form biofilm and adsorbing heavy metals including $Cu^{2+}$, $Zn^{2+}$, $Cd^{2+}$, and $Pb^{2+}$. It has been observed that C–OH and P=O groups related to polysaccharides showed a significant role in heavy metal adsorption and immobilization [135]. A biofilm forming cadmium tolerant PGPR, *Aeromonas* sp enhanced the root length and shoot height of augmented plant by 21.4 and 17.36%, respectively, as compared to the non-augmented plants. It was also noticed that bioaugmentation of *Aeromonas* sp. in the rhizosphere of *Vetiveria zizanioides* increased cadmium uptake by 67.7% in the soil treated with 15 mg/kg of Cd [136].

## *Rhizobacterial Biosurfactant Mediated Heavy Metal Remediation*

Biosurfactant-mediated metal remediation from metal-polluted soils is considered a promising environmental green technology [137]. Biosurfactants are surface-active molecules that reduce the surface tension between liquid and liquid or liquid and solid [138]. Several microorganisms such as bacteria, yeast, and fungi have been reported to be capable of producing biosurfactants. These biosurfactants are commonly used for remediation of heavy metals such as cadmium, lead and zinc [139]. Several bacterial isolates within the genus *Pseudomonas*, *Bacillus*, *Micrococcus*, *Arthrobacter*, and *Rahnella* have been reported as potent producers of biosurfactants [140]. Endophytic *Rahnella* sp. JN6 significantly enhanced the phytoremediation efficacy in cadmium, lead and zinc contaminated soil [141]. Rhizobacteria produce biosurfactants that not only contribute to metal bioavailability but also promote plant growth. Biosurfactants are composed of polysaccharides, proteins, lipoproteins, lipopolysaccharides, or complex mixtures. Many species of *Acinetobacter* have produced high-molecular weight emulsifiers [77, 138]. However, rhamnolipids are the major class of biosurfactants produced by *P. aeruginosa* and other several microorganisms [139].

A potential of biosurfactant producing the endophytic *Pseudomonas* sp. Lk9 was tested for cadmium uptake and growth promotion of *Solanum nigrum* L. Researcher has found that *Solanum nigrum* L inoculated by *Pseudomonas* sp. Lk9 increases the cadmium availability, increases shoot dry biomass by 14% and total Cd accumulated in the shoot by 46.6% mg/kg [142]. Similarly, *Miscanthus sinensis* inoculation with the biosurfactant-producing multimetal-tolerant endophytic *P. koreensis* AGB-1 improved plant biomass by 54% and also increased metals content in roots and shoots [143]. Further study has been performed on the metal speciation by biosurfactant-producing *B. subtilis*, *P. aeruginosa*, and *P. fluorescence*. This study showed that *P. aeruginosa* strain has high metal exchangeable fraction concentrations compared to other strains [144].

## Conclusion

Restoring soil contaminated with toxic metals is a major challenge. Several physico-chemical methods are available for treating metal-contaminated soil. These methods have several disadvantages. Therefore, searching an alternative method is of high priority. A biological approach that fascinates many scientists because it has many advantages over traditional methods. Microbial remediation of heavy metal-polluted environment has emerged as an efficient green technology. There are several reports available on bioremediation of heavy metal-polluted soil by PGPR.

It has been investigated that PGPR is a promising agent for remediation of heavy metal-contaminated soils. There are various strategies like biosorption, bioaccumulation, EPS-assisted, bioleaching, biosurfactant-assisted, and biofilm-based techniques that have been used for restoration purposes. In the future, further research is needed to improve the bioremediation process with PGPR. Heavy metal tolerance in PGPR needs to be understood in detail, and genes responsible for metal tolerance need to be thoroughly studied in the future. Since the bioaccumulation of heavy metals by PGPR has not been sufficiently studied, it is very important to carry out the research work in detail. In order to develop efficient green technology in the future, it is necessary to study the interaction between PGPR and heavy metals at the molecular level. PGPR-metal interactions need to be study at molecular level in order to develop efficient green technology in future. Further genetic modification in PGPR is of high importance to improve efficacy of bioremediation process. Another genetic manipulation in PGPR is very important for improving the efficiency of the bioremediation process.

**Acknowledgements** All authors are thankful to DST, New Delhi, India for financial assistance in the form of a major project (DST-SERB file no. EEQ/2018/001202).

## References

1. Ersoy A, Yunsel TY, Cetin M et al (2004) Characterization of land contaminated by past heavy metal mining using geostatistical methods. Arch Environ Contam Toxicol 46:162–175
2. Bankar A, Kumar A, Zinjarde S et al (2009a) Removal of chromium (VI) ions from aqueous solution by adsorption onto two marine isolates of *Y. lipolytica*. J Hazard Mater 170(1):487–494
3. WHO (2000) Safety evaluation of certain food additives and contaminants. In: Fifty-Third Meeting of the Joint FAO/WHO Expert Committee on Food Additives, Food Additives Series No 830. WHO, Geneva
4. Yahaghi Z, Shirvani M, Nourbakhsh F, de la Peña TC, Pueyo JJ, Talebi M et al (2018) Isolation and characterization of Pb-solubilizing bacteria and their effects on Pb uptake by *Brassica juncea*: implications for microbe-assisted phytoremediation. J Microbiol Biotechnol 28:1156–1167
5. Bankar AV, Kumar AR, Zinjarde SS et al (2009) Environmental and industrial applications of *Yarrowia lipolytica*. Appl Microbiol Biotechnol 84(5):847–865
6. Rajendran S, Priya TAK, Khoo KS, Hoang TKA, Ng HS, Munawaroh HSH, Rajkumar M, Freitas H et al (2008) Effects of inoculation of plant growth promoting bacteria on Ni uptake by Indian mustard. Bioresour Technol 99:3491–3498
7. Emenike CU, Jayanthi B, Agamuthu P, Fauziah SH et al (2018) Biotransformation and removal of heavy metals: a review of phytoremediation and microbial remediation assessment on contaminated soil. Environ Rev 26:156–168
8. Sandaa RA, Torsvik V, Enger O et al (2001) Influence of long term heavy-metal contamination on microbial communities in soil. Soil Biol Biochem 33:287–295
9. Zafar S, Aqil F, Ahmad I et al (2007) Metal tolerance and biosorption potential of filamentous fungi isolated from metal contaminated agriculture soil. Bioresour Technol 98:2557–2561
10. Jan AT, Azam M, Siddiqui K, Ali A, Choi I, Haq QM et al (2015) Heavy metals and human health: mechanistic insight into toxicity and counter defense system of antioxidants. Int J Mol Sci 16:29592–29630
11. Balali-Mood M, Naseri K, Tahergorabi Z, Khazdair MR, Sadeghi M et al (2021) Toxic mechanisms of five heavy metals: mercury, lead, chromium, cadmium, and arsenic. Front Pharmacol 12:643972
12. Brooks PR, Crowe TP (2019) Combined effects of multiple stressors: new insights into the influence of timing and sequence. Front Ecol Evol 7:387
13. Tang J, Zhang J, Ren L, Zhou Y, Gao J, Luo L, Yuan Y, Qinghui P, Hongli H, Anwei C et al (2019) Diagnosis of soil contamination using microbiological indices: a review on heavy metal pollution. Environ Manage 242:121–130
14. Nagajyoti PC, Lee KD, Sreekanth TVM et al (2010) Heavy metals, occurrence and toxicity for plants: a review. Environ Chem Lett 8:199–216
15. McLaren RG, Clucas LM, Taylor MD, Hendry T et al (2004) Leaching of macronutrients and metals from undisturbed soils treated with metal-spiked sewage sludge. 2. Leaching of metals. Soil Res 42(4):459–471
16. Baker AJM, McGrath SP, Sidoli CMD, Reeves RD et al (1994) The possibility of in-situ heavy-metal decontamination of polluted soils using crops of metal-accumulating plants. Resour Conserv Recycl 11:41–49
17. Liu S, Yang B, Liang Y, Xiao Y, Fang J et al (2020) Prospect of phytoremediation combined with other approaches for remediation of heavy metal-polluted soils. Environ Sci Pollut Res 27:16069–16085

18. Ojuederie OB, Babalola OO (2017) Microbial and plant-assisted bioremediation of heavy metal polluted environments: a review. Int J Environ Res Public Health 14:1504
19. Hadiani MR, Darani KK, Rahimifard N, Younesi H et al (2018) Biosorption of low concentration levels of lead (II) and cadmium (II) from aqueous solution by *Saccharomyces cerevisiae*: Response surface methodology. Biocatal Agric Biotechnol 15:25–34
20. Khan I, Aftab M, Shakir SU, Ali M, Qayyum S, Rehman MU, Haleem KS, Touseef I et al (2019) Mycoremediation of heavy metal (Cd and Cr)-polluted soil through indigenous metallotolerant fungal isolates. Environ Monit Assess 191:585
21. Lopes CSC, Teixeira DB, Braz BF, Santelli RE, de Castilho LVA, Gomez JGC, Castro RPV, Seldin L, Freire DMG et al (2021) Application of rhamnolipid surfactant for remediation of toxic metals of long- and short-term contamination sites. Int J Environ Sci Technol 18:575–588
22. Chang J, Si G, Dong J, Yang Q, Shi Y, Chen Y, Zhou K, Chen J et al (2021) Transcriptomic analyses reveal the pathways associated with the volatilization and resistance of Mercury (II) in the fungus *Lecythophora* sp. DC-F1. Sci Total Environ 752:42172
23. Dobrowolski R, Szczé SA, Czemierska M, Jarosz-Wikołazka A et al (2017) Studies of Cadmium (II), Lead (II), Nickel (II), Cobalt (II) and Chromium (VI) sorption on extracellular polymeric substances produced by *Rhodococcus opacus* and *Rhodococcus rhodochrous*. Bioresour Technol 225:113–120
24. Nayak AK, Panda SS, Basu A, Dhal NK et al (2018) Enhancement of toxic Cr (VI), Fe, and other heavy metals phytoremediation by the synergistic combination of native *Bacillus cereus* strain and *Vetiveria zizanioides* L. Int J Phytoremediat 20:682–691
25. Ashraf S, Ali Q, Zahir ZA, Ashraf S, Asghar HN et al (2019) Phytoremediation: environmentally sustainable way for reclamation of heavy metal polluted soils. Ecotoxicol Environ Saf 174:714–727
26. Raklami A, Meddich A, Oufdou K, Baslam M et al (2022) Plants-microorganisms-based bioremediation for heavy metal cleanup: recent developments, phytoremediation techniques, regulation mechanisms, and molecular responses. Int J Mol Sci 23:5031
27. Tank N, Saraf M (2009) Enhancement of plant growth and decontamination of nickel spiked soil using PGPR. J Basic Microbiol 49:195–204
28. Vessey JK (2003) Plant growth promoting rhizobacteria as biofertilizers. Plant Soil 255:571–586
29. Wani PA, Khan MS, Zaidi A et al (2008) Effect of heavy metal toxicity on growth, symbiosis, seed yield and metal uptake in pea grown in metal amended soil. Bull Environ Contam Toxicol 81:152–158
30. Wani PA, Khan MS, Zaidi A et al (2008) Chromium reducing and plant growth promoting Mesorhizobium improves chickpea growth in chromium amended soil. Biotechnol Lett 30:159–163
31. Mamaril JC, Paner ET, Alpante BM et al (1997) Biosorption and desorption studies of chromium (iii) by free and immobilized Rhizobium (BJVr 12) cell biomass. Biodegradation 8:275–285
32. Kumar P, Deshwal VK (2013) Effect of heavy metals on growth and PGPR activity of *Pseudomonads*. J Acad Ind Res 2:286–290
33. Gao J, Wu S, Liu Y, Wu S, Jiang C, Li X, Wang R, Bai Z, Zhuang G, Zhuang X et al (2020) Characterization and transcriptomic analysis of a highly Cr (VI)-resistant and-reductive plant-growth-promoting rhizobacterium *Stenotrophomonas rhizophila* DSM14405T. Environ Pollut 263:114622
34. Patel M, Patel K, Al-Keridis LA, Alshammari N, Badraoui R, Elasbali AM, Al-Soud WA, Hassan MI, Yadav DK, Adnan M et al (2022) Cadmium-tolerant plant growth-promoting bacteria *Curtobacterium oceanosedimentum* improves growth attributes and strengthens antioxidant system in Chili (*Capsicum frutescens*). Sustainability 14:4335
35. Khanna K, Jamwal VL, Gandhi SG, Ohri P, Bhardwaj R et al (2019) Metal resistant PGPR lowered Cd uptake and expression of metal transporter genes with improved growth and photosynthetic pigments in *Lycopersicon esculentum* under metal toxicity. Sci Rep 9:5855

36. Tirry N, Kouchou A, El Omari B, Ferioun M, El Ghachtouli N et al (2021) Improved chromium tolerance of *Medicago sativa* by plant growth-promoting rhizobacteria (PGPR). J Genet Eng Biotechnol 19:149
37. Mazhar R, Ilyas N, Arshad M, Khalid A, Hussain M et al (2020) Isolation of heavy metal-tolerant PGPR strains and amelioration of chromium effect in wheat in combination with biochar. Iran J Sci Technol Trans Sci 44:1–12
38. Zaidi S, Usmani S, Singh BR, Musarrat J et al (2006) Significance of *Bacillus subtilis* strain SJ 101 as a bioinoculant for concurrent plant growth promotion and nickel accumulation in *Brassica juncea*. Chemosphere 64:991–997
39. Pinter IF, Salomon MV, Berli F, Bottini R, Piccoli P et al (2017) Characterization of the As (III) tolerance conferred by plant growth promoting rhizobacteria to in vitro-grown grapevine. Appl Soil Ecol 109:60–68
40. Hao X, Taghavi S, Xie P, Orbach MJ, Alwathnani HA, Rensing C et al (2014) Phytoremediation of heavy and transition metals aided by legume-rhizobia symbiosis. Int J Phytoremediat 16:179–202
41. Rangel WM, Thijs S, Janssen J, Oliveira Longatti SM, Bonaldi DS, Ribeiro PR et al (2017) Native rhizobia from Zn mining soil promote the growth of *Leucaena leucocephala* on contaminated soil. Int J Phytoremediat 19:142–156
42. He J, Zhang Q, Achal V et al (2020) Heavy metals immobilization in soil with plant-growth promoting precipitation in support of radish growth. Microbiol Biotechnol Lett 48:223–229
43. Wani PA, Khan S, Zaidi A et al (2008) Effect of metal-tolerant plant growth-promoting Rhizobium on the performance of pea grown in metal-amended soil. Arch Environ Contam Toxicol 55:33–42
44. Wang Q, Xiong D, Zhao P, Yu X, Tu B, Wang G et al (2011) Effect of applying an arsenic-resistant and plant growth-promoting rhizobacterium to enhance soil arsenic phytoremediation by Populus deltoides LH05-17. J Appl Microbiol 111:1065–1074
45. Di Gregorio S, Barbafieri M, Lampis S, Sanangelantoni AM, Tassi E, Vallini G et al (2006) Combined application of Triton X-100 and *Sinorhizobium* sp. Pb002 inoculum for the improvement of lead phytoextraction by *Brassica juncea* in EDTA amended soil. Chemosphere 63(2):293–299
46. Oubohssaine M, Sbabou L, Aurag J et al (2022) Native heavy metal-tolerant plant growth promoting rhizobacteria improves *Sulla spinosissima* (L.) growth in post-mining contaminated soils. Microorganisms 10(5):838
47. Gupta DK, Rai UN, Sinha S, Tripathi RD, Nautiyal BD, Rai P, Inouhe M et al (2004) Role of Rhizobium (CA-1) inoculation in increasing growth and metal accumulation in *Cicer arietinum* L. growing under fly-ash stress condition. Bull Environ Contam Toxicol 73:424–431
48. Wani PA, Khan MS, Zaidi A et al (2007) Impact of zinc-tolerant plant growth promoting rhizobacteria on lentil grown in zinc-amended soil. Agron Sustain Dev 28:449–455
49. Zaidi A, Khan MS (2006) Co-inoculation effects of phosphate solubilizing microorganisms and *Glomus fasciculatum* on greengram-Bradyrhizobium symbiosis. Turk J Agric For 30:223–230
50. Zaidi A, Khan MS, Aamil M et al (2004) Bioassociative effect of rhizospheric microorganisms on growth, yield and nutrient uptake of greengram. J Plant Nutr 27:599–610
51. Maier RM, Pepper IL, Gerba CP (2009) Introduction to environmental microbiology. In: Environmental microbiology. Academic Press, pp 3–7
52. Glick BR (2010) Using soil bacteria to facilitate phytoremediation. Biotechnol Adv 28:367–374
53. Belimov AA, Hontzeas N, Safronova VI, Demchinskaya SV, Piluzza G, Bullitta S, Glick BR et al (2005) Cadmium-tolerant plant growth promoting rhizobacteria associated with the roots of Indian mustard (*Brassica juncea* L. Czern.). Soil Biol Biochem 37:241–250
54. Uchiumi T, Oowada T, Itakura M, Mitsui H, Nukui N, Dawadi P, Kaneko T, Tabata S, Yokoyama T, Tejima T, Saeki K, Oomori H, Hayashi M, Maekawa T, Sriprang R, Murooka Y, Tajima S, Simomura K, Nomura M, Suzuki A, Shimoda S, Sioya K, Abe M, Minamisawa K et al (2004) Expression islands clustered on symbiosis island of mesorhizobium loti genome. J Bacteriol 186:2439–2448

55. Roane TM, Pepper IL (2000) Microorganisms and metal pollution. In: Maier RM, Pepper IL, Gerba CB (eds) Environmental microbiology. Academic Press, London, p 55
56. Zubair M, Shakir M, Ali Q, Rani N, Fatima N, Farooq S, Nasir IA et al (2016) Rhizobacteria and phytoremediation of heavy metals. Environ Technol Rev 5:112–119
57. Yang J, Kloepper JW, Ryu CM et al (2009) Rhizosphere bacteria help plants tolerate abiotic stress. Trends Plant Sci 14:1–4
58. Zhuang X, Chen J, Shim H, Bai Z et al (2007) New advances in plant growth-promoting rhizobacteria for bioremediation. Environ Int 33:406–413
59. Valls M, De Lorenzo V (2002) Exploiting the genetic and biochemical capacities of bacteria for the remediation of heavy metal pollution. FEMS Microbiol Rev 26:327–338
60. Alotaibi F, Hijri M, St-Arnaud M et al (2021) Overview of approaches to improve rhizoremediation of petroleum hydrocarbon contaminated soils. Appl Microbiol 1:329–351
61. Guo JK, Ding YZ, Feng RW, Wang RG, Xu YM, Chen C, Wei XL, Chen WM et al (2015) *Burkholderia metalliresistens* sp. nov., a multiple metal-resistant and phosphate-solubilising species isolated from heavy metal-polluted soil in Southeast China. Antonie Leeuwenhoek Int J G 107:1591–1598
62. Ranjard L, Nazaret S, Cournoyer B et al (2003) Freshwater bacteria can methylate selenium through the thiopurine methyltransferase pathway. Appl Environ Microbiol 69(7):3784–3790
63. Lovley DR, Holmes DE, Nevin KP et al (2004) Dissimilatory Fe(iii) and Mn (iv) reduction. Adv Microbial Physiol 49:219–286
64. Lasat MM (2002) Phytoextraction of toxic metals. J Environ Qual 31:109–120
65. Whiting SN, de Souza MP, Terry N et al (2001) Rhizosphere bacteria mobilize Zn for hyperaccumulation by *Thlaspi caerulescens*. Environ Sci Technol 35:3144–3150
66. Khan MS, Zaidi A, Wani PA et al (2007) Role of phosphate-solubilizing microorganisms in sustainable agriculture-a review. Agron Sustain Dev 27:29–43
67. Ledin M, Krantz-Rulcker C, Allard B et al (1996) Zn, Cd and Hg accumulation by microorganisms, organic and inorganic soil components in multicompartment system. Soil Biol Biochem 28:791–799
68. Madhaiyan M, Poonguzhali S, Sa T et al (2007) Metal tolerating methylotrophic bacteria reduces nickel and cadmium toxicity and promotes plant growth of tomato (*Lycopersicon esculentum* L.). Chemosphere 69:220–228
69. Mishra A, Malik A (2013) Recent advances in microbial metal bioaccumulation. Crit Rev Environ Sci Technol 43:1162–1222
70. Gorfer M, Persak H, Berger H, Brynda S, Bandian D, Strauss J et al (2009) Identification of heavy metal regulated genes from the root associated ascomycete *Cadophora finlandica* using a genomic microarray. Mycol Res 113:1377–1388
71. Italiano F, Buccolieri A, Giotta L, Agostiano A, Valli L, Milano F, Trotta M et al (2009) Response of the carotenoidless mutant *Rhodobacter sphaeroides* growing cells to cobalt and nickel exposure. Int Biodeterior Biodegrad 63:948–957
72. Choi DH, Kwon YM, Kwon KK, Kim SJ et al (2015) Complete genome sequence of *Novosphingobium pentaromativorans* US6-1(T). Stand Genom Sci 10(1):1–8
73. Shi B, Huang Z, Xiang X, Huang M, Wang WX, Ke C et al (2015) Transcriptome analysis of the key role of GAT2 gene in the hyper-accumulation of copper in the oyster *Crassostrea angulata*. Sci Rep 5:1–12
74. Stadnicka J, Schirmer K, Ashauer R et al (2012) Predicting concentrations of organic chemicals in fish by using toxicokinetic models. Environ Sci Technol 46:3273–3280
75. Hong SH, Ryu HW, Kim J, Cho KS et al (2011) Rhizoremediation of diesel-contaminated soil using the plant growth-promoting rhizobacterium *Gordonia* sp S2RP-17. Biodegradation 22:593–601
76. Khan MS, Zaidi A, Wani PA, Oves M et al (2009) Role of plant growth promoting rhizobacteria in the remediation of metal contaminated soils. Environ Chem Lett 7:1–19
77. Bankar A, Patil S (2021) Microbial cell factories for treatment of soil polluted with heavy metals: a green approach. In: Microbiome stimulants for crops. Woodhead Publishing, pp 315–332

78. Khan M, Asghar H, Jamshaid M, Akhtar M, Zahir Z et al (2013) Effect of microbial inoculation on wheat growth and phytostabilization of chromium contaminated soil. Pak J Bot 45:27–34
79. Sorour AA, Khairy H, Zaghloul EH, Zaghloul HA et al (2022) Microbe-plant interaction as a sustainable tool for mopping up heavy metal contaminated sites. BMC Microbiol 22:1–13
80. Jiang J, Pan C, Xiao A, Yang X, Zhang G et al (2017) Isolation, identification, and environmental adaptability of heavy-metal-resistant bacteria from ramie rhizosphere soil around mine refinery. 3 Biotech 7:5
81. Chowdhury SK (2021) Application of heavy metal tolerance plant growth promoting bacteria for remediation of metalliferous soils and their growth efficiency on maize (*Zeamays* L.). Plant Isol Sci J Biol 4(1):039–050
82. Wang Q, Zhang WJ, He LY, Sheng XF et al (2018) Increased biomass and quality and reduced heavy metal accumulation of edible tissues of vegetables in the presence of Cd-tolerant and immobilizing Bacillus megaterium H3. Ecotoxicol Environ Saf 148:269–274
83. Mousavi SM, Motesharezadeh B, Hosseini HM, Alikhani H, Zolfaghari AA et al (2018) Root-induced changes of Zn and Pb dynamics in the rhizosphere of sunflower with different plant growth promoting treatments in a heavily contaminated soil. Ecotoxicol Environ Saf 147:206–216
84. Arunakumara KKIU, Walpola BC, Yoon MH et al (2015) Bioaugmentation-assisted phytoextraction of Co, Pb and Zn: an assessment with a phosphate-solubilizing bacterium isolated from metal-contaminated mines of Boryeong area in South Korea. Biotechnol Agron Soc Environ 19:143–152
85. Chakraborty S, Das S, Banerjee S, Mukherjee S, Ganguli A, Mondal S et al (2021) Heavy metals bio-removal potential of the isolated *Klebsiella* Sp TIU20 strain which improves growth of economic crop plant (*Vigna radiata* L.) under heavy metals stress by exhibiting plant growth promoting and protecting traits. Biocatal Agric Biotechnol 38:102204
86. Mallick I, Bhattacharyya C, Mukherji S, Dey D, Sarkar SC, Mukhopadhyay UK, Ghosh A et al (2018) Effective rhizoinoculation and biofilm formation by arsenic immobilizing halophilic plant growth promoting bacteria (PGPB) isolated from mangrove rhizosphere: a step towards arsenic rhizoremediation. Sci Total Environ 610:1239–1250
87. Fomina M, Gadd GM (2014) Biosorption: current perspectives on concept, definition and application. Bioresour Technol 160:314
88. Aryal M, Liakopoulou-Kyriakides M (2015) Bioremoval of heavy metals by bacterial biomass. Environ Monit Assess 187(1):1–26
89. Saba RY, Ahmed M, Sabri AN et al (2019) Potential role of bacterial extracellular polymeric substances as biosorbent material for arsenic bioremediation. Bioremediat J 23:2–81
90. Bankar A, Geetha N (2018) Recent trends in biosorption of heavy metals by Actinobacteria. In: Singh B, Gupta V, Passari A (eds) Actinobacteria: diversity and biotechnological applications. Elsevier, pp 257–275
91. Alloway BJ (1995) Heavy metals in soils, 2nd ed. Springer, Dordrecht
92. Bankar A, Zinjarde S, Shinde M, Gopalghare G, Ravikumar A et al (2018) Heavy metal tolerance in marine strain of *Yarrowia lipolytica*. Extremophiles 22(4):617–628
93. Bankar A, Zinjarde S, Telmore A, Walke A, Ravikumar A (2018) Morphological response of *Yarrowia lipolytica* under stress of heavy metals. Can J Microbiol 64(8):559–566
94. Wang Y, Guo J, Liu R et al (2001) Biosorption of heavy metals by bacteria isolated from activated sludge. Appl Biochem Biotechnol 91:171–184
95. Vijayaraghavan K, Yun YS (2008) Bacterial biosorbents and biosorption. Biotechnol Adv 26:266–291
96. Brady D, Duncan JR (1994) Cation loss during accumulation of heavy metal cations by *Saccharomyces cerevisiae*. Biotechnol Lett 16:543–548
97. Sarret G, Manceau A, Spadini L, Roux JC, Hazemann JL, Soldo Y, Eybert-BÉrard L, Menthonnex J et al (1998) Structural determination of Zn and Pb binding sites in *Penicillium chrysogenum* cell walls by EXAFS spectroscopy. Environ Sci Technol 32:1648–1655
98. Van Hullebusch ED, Zandvoort MH, Lens PN et al (2003) Metal immobilisation by biofilms: mechanisms and analytical tools. Rev Environ Sci Biotechnol 2:9–33

99. Ahluwalia SS, Goyal D (2007) Microbial and plant derived biomass for removal of heavy metals from wastewater. Bioresour Technol 98:2243–2257
100. Abdi O, Kazemi M (2015) A review study of biosorption of heavy metals and comparison between different biosorbents. J Mater Environ Sci 6(5):1386–1399
101. Tsezos M, Remoundaki E, Hatzikioseyian A et al (2014) Biosorption - principles and applications for metal immobilization from waste-water streams. Clean Prod Nano Technol 23–33
102. Ilyas S, Kim MS, Lee JC, Jabeen A, Bhatti HN et al (2017) Bio-reclamation of strategic and energy critical metals from secondary resources. Metals 7:1–17
103. Volesky B (2007) Biosorption and me. Water Res 41:4017–4029
104. Tangaromsuk J, Pokethitiyook P, Kruatrachue M, Upatham E et al (2002) Cadmium biosorption by *Sphingomonas paucimobilis* biomass. Bioresour Technol 85:103–105
105. Xu C, He S, Liu Y, Zhang W, Lu D et al (2017) Bioadsorption and biostabilization of cadmium by *Enterobacter cloacae* TU. Chemosphere 173:622–629
106. Huang FZ, Dang CL, Guo GN, Lu RR, Gu HJ, Liu HZ et al (2013) Biosorption of Cd (II) by live and dead cells of Bacillus cereus RC-1 isolated from cadmium-contaminated soil. Colloids Surf B 107:11–18
107. Guo J, Zheng XD, Chen QB, Zhang L, Xu XP et al (2012) Biosorption of Cd (II) from aqueous solution by *Pseudomonas plecoglossicida*: kinetics and mechanism. Curr Microbiol 65(4):350–355
108. Li X, Li D, Yan Z, Ao Y et al (2018) Adsorption of cadmium by live and dead biomass of plant growth-promoting rhizobacteria. RSC Adv 8:33523–33533
109. Titah HS, Abdullah SRS, Idris M, Anuar N, Basri H, Mukhlisin M, Tangahu BV, Purwanti IF, Kurniawan SB et al (2018) Arsenic resistance and biosorption by isolated rhizobacteria from the roots of *Ludwigia octovalvis*. Int J Microbiol e3101498
110. Li X, Li D, Yan Z, Ao Y et al (2018) Biosorption and bioaccumulation characteristics of cadmium by plant growth-promoting rhizobacteria. RSC Adv 8(54):30902–30911
111. Ayangbenro AS, Babalola OO (2017) A new strategy for heavy metal polluted environments: a review of microbial biosorbents. Int J Environ Res Public Health 14(1):94
112. Ueda M (2016) Establishment of cell surface engineering and its development. Biosci Biotechnol Biochem 80:1243–1253
113. Samuelson P, Wernérus H, Svedberg M, Stahl S et al (2000) *Staphylococcal* surface display of metal-binding polyhistidyl peptides. Appl Environ Microbiol 66:1243–1248
114. Li Q, Yu Z, Shao X, He J, Li L et al (2009) Improved phosphate biosorption by bacterial surface display of phosphate-binding protein utilizing ice nucleation protein. FEMS Microbiol Lett 299:4452
115. Krishnaswamy R, Wilson DB (2000) Construction and characterization of an *Escherichia coli* strain genetically engineered for Ni (II) bioaccumulation. Appl Environ Microbiol 66:53835386
116. Mowell JL, Gadd GM (1984) Cadmium uptake by *Aureobasidium pullulans*. J Gen Microbiol 130:279–284
117. Mosa KA, Saadoun I, Kumar K, Helmy M, Dhankher OP et al (2016) Potential biotechnological strategies for the cleanup of heavy metals and metalloids. Front Plant Sci 7:303
118. Sessitsch A, Kuffner M, Kidd P, Vangronsveld J, Wenzel WW, Fallmann K, Puschenreiter M et al (2013) The role of plant-associated bacteria in the mobilization and phytoextraction of trace elements in contaminated soils. Soil Biol Biochem 60:182–194
119. Saier MH (2016) Transport protein evolution deduced from analysis of sequence, topology and structure. Curr Opin Struct Biol 38:9–17
120. Bankar AV, Zinjarde SS, Kapadnis BP et al (2012) Management of heavy metal pollution by using yeast biomass. In: Satyanarayana T, Johri BN, Prakash A (eds) Microorganisms in environmental management. Springer, pp 335–363. ISBN 978-94-007-2228-6
121. Bankar A, Winey M, Prakash D, Kumar AR, Gosavi S, Kapadnis B, Zinjarde S et al (2012) Bioleaching of fly ash by the tropical marine yeast, *Yarrowia lipolytica* NCIM 3589. Appl Biochem Biotechnol 168(8):2205–2217

122. Limcharoensuk T, Sooksawat N, Sumarnrote A, Awutpet T, Kruatrachue M, Pokethitiyook P, Auesukaree C et al (2015) Bioaccumulation and biosorption of $Cd^{2+}$ and $Zn^{2+}$ by bacteria isolated from a zinc mine in Thailand. Ecotox Environ Saf 122:322–330
123. Alam MZ, Ahmad S (2013) Multi-metal biosorption and bioaccumulation by Exiguobacterium sp. ZM-2. Ann Microbiol 63(3):1137–1146
124. Das S, Elavarasi A, Lyla PS, Khan SA et al (2009) Biosorption of heavy metals by marine bacteria: potential tool for detecting marine pollution. Environ Health 9:38–43
125. Gupta P, Diwan B (2017) Bacterial exopolysaccharide mediated heavy metal removal: a review on biosynthesis, mechanism and remediation strategies. Biotechnol Rep 13:58–71
126. Rasulov BA, Yili A, Aisa HA et al (2013) Biosorption of metal ions by exopolysaccharide produced by 1025 *Azotobacter chroococcum* XU1. J Environ Prot 4(09):989
127. Sheng GP, Yu HQ, Li XY et al (2010) Extracellular polymeric substances (EPS) of microbial aggregates in biological wastewater treatment systems: a review. Biotechnology 28:882–894
128. Vanderlinde EM, Harrison JJ, Muszynski A, Carlson RW, Turner RJ, Yost CK et al (2010) Identification of a novel ABC transporter required for desiccation tolerance and biofilm formation in *Rhizobium leguminosarum* bv. viciae 3841. FEMS Microbiol Ecol 71:327–340
129. Shukla SK, Mangwani N, Karley D, Rao TS et al (2017) Bacterial biofilms and genetic regulation for metal detoxification. In: Handbook of metal-microbe interactions and bioremediation. CRC Press, pp 317–332
130. Zhang D, Pan X, Mostofa KM, Chen X, Mu G, Wu F, Liu J, Song W, Yang J, Liu Y, Fu Q et al (2010) Complexation between Hg (II) and biofilm extracellular polymeric substances: an application of fluorescence spectroscopy. J Hazard Mater 175:359–365
131. Joshi PM, Juwarkar AA (2009) In vivo studies to elucidate the role of extracellular polymeric substances from Azotobacter in immobilization of heavy metals. Environ Sci Technol 43:5884–5889
132. Wei X, Fang L, Cai P, Huang Q, Chen H, Liang W, Rong X et al (2011) Influence of extracellular polymeric substances (EPS) on Cd adsorption by bacteria. Environ Pollut 159:1369–1374
133. Meliani A, Bensoltane A (2016) Biofilm-mediated heavy metals bioremediation in PGPR *Pseudomonas*. J Bioremed Biodegrad 7:370
134. Nocelli N, Bogino PC, Banchio E, Giordano W et al (2016) Roles of extracellular polysaccharides and biofilm formation in heavy metal resistance of rhizobia. Materials 9:418
135. Xing Y, Tan S, Liu S, Xu S, Wan W, Huang Q, Chen W et al (2022) Effective immobilization of heavy metals via reactive barrier by rhizosphere bacteria and their biofilms. Environ Res 207:112080
136. Itusha A, Osborne WJ, Vaithilingam M et al (2019) Enhanced uptake of Cd by biofilm forming Cd-resistant plant growth promoting bacteria bioaugmented to the rhizosphere of *Vetiveria zizanioides*. Int J Phytoremediat 21:487–495
137. Lal S, Ratna S, Said OB, Kumar R et al (2018) Biosurfactant and exopolysaccharide-assisted rhizobacterial technique for the remediation of heavy metal contaminated soil: an advancement in metal phytoremediation technology. Environ Technol Innov 10:243–263
138. Ron EZ, Rosenberg E (2001) Natural roles of biosurfactants. Environ Microbiol 3:229–236
139. Maier RM, Soberón-Chávez G (2000) *Pseudomonas aeruginosa* rhamnolipids: biosynthesis and potential applications. Appl Microbiol Biotechnol 54:625–633
140. Saikia RR, Deka S, Deka M, Sarma H et al (2012) Optimization of environmental factors for improved production of rhamnolipid biosurfactant by *Pseudomonas aeruginosa* RS29 on glycerol. J Basic Microbiol 52:446–457
141. He HD, Ye ZH, Yang DJ, Yan JL, Xiao L, Zhong T, Yuan M, Cai XD, Fang ZQ, Jing YX et al (2013) Characterization of endophytic *Rahnella* sp. JN6 from *Polygonum pubescens* and its potential in promoting growth and Cd, Pb, Zn uptake by *Brassica napus*. Chemosphere 90:1960–9165
142. Chen L, Luo S, Li X, Wan Y, Chen J, Liu C et al (2014) Interaction of Cd-hyperaccumulator *Solanum nigrum* L. and functional endophyte *Pseudomonas* sp. Lk9 on soil heavy metals uptake. Soil Biol Biochem 68:300–308

143. Babu AG, Shea PJ, Sudhakar D, Jung IB, Oh BT et al (2015) Potential use of *Pseudomonas koreensis* AGB-1 in association with *Miscanthus sinensis* to remediate heavy metal (loid)-contaminated mining site soil. J Environ Manage 151:160–166
144. Braud A, Jezequel K, Vieille E, Tritter A, Lebeau T et al (2006) Changes in extractability of Cr and Pb in a polycontaminated soil after bioaugmentation with microbial producers of biosurfactants, organic acids and siderophores. Water Air Soil Pollut 6:261–279

# Chapter 19
# The Utilization of Arbuscular Mycorrhiza to Support Revegetation on Degraded Tropical Peatland of Central Kalimantan

**Tri Wira Yuwati and Safinah Surya Hakim**

**Abstract** Tropical peatland in Indonesia especially in Central Kalimantan has been degraded due to various factors including repeating fires, illegal logging, and conversion into other land use and inappropriate drainage such as the ex-Mega Rice Project. Efforts to revegetate this area have encountered many obstacles due to nutrient poor peat soil characteristic. Arbuscular mycorrhiza is one of potential soil microbes that can be utilized as plant growth-booster in bio-rehabilitation technology particularly in degraded land. However, this bio-rehabilitation-technology has not been utilized intensively to support revegetation of degraded tropical peatland. This paper aimed to summarize the recent progress on the utilization of arbuscular mycorrhiza fungi in supporting the plant's growth of the peatland revegetation efforts. The result showed that arbuscular mycorrhiza application significantly increased plant's growth and survival rates especially in the nursery stage. However, compatibility between arbuscular mycorrhiza fungal species and host plants was an important factor that determines the success of colonization and its contribution to plant's growth performance. Appropriate combination of indigenous mycorrhizal fungal species and native peatland plant species needs to be considered for the success of this bio-rehabilitation technology in revegetating degraded tropical peatland.

## Introduction

Eleven percent or approximately 44 million ha of the world's peatland is tropical peatland [1]. The tropical peatland in Indonesia covers an area of 13.43 million ha distributed across four main islands namely Sumatera (5.85 Mha), Kalimantan (4.54 Mha), and Papua (3.01 Mha) and Sulawesi (0.024 Mha) [2]. The tropical

T. W. Yuwati (✉)
Research Center for Ecology and Ethnobiology, National Research and Innovation Agency (BRIN), Bogor, Indonesia
e-mail: tri.wira.yuwati@brin.go.id

S. S. Hakim
Research Center for Applied Microbiology, National Research and Innovation Agency (BRIN), Bogor, Indonesia



J. A. Parray (ed.), *Climate Change and Microbiome Dynamics*, Climate Change Management, https://doi.org/10.1007/978-3-031-21079-2_19

peatland has ecologically important roles as carbon storage, hydrological control, and habitat of flora fauna and microbes. After 2015, the tropical peatland in South East Asia experienced massive changes and only 6% remains in pristine condition [3]. Fire, logging, drainage, and conversion into other land uses such as agriculture, oil palm plantation, Acacia plantations and smallholdings were causes of those massive changes [4–6]. Two thousand and fifteen fire has burnt 2.6 million ha of Indonesia's tropical peatland and in order to restore it, the government of Indonesia established The Peatland Restoration Agency or Badan Restorasi Gambut in early 2016 [7]. The restoration policy includes rewetting, revegetation, and revitalization of local livelihood, known as the 3R approach. As of the end of 2019, the Peatland Restoration Agency claimed to construct 713 revegetation demonstration plots across various provinces of Indonesia namely Riau, Jambi, South Sumatera, South Kalimantan, and Central Kalimantan. The direct barriers of peatland revegetation includes physical, hidrological and biological constrains [8]. Most of Indonesia's peatland characterised as lowland ombrotrophic meaning that it has low nutrient and acidic conditions [9]. Moreover, according to [8] the dense shrubs and ferns communities has caused the increased competition for nutrients and makes it difficult for indigenous plant species to survive.

Soil microorganisms such as ectomycorrhizal and arbuscular mycorrhiza fungi (AMF) have the potential to be used as growth stimulators and bioremediation agents for degraded or polluted lands; but this potential has not been recognized or utilized until recently. There is a number of benefits potentially obtained from tree-mycorrhizal associations, such as increased seedling and mature plant growth, increased uptake of phosphorus (P) and other nutrients, increased root longevity, increased disease resistance, increased resistance to water stress and increased resistance to toxic elements [10].

A preliminary study to determine the presence of mycorrhizal association in peat swamp forest species found that there were mycorrhizal associations in the roots of peat swamp forest species such as *Shorea balangeran* (Balangeran), *Gonistylus bancanus* (Ramin), *Cratoxylon arborescens* (Gerunggang) and *Calophyllum soulattri* (Kapur Naga) [11, 12]. The effect of *Glomus clarum* and *Gisgasora decipiens* inoculation on *Dyera polyphylla* and *Aquilaria filaria* under green house conditions was investigated [13]. The result showed that plant height, diameter and shoot and root dry weight of *D. polyphylla* and *A. filaria* increased after innoculation. The positive effect of *G. clarum* and *G. aggregatum* on *Ploiariuum alternifolium* and *Calophyllum hosei* was also reported [14].

This paper presented the result of AMF species innoculation research on several local peatland plant species namely *Alstonia pneumatophora*, *Gonistylus bancanus*, *Stemonurus scorpioides*, *Callophylum soulattri*, *Tetramerista glabra* and *Palaquium* sp in the nursery condition and also reporting the growth and survival after being planted in the field. The spores of *Glomus clarum*, *Gigaspora decipiens* and *Enthrophospora* sp. were isolated from degraded tropical peatland in Kalampangan, Central Kalimantan province. The spores were mass-produced by using *Pueraria javanica* as host plant in a pot culture with zeolite as the growth medium. The seedlings media were autoclaved-sterilized (121 °C for 15 min). The seedlings and cuttings were surfaced-sterilized with $H_2O_2$ 5% for 5 min and rinsed with tap-water before sowing. Ten milligrams of the inoculums were applied. The plants were grown in the nursery for 24 weeks (6 months) and transplanted to the field. The growth performance and survival rate were recorded periodically.

## The Growth of Tropical Peatland Plant Species After AMF Innoculation

The growth performance, number of leaves and survival rate of the tropical peatland plant species in the nursery and the field is presented in Table 19.1. The early height and diameter growth of *C. soulattri* increased after inoculated with *G. clarum*, *G. decipiens* and *Enthrophospora* sp. five months after inoculation in the nursery [15]. Moreover, for *A. pneumatophora*, inoculation with *G. clarum* significantly effect the height and diameter growth 6 months after transplanted in the field; while *T. glabra* and *G. bancanus* did not show any significant growth effect after AMF inoculation in the nursery and transplant to the field [16]. The study of [16] showed interesting result for *S. scorpioides* after AMF innoculation. There was no significant different in term of height and diameter growth in the nursery, however ten months after transplant to the field innoculated seedlings were shown better growth. Another study [17] showed a consistent effect of *Gigaspora decipiens* inoculation on *A. pneumatophora* 24 weeks in the nursery and 5 years after transplanted in the field.

Moreover, the study [16] showed that *G clarum* is significantly effect height growth of *C. rotundatus* 9 months after inoculation in the nursery.

The effect of AMF innoculation varied between treatments and control both on the height and diameter. It was considered that the growth response to AMF colonization appeared more than 24 weeks after innoculation because the growth of peat swamp species was slow [13]. The compatibility of AMF to the host plant was also considered. The AMF which was not compatible to the host plant would not result in positive symbiosis. This will lead to limited P-available absorption to the plant root. The height, diameter and survival rate of peat swamp plant species varied in the nursery and the field. The growth response to AMF colonization appeared longer after inoculation because of the slow groath of peat swamp plant species. The compatibility between AMF and its host plant should also be taken into consideration. Indigenous mycorrhiza exploration and field trials of inoculated peat swamp plant species were needed to support the revegetation of degraded peatland especially in Central Kalimantan. It was expected that AMF application increase the growth and survival rate of seedlings in the nursery and the field.

**Table 19.1** Height, diameter growth, number of leaves and survival rates of tropical peatland plant species after AMF innoculation

| No | Plant species | AMF species | Height (cm) | Diameter (cm) | Leaf number | Survival rate (%) | Age | | Reference |
|---|---|---|---|---|---|---|---|---|---|
| | | | | | | | Nursery | Field | |
| 1 | *Alstonia pneumatophora* (Pulai Rawa) | Control | 40 a | 0.7 a | | 40 | | 6 months | [15] |
| | | *Glomus clarum* | 55 b | 1.0 b | | 40 | | 6 months | |
| | | *Gigaspora* sp. | 45 a | 0.9 b | | 30 | | 6 months | |
| 2 | *Callophylum soulattri* (Kapur Naga) | Control | 7.5 a | 2.0 a | | 78 | 5 months | | |
| | | *Glomus* sp. | 11 b | 5.0 b | | 100 | 5 months | | |
| | | *Gigaspora* sp. | 10 b | 4.0 b | | 80 | 5 months | | |
| | | *Enthrophospora* sp. | 12.5 b | 5.0 b | | 100 | 5 months | | |
| 3 | *Tetramerista glabra* (Punak) | Control | 13 a | 0.6 a | 3.0 a | | 6 months | | |
| | | *G. clarum* | 14 a | 0.5 b | 5.0 b | | 6 months | | |
| | | *Gigaspora decipiens* | 14 a | 0.7 a | 4.0 a | | 6 months | | |
| 4 | *Gonistylus bancanus* (Ramin) | Control | 30 ab | 0.4 a | 5.0 a | 40 | | 2 years | |
| | | *G. clarum* | 25 a | 0.4 a | 5.0 a | 40 | | 2 years | |
| | | *G. decipiens* | 50 b | 0.7 b | 6.0 a | 50 | | 2 years | |
| 5 | *Stemonurus scorpioides* (Medang telur) | *Control* | 9.0 ab | 0.28 ab | | | 24 weeks | | [16] |
| | | *Glomus clarum* | 8.5 a | 0.28 ab | | | 24 weeks | | |
| | | *G. decipiens* | 9.0 ab | 0.30 a | | | 24 weeks | | |
| | | *Enthrophospora sp.* | 10.0 b | 0.30 a | | | 24 weeks | | |
| | | *Mix* | 11.0 b | 0.25 b | | | 24 weeks | | |

(continued)

**Table 19.1** (continued)

| No | Plant species | AMF species | Height (cm) | Diameter (cm) | Leaf number | Survival rate (%) | Age | | Reference |
|---|---|---|---|---|---|---|---|---|---|
| | | | | | | | Nursery | Field | |
| 6 | *Palaquium sp.* (Nyatoh) | *Control* | 7.0 a | 0.2 a | | | 24 weeks | | |
| | | *Glomus clarum* | 10.0 b | 0.2 a | | | 24 weeks | | |
| | | *G. decipiens* | 7.0 a | 0.2 a | | | 24 weeks | | |
| | | *Enthrophospora sp.* | 7.5 a | 0.2 a | | | 24 weeks | | |
| | | *Mix* | 7.5 a | 0.2 a | | | 24 weeks | | |
| 7 | *T. glabra* (Punak) | *Control* | 12.0 a | 0.6 a | | | 24 weeks | | |
| | | *Glomus clarum* | 12.0 a | 0.4 b | | | 24 weeks | | |
| | | *Gigaspora sp.* | 12.0 a | 0.8 a | | | 24 weeks | | |
| 8 | *S. scorpioides* (Medang telur) | *Control* | 15 a | 0.35 ab | | | | 10 months | |
| | | *Glomus clarum* | 18 ab | 0.32 a | | | | 10 months | |
| | | *G. decipiens* | 22 b | 0.35 ab | | | | 10 months | |
| | | *Enthrophospora sp.* | 22 b | 0.40 b | | | | 10 months | |
| | | *Mix* | 18 ab | 0.40 b | | | | 10 months | |
| 9 | *Palaquium sp.* (Nyatoh) | *Control* | 10.0 a | 0.20 a | | | | 10 months | |
| | | *Glomus clarum* | 14.0 a | 0.20 a | | | | 10 months | |
| | | *G. decipiens* | 13.0 a | 0.20 a | | | | 10 months | |
| | | *Enthrophospora sp.* | 14.0 a | 0.20 a | | | | 10 months | |
| | | *Mix* | 13.0 a | 0.20 a | | | | 10 months | |
| 10 | *T. glabra* (Punak) | *Control* | 25.0 a | 0.65 ab | | | | 10 months | |
| | | *Glomus clarum* | 20.0 a | 0.60 a | | | | 10 months | |
| | | *Gigaspora sp.* | 25.0 a | 0.80 b | | | | 10 months | |

(continued)

**Table 19.1** (continued)

| No | Plant species | AMF species | Height (cm) | Diameter (cm) | Leaf number | Survival rate (%) | Age | | Reference |
|---|---|---|---|---|---|---|---|---|---|
| | | | | | | | Nursery | Field | |
| 11 | *A. pneumatophore* | *Control* | 18.0 a | 4.0 a | | | 24 weeks | | [17] |
| | | *Glomus clarum* | 25.0 b | 6.0 b | | | 24 weeks | | |
| | | *Gigaspora sp.* | 25.0 b | 6.0 b | | | 24 weeks | | |
| 12 | *G. bancanus* | *Control* | 28.0 a | 5.0 a | | | 6 months | | |
| | | *Glomus clarum* | 25.0 b | 5.0 a | | | 6 months | | |
| | | *Gigaspora sp.* | 32.0 a | 6.0 b | | | 6 months | | |
| 13 | *A. pneumatophora* | *Control* | 300 a | 10.0 a | | 78 | | 5 years | |
| | | *Glomus clarum* | 300 a | 15.0 a | | 80 | | 5 years | |
| | | *Gigaspora sp.* | 250 b | 30.0 b | | 76 | | 5 years | |
| 14 | *G. bancanus* | *Control* | 80 a | 5.0 a | | 83 | | 3 years | |
| | | *G. clarum* | 85 ab | 5.0 a | | 80 | | 3 years | |
| | | *G. decipiens* | 85 ab | 7.0 b | | 89 | | 3 years | |
| | | *Enthrophospora sp.* | 120 c | 5.5 ab | | 100 | | 3 years | |
| | | *Mix* | 100 bc | 5.5 ab | | 100 | | 3 years | |
| 15 | *Cratoxylon arborescens* (Gerunggang) | *Control* | 2.0 a | 0.4 a | 1.5 a | | 3 months | | [18] |
| | | *Glomus sp.1* | 3.0 ab | 0.35 a | 2.0 ab | | 3 months | | |
| | | *Glomus sp.2* | 2.0 a | 0.35 a | 2.0 ab | | 3 months | | |
| | | *Glomus sp. 5* | 4.0 ab | 0.4 a | 3.5 ab | | 3 months | | |
| | | *Gigaspora sp.* | 5.0 b | 0.3 a | 4.0 b | | 3 months | | |
| 16 | *Combretocarpus rotundatus* (Merapat) | *Control* | 20.0 a | 0.6 a | | 60 | 9 months | | [16] |
| | | *G. clarum* | 10.0 b | 0.65 a | | 100 | 9 months | | |
| | | *Gigaspora sp.* | 25.0 a | 0.6 a | | 70 | 9 months | | |

## The Root Colonization of Tropical Peatland Plant Species After AMF Innoculation

The root colonization after AMF innoculation is presented in Table 19.2. From Table 19.2 we can see that root colonization was higher for inoculated seedlings compared with control. AMF colonization on plant's roots showed variation from negative (parasite endophyte) to positive (mutualistic) [19]. High number of colonization does not always correlate with the benefit obtained by the host plants [20]. The AMF symbiosis is said to be effective when providing positive effect on the host plant and its environment. Those positive reaction is determined by various factors such as AMF species, soil types, the age of host plant and time needed for symbiosis to happen [21]. Inoculums, which could effectively colonize the roots, is potential to be utilized as inoculum source. However, each AMF genus owns different infection characteristics and various sporulation in different environmental condition [21].

Arbuscular mycorrhizal fungi are obligate symbiotic fungi that have been known to have a positive effect on plant growth. Arbuscular Mycorrhiza Fungi has four functional roles [21], namely: (1) as a bioprocessor, able to help the absorption of nutrients and water in plants from locations that are not reached by hair roots; (2) as

**Table 19.2** Root colonization of tropical peatland species after AMF innoculation

| No | Plant species | AMF species | Root colonization (%) | Age | | Reference |
|---|---|---|---|---|---|---|
| | | | | Nursery | Field | |
| 1 | *Alstonia pneumatophora* (Pulai Rawa) | Control | 5.0 a | | 6 months | [15] |
| | | *Glomus clarum* | 75 b | | 6 months | |
| | | *Gigaspora* sp. | 70 b | | 6 months | |
| 2 | *Callophylum soulattri* (Kapur Naga) | Control | 0.0 a | 5 months | | |
| | | *Glomus* sp. | 10.0 a | 5 months | | |
| | | *Gigaspora* sp. | 2.0 a | 5 months | | |
| | | *Enthrophospora* sp. | 3.0 a | 5 months | | |
| 3 | *Tetramerista glabra* (Punak) | Control | 0.0 a | 6 months | | |
| | | *G. clarum* | 100 b | 6 months | | |
| | | *Gigaspora decipiens* | 45 b | 6 months | | |
| 4 | *A. pneumatophore* | *Control* | 0.0 a | 24 weeks | | [17] |
| | | *Glomus clarum* | 80.0 b | 24 weeks | | |
| | | *Gigaspora sp.* | 80.0 b | 24 weeks | | |
| 5 | *G. bancanus* | *Control* | 15.0 a | 6 months | | |
| | | *Glomus clarum* | 10.0 a | 6 months | | |
| | | *Gigaspora sp.* | 100.0 b | 6 months | | |

(continued)

**Table 19.2** (continued)

| No | Plant species | AMF species | Root colonization (%) | Age | | Reference |
|---|---|---|---|---|---|---|
| | | | | Nursery | Field | |
| 6 | *Cratoxylon arborescens* (Gerunggang) | *Control* | 0.0 a | 3 months | | [18] |
| | | *Glomus sp. 1* | 20.0 a | 3 months | | |
| | | *Glomus sp.* 2ara> | 60.0 b | 3 months | | |
| | | *Glomus sp. 5* | 65.0 b | 3 months | | |
| | | *Gigaspora sp.* | 70.0 b | 3 months | | |
| 7 | *Combretocarpus rotundatus* (Merapat) | *Control* | 0 | 9 months | | [16] |
| | | *G. clarum* | 29 | 9 months | | |
| | | *Gigaspora sp.* | 65 | 9 months | | |

a bioprotector, capable of protecting plants from biotic stresses such as pathogens, pests and weeds as well as biotic stresses such as temperature, soil moisture, soil density and heavy metals; (3) as a bioactivator, able to help increase carbon storage in the rhizosphere so that the activity of microorganisms increases and (4) as a bioaggregator, able to increase soil aggregation.

In the forestry sector, this AMF is widely recommended as a stimulant to accelerate plant growth (biofertilizer) in restoration activities of degraded land [22]. The use of AMF in the forestry sector can be seen from several related studies that have been carried out. AMF inoculation and composting increased the growth of teak seedlings on planting media from limestone ex-mining soil [23]. Provision of compost by inoculation of several doses of AMF on ultisol soil media also affects the increase in stem diameter of Surian seedlings [24]. Local AMF inoculum proved to be quite effective in increasing growth, biomass and nutrient uptake of nail wood seedlings *Pericopsis mooniana* [25].

## Conclusion

AMF application in tropical peatland plant species at nursery level showed varying effects on height, diameter growth and survival rate. Field test results showed that the application of mycorrhizae on tropical peatland plants could increase the diameter and number of leaves of the plants. AMF has prospects to be developed in order to support the revegetation of degraded peatlands in Central Kalimantan, but there are still challenges to be faced, namely the suitability of AMF with host plants and plant survival rate which is still low in the field. Further AMF exploration in peat swamp forest needs to be carried out to obtain new isolates that are compatible with the plants to be developed.

**Acknowledgements** This study was financially supported by the Banjarbaru Environment and Forestry Research and Development Institute, South Kalimantan.

**Contributorship** Yuwati, T.W. contributed to the design and implementation of the research, to the analysis of the results and to the writing of the manuscript. Hakim, S.S. contributed to the writing of the manuscript.

## References

1. Page SE, Rieley JO, Banks CJ (2011) Global and regional importance of the tropical peatland carbon pool. Glob Chang Biol 17(2):798–818
2. Anda M, Ritung S, Suryani E, Sukarman, Hikmat M, Yatno E (2021) Revisiting tropical peatlands in Indonesia: semi-detailed mapping, extent and depth distribution assessment. Geoderma 402:115235
3. Miettinen J, Shi C, Liew SC (2016) Land cover distribution in the peatlands of Peninsular Malaysia, Sumatra and Borneo in 2015 with changes since 1990. Glob Ecol Conserv 6:67–78
4. Silvius BM, Diemont H (2007) Peatlands, climate change, poverty, biofuels, pulp and reduced emissions from deforestation and degradation. Wetland International, Wageningen
5. Miettinen J, Liew SC (2016) Status of peatland degradation and development in Sumatra and Kalimantan. Ambio 39(5):394–401
6. Dohong A, Aziz AA, Dargusch P (2017) A review of the drivers of tropical peatland degradation in South-East Asia. Land Use Policy 69:349–360
7. BRG (2019) Three years of peatland restoration in Indonesia. Technical Report, Badan Restorasi Gambut Republik Indonesia (the Peatland Restoration Agency of the Republic of Indonesia)
8. Dohong A, Abdul AA, Dargusch P (2018) A review of techniques for effective tropical peatland restoration. Wetlands 38(2):275–292
9. Page S, Hosciło A, Wösten H, Jauhiainen J, Silvius M, Rieley J (2009) Restoration ecology of lowland tropical peatlands in Southeast Asia: current knowledge and future research directions. Ecosystems 12(6):888–905
10. Brundrett MC, Bougher N, Dell B, Grove T, Malajczuk N (1996) Working with mycorrhizas in forestry and agriculture. In: International mycorrhizal workshop in Kaiping – China, vol 32. Australian Center for International Agricultural Research, p 374
11. Tawaraya K, Takaya Y, Turjaman M, Tuah SJ, Limin SH, Tamai Y (2003) Arbuscular mycorrhizal colonization of tree species grown in peat swamp forests of Central Kalimantan, Indonesia. For Ecol Manag 182(1–3):381–386
12. Yuwati TW (2003) Keberadaan mikoriza asli setempat pada hutan rawa gambut pasca kebakaran Tumbang Nusa, Kalimantan Tengah. Bul Tekno Hutan Tanam 1(1)
13. Turjaman M, Tamai Y, Santoso E, Osaki M, Tawaraya K (2006) Arbuscular mycorrhizal fungi increased early growth of two nontimber forest product species *Dyera polyphylla* and *Aquilaria filaria* under greenhouse conditions. Mycorrhiza 16(7):459–464
14. Turjaman M, Tamai Y, Sitepu IR, Santoso E, Osaki M, Tawaraya K (2008) Improvement of early growth of two tropical peat-swamp forest tree species *Ploiarium alternifolium* and *Calophyllum hosei* by two arbuscular mycorrhizal fungi under greenhouse conditions. New For 36(1):1–12
15. Yuwati TW, Hermawan B (2011) Inokulasi Mikoriza Tanah Lokal pada Jenis Tanaman Rawa Gambut untuk Peningkatan Pertumbuhannya. In: Arifin YF, Savitri E, Akbar A (eds) Prosiding Ekspose Hasil Penelitian "Dukungan BPK Banjarbaru dalam Pembangunan Kehutanan di Kalimantan", 25–26 Okt 2011. BPK Banjarbaru, pp 13–21
16. Yuwati TW , Santosa PB, Hermawan B (2007) Arbuscular mycorrhiza fungi application for rehabilitation of degraded peat swamp forest in Central Kalimantan. In: Rieley JO, Banks CJ,

Radjagukguk B (eds) Carbon – climate human interaction on tropical peatland. Proceedings of the international symposium and workshop on tropical peatland, Yogyakarta, 27–29 Aug 2007, EU CARBOPEAT and RESTORPEAT partner. Gadjah Mada University and University of Leicester

17. Yuwati TW, Rachmanadi D (2010) Challenges and opportunities of soil microbes utilization to support plantation forest establishment in degraded peat swamp forest of Central Kalimantan. In: Rimbawanto A, Febrianto FKE (eds) Proceedings international seminar research on plantation forest management: challenges and opportunities, Bogor, Indonesia, 5–6 Nov 2009. Centre for Plantation Forest Research and Development, Bogor, p 8
18. Yuwati TW, Hakim SS, Alimah D (2018) Pengaruh Aplikasi Mikoriza Arbuskula terhadap Pertumbuhan Gerunggang (*Cratoxylon arborescens*) di Persemaian. J Hutan Trop 6(2):170
19. Brundrett M (2004) Diversity and classification of mycorrhizal associations. Biol Rev Camb Philos Soc 79(3):473–495
20. Corkidi L, Allen EB, Merhaut D, Allen MF, Downer J, Bohn J (2004) Assessing the infectivity of commercial mycorrhizal inoculants in plant nursery conditions. J Environ Hortic 22(3):149–154
21. Nusantara AD, Irdika M (2012) Bekerja dengan Fungi Mikoriza Arbuskula. SEAMEO-BIOTROP, pp 978–979
22. Yuwati TW, Hakim SS, Alimah D, Hermawan B, Musthofa AA (2017) Keanekaragaman spora mikoriza arbuskula di hutan rawa gambut Kalimantan Tengah. In: Diana R, Sulistioadi YB, Karyati Sarminah S, Widiati KY, Kuspradini H, Sari DR, Mulyadi R (eds) Prosiding Seminar Nasional Silvikultur IV, 19–20 Juli 2016. Universitas Mulawarman, p 493
23. Prayudyaningsih R, Sari R (2016) Aplikasi Fungi Mikoriza Arbuskula (FMA) dan Kompos untuk Meningkatkan Pertumbuhan Semai Jati (*Tectona grandis* Linn.f.) pada Media Tanah Bekas Tambang Kapur (The application of arbuscular mycorrhizal fungi (AMF) and compost to improve the growth of Jati). J Penelit Kehutan Wallacea 5(1):37–46
24. Zulya F, Noli ZA, Maideliza T (2016) Respon Bibit Surian (*Toona sinensis* (Juss,) M. Roem.) Terhadap Inokulasi Beberapa Dosis Fungi Mikoriza Arbuskula pada Media Tanah Ultisol yang Dicampur Pupuk Kompos. Al-Kauniyah J Biol 9(1):10–18
25. Husna, Wilarso S, Mansur I, Kusmana C (2015) Respon pertumbuhan bibit kayu kuku (*Pericopsis mooniana*). J Pemuliaan Tanam Hutan 9(3):131–148

GPSR Compliance

*The European Union's (EU) General Product Safety Regulation (GPSR) is a set of rules that requires consumer products to be safe and our obligations to ensure this.*

*If you have any concerns about our products, you can contact us on ProductSafety@springernature.com*

In case Publisher is established outside the EU, the EU authorized representative is:

Springer Nature Customer Service Center GmbH
Europaplatz 3
69115 Heidelberg, Germany

**Batch number: 10756376**

Printed by Printforce, the Netherlands